The Jurassic of the Gulf Rim

PROCEEDINGS OF THE
THIRD ANNUAL RESEARCH CONFERENCE
GULF COAST SECTION
SOCIETY OF ECONOMIC PALEONTOLOGISTS AND
MINERALOGISTS FOUNDATION

Edited by

William P. S. Ventress
Don G. Bebout
Bob F. Perkins
and
Clyde H. Moore

1984

Published March 25, 1984

Distributed by
Earth Enterprises, Inc.
P.O. Box 672
Austin, Texas 78767

FOREWORD

The Third Annual Research Conference, sponsored by the Gulf Coast Section Society of Economic Paleontologists and Mineralogists Foundation, was held in Baton Rouge, Louisiana on November 28 to December 1, 1982. The subject - THE JURASSIC OF THE GULF RIM - was addressed through 24 oral presentations and 10 core displays. The abstracts of these presentations were published in the Program and Abstracts of the Conference. The enthusiastic response of the participants during and after the Conference was gratifying and served as stimulus for the editors to compile this volume of the full-length papers. Each of the speakers and core presenters (total of 27) was invited to submit a paper; 22 responded positively and their papers are included in this volume. Subsequent to the meeting another paper on "Significant Fossils of the Knowles Limestone, Lower Cretaceous, Texas" was contributed by R. W. Scott.

This volume is not a collection of reprints of previously published papers nor a synthesis of Jurassic studies of the past. Most of the papers present new data and interpretations while using as supporting data the many excellent papers published on the Jurassic sediments during the past 30 years. All of the papers are applicable to the exploration and production of hydrocarbons from the Jurassic sediments, and many of the authors have emphasized this relationship.

The pattern of Jurassic carbonate and sandstone deposition around the Gulf of Mexico is complex. Jurassic deposystems not only were controlled by bathmetry but also by local topography of the sea floor related to early salt movement. The nature of the rapid horizontal facies changes characteristic of the Jurassic is demonstrated by the several important papers which are based on detailed study of the well-exposed out-cropping Jurassic of northern Mexico. Regional and detailed field studies along the northern Gulf of Mexico extend these depositional patterns into the subsurface. Finally, complex diagenetic patterns controlled by both depositional facies and subsequent burial history alter the original depositional porosity. The papers on diagenesis and geochemistry attest to the difficulty in predicting the location of significant porosity in the Jurassic rocks.

It is felt that the papers included in this volume will contribute to the understanding of the nature and distribution of Jurassic rocks in the Gulf Coast.

D. G. Bebout
C. H. Moore, Jr.

GULF COAST SECTION
SOCIETY OF ECONOMIC PALEONTOLOGISTS AND MINERALOGISTS FOUNDATION

ACKNOWLEDGEMENTS

The Jurassic of the Gulf Rim Proceedings represents the efforts of many people. However, certain individuals warrant specific recognition for their contributions which made the publication of this proceedings a reality.

Bob F. Perkins, The University of Texas at Arlington, guided the publication in every phase, from initial planning to final printing. His patience, editorial expertise, and dedication to the Foundation in this endeavor are gratefully acknowledged.

Also, thanks are extended to Beverly J. Tyler, The University of Texas at Arlington, for typesetting the book and for handling many details of the publication effort.

Special appreciation is extended to Dennis J. Greig, Chevron USA, Inc., for his exceptional efforts in overseeing the redrafting and reformatting of many of the illustrations. Chevron USA, Inc. is acknowledged for providing equipment, facilities, and manpower support throughout the preparation of the proceedings.

W. P. S. Ventres
D. G. Bebout

CONTRIBUTORS
TO THE
GULF COAST SECTION
SOCIETY OF ECONOMIC PALEONTOLOGISTS AND MINERALOGISTS FOUNDATION

The Gulf Coast Section of the Society of Economic Paleontologists and Mineralogists Foundation extends sincere appreciation and gratitude to the following companies, organizations, and individuals whose financial support made the Annual Research Conference series and publication of these Proceedings possible.

Aminoil USA, Inc.
Amoco Production Company (USA)
Anadarko Production Company
ARCO Oil and Gas Company
Atlantic Richfield Foundation
Sheila C. Barnette
Bass Enterprises Production Company
Bill Forney Inc.
Jannette Boyer
Brian O'Brien
C & R Exploration Inc.
Central Louisiana Energy Corporation
Champlin Petroleum Company
Chevron U.S.A. Inc.
Cities Service Oil Company
Conoco Inc.
Charles Copeland, Jr.
Crosbie-Macomber Inc.
Dapsco Inc.
Data Log Inc.
Delta Drilling Company
Dresser Industries Inc.
John B. Dunlap, Jr.
Virginia C. Elias
Albert D. Ellis, Jr.
Exxon Company U.S.A.
Franks Petroleum, Inc.
Getty Oil Company
Gulf Oil Corporation
Robert Hennington
S. Hixon
J. M. Huber Corporation
Barry Kohl
Marathon Oil Company
Gene B. Martin
Martin Exploration Company
Mesa Petroleum Company
Mobil Exploration Company
Monarch Logging Company, Inc.
Monsanto Oil Company
Robert Mullins
NL Baroid Petroleum Company
Paleo Data, Inc.
Pend Oreille Oil & Gas Company
Bob F. Perkins
Petroleum Information
Phillips Petroleum Company
Edward B. Picou, Jr.
Placid Oil Company
Prairie Producing Company
Claude M. Quigley, Jr.
M. L. Randall Explorations Inc.
Sanchez-O'Brien Oil and Gas Corporation
J. Frederick Sarg
Schares Instrument Corporation
Schlumberger Well Services
Shell Oil Company
Sohio Petroleum Company
Spooner Petroleum Company
Teledyne
Texaco
Transco Energy Company
Union Oil of California
C. G. Ventress
H. Vaughan Watkins, Jr.

CONTENTS

POROSITY DEVELOPMENT AND RESERVOIR CHARACTERISTICS OF THE SMACKOVER FORMATION IN SOUTHWEST ALABAMA

D. JOE BENSON

Department of Geology
The University of Alabama
University, Alabama 35486

ERNEST A. MANCINI

Geological Survey of Alabama
State Oil and Gas Board
University, Alabama 35486

ABSTRACT

Deposition of the Upper Jurassic Smackover Formation in southwest Alabama was significantly influenced by a number of topographic highs that were present during Smackover time. These highs, which are extensions of the Appalachian structural trend, salt diapiric features, or fault-related structures, include the Wiggins Uplift and the Conecuh and Pensacola Ridges. These features essentially divided the Smackover depositional basin in southwest Alabama into three smaller subbasins: the Conecuh Embayment, Manila Embayment, and eastern extension of the Mississippi Interior Salt Basin. These subbasins and the paleotopographic highs that separated them were responsible for a very irregular pattern of lithofacies distribution. Because of this widely varied depositional setting, the Smackover in southwest Alabama exhibits significant variation in lithologies, porosity types, and reservoir characteristics.

Examination of porosity types and reservoir characteristics in four of the major Smackover-producing areas in southwest Alabama indicates that Smackover reservoirs vary dramatically from area to area. This variation is a result of differences in principal lithologies among areas and in the diagenetic sequence produced by different depositional settings.

In the eastern portion of the Mississippi Interior Salt Basin, principally in Washington and Choctaw Counties, Smackover reservoirs are partially dolomitized oolitic and oncolitic grainstones; and porosity is a mixture of primary interparticulate and secondary grain moldic (oomoldic) and dolomitic intercrystalline porosity. Reservoir porosities range from 6 to 30% with permeabilities of 0.5 to several hundred md.

In the Manila Embayment in Monroe County, facies indicative of high energy are generally lacking and are replaced in the upper Smackover by those deposited in lower energy, restricted-marine environments. Porosity is dominantly secondary and includes intercrystalline and fracture porosity in dolostone lithologies. Reservoir porosities range from 8 to 20% with 10 to several thousand md permeability.

In the Conecuh Embayment in Escambia County, the Smackover reservoirs are characterized by dolomitized peloidal wackestones and packstones and laminated mudstones all indicative of deposition in lower energy environments. Porosity is secondary and is mainly dolomitic intercrystalline and grain moldic. Porosities in producing horizons range from 8 to 30% with permeabilities of 1 to 50 md.

In the Hatter's Pond Field in Mobile County, situated on the northwestern flank of the Wiggins Uplift, the Smackover consists of highly dolomitized oolitic, oncolitic, and peloidal packstones and grainstones. Porosity is mainly secondary and includes dolomitic intercrystalline, grain moldic (oomoldic), and vuggy types. Reservoir grade porosity (which ranges from 10 to 20%) and permeability (1 to 100 md) are facies selective and occur in the dolomitized grainstones.

Because of the wide variation in depositional and diagenetic settings, reservoir characteristics for the Smackover Formation of southwest Alabama cannot be integrated into a single exploration model. Instead, separate models must be developed for individual areas, and exploration techniques successful in one area may not necessarily be successful in another.

INTRODUCTION

Petroleum exploration has increased rapidly in southwest Alabama since the discovery of petroleum in Gilbertown Field in Choctaw County in 1944. The primary exploration target has been, and continues to be, the Upper Jurassic Smackover and Norphlet Formations (Fig. 1). These two formations are the most productive hydrocarbon reservoirs in the state. Cumulative production for the Smackover and Norphlet in Alabama includes more than 227 million barrels of oil, 59 million barrels of condensate, and 513 billion cubic feet of gas. To date, more than 750 wells have been drilled in search of oil and gas in southwest Alabama and have resulted in the discovery of 37 Upper Jurassic fields (Fig. 2).

SYSTEM	GROUP OR FORMATION
JURASSIC	COTTON VALLEY GROUP
	HAYNESVILLE FORMATION
	SMACKOVER FORMATION
	NORPHLET FORMATION
	LOUANN SALT

Figure 1. Jurassic section in southwest Alabama.

GEOLOGIC SETTING

Smackover deposition in Alabama closely approximated present-day sedimentation in the Persian Gulf. Louann Salt was probably responsible for forming a ramp surface. Variations in carbonate deposition are related to the regional paleogeography during the late Jurassic for southwest Alabama. One basin, the Mississippi Interior Salt, and two embayments, the Manila and Conecuh, are recognized (Fig. 3). Lithofacies trends similar to those of the Persian Gulf are characteristic of the Mississippi Interior Salt Basin. Higher energy conditions occurred in updip nearshore areas and lower energy facies, in downdip offshore environments. Predicted lithofacies distributions, however, have not been observed for the Manila and Conecuh Embayments. The Manila Embayment is apparently separated from the Mississippi Interior Salt Basin by a salt ridge. This ridge probably acted as a barrier inhibiting ocean circulation in the area (Mancini and Benson, 1980). Lithofacies typical of lower energy, restricted environments characterize the Manila Embayment area. The Wiggins-Conecuh trend separates the Manila and Conecuh Embayments. This trend probably also contributed to producing restrictive marine conditions in the Manila Embayment. The Conecuh Embayment was protected to the north by the Conecuh Ridge and to the southeast by the Pensacola Ridge. Predominantly lower energy characterized the restricted lagoonal and intertidal environments in this embayment.

In southwest Alabama, moderate- to high-energy conditions prevailed downdip in offshore areas as well as updip in nearshore environments. The presence of updip grainstones is expected using the Persian Gulf model (Ahr, 1973). In addition, employing Purser's (1973) findings for the Persian

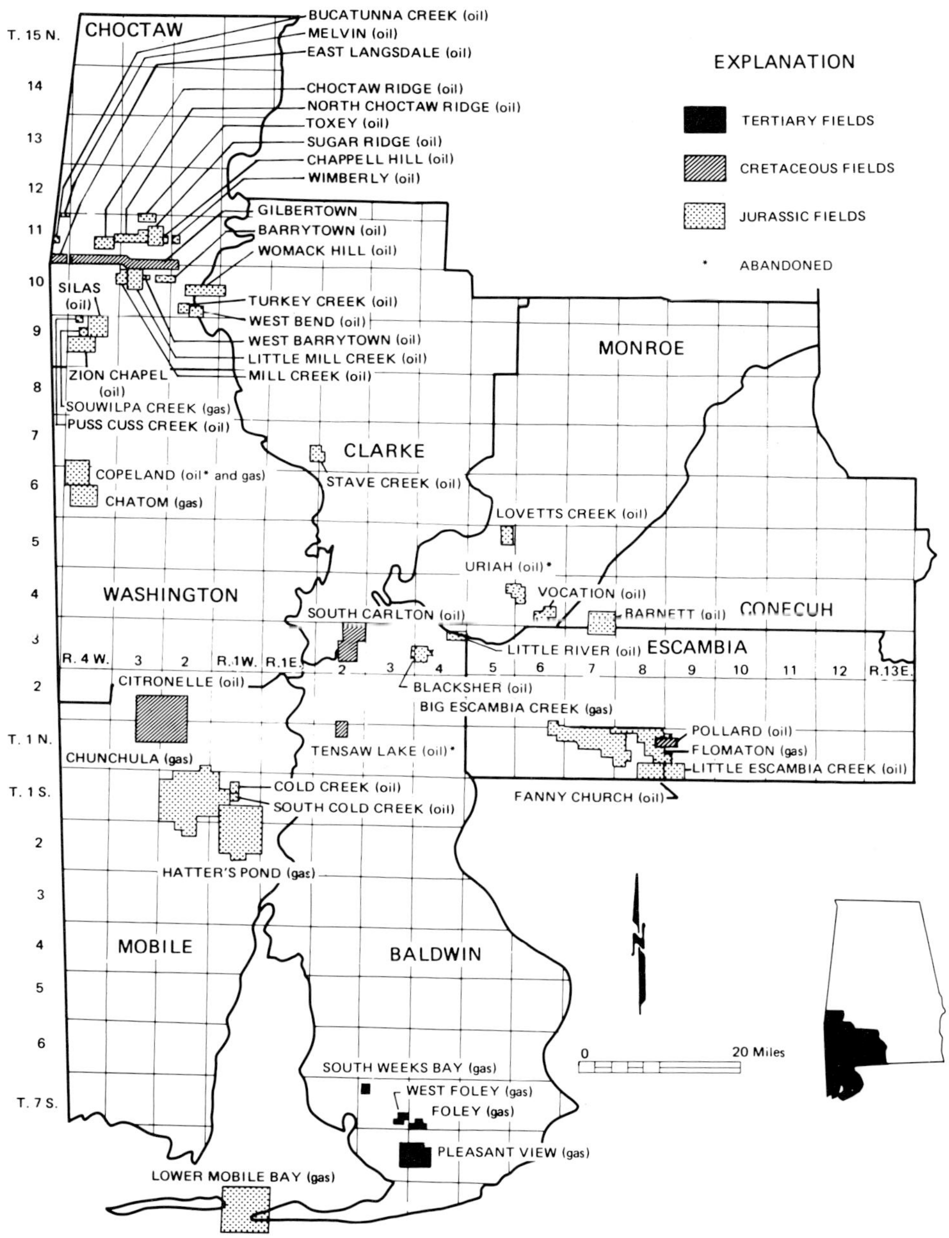

Figure 2. Oil and gas fields in southwest Alabama (Masingill and Bolin, 1982).

Gulf, downdip moderate- to high-energy conditions were also present on the crest and flank of basement paleohighs and salt ridges. These features, including the Wiggins-Conecuh Trend, domal paleohighs, and salt diapirs, acted as basinward highs during Smackover time.

The variability in depositional setting had a significant effect on diagenetic history of the Smackover in southwest Alabama. Because diagenetic history plays a significant role in reservoir development in the Smackover throughout the Gulf Coast, geographic variation in depositional setting is accompanied by geographic variation in reservoir types and characteristics. In order to illustrate the geographic variation in reservoir characteristics across southwest Alabama, four separate areas will be described: the northern portion of the Mississippi Interior Salt Basin, the Manila Embayment, the Conecuh Embayment, and along the Wiggins Uplift (Fig. 3).

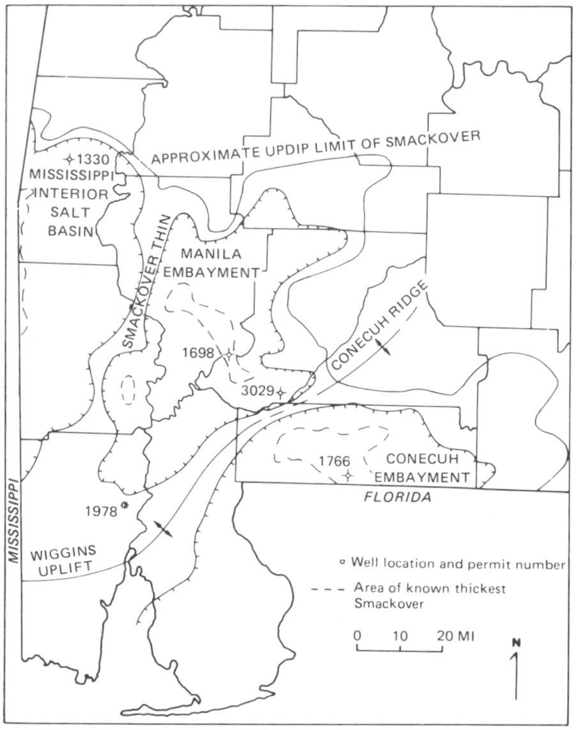

Figure 3. Paleogeographic map of southwest Alabama during Smackover time (Mancini and Benson, 1980).

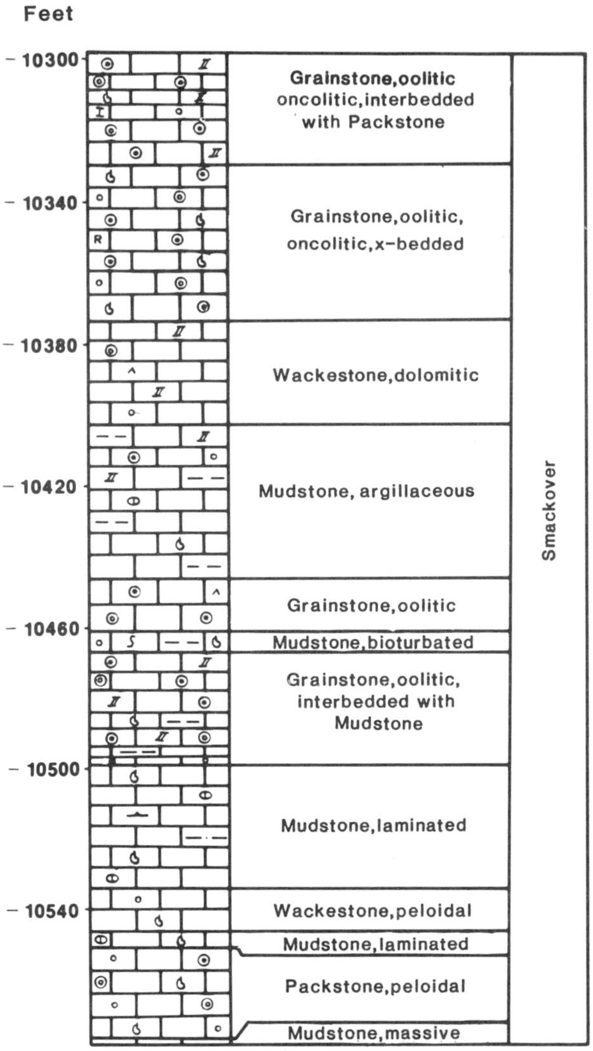

Figure 4. Core description of the Placid Oil Jackson No. 1 well (Permit No. 1330, Fig. 3), 4 miles north of Toxey, Choctaw County, Alabama.

Figure 5. Photomicrographs showing porosity types in the Smackover Formation in the Placid Oil Jackson No. 1, southwest Alabama; bar scales equal 0.5 mm.

a. Solution-enlarged interparticulate porosity (i) in oolitic grainstone; 10,344 ft.

b. Intercrystalline porosity in dolomitzed oolitic grainstone; ooids are replaced by finely-crystalline dolomite and cemented by coarse, isopachous dolomite mosaic (c); 10,313 ft.

c. Dolomitized oolitic grainstone; ooids (o) have been replaced by finely-crystalline dolomite; interparticulate cement (g) is granular calcite; 10,494 ft.

d. Compacted oolitic grainstone; notice stylolitized (s) and interpenetrating grain boundaries; 10,348 ft.

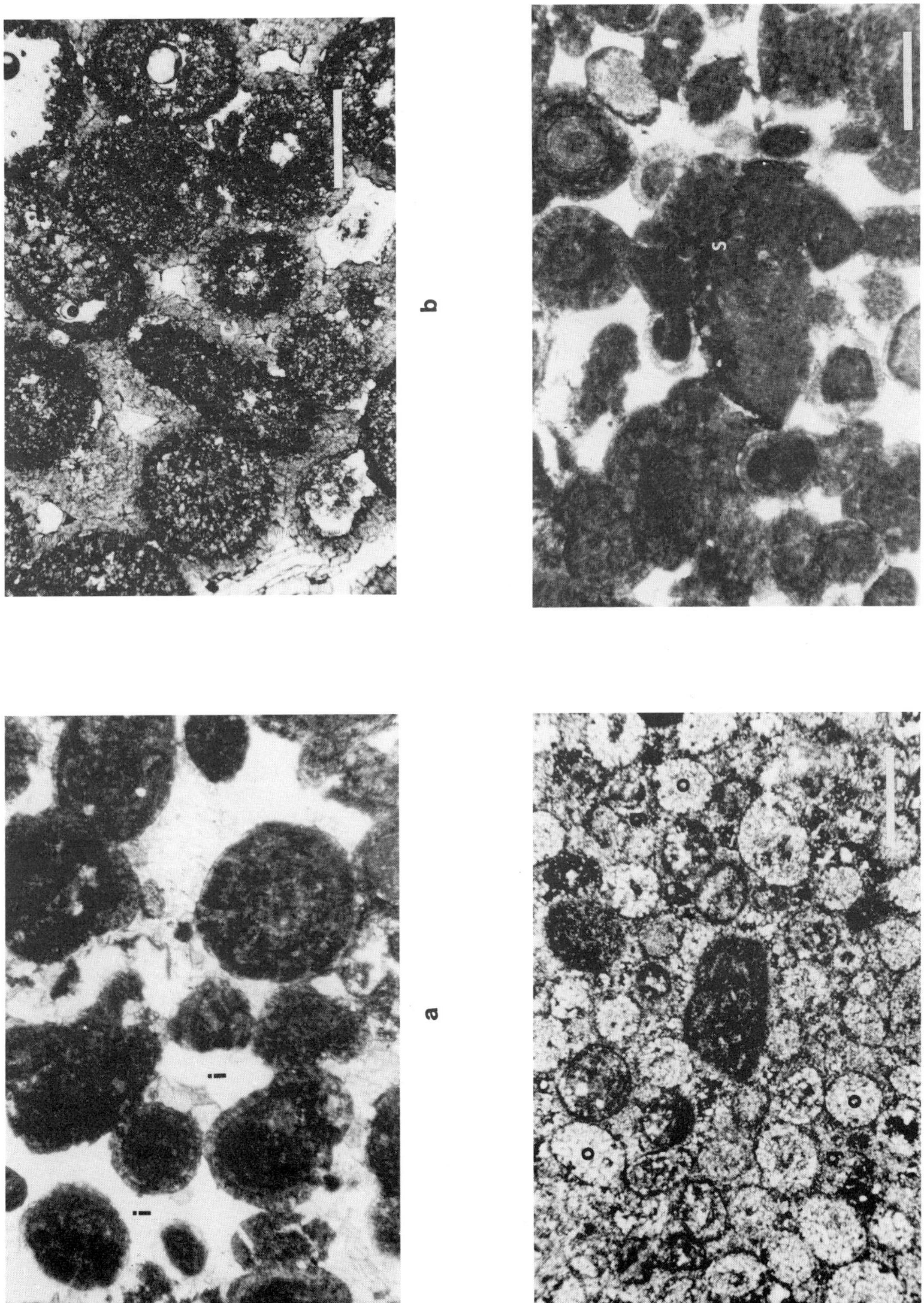

Figure 5

POROSITY DEVELOPMENT AND RESERVOIR CHARACTERISTICS

Mississippi Interior Salt Basin

The Smackover in the northern portions of the Mississippi Interior Salt Basin, principally in Choctaw and Washington Counties, Alabama, is lithologically very similar to the Smackover throughout much of the Gulf Coast. The lower Smackover, commonly called the "brown dense," consists of laminated mudstones deposited in a low-energy subtidal setting. The upper Smackover, which is well represented in the Placid Oil Jackson No. 1 well (Fig. 4), consists of a series of shoaling upward cycles that grade from peloidal wackestone/packstone and laminated mudstone into oncolitic and oolitic grainstone deposited in a high-energy shoal environment. These ooid shoals were periodically exposed and the deposits partially dolomitized.

Reservoir grade porosity in this area is restricted to the oolitic grainstones and is a mixture of solution-enlarged, primary interparticulate (Fig. 5a), moldic, and dolomitic intercrystalline porosity (Fig. 5b). Reservoir grade porosity ranges from 6 to 30% with permeabilities from 0.5 to several hundred md. A porosity-permeability cross plot for the Midroc and Watkins Bolinger 26-7 well from Chappell Hill Field (Fig. 6) is representative of porosity and permeability relations in the area.

The effect of diagenesis on reservoir development in southwest Alabama is perhaps best illustrated in the Mississippi Interior Salt Basin where the upper Smackover deposits have undergone a complex history of diagenetic alteration (Fig. 7). Four separate processes have had a significant effect on reservoir development.

Dolomitization. Dolomite occurs in the upper Smackover as fine- to medium-crystalline, hypidiotopic rhombs, which commonly replace aragonitic ooids (Fig. 5c), or as a bladed, isopachous cement (Fig. 5b). This dolomite is a product of mixing-zone dolomitization associated with periodic exposure of the ooid shoals or with later Bucknew regression. Dolomitization created small amounts of intercrystalline porosity and, where the dolomitization process was incomplete, provided a compositional distinction that was important during later dissolution.

Cementation. Three generations of cement are represented in the oolitic grainstones. The first of these is a marine phreatic, isopachous, fibrous cement which occurs as rims 20 to 50 microns in thickness. The fibrous cement is a first generation cement wherever found, but has a highly variable distribution and is frequently lacking.

Meteoric phreatic, granular cements occur as equant to loaf-shaped crystals 20 to 50 microns in diameter. They are isopachous, but coarsen away from grain substrates as mosaic pore filling. Where the fibrous cements are present, the granular cements are second

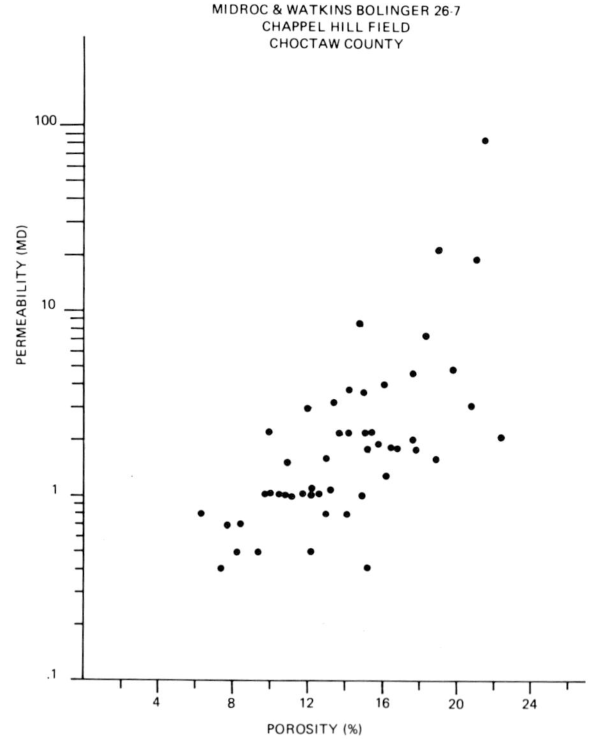

Figure 6. Porosity-permeability cross-plot of potential reservoir zones in the Midroc and Watkins Bolinger 26-7 well, Chappell Hill Field, Choctaw County, Alabama.

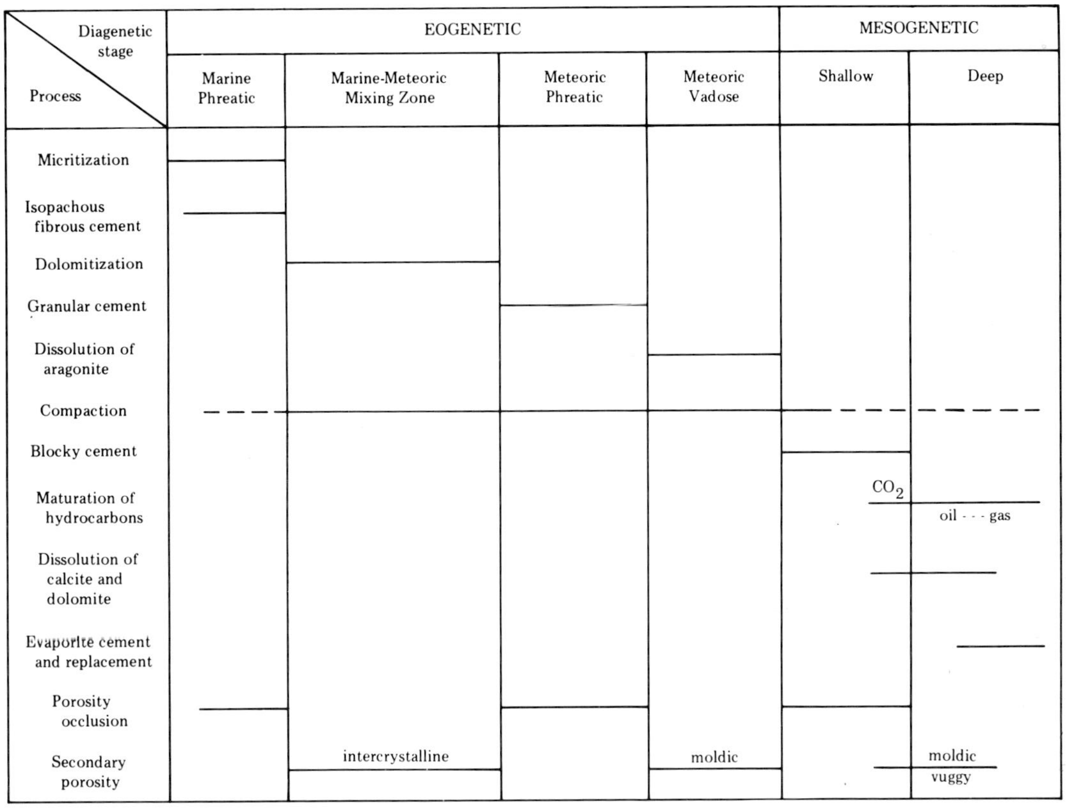

Figure 7. Generalized paragenetic sequence for upper Smackover grainstones in southwest Alabama.

generation cements; where the fibrous cements are lacking, the granular cements are first generation.

The third cement type is a very coarse-grained, poikilotopic, blocky cement which fills entire pores and includes adjacent allochemical grains. Paragenetically, this is a late-stage mesogenetic cement that post-dates compaction (Fig. 7). It is very similar to the "calcspar cement" described by Moore and Druckman (1981).

All three cement types have the same influence on reservoir development because they each occlude porosity. Volumetrically, the fibrous cements are minor, but the granular and blocky cements can each account for up to 25% of the total rock volume.

Compaction. Compaction has also had a significant effect on porosity occlusion. Deposits which were not cemented early show strong evidence of grain-to-grain compaction (Fig. 5d). Compaction is indicated by stylolitic and embayed grain contacts and high grain densities (70 to 90%). In many cases, deposits with isopachous granular cement rims also show evidence of compaction in the form of highly distorted grains, which show very small numbers of grain-to-grain contacts.

Leaching. Perhaps the most important process in reservoir evolution in the Smackover in the Alabama portion of the Mississippi Interior Salt Basin and throughout southwest Alabama is leaching. Most of the porosity in the Smackover shows evidence of solution enhancement of primary porosity or development of solution-derived secondary porosity.

Dissolution occurred at two separate stages in the paragenetic sequence (Fig. 7). An early stage of vadose dissolution occurred asociated with exposure of the ooid shoals

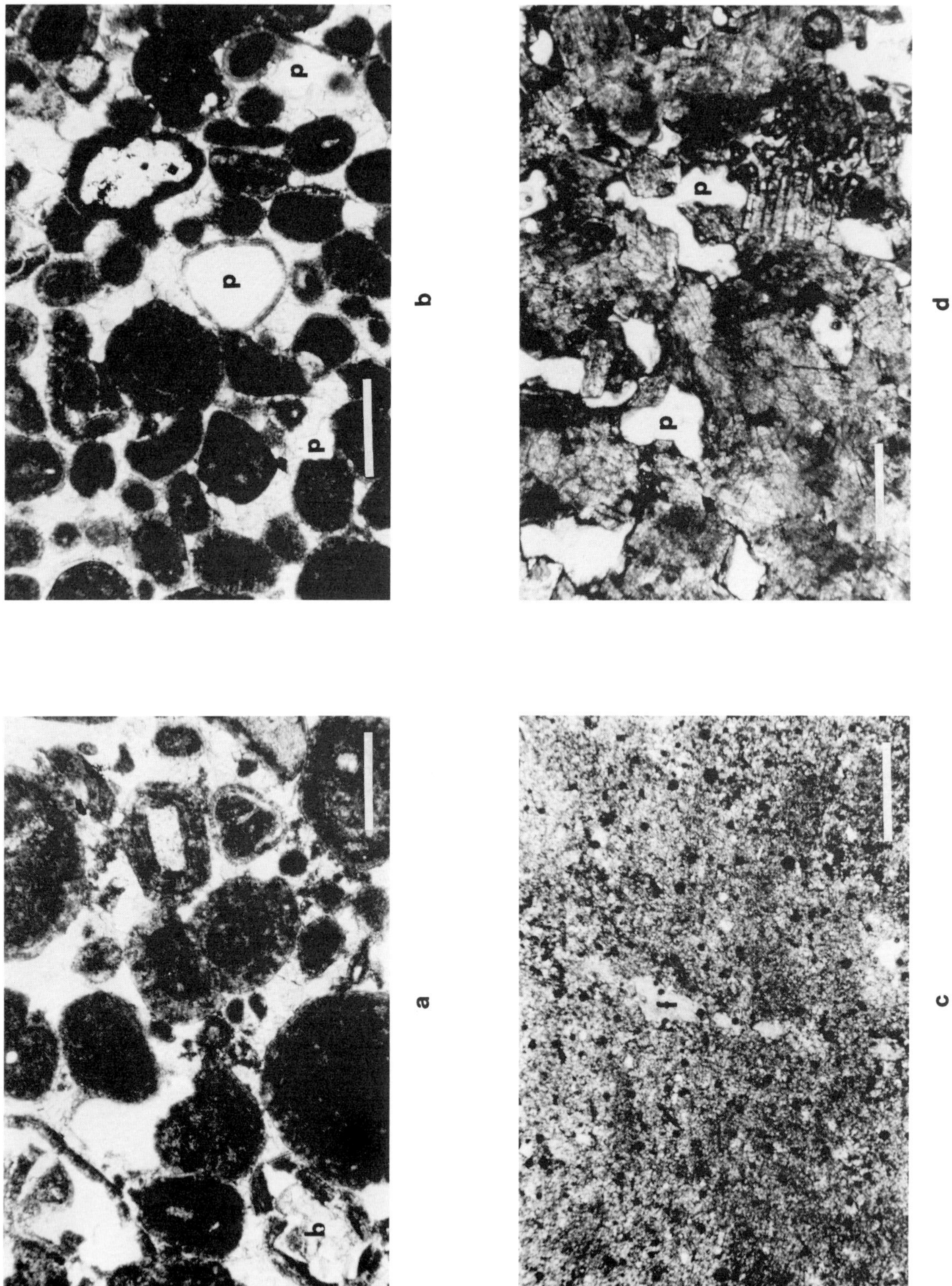

Figure 8

and/or Buckner regression. The dissolution is precompaction as indicated by crushed, leached ooids and pre-precipitation of the late-stage, blocky cement as indicated by blocky cement filling leached ooids (Fig. 8a). This early stage of dissolution produced moldic and solution enlarged interparticulate porosity.

The second episode of leaching was a late-stage, mesogenetic event initiated perhaps by the introduction into the reservoir of CO_2-charged fluids generated during the early stages of hydrocarbon maturation. This period of dissolution is post-compaction and post-precipitation of the blocky cements indicated by leacing of the cements (Fig. 8b).

The formation of porosity by late-stage, mesogenetic dissolution in carbonates is a relatively new concept (Moore and Druckman, 1981; Loucks and Budd, 1981; Benson and Mancini, 1982; and Loucks and Budd, 1982), and it is still the subject of active debate. Some authors have rejected the idea that dissolution can occur in carbonates at depth, reserving the mesogenetic realm for compaction and cementation (Longman, 1982). This rejection seems inconsistent, however, in light of the wide acceptance that the concept of mesogenetic, secondary porosity development in clastics has received. Furthermore, much of the secondary porosity in clastics is the result of dissolution of calcium carbonate cements. If mesogenetic fluids can dissolve calcium carbonate in clastics, it seems reasonable that the same process would occur in carbonate rocks.

The significance of late-stage porosity development in Smackover reservoirs of southwest Alabama is unknown. Most productive reservoirs show some evidence of late-stage dissolution. The proportion of late-stage, secondary porosity to total porosity is difficult to determine. In some cases, the late- stage porosity appears to comprise most of the total porosity; in others, it appears to be minor with the majority of the porosity developed in the eogenetic realm. Further work is ongoing in an effort to determine the overall significance of late-stage porosity.

A second important question concerns the controls on secondary porosity development; little is known on the subject. It seems obvious that migration paths of mesogenetic fluids are controlled by inherent porosity and permeability within the rock. Migration would be most significant and dissolution most rapid in zones that possessed the greatest porosity and permeability. Thus, there may be a direct correlation between early formed eogenetic porosity and mesogenetic dissolution. Additional work is needed to define the controls on mesogenetic dissolution more accurately.

Manila Embayment

Depositional conditions in the Manila Embayment are very different from those found

Figure 8. Photomicrographs showing porosity types from the Smackover Formation of southwest Alabama; bar scales equal 0.5 mm.

a. Early moldic porosity; notice leached ooid which has been compacted and then had the porosity occluded by precipitation of late-stage, blocky cement (b). This dissolution preceeded compaction and precipitation of the blocky cement. Placid Oil Jackson No. 1, 10,344 ft.

b. Late-stage porosity (p); a mixture of solution enlarged interparticulate and moldic porosity; note delicate rim on leached ooid. Dissolution postdated compaction and precipitation of the blocky cement which fills interparticulate pores. Placid Oil Jackson No. 1, 10,348 ft.

c. Intercrystalline and fracture (f) porosity in dolomitized mudstone. Getty C.D. Broughton No. 3, 13,768 ft.

d. Late-stage solution-enlarged intercrystalline porosity (p). Reef interval, Texaco Coley No. 1, 14,119 ft.

in the Mississippi Interior Salt Basin to the west. High-energy oolitic grainstone deposits, which characterize the upper Smackover in the Mississippi Interior Salt Basin, are minor constituents of the Smackover throughout much of the Manila Embayment. Low energy and restricted circulation prevailed because of paleobathymetric highs (Wiggins-Conecuh trend and the unnamed salt ridge) which encircle the embayment.

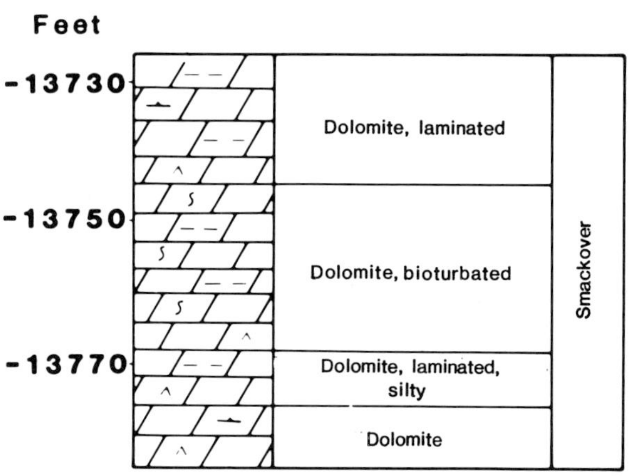

Figure 9. Core description of the Getty C.D. Broughton No. 3 core (Permit No. 1698, Fig. 3), 7.5 miles northwest of Uriah, Monroe County, Alabama.

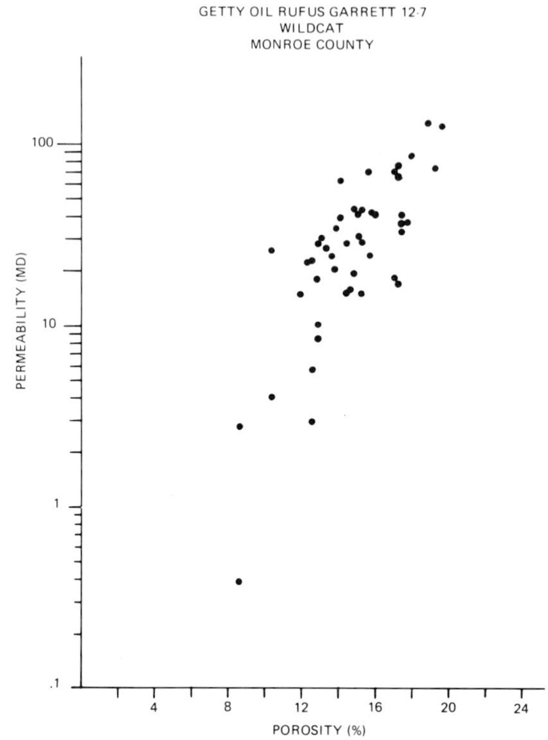

Figure 10. Porosity-permeability cross-plot of potential reservoir zones in the Getty Oil Rufus Garrett 12-7 well, Uriah Field, Monroe County, Alabama.

The upper Smackover, as exemplified by the Getty C. D. Broughton No. 3 well (Fig. 9), consists predominantly of laminated or bioturbated, silty dolomite. Thin zones of litharenitic sandstone are interbedded within the sequence indicating episodic regression accompanied by influx of clastics from upland areas to the north and northeast. Peloids are the dominant allochems, whereas invertebrate fossils are minor constituents, probably reflecting restrictive conditions in the embayment.

With high-energy conditions generally lacking throughout much of the embayment, reservoir grade porosity is developed only through dolomitization of portions of the section. Porosity is secondary and is a mixture of solution enlarged dolomitic intercrystalline and fracture porosity (Fig. 8c). The intercrystalline porosity is a product of early mixing-zone dolomitization, while the fracture porosity is a late-stage feature produced by compactional and/or tectonic activity. Most of this porosity shows evidence of late-stage mesogenetic dissolution.

Reservoir-grade porosity ranges from 8 to 20% with permeabilities of from 10 to several thousand md. Typical permeabilities range from 10 to several hundred md (Fig. 10) with higher permeabilities in highly fractured zones. These porosities are typically lower than in the Mississippi Interior Salt Basin but permeabilities are significantly higher, particularly in the fractured reservoirs.

To the southwest in the Manila Embayment, on the flank of the Conecuh Ridge, a different lithologic sequence is encountered. In the Texaco No. 1 Coley well (Fig. 11) in Vocation Field, subtidal, low-energy conditions in the lower Smackover resulted in deposition of carbonate mudstones overlain by an algal boundstone "reef" sequence. Stromatolitic algae are the dominant organic constituents of the "reef" with lesser amounts of red algae, sponges, echinoid spines, pelecypod fragments, and foraminifera (Baria *et al.*, 1982). The "reef" is overlain by a series of grainstone cycles which grade upward from peloidal packstones and grainstones into ooid grainstones.

The appearance of organic accumulations and occurrence of grainstones in the upper Smackover are a reflection of higher energy and more open-marine conditions along the crest of the Conecuh Ridge. The Smackover thins appreciably across this feature, suggesting that the ridge was a bathymetric high during Smackover deposition. The high-energy deposits are a reflection of the shallow-water nature of deposition across this paleohigh. The entire upper Smackover sequence in the Coley well is extensively dolomitized and shows signs of periodic exposure, indicating a very shallow-water environment of deposition.

Porosity in the Coley well ranges from 6 to 15% and is a mixture of solution-enlarged, dolomitic intercrystalline (Fig. 8d) in the "reef" interval and solution-enlarged, interparticulate, dolomitic intercrystalline and grain moldic porosities (Fig. 13a) in the

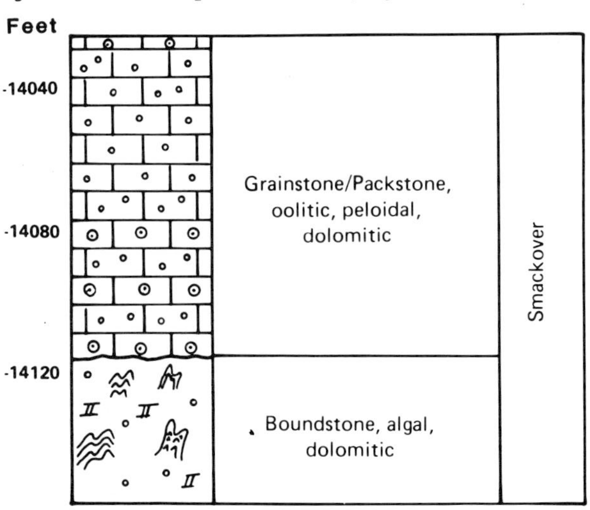

Figure 11. Core description of the Texaco Coley No. 1 core (Permit No. 3029, Fig. 3), Vocation Field, Monroe County, Alabama.

packstones and grainstones. The well tested 105 BCPD and 245 MCFD from a 14-foot zone in the upper grainstone interval.

Conecuh Embayment

Depositional conditions in the Conecuh Embayment show significant evidence of restriction. High-energy conditions were uncommon in the embayment. They were lacking in the central portion as illustrated in the Exxon L & N Railroad well (Fig. 12). The

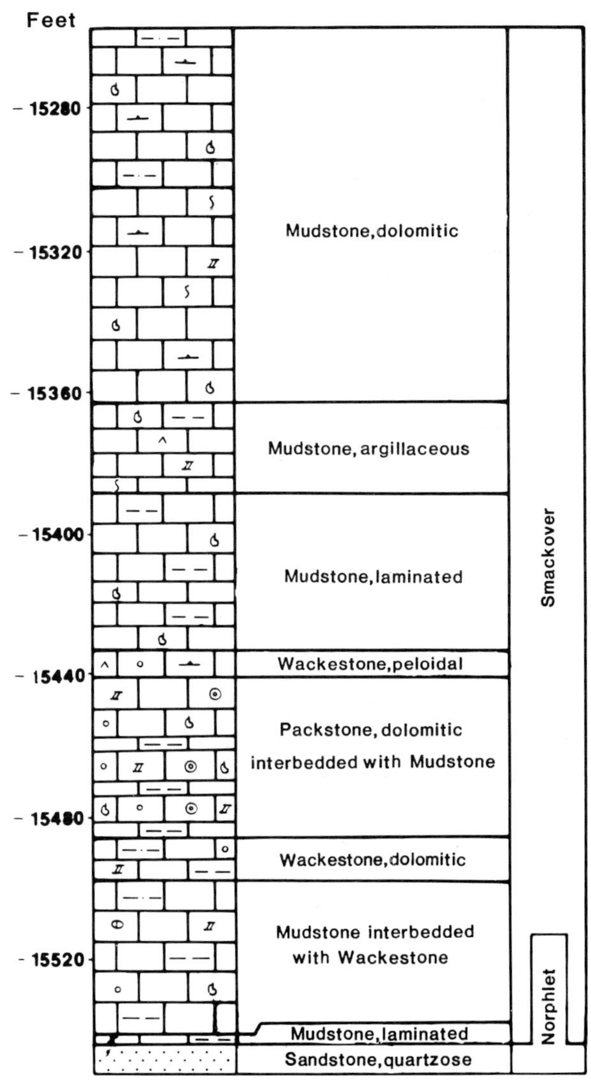

Figure 12. Description of the Exxon L&N Railroad No. 1 core (Permit No. 1766, Fig. 3), 1.5 miles west of Pollard, Escambia Co., Ala.

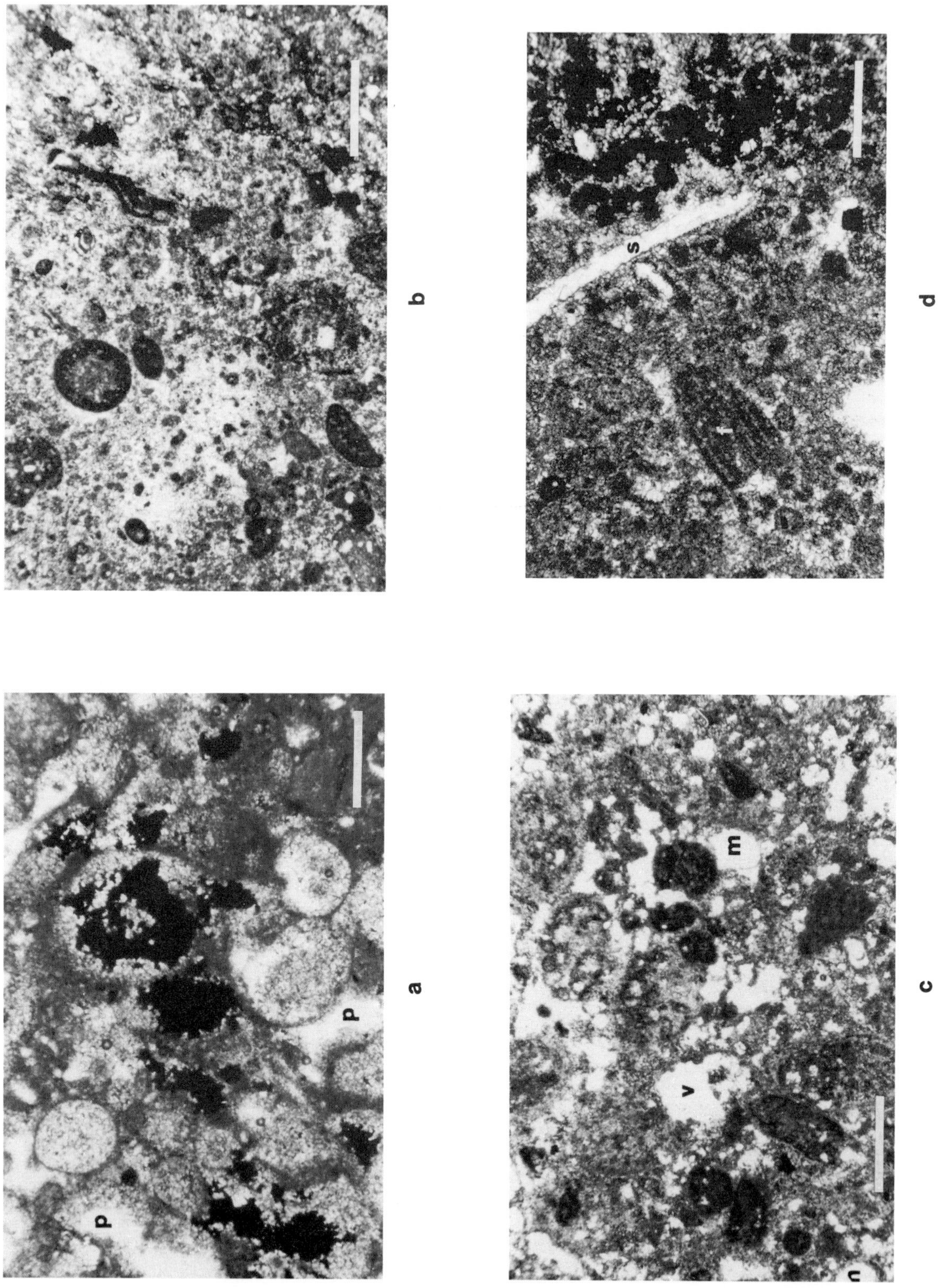

Figure 13

lower Smackover interval consists of oncolitic and peloidal wackestone/packstone interbedded with laminated mudstone. This sequence grades upward into dark-colored, laminated mudstones, which are overlain by peloidal wackestones and packstones of the upper portion of the Smackover interval (Fig. 13b). Shoal-water grainstone deposits are lacking across much of the embayment. Ooid grainstones do occur locally, in many cases associated with paleobathymetric highs (Bradford, 1982; Vinet, 1982). These grainstones or the peloidal wackestone/packstone are locally dolomitized, which is an important process in reservoir development.

Reservoir-grade porosity is restricted to the upper Smackover peloidal or oolitic grainstone/packstone deposits and is greatest where these deposits are dolomitized. The importance of dolomitization to overall reservoir development is well illustrated in the Big Escambia Creek-Flomaton Field area where upper Smackover carbonates are dolomitized and gas productive on top of the structure and nondolomitized and nonproductive off the structure (Bradford, 1982).

Porosity types include solution-enlarged intercrystalline, vuggy and grain moldic (Fig. 13c). Reservoir-grade porosity ranges from 8 to 30% with permeabilities of from 1 to 50 md. A porosity-permeability cross plot shows typical porosity-permeability relations for the area (Fig. 14).

Wiggins Uplift

In Hatter's Pond Field on the northwest flank of the Wiggins Uplift in Mobile County (Fig. 3), there is perhaps the best evidence of the control of pre-Smackover depositional topography on facies distribution and diagenesis. Because of the downdip position of the Hatter's Pond Field, it would be reasonable to expect the Smackover to consist of a thick section of lower Smackover basinal facies. However, this is not the case. The Smackover is very thin (less than 150 feet) and the facies present are high-energy upper Smackover facies (Benson and Mancini, 1982). Typical lower Smackover deposits are lacking and lower Smackover deposition is represented by time-equivalent eolian and marine-reworked sands of the upper portion of the Norphlet Formation.

The Smackover sequence is heavily dolomitized and appears to consist of a mixture of fine- to medium-crystalline dolomite interbedded with coarse- to very coarse-crystalline dolomite. This is well illustrated in the Getty Peter Klein No. 1 well, the discovery well for the Hatter's Pond Field (Fig. 15). Nondolomitized rocks only occur in the uppermost portions of the Smackover interval. These nondolomitized rocks include oolitic, skeletal, and peloidal grainstones and packstones.

In thick thin-sections, however, the

Figure 13. Photomicrographs showing porosity types in the Smackover Formation of southwest Alabama; bar scales equal 0.5 mm.

a. Solution-enlarged interparticulate (p) and oomoldic porosity in dolomitized oolitic grainstone; note dead hydrocarbons in oomolds; Texaco Coley No. 1 well, 14,025 ft.

b. Skeletal-peloidal wackestone typical of lower energy upper Smackover deposits in the Conecuh Embayment; Exxon L&N Railroad No. 1 well, 15,451 ft.

c. Vuggy (v) and grain moldic (m) porosity in dolomitized and leached upper Smackover peloidal wackestone; Exxon L&N Railroad No. 1 well, 15,454 ft.

d. Dolomitized skeletal-peloidal grainstone-packstone. Note the dolomitized *Faverina* pellet and the skeletal fragment (s); Getty Oil Creola Minerals 35-11 No. 2 well, Hatter's Pond Field, 18,090 ft. Note that this thin-section was made thicker than normal in order to better view the fabric.

true nature of the Smackover dolomites in Hatter's Pond is observed. The finely crystalline dolomites are peloidal grainstone and packstone that contain common *Faverina* pellets and scattered skeletal grains (Fig. 13d). The coarsely crystalline dolomites show definite ghosts of ooids and represent dolomitized oolitic grainstones (Fig. 16a). The oolitic grainstones represent ooid shoals, while the peloidal grainstones and packstones represent intershoal flats. The sequence has been subsequently dolomitized by mixing-zone dolomitization to produce the alternating finely and coarsely crystalline dolomite units.

Reservoir grade porosity in the Smackover in Hatter's Pond is restricted to the dolomitized ooid grainstones (coarsely crystalline dolomite). Porosity types include oomoldic, dolomitic intercrystalline, and substantial amounts of vuggy porosity (Fig. 16b). The vuggy porosity is primarily solution-enlarged moldic or intercrystalline and is a product of late-stage dissolution.

Porosities in the Hatter's Pond Field reservoir range from 10 to 20% with permeabilities of from 1 to 100 md. A porosity-permeability cross-plot for the Getty Peter Klein No. 1 well (Fig. 17) illustrates representative porosity-permeability relations in the field.

CONCLUSIONS

1. The depositional setting of the Smackover Formation in southwest Alabama can-

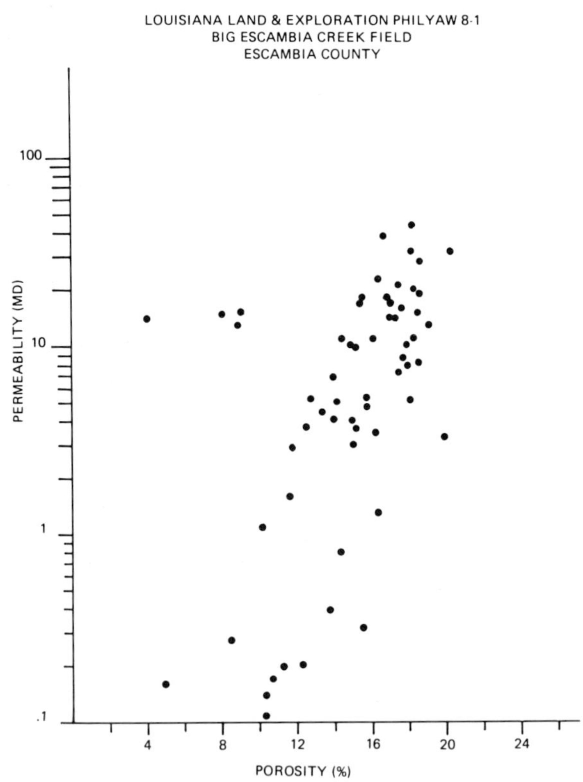

Figure 14. Porosity-permeability cross plot of potential reservoir zones in the Louisiana Land & Exploration Philyaw 8-1 well, Big Escambia Creek Field, Escambia County, Alabama.

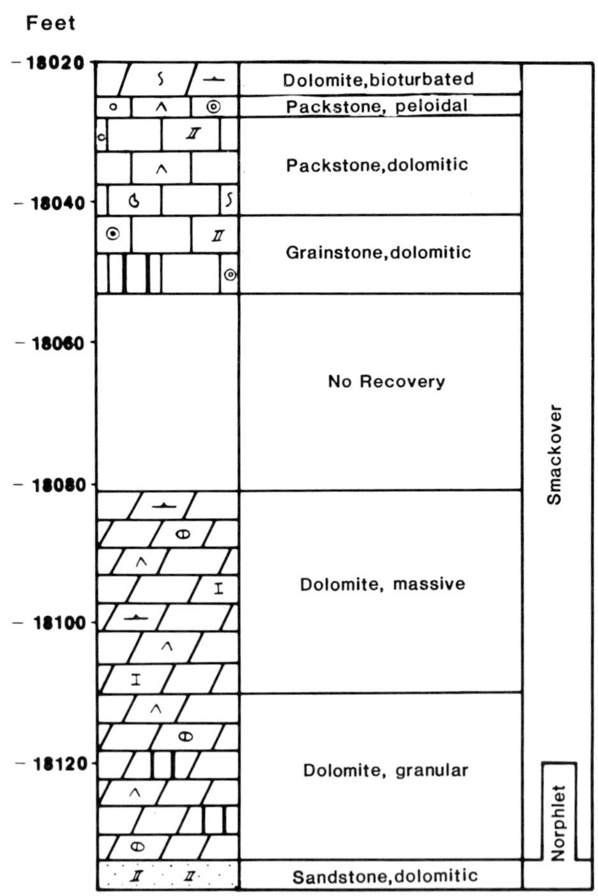

Figure 15. Core description of the Getty Oil Peter Klein No. 1 core (Permit No. 1978, Fig. 3), 12 miles northeast of Mobile, Mobile County, Alabama.

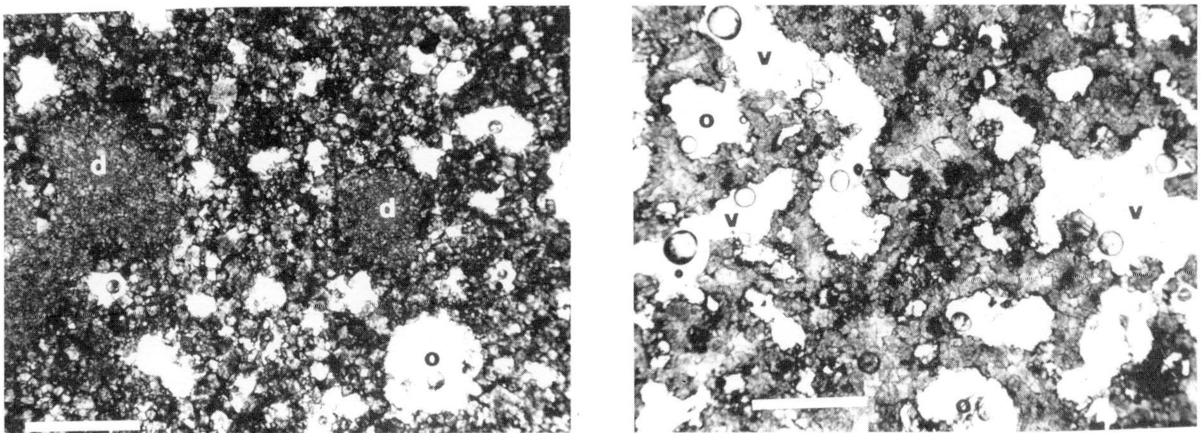

a b

Figure 16. Photomicrographs of porosity types in the Smackover Formation, Hatter's Pond Field, southwest Alabama; bar scales equal 0.5 mm.

a. Dolomitized oolitic grainstone; note the ghosts of micritized ooids represented by 0.2 to 0.5 mm patches of finely crystalline dolomite (d) included within the coarser crystalline dolomite mosaic and the spherical oomolds (o); Getty Oil Creola Minerals 35-11 No. 2 well, 18,117 ft.

b. Vuggy (v) and oomoldic (o) porosity; Getty Oil G.W. Radcliff 9-6 well, 18,444 ft.

not be adequately explained using the standard ramp model in use throughout the Gulf Coast. The Smackover depositional surface was significantly modified by a number of pre-Smackover paleobathymetric highs, which are reflections of buried Paleozoic ridges, salt diapiric structures, or pre-Smackover faults. These highs divided the Smackover depositional basin in Alabama into three separate basins or embayments, each characterized by a unique facies sequence.

2. The paleohighs had significant control on Smackover deposition in that most were fringed by higher energy deposits. Off the highs, lower energy conditions prevailed and organic-rich sediments with potential as hydrocarbon source rocks were deposited. The paleohighs are also significant in that they provide structural closure in many cases.

3. Porosity types are closely related to facies. Primary interparticulate porosity is only significant in areas where high- energy conditions were common in the upper Smackover such as in the northern portion of the Mississippi Interior Salt Basin and over the paleohighs. In areas where the upper Smackover sediments were deposited under low-energy conditions such as in the Manila and Conecuh Embayments, primary interparticulate porosity is minimal and development of productive reservoirs is dependent upon the formation of secondary porosity through diagenetic alteration.

4. Diagenesis has played a major role in reservoir development in the Smackover in southwest Alabama. Dolomitization is responsible for the formation of appreciable amounts of intercrystalline porosity, the dominant type of reservoir porosity in many areas. There is strong evidence that suggests dolomitization is controlled by paleobathymetry. Dolomitization is most extensive over paleohighs with the degree of dolomitization decreasing off the highs. Both early- and late-stage dissolution are also important processes in reservoir development, with moldic and vuggy porosity being the major porosity types in many of the reservoirs.

5. Evidence of late-stage dissolution is widespread in many of the Smackover reservoirs in southwest Alabama. In some areas, such as the Hatter's Pond Field area, this late-stage dissolution was a principal factor in reservoir development. In areas where

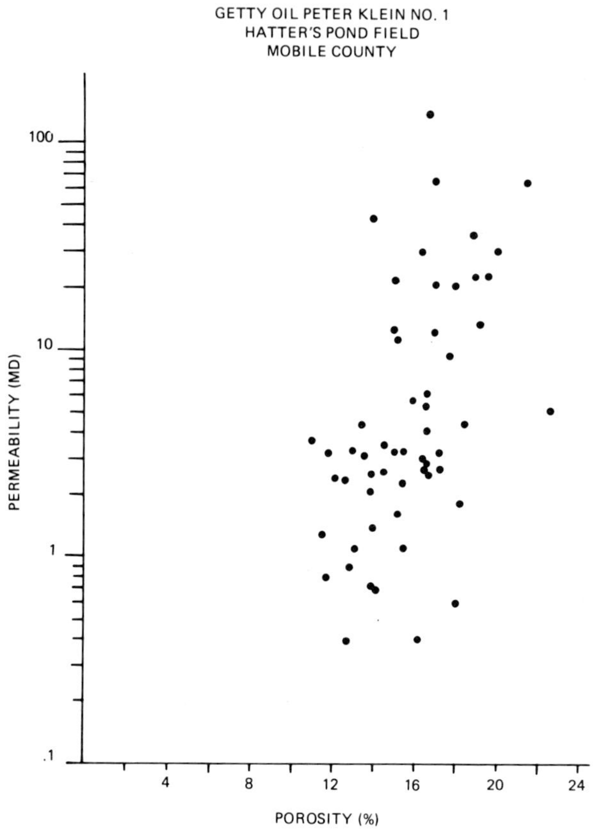

Figure 17. Porosity-permeability cross plot of potential reservoir zones in the Getty Oil Peter Klein No. 1 well, Hatter's Pond Field, Mobile County, Alabama.

eogenetic diagenesis was widespread, however, it is difficult to assess the volumetric importance of the late-stage dissolution. Further study is necessary to determine the significance of late-stage mesogenetic porosity in the Smackover of southwest Alabama and to determine its relation to earlier, eogenetically derived porosity.

6. Because of the wide variation in depositional and diagenetic settings, reservoir characteristics for the Smackover Formation in southwest Alabama cannot be integrated into a single exploration model. Instead, separate models must be developed for an individual area, and exploration techniques successful in one area may not necessarily be successful for all of southwest Alabama.

REFERENCES

Ahr, W. M., 1973, The carbonate ramp: an alternative to the shelf model: Gulf Coast Assoc. Geol. Soc. Trans., v.23, p. 221-225.

Baria, L. R., D. L. Stoudt, P. M. Harris, and P. D. Crevello, 1982, Upper Jurassic reefs of the Smackover Formation, U.S. Gulf Coast: Am. Assoc. Petroleum Geol. Bull., v. 66, p. 1449-1482.

Benson, D. J., and Mancini, E. A., 1982, Petrology and reservoir characteristics of the Smackover Formation, Hatter's Pond Field; implications for Smackover exploration in southwestern Alabama: Gulf Coast Assoc. Geol. Soc. Trans., v. 32, p. 67-75.

Bradford, C. A., 1982, Transgressive-regressive carbonate facies of the Jurassic Smackover Formation, Big Escambia Creek and Flomaton Fields, Escambia County, Alabama *in* The Jurassic of the Gulf Rim: Program and Abstracts, Third Annual Research Conf., Gulf Coast Section, Soc. Econ. Paleo. and Min., p. 15-17.

Longman, M. W., 1982, Carbonate diagenesis as a control on stratigraphic traps: Am. Assoc. Petroleum Geol., Ed. Course Note Ser. 21, 159 p.

Loucks, R. G., and D. A. Budd, 1981, Diagenesis and reservoir potential of the Upper Jurassic Smackover Formation of south Texas: Gulf Coast Assoc. Geol. Soc. Trans., v. 31, p. 339-346.

______ 1982, Diagenesis and subsurface-derived secondary oomoldic porosity in Upper Jurassic Smackover Formation, south Texas *in* The Jurassic of the Gulf Rim: Program and Abstracts, Third Annual Research Conf.

Gulf Coast Section, Soc. Econ. Paleo. and Min., p. 51-53.

Mancini, E. A., and D. J. Benson, 1980, Regional stratigraphy of Upper Jurassic Smackover carbonates of southwest Alabama: Gulf Coast Assoc. Geol. Soc. Trans., v. 30, p. 151-165.

Masingill, J. H., and D. E. Bolin (editors), 1982, The petroleum industry in Alabama, 1981: Alabama Geol. Survey, Oil and Gas Report 3E, 47p.

Moore, C. H., and Y. Druckman, 1981, Burial diagenesis and porosity evolution, Upper Jurassic Smackover, Arkansas and Louisiana: Am. Assoc. Petroleum Geol. Bull., v. 65, p. 597-628.

Purser, B. H., 1973, Sedimentation around bathymetric highs in the southern Persian Gulf *in* Purser, B. H. (ed.), The Persian Gulf, Springer-Verlag, New York, p. 157-178.

Vinet, M. J., 1982, Stratigraphy and dolomitization of the Smackover and Buckner Formations (Upper Jurassic) in the Jay-Big Escambia Creek Field area, southern Alabama-western Florida *in* The Jurassic of the Gulf Rim: Program and Abstracts, Third Annual Research Conference, Gulf Coast Section, Soc. Econ. Paleo. and Min., p. 110-113.

ENVIRONMENTS OF DEPOSITION AND EARLY DIAGENESIS, LA JOYA FORMATION, HUIZACHAL GROUP RED BEDS, NORTHEASTERN MEXICO

BRYAN BRACKEN

Department of Geosciences
The University of Texas at Austin
Austin, Texas 78712

ABSTRACT

During the Middle Jurassic, continental basins formed in northeastern Mexico that were complexly related to the initial stages of the opening of the Gulf of Mexico. The La Joya Formation is one of the many sequences of red beds that partially filled these basins. It is composed of terrigenous clastics and freshwater limestones. The vertical sequence of facies, from the base up, are alluvial-fan clastics, lacustrine carbonates and clastics, meandering-fluvial and low-sinousity, ephermeral fluvial clastics. Early diagenetic modifications include pedoturbation and pedogenic calcite precipitation, infiltration of clay, dissolution of unstable framework grains, and precipitation of authigenic phases from the released elements. Infiltered clay was a major porosity reducing process and is, in part, facies controlled. This early diagenesis significantly modified the compositional and textural characteristics and porosity of the La Joya sandstones and contributed to the poor reservoir quality.

INTRODUCTION

The quest for geological information concerning the initial opening of the Gulf of Mexico Basin in the early Mesozoic is both academically and economically motivated. Rocks deposited during this early stage of basin formation are deeply buried in the northern and southern part of the basin (Scott *et al.*, 1961; Vernon, 1971; Woods and Addington, 1973). The outcrop on the western margin, however, provides the only opportunity for gathering more detailed sedimentological and stratigraphic information.

GEOLOGICAL SETTING AND STRATIGRAPHY

The La Joya Formation crops out in the southwestern corner of the state of Tamaulipas, west of Cuidad Victoria (Fig. 1). Exposures are located in canyons deeply incised

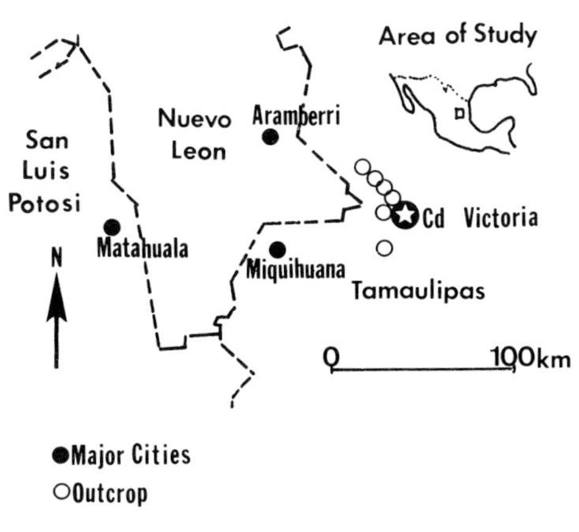

Figure 1. Map showing the location of outcrops of the La Joya Formation.

into the flanks of the Huizachal-Peregrina Anticlinorium, which is a northwest-trending structural element of the eastern front of the Sierra Madre Oriental (Fig. 2).

The La Joya Formation is the younger of two Early Mesozoic,Huizachal Group red-bed sequences (Mixon, *et al.*, 1959). Because of the lack of age definitive fossils, this se-

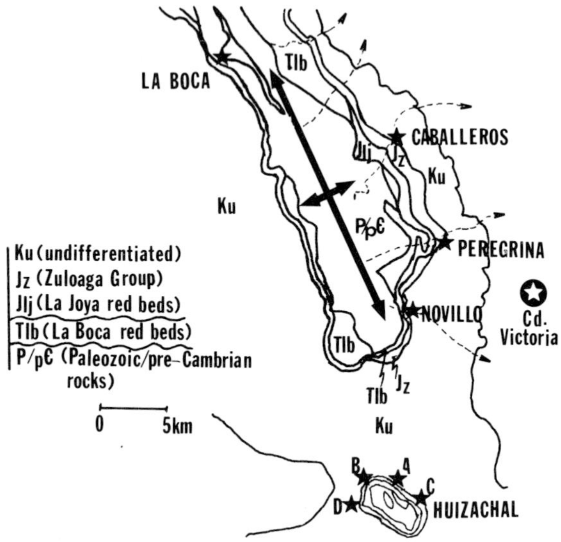

Figure 2. Geological map of Huizachal-Peregrina Anticlinorium. Stars indicate outcrops of the La Joya Formation; after Lopez Ramos, 1980.

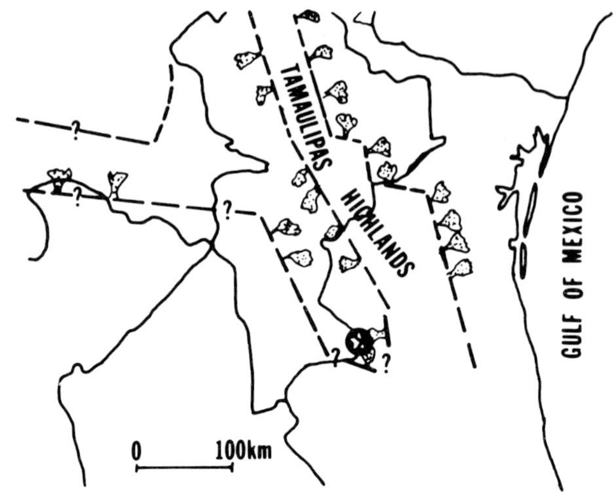

Figure 4. Configuration of continental basins and location of red beds of Late Triassic to Middle Jurassic age; after Padilla, 1982.

quence of rocks can be dated only as older than the Oxforidan-Kimmeridgian Novillo Formation of the Zuloaga Group (Humphrey and Diaz, 1956) and younger than the La Boca Formation red beds (Late Triassic?) (Fig. 3). At all localities, the La Joya Formation is overlain conformably by limestones and dolostones of the Novillo Formation and unconformably overlies the La Boca Formation and, less commonly, altered felsic volcanics and pre-Cambrian medium- to high-grade metamorphic rocks. The La Joya may only be positively identified by its association with the La Boca Formation. With present levels of information about the geology of northern d Mexico and the inability to correlate red-bea sequences throughout this region, the name La Joya Formation must be restricted to the are of this study.

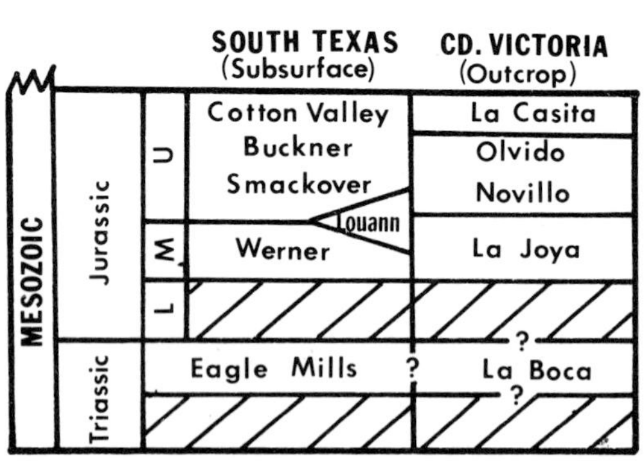

Figure 3. Stratigraphic column of the study area and south Texas.

The La Joya Formation consists of 55 to-120 meters of terrigenous clastics and fresh water limestones deposited in a northwest-trending basin prior to the transgression of the Oxfordian sea (Fig. 4). In general, these rocks represent a period of sedimenta-r tion in a peripheral, continental basin prio to, and in part synchronous with deposition of carbonates and evaporites of the transgressive Zuloaga Group.

LITHOFACIES AND INTERPRETATION

More than three quarters of the La Joya Formation consists of mudstone, sandstone, and conglomerate, with fine-grained sandstone the most common. The remainder of the formation consists of laminated carbonate and calcareous mudstone. The sequence is predominantly red in color, except the top 5 to 10 meters of each section which are a drab, green/gray. Regionally, all exposures have the same vertical variation in grain size. There is an upward fining in grain size followed by a coarsening sequence that corresponds to each of the four recognized depositional environments. The following facies occur in the La Joya, from bottom to top: (1) conglomeratic, alluvial-fan deposits; (2) lacustrine carbonates and fine-grained clastics; (3) mixed-load, high-sinuosity fluvial clastics; and (4) low-sinuosity fluvial clastics.

Alluvial-Fan Facies

The alluvial-fan facies is composed of 0.75 to 14.5 meters of poorly sorted boulder to pebble conglomerates and coarse-grained sandstones. Individual beds are 0.5 to 0.75 meters thick, structureless, have clast- and matrix-supported textures, lack channel-shaped lower boundries, and commonly are graded. Thin lenses of coarse- to medium-grained sandstone overlie the conglomeratic beds. Vertically, the average grain size and abundance of matrix-supported textures decreases.

The poor sorting, upward fining in grain size, and the mixture of matrix- and clast-supported textures suggest that these are debris- and poorly confined stream-flow deposits, which are characteristic of deposition on steep alluvial surfaces. The overall "fan-shaped" geometry for the Middle Jurassic red-bed sequences of northeastern Mexico (Padilla, 1982) supports the interpretation of deposition as alluvial fans along nearby highlands (Fig. 4).

Lacustrine Facies

Abruptly overlying and interfingering with the upper part of the conglomeratic deposits are laminated limestones and calcareous and non-calcareous mudstones (Fig. 5). The lacustrine facies, which is 1.5 to 25 meters thick, mainly consists of 1 to 5 meter-thick beds of thinly interlaminated (laminae are 2 to 3 mm thick), fine-grained, terrigenous clastics and micritic to microsparric calcite. The limestone units have soft-sediment deformational features, and contain no organic matter or minerals to suggest reducing conditions. Overlying each limestone unit is a thin (less than 1 m thick) struc-

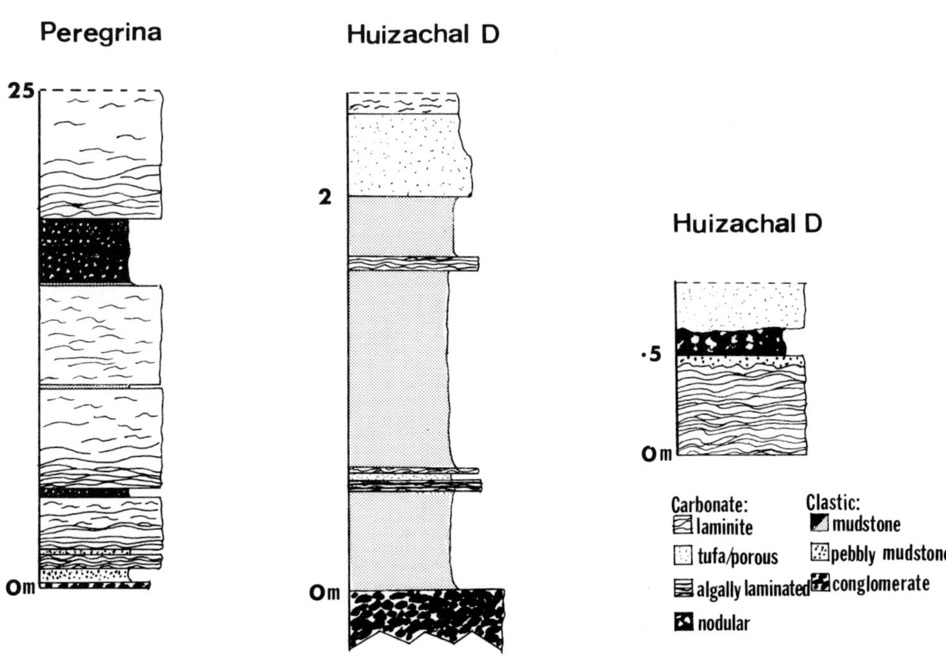

Figure 5. Lithic column of the lacustrine facies from two outcrop localities. Note the variation in thickness between the two outcrops.

tureless, red mudstone which contains dispersed calcitic nodules. The upper surfaces of the limestone beds are irregular and brecciated. Recrystallized textures are also common immediately beneath the irregular upper surfaces. Most of the features of the limestone suggest reworking during periods of subaerial exposure, and the development of paleosols (Freytet, 1973).

The mineralogy and texture and structures of the rocks of this facies suggest deposition in a low-energy, carbonate-precipitating lake adjacent to the coarse-grained, alluvial-fan environment. Sediment may have accumulated by a cyclic mechanism of carbonate and clastic deposition similar to the seasonal couplets that develop in a thermally stratified lake ("laminites"). There is no evidence to indicate depth of water. The repeated episodes of subaerial exposure and thoroughly oxidized condition of the lacustrine rocks suggest a shallow water depth which more likely would be affected by water-level fluctuations.

Fluvial Facies

The upper and major part of the La Joya red beds are silty mudstones and conglomeratic, fine- to medium-grained sandstone which comprise an upward-coarsening sequence up to the abrupt contact with the overlying Novillo Formation. The sequence changes from mudstone, overlying the lacustrine rocks, to fine- to medium-grained sandstone of the lower fluvial facies; these, in turn, are overlain by pebbly, fine- to medium-grained sandstone of the upper fluvial facies.

High-sinuosity fluvial facies. The one to three meters thick sandstone units within the lower part of the fluvial sequence are characterized by an upward-fining grain size, an erosional base with or without a lag deposit, and a sequence of sedimentary structures suggestive of deposition on a point bar of a high-sinuosity fluvial stream (Fig. 6). Exposures of this facies in Caballeros Canyon (Fig. 2) show accretionary bar deposits with 1.1 to 2.0 m thick sets of "epsilon" cross-bedding (Allen, 1965; 1970b), indicating from empirical width/depth relations a channel width of 2 to 25 m at a 95% confidence level (Leeder, 1973).

Thinly-bedded, silty mudstones and very fine- and fine-grained sandstone that encase the channel sandstones commonly have pedogenetically disrupted, or "wispy" stratification, planar lamination, and climbing ripple cross-stratification. Structures and textures suggest deposition of a suspension and bedload transported sediment during overbank-

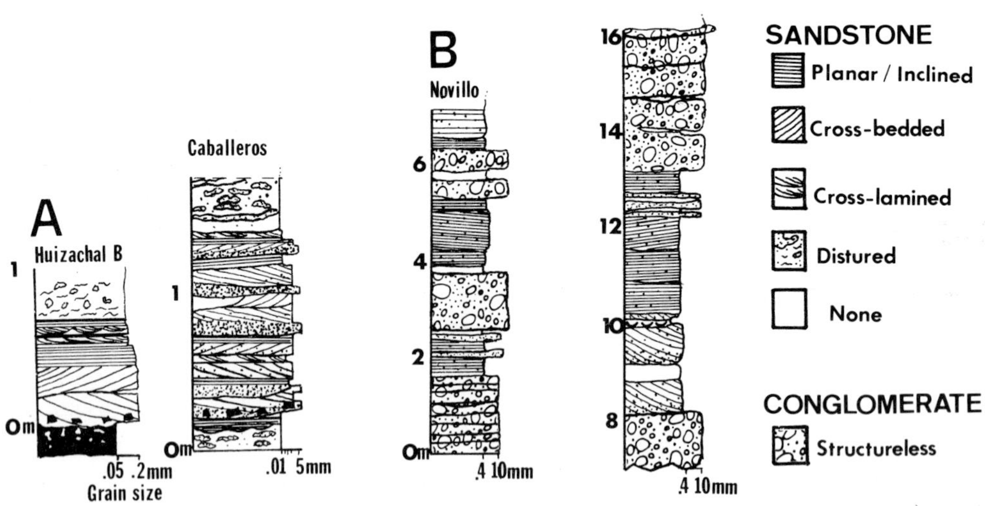

Figure 6. Fluvial facies, (A) high-sinuosity and (B) low-sinuosity, channel sandstsones.

ing of floodwater. Extended periods of weathering (non-deposition) are indicated by relict pedogenic and diagenetic features. Some of the results of the weathering are discussed in the following section on diagenetic modifications of the sandstones.

Low-sinuosity, fluvial facies. The uppermost facies is composed predominantly of pebbly, fine- to medium-grained sandstone, with pebble conglomerates a minor component (Fig. 6). These rocks abruptly overlie the high-sinuosity, fluvial rocks and lack their characteristic sequence of structures and grain sizes. Bedding ranges from 0.5 m to 0.75 m thick, and internally is dominated by planar or slightly inclined laminations. Rarely 0.5 m thick, tangentially-planar, cross-stratified units are present. Thin mud strips with dessication polygons overlie many of the sandstone beds.

The only modern environment which is dominated by planar stratification is a sandy, ephemeral-fluvial stream. High-energy stream flows or sheet floods are the dominant depositional mechanism (McKee *et al.*, 1967; Scott *et al.*,1969; Williams, 1971). Abundant parallel laminations are the dominant type of stratification characteristic of high-energy flows common to regions of sporadic rainfall and sparse vegetation (*i.e.*, relatively arid climates). Evaporites that overlie similar red beds to the north near Aramberri (Fig. 1; Padilla, 1978) and evaporites in the Olvido Formation in El Huizachal Valley support the interpretation of an arid climate during the deposition of the uppermost facies.

EARLY DIAGENETIC MODIFICATIONS

Thin-section analysis, X-ray diffractometry, and scanning electron microscopy were utilized to determine: (1) source-rock composition and (2) early diagenetic modification of sandstone porosity in the La Joya Formation. Porosity evolution and distribution were then determined in order to evaluate the reservoir potential of the sandstones.

The sandstones are volcanic-litharenite and feldspathic volcanic-litharenites (Folk, 1974) with a lesser contribution from high- to low-grade metamorphic and granitic rocks. The abundance of volcanic rocks in the source areas of the La Joya Formation presents an enigma, since insufficient quantities of pre-Middle Jurassic volcanic rocks have been identified. This problem illustrates the present lack of understanding of Mesozoic volcanism and tectonism in northeastern Mexico.

The original depositional fabric of the sandstones was modified by the following early diagenetic processes: (1) pedoturbation, (2) clay infiltration, (3) pedogenic calcite precipitation, (4) dissolution of unstable framework grains, and (5) the precipitation of ferric oxide, mixed-layer smectitic/illitic clay minerals, quartz, and feldspar cements. Petrographic textural relationships indicate that these five modifications occurred before significant burial compaction or cementation.

Periods of weathering (slow or non-deposition) modified the sorting, textural and compositional maturity of sandstones. Pedoturbation (wet/dry alternations) disrupted floodplain sediments and commonly produced disturbed, poorly sorted, structureless beds. In addition, small amounts of finely disseminated nodular masses of pedogenic calcite crystals (less than 5% of the volume of a sample) reduced porosity. These modifications were most active in the floodplain environment, but reduced the depositional porosity the least of all modifying processes acting on the La Joya sandstones. The infiltration of clay into pore spaces created argillans (Brewer, 1964) on sand grains. This occurred in all facies (Fig. 7). Clay is infiltered either by pedogenic processes or by the influent seepage of floodwater laden with suspended material (Crone, 1975). The volume of matrix in the sandstones is facies controlled (Fig. 8). Floodplain and meandering-channel sandstones have the highest amount of matrix ($\bar{x}$15% and 14%, respectively) as is expected with the greater availability of clay-sized particles and the

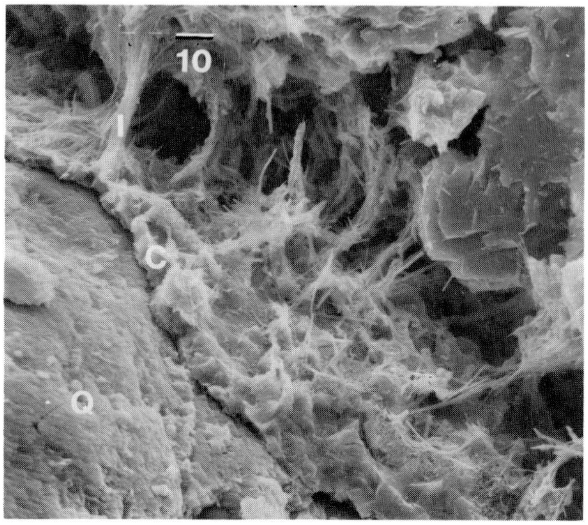

Figure 7. S.E.M. photomicrograph of argillaceous cutan (C), rounded quartz grain (Q) and later stage fibrous illitic cement (I). Bar in microns.

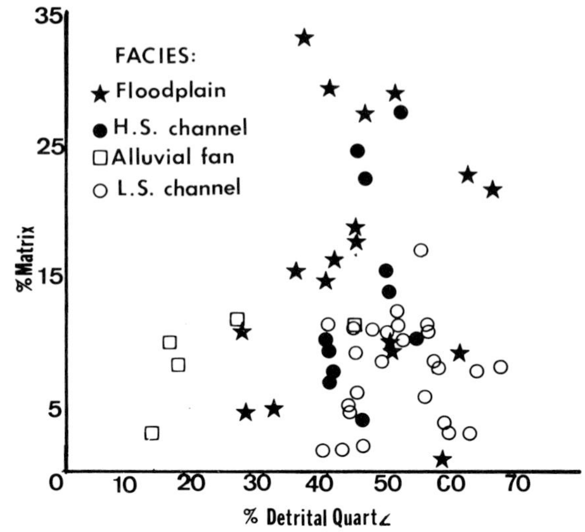

Figure 8. Graph of infiltered matrix vs. amount of detrital quartz grains. Samples from the high-sinuosity facies contain more matrix.

increased pedogenic activity in the floodplain environment. Sandstones from the other facies contain on the average less than 9% infiltered matrix. The addition of clay is the most effective porosity-reducing process during early diagenesis. In fact, infiltered clay occludes more porosity than all other diagenetic modifications, except later stage calcite cement.

Dissolution occurred at the same time as the pedogenic modifications and clay infiltration. The exposure of sand grains to meteoric water had two important influences on all sandstones. The first was the dissolution of unstable framework grains, and second was the precipitation of authigenic phases from the released elements. La Joya sandstones contain only ultrastable heavy minerals (*e.g.*, zircon and tourmaline), which in the light of the first-cycle nature of the sandstones, suggest removal and alteration of the less stable grains. Oxidizing conditions in this early diagenetic stage formed ferric hydroxide (red pigment precursor) from the iron released from altered framework grains (Fig. 9). At the same time, less mobile elements precipitated as mixed-layer smetitic/illitic clay minerals (Fig. 9), but both form, on the average, less than 1% of a sample. Quartz, plagioclase, and potassium feldspar cements also were precipitated, but are volumetrically only a minor component of the sandstones.

SUMMARY AND CONCLUSIONS

The continental basins in northeastern Mexico, which formed in the Early and Middle Mesozoic, were the result of complex, poorly understood tectonic activity. Into these basins in the southwestern corner of the state of Tamaulipas, a thick sequence of red, terrigenous clastics and freshwater limestones accumulated. Deposition occurred both later than and partly synchronous with the deposition of the Oxfordian-Kimmeridgian aged, transgressive carbonates and evaporites of the Zuloaga Group.

The sequence of facies and their structures and textures indicate that alluvial-fan clastics and lacustrine carbonates and clastics were the initial sediments deposited (Fig. 10). The remainder of the basin fill

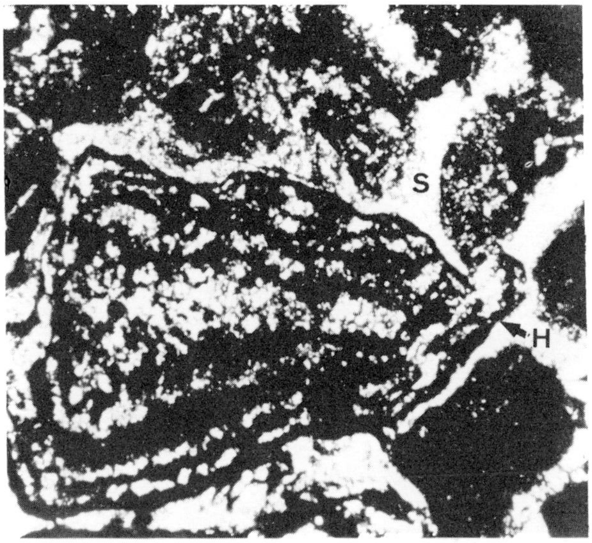

Figure 9. Photomicrograph of relict hematite (H) rims in smetitic clay (S). Note the concentric dissolution fronts (rims) with a jagged pattern. Width = 0.65 mm.

consists of fluvial clastics deposited first by mixed-load, high-sinuosity streams and then by low-sinuosity, ephermeral-flow streams.

Depositional fabric and porosity of the La Joya sandstones were significantly modified during early diagenesis. Pedogenic processes (pedoturbation and pedogenic precipitation of calcite), which are of primary importance in the floodplain environment, decreased sorting and porosity. Clay infiltration (*i.e.*, infiltered matrix) reduced significant amounts of porosity ($\bar{x}$=14% to 15%) in all sandstones of the meandering fluvial facies and to a lesser extent in sandstone from the other facies (less than 9%). Dissolution under the influence of meteoric water removed and altered unstable framework grains and provided a source of elements for the precipitation of iron oxide, mixed-layer smectitic/ illitic clay minerals, quartz and feldspar cements. Although the composition and color of the sediment were drastically altered by the dissolution of framework grains and precipitation of early authigenic phases, there was only a minor effect on porosity. The combination of infiltered clay and small amounts of cement in combination with later diagenetic phases (*i.e.*, illitic

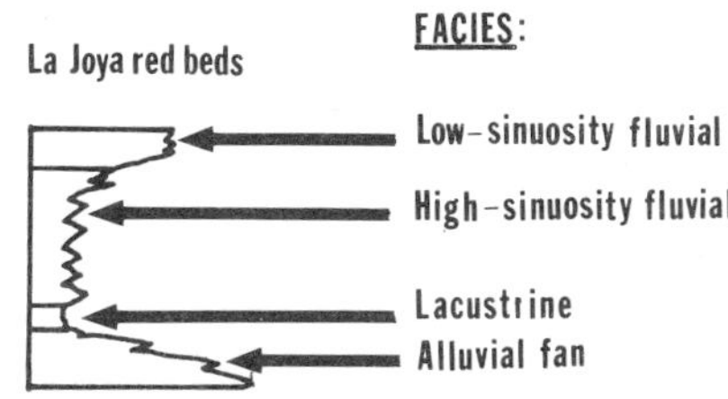

Figure 10. Graphic summary of vertical variation in facies of the La Joya Formation.

clay and calcitic cements) creates poor reservoir quality in the sandstone of the La Joya Formation.

ACKNOWLEDGMENTS

This paper presents the results of Master's thesis research performed at The University of Texas at Austin on the sedimentological and diagenetic history of the Middle Jurassic La Joya Formation in northeastern Mexico. I would like to thank Shell Oil Company, New Orleans, for the major part of the funding of the thesis research and the Geology Foundation, University of Texas at Austin, for additional funds. Melissa Sandstrom's guidance and companionship made the months of field work enjoyable and rewarding.

REFERENCES

Allen, J. R. L., 1965a, Fining-upward cycle in alluvial sedimentation: Geol. Jour., v. 4, p. 229-246.

__________, 1970, A quantitative model of grain size and sedimentary structures in lateral deposits: Geol. Jour., v. 7, p. 129-146.

Brewer, R., 1964, Fabric and mineral analysis of soils: Wiley, N.Y., N.Y., 259p.

Crone, A. J., 1975, Laboratory and field studies of mechanically infiltered matrix clay in arid fluvial deposits: Ph.D. thesis, Boulder, U. of Colorado, 162p.

Folk, R. L., 1974, Petrology of sedimentary

rocks: Hemphill Pub. Co., Austin Texas, 182p.
Freytet, P., 1973, Petrography and paleoenvironment of continental carbonate deposits with particular references to Upper Cretaceous and Eocene of Languedoc (southern) France: Sed. Geol., v. 10, p. 26-60.
Humphrey, W. E. and T. Diaz, 1956, Tectonic framework of northeast Mexico: unpublished manuscript.
Leeder, M. R., 1973, Fluviatile fining-upwards cycles and the magnitude of paleochannel: Geol. Mag., v. 110, p. 265-276.
McKee, E. D., E. J. Crosby, and H. L. Berryhill, 1967, Flood deposits, Bijou Ck., Colorado, June, 1965: Jour. Sed. Pet., v. 37, p. 829-851.
Mixon, R. B., G. E. Murray, and T. Diaz, 1959, Age and correlation of Huizachal Group (Mesozoic), State of Tamaulipas, Mexico: Am. Assoc. Pet. Geol. Bull., v. 43, p. 757-771.
Padilla y Sanchez, R. L., 1978, Bosquejo geologico-estructural de la Sierra Madre Oriental en el area Linares-Galeana-San Roberto, Estado de Nuevo Leon: Univ. Nac. Auton. Mexico, Inst. Geologia, Revista, v. 2, no. 1, p. 45-54.
__________, 1982, Geological evolution of the Sierra Madre Oriental between Linares, Conception del Oro, Saltillo, and Monterrey, Mexico: Dissertation, U. of Texas, Austin, 1982.
Scott, A. J., R. A. Hoover and J. McGowen, 1969, Effects of Hurricane "Beulah", 1967, on Texas coastal lagoons and barriers: U.N.E.S.C.O., Simposia Intern. Sobre Lagunas Costeras, p. 221-236.
Scott, K. R., W. E. Hayes and R. P. Fietz, 1961, Geology of the Eagle Mills Formation: Trans, Gulf Coast Assoc. Geol. Soc., V. 11, p. 1-14.
Vernon, R. G., 1971, Possible future petroleum potential of the pre-Jurassic western Gulf Basin: *in,* Cram, I. H., (ed.), Future petroleum provinces of the U. S.- Their geology and potential, v. 2, p. 954-979.
Williams, G. E., 1971, Flood deposits of the sand-bed ephemeral streams of central Australia: Sed., v. 17, p. 1-40.
Woods, R. D. and J. W. Addington, 1973, Pre-Jurassic geologic framework, northern Gulf Basin: Trans. Gulf Coast Assoc. Geol. Soc., v. 23, p. 92-108.

TRANSGRESSIVE-REGRESSIVE CARBONATE FACIES OF THE SMACKOVER FORMATION, ESCAMBIA COUNTY, ALABAMA

CYNTHIA A. BRADFORD

Mobil Oil Exploration and Producing Southeast, Inc.
1250 Poydras Street
New Orleans, Louisiana 70113

ABSTRACT

The Smackover Formation in Escambia County, Alabama is divided into ten facies. The transgressive facies of the basal Smackover were deposited in a lower supratidal to upper subtidal environment and consist of interbedded algal-laminated mudstone and oncolite-pellet packstone. A coated fossil packstone facies was also deposited in the lower unit of the Smackover, but in more open-marine conditions. During the deposition of the middle Smackover this facies acted as a partial barrier and restricted the energy and salinity in the landward area. These restrictions resulted in the deposition of a fossiliferous mudstone facies, which alternated with a distinctive, black laminated mudstone devoid of preserved fauna or bioturbation. The regressive upper Smackover sequence is dominated by a lagoonal pellet wackestone/mudstone. This lagoonal mudstone was deposited behind an oolite shoal which developed seaward on the slight topographic relief formed by salt movement during middle Smackover time. Landward of the lagoon, a sabkha developed which eventually prograded over the entire facies tract.

Dolomitization of the oolite grainstone and underlying lagoonal mudstone created porosity in the upper Smackover. This porosity, along with the presence of favorable structure and dense anhydrite seal provided by the overlying Buckner Formation, combined to produce an effective trap for hydrocarbon accumulation.

INTRODUCTION

The Late Jurassic Smackover Formation, a major producer of oil and gas, is present in the subsurface along the northern Gulf Coast Basin from Texas to Florida. Variations in lithofacies, porosity types, and hydrocarbon trapping mechanisms are present along this trend due to differences in depositional environments, diagenesis, and local structural features (Mancini and Benson, 1980; Moore and Druckman, 1981; Loucks and Budd, 1981).

This study concentrates on the Smackover Formation in Escambia County, Alabama (Fig. 1). The major Smackover fields in this area are Big Escambia Creek and Jay Fields, while Flomaton Field produces from the underlying Norphlet Sandstone. The objectives of this paper are to describe the major Smackover facies, interpret the depositional environments in which they formed, and to relate the environments, structure, and diagenesis to hydrocarbon accumulations.

STRATIGRAPHY AND REGIONAL SETTING

In Alabama and Florida the Smackover Formation overlies the Norphlet Formation, which is interpreted as alluvial braided-stream deposits overlain by eolian and intertidal quartzose sandstone (Wilkerson, 1981). The Norphlet, in turn, overlies the Louann Salt. Above the Smackover is the Buckner Formation which consists of anhydrite, salt, and shale. The Buckner is interpreted as a

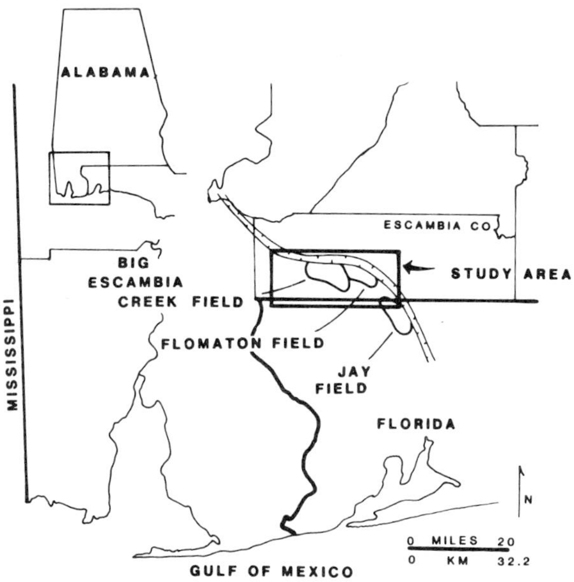

Figure 1. Location of major oil and gas fields in the study area.

sabkha brine-pan complex (Sigsby, 1976), and is a continuation of the regressive phase of the upper Smackover carbonates.

Smackover deposition in southwest Alabama was significantly affected by three Paleozoic structural elements: the Conecuh Ridge, Pensacola Arch, and Wiggins Uplift (Ottman *et al.*, 1973) (Fig. 2). The Conecuh Ridge and Pensacola Arch formed the updip limits to Smackover deposition. The Pensacola Arch also provided a partial barrier to waves and currents from the east and southeast. The Wiggins Uplift partly restricted the Manila and Conecuh Embayments. The effects of the restrictions are seen in the paucity of fossils in the Smackover, and high percentage of low-energy facies found throughout the Smackover. The low energy level of the system was probably also caused by the broad, gently dipping nature of the ramp surface, especially in the Conecuh Embayment.

A series of grabens and normal faults called the Pickens-Gilbertown Fault System are present in the study area. Similar fault systems surround the margin of the Gulf of Mexico Basin and are believed to have been caused by salt flowage beginning in the Jurassic (Hughes, 1968). Studies by Sigsby (1976) indicate that in the Alabama-Florida area the graben-system faults were initiated during the Kimmeridgian (post-Smackover), but that earlier basement faults produced some structures affecting Smackover deposition. The main Jurassic oil and gas fields in the study area are bounded to the north and east by this fault system (Fig. 1).

FACIES DESCRIPTIONS AND DISTRIBUTIONS

Methods

Gross correlation of the Jurassic sedimentary section in Escambia County was achieved through the use of resistivity logs from 80 wells which penetrated all or part of the Smackover Formation. Slabbed cores and thin-sections from 20 of these wells (Table 1), totaling 4,000 feet of section, were described in order to define lithofacies. Density-neutron logs from 40 wells were used to determine lithology and to correlate core data with uncored wells. Locations of these

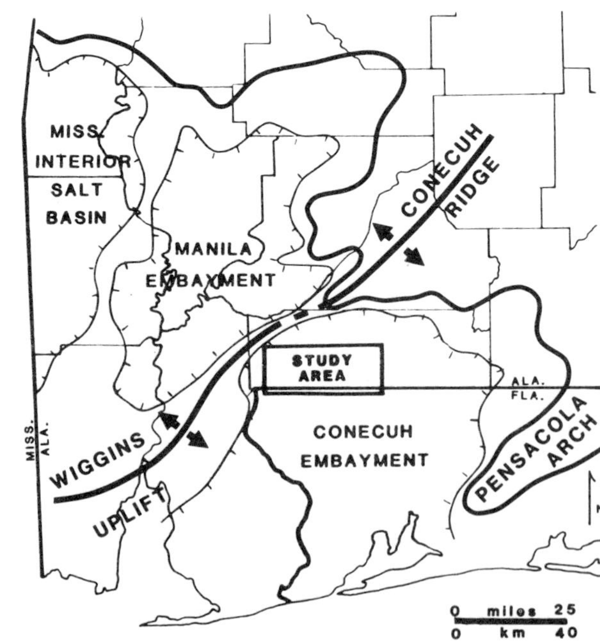

Figure 2. Regional paleogeographic features present during deposition of the Smackover Formation in southwestern Alabama (from Mancini and Benson, 1980).

Table 1. Core Used in Study

Core No.	Well Names	Location	Core Depth feet
1	Exxon #26-11 Scott Paper	26-2N-5E	15,771-15,853
2	Placid Digman	31-2N-6E	15,735-16,011
3	Exxon St. Lease 338	28-2N-6E	15,420-15,661
4	Exxon Scott Paper #2	2-1N-6E	15,520-15,615
5	Exxon Helton	1-1N-6E	15,485-15,644
6	Exxon Turner	6-1N-7E	15,409-15,691
7	Exxon Scott #5	5-1N-7E	15,322-15,661
8	Exxon Scott Paper #1	1-1N-7E	15,107-15,442
9	Exxon Steward	11-1N-8E	15,002-15,067
10	La. Land and Expl. Philyaw	8-1N-7E	15,490-15,659
11	Exxon Scott Paper #17	17-1N-7E	15,605-15,722
12	La. Land and Expl. Hall	30-1N-7E	15,822-15,978
13	Exxon McCurdy	12-1N-7E	15,254-15,581
14	Exxon Scott Paper #15	15-1N-7E	15,446-15,580
15	Exxon Scott Paper #23	23-1N-7E	15,615-15,703
16	Exxon Scott Paper #25	25-1N-7E	15,545-15,595
17	Exxon Hart	26-1N-8E	15,395-15,697
18	Exxon Bush	25-1N-8E	15,297-15,580
19	Exxon L. and N. RR	30-1N-9E	15,257-15,544
20	La. Land and Expl. Mill	32-1N-9E	15,152-15,498

wells and cores and two cross-sections which illustrate the relationships of the ten Smackover facies are shown in Figure 3. In order to describe these facies relationships the Smackover was divided into lower, middle, and upper units.

Lower Smackover (Fig. 4)

The lowermost Smackover facies is an algal-laminated mudstone. This facies is light brown in color, almost totally dolomitized, and consists of continuous wavy to parallel laminations of alternating dolomicrite and algal material. Peloids are rare and the algal laminations are occasionally disrupted by burrows. The laminated mudstone is interbedded with an oncolite-pellet packstone/wackestone. This facies is also well dolomitized and commonly contains anhydrite nodules and pyrite. Allochems consist of abundant and ubiquitous *Favreina* fecal pellets, scattered, but common, oncolites and intraclasts of algal mats, and minor amounts of high-spired gastropods. The facies is frequently burrowed. The contact between the laminated mudstone and the overlying packstone is sometimes sharp with reworked algal fragments incorporated into the overlying pellet packstone. In other areas it is gradational, with the laminations becoming more burrowed and less distinct upward.

The algal-laminated facies is thickest toward the northeast part of the study area and laterally forms a narrow band that trends in a northwest-southeast direction. The associated oncolite-pellet facies thins in the area of maximum algal mudstone, but it is generally constant in thickness and averages 60 feet (18 meters) in the rest of the study area. Figure 5 is a cross-section of the en-

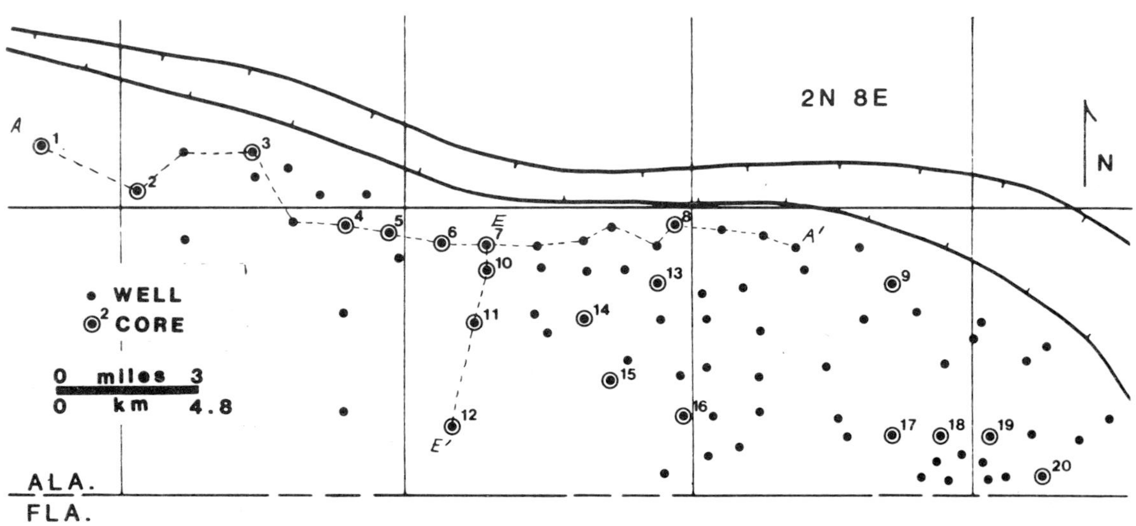

Figure 3. Well and core control and cross-section locations.

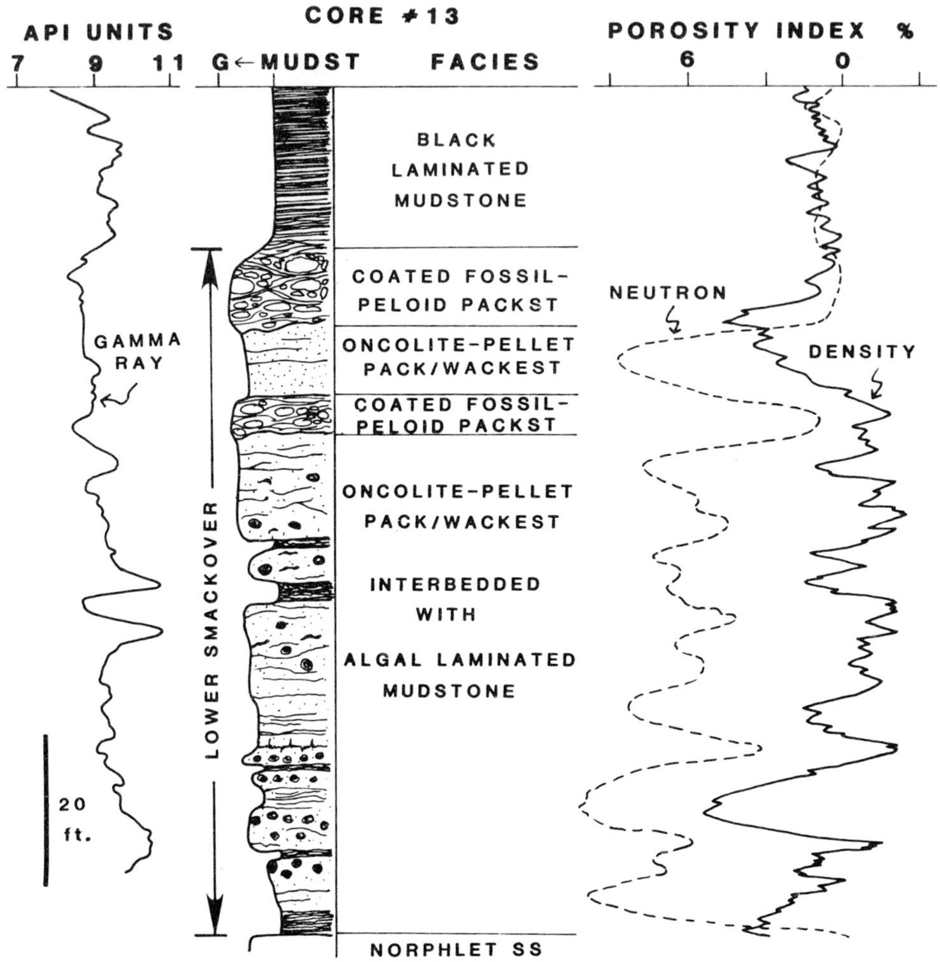

Figure 4. Core description, gamma-ray log, and density-neutron log of the lower unit of the Smackover Formation. (Exxon McCurdy, core #13, 15,470-15,581 feet. See Figure 3 for location.)

tire Smackover Formation and shows the lateral continuity of the oncolite-pellet packstone facies. Individual algal laminated beds were more difficult to correlate from well to well and are shown by wavy lines only where core material was available.

The uppermost facies in the lower Smackover is the coated fossil-peloid packstone. This is a light to medium gray limestone, which is rarely dolomitized and is commonly burrowed. The major allochems are peloids (smaller in size and not ornamented as the *Favreina* pellets) and coated fossil fragments. Minor allochems include pellets, thin-shelled pelecypod fragments, echinoids and benthic foraminifers.

In the east and northeast part of the study area, this facies occurs in two well-defined beds which are separated by the oncolite-pellet packstone (Figs. 4 and 5). To the west, these beds become thicker with a few less distinct and thinner beds occurring below them. In the most seaward (southwest) well, no. 12, the top bed of this facies greatly increases in thickness and persists into the middle and upper Smackover (Fig. 6).

Middle Smackover (Figs. 5 and 7)

The fossiliferous mudstone of the middle

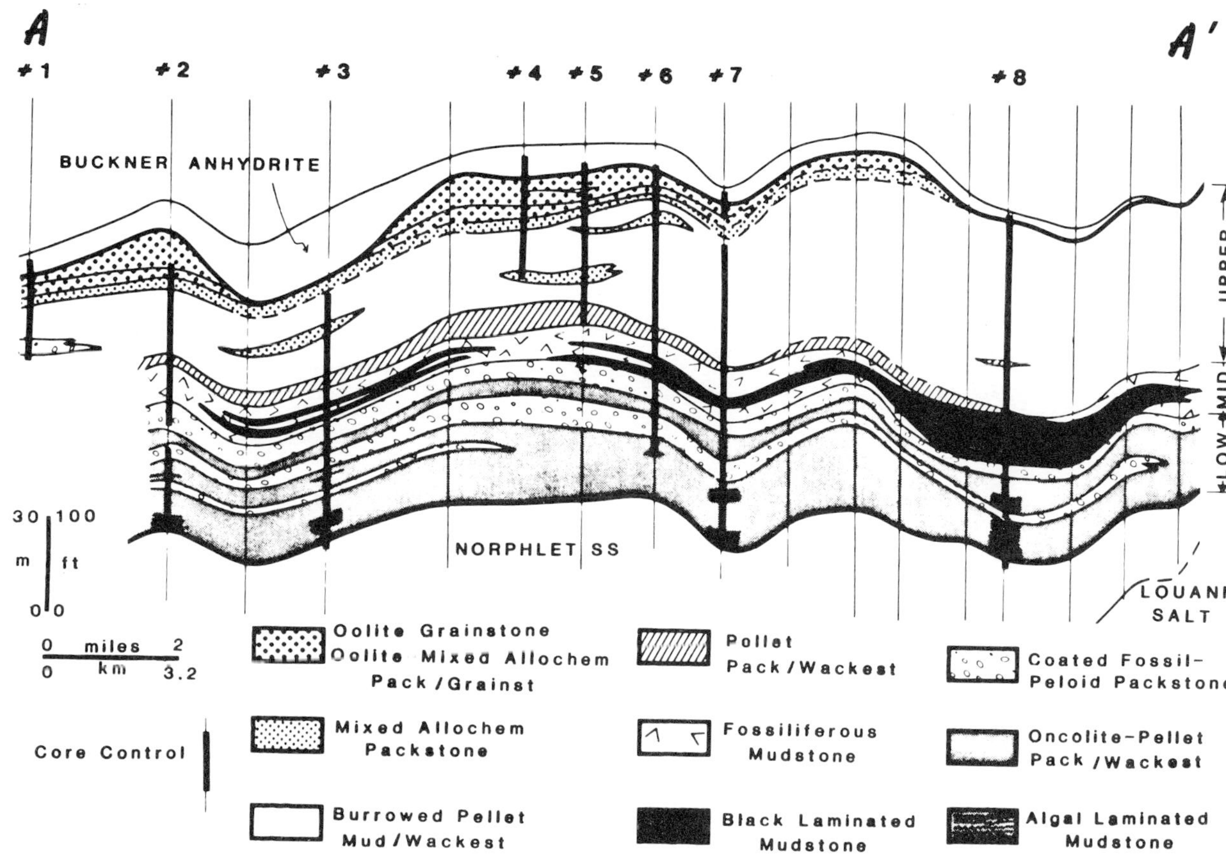

Figure 5. Facies cross section A-A'.

Smackover is a medium to dark gray limestone. Pelecypods, which are usually complete specimens, are the major allochems present, but constitute only about 5 to 10 percent of this facies. Pellets and peloids are also present, but are less common. Interbedded with the fossiliferous mudstone is a black laminated mudstone. This limestone consists of alternating laminae of gray micrite and black organic material with minor amounts of quartz silt. Burrows are lacking and fossils and pellets are rare. Overlying these two facies is a pellet packstone/wackestone, which is dolomitized, burrowed, and contains abundant fecal pellets.

As shown in Figure 5, the facies of the middle Smackover are more variable in thickness and continuity than those in the lower Smackover. The main area of concentration of the black laminated mudstone facies is to the northeast part of the study area. Here, thicknesses of up to 50 feet (15 meters) are attained in well no.8 (Fig. 5). This facies thins to the southwest and becomes interbedded with the fossiliferous mudstone, which, in turn, thickens to the southwest. Both of these facies are lacking in well no. 12 where the coated fossil packstone dominates (Fig. 6). The pellet packstone occurs above these two mudstone facies in a limited area and is not present to the southeast or west.

Upper Smackover (Figs. 5 and 8)

The burrowed, pellet mudstone/wackestone facies is a medium brown to gray, partially dolomitized limestone. The facies has abundant burrows and minor amounts of pellets, pelecypods, gastropods, and echinoids. Peloids and intraclasts are also present but occur only locally. The highly dolomitized packstones to grainstones that overlie this mudstone/wackestone have been subdivided according to the percentage of allochems pres-

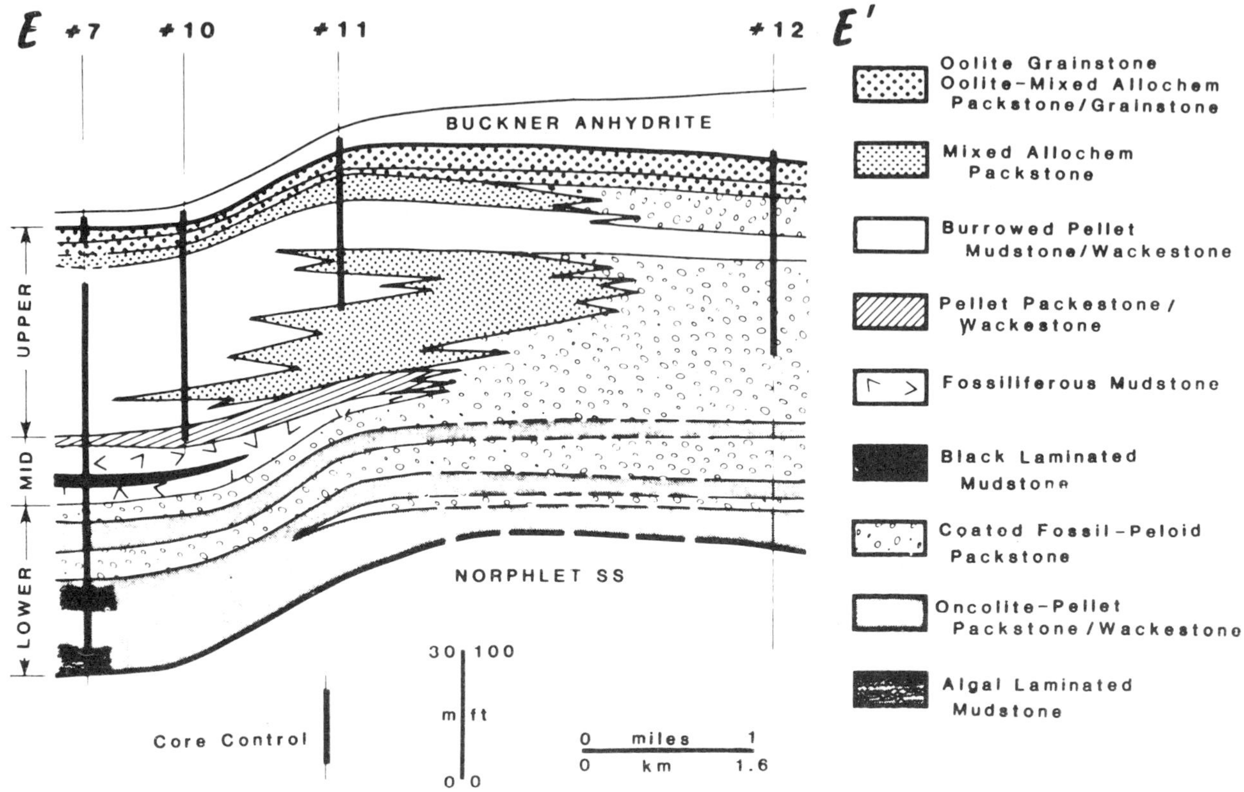

Figure 6. Facies cross section E-E'.

ent. The mixed-allochem packstone contains equal amounts of pellets, peloids and intraclasts. This packstone grades upward into an oolite-mixed allochem packstone/grainstone, which contains up to 50 percent ooids, along with varying amounts of pellets, peloids and intraclasts. Upward, this facies grades into an oolite grainstone in which ooids comprise more than 50 percent of the rock. In places the grainstone is cross-bedded and coarsens upward.

The burrowed, pellet mudstone/wackestone, which dominates the upper Smackover (Figs. 5 and 6), is thickest in the northeast part of the study area and thins gradually to the southwest. In general vertical sequence (Fig. 8), the burrowed mudstone is overlain by the mixed allochem facies, which grades upward into first the oolite-mixed allochem and then into the oolite grainstone facies. To the east no grainstone is present and core no. 8 is capped by a thin bed of algal laminated mudstone. To the west in core no. 12, the oolite-mixed allochem facies caps the coated-fossil facies. An isopach of the ooid grainstone shows that the main axis of this belt trends northwest-southeast (Fig. 9). Finally, the Buckner anhydrite overlies the entire sequence. This anhydrite thins to the northeast (Fig. 5) where no grainstone facies is present. Here, massive salt is common.

INTERPRETATION AND MODELS OF ENVIRONMENTS

Lower Smackover

The lower Smackover is a transgressive sequence which was deposited over the eolian and intertidal sandstones of the Norphlet Formation. The lowermost algal-laminated mudstone is interpreted to have been deposited in a protected, low-energy, upper intertidal to lower supratidal environment (Fig. 10). Work on recent algal-laminated sedi-

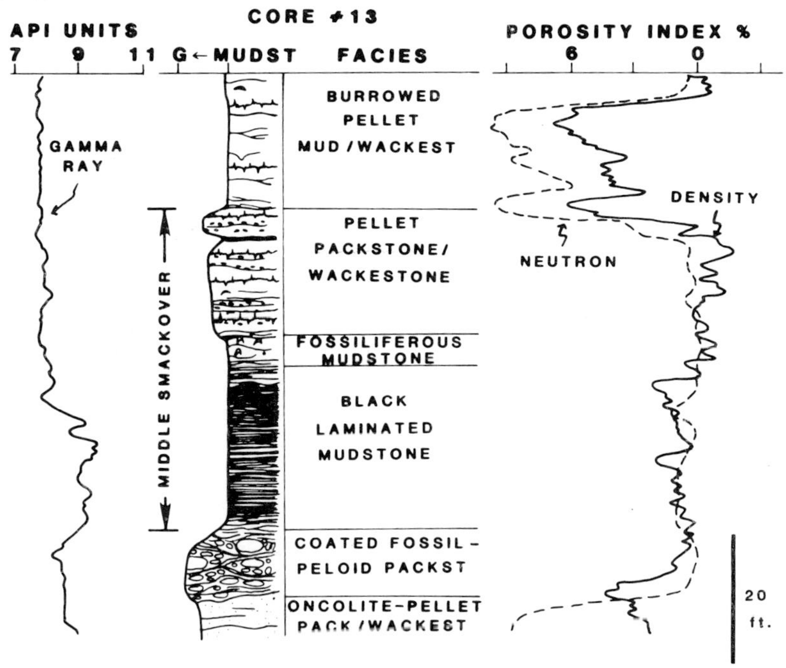

Figure 7. Core description, gamma-ray log, and density-neutron log of the middle Smackover; Exxon McCurdy, core No. 13, 15,420-15,507 feet.

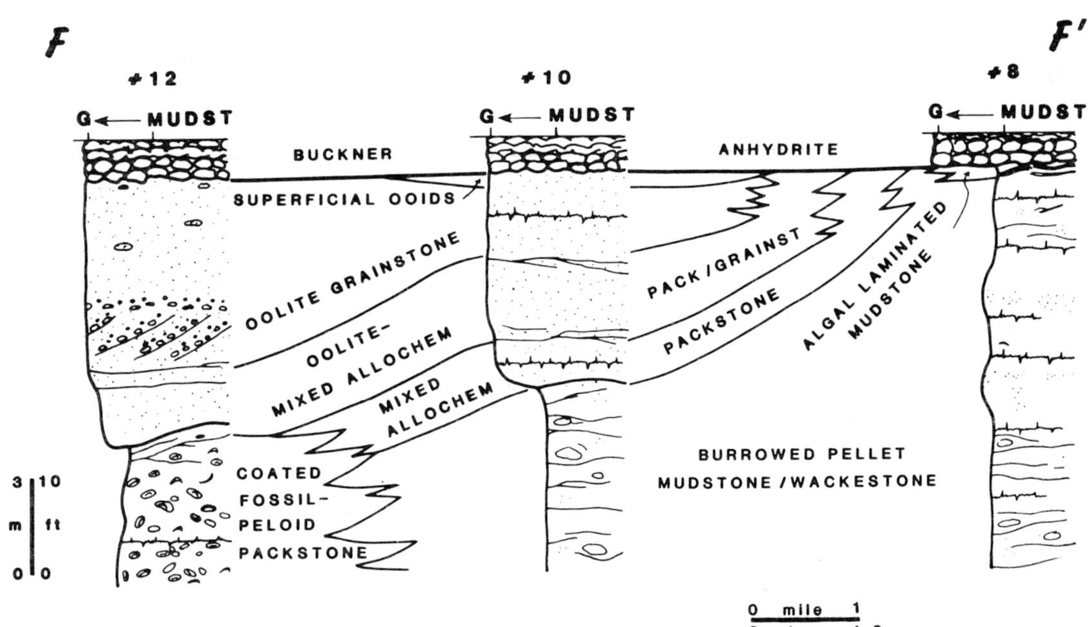

Figure 8. Facies cross section F-F' of the upper Smackover. (From left to right: La. Land and Expl. Hall, core No. 12, 15,819-15,861 feet; La. Land and Expl. Philyaw, core No. 10, 15,488-15,530 feet; Exxon No. 1 Scott, core No. 8, 15,103-15,145 feet. See Figure 9 for location.)

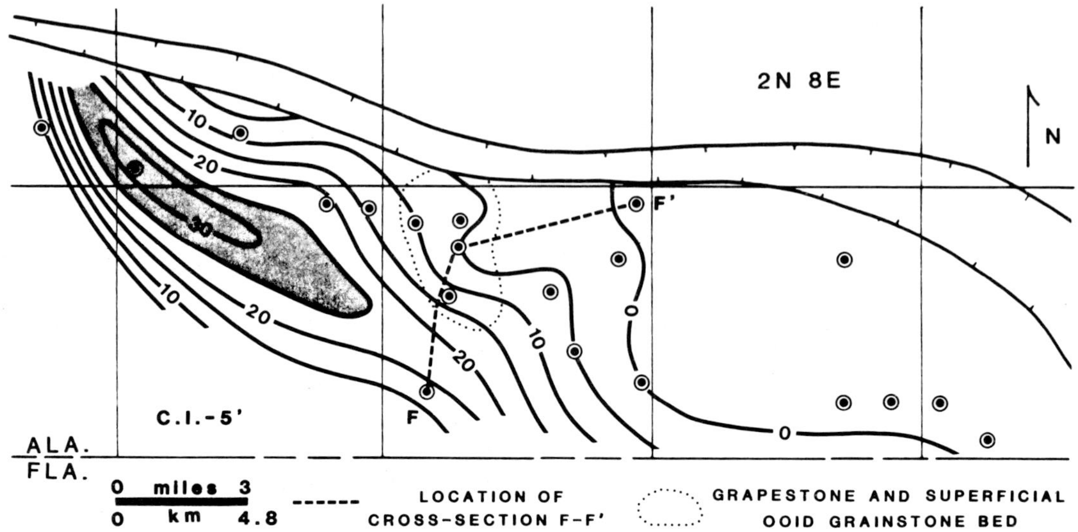

Figure 9. Isopach map of the oolite grainstone facies.

ments in the Persian Gulf (Kendall and Skipwith, 1969; Purser and Evans, 1973) and Shark Bay, Western Australia (Davies, 1970a, b; Logan and Cebulski, 1970) indicates that significant thicknesses of algal mats develop only in hypersaline areas where few browsing gastropods and burrowing animals are able to live. In Abu Dhabi, these browsing and burrowing organisms and their pellets are abundant in the lower intertidal and upper subtidal zones where the water is less saline. The oncolite-pellet packstone/wackestone facies is interpreted to have been deposited in a similar lower intertidal environment. Some of the oncolites in this facies are irregular in shape with distinctive algal growths projecting in one direction, which indicates that the oncolites were only periodically moved by waves and currents, perhaps due to storms (Logan et al., 1964). Intraclasts of algal-mat fragments found in this facies were probably also formed during periodic storms. The well-preserved fecal pellets provide more evidence that low-energy conditions existed because violent waves or currents tend to disrupt the soft pellets before they have a chance to be cemented and hardened (Kendall and Skipwith, 1969).

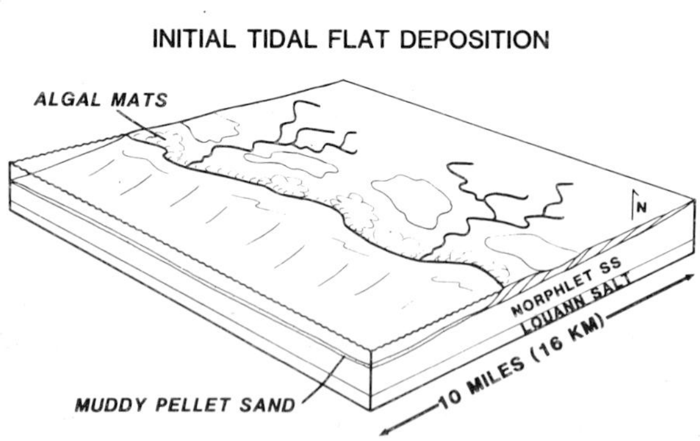

Figure 10. Depositional model of the basal Smackover in the study area.

In southwestern Alabama Paleozoic highs composed of igneous and metamorphic rocks to the north and east and eolian dunes to the west are thought to have partially restricted the Conecuh Embayment. In the Smackover, the vertical cyclic sequences of alternating algal mudstones and oncolite-pellet packstones are believed to be prograding shoaling-upward cycles capped by the algal laminated sediments (Fig. 11). Continued subsidence and/or minor transgression could have accounted for the formation of these cycles.

The presence of the coated fossil-peloid packstone facies in the upper part of the lower Smackover indicates more normal-marine and moderate-energy conditions. Skeletal fragments of echinoderms, molluscs, and foraminifers are relatively numerous compared to

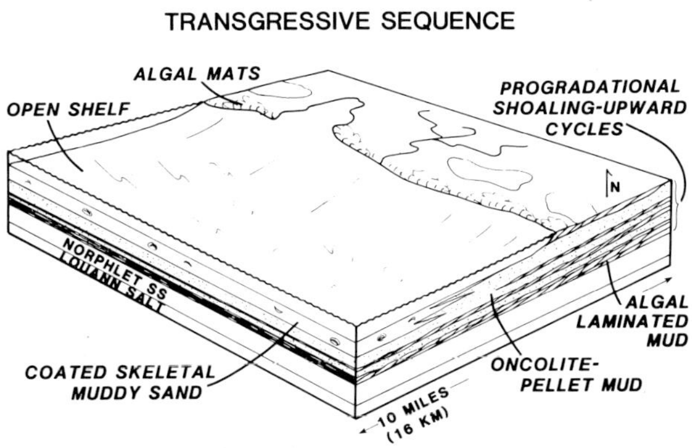

Figure 11. Depositional model of the progradational shoaling-upward cycles in the overall transgressive lower Smackover.

their numbers in the lower two facies; these fragments usually have evenly laminated coatings. In the more protected landward areas to the northeast, skeletal fragments are still numerous, but not as many are coated. This coated-fossil packstone facies is interpreted to have formed in the subtidal zone near wave base (Fig. 11). Similar rock types described by Wilson (1975) are categorized as shelf facies with open-marine circulation. This facies, like the algal mudstone, is also interbedded with the oncolite-pellet facies. The interbedded coated-fossil and oncolite-pellet facies can be interpreted as prograding, shallowing-upward cycles (capped by the oncolite-pellet packstone) in an overall transgressive sequence.

Middle Smackover

The middle Smackover is characterized by three distinct facies which form a facies tract (Fig. 12) from the northeast (landward) to the southwest (seaward). The well farthest seaward (no. 12) contains a thick vertical continuation of the coated fossil-peloid facies. To the northeast a fossiliferous mudstone which contains an infauna is present and is interpreted to represent a protected shallow-water subtidal deposit. Wilson (1975) described a similar facies type as a quiet-water facies deposited below wave base on either an open shelf or open platform. I believe the presence of the coated-fossil facies in a seaward direction provided a protective barrier against any high-energy waves invading the area, and allowed whole pelecypods to be preserved in place. Interbedded with the fossiliferous mudstone is a black, laminated mudstone which becomes thickest in a landward direction. The lack of preserved fauna or burrowing in this facies and high organic content indicate deposition in a restricted, anoxic environment (Demison and Moore, 1980).

The formation of the moderate-energy, coated-fossil facies to the southwest was probably initiated on low-relief bottom highs formed by the underlying salt during Smackover time. During the rest of Smackover deposition, the area to the northeast continued to be a region of restricted lagoonal deposits - an environment that persisted during Buckner time when massive salt was deposited in the area.

A thin, pellet packstone facies is also present in the middle Smackover above these two facies and may represent an influx of more normal marine water flowing into the lagoon.

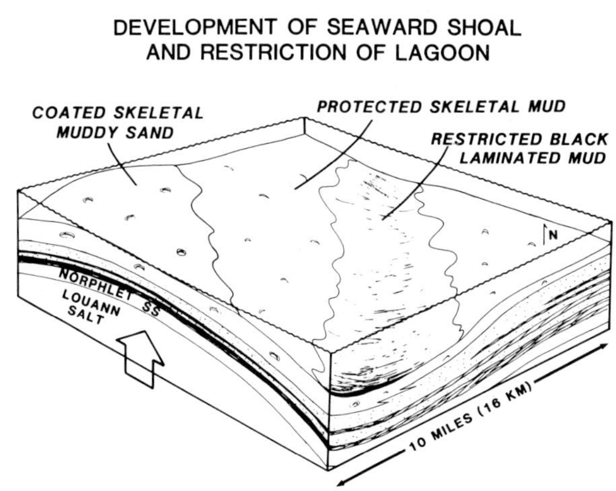

Figure 12. Depositional model of the restricted-lagoon facies and seaward shoal of the middle Smackover. Initial development of the shoal is believed to have been caused by salt withdrawal and associated salt-ridge formation.

Upper Smackover

In the upper Smackover deposition of the coated fossil-peloid facies continued in the southwest and a burrowed pellet mudstone/wackestone was deposited in the northeast. The mudstone/wackestone facies, with its abundant burrows and low species diversity, is interpreted to be a partially restricted lagoonal deposit. This lagoonal deposit grades upward into a mixed allochem packstone/grainstone and finally into an oolite grainstone. These rocks represent a shoaling upward sequence, from very low-energy deposits to high-energy grainstones, similar to sequences found in the oolite sand belts of the Bahamas (Newell *et al.*, 1960; Ball, 1967). One main difference is that these ooid shoals were not developed on a shelfedge break as are those in the Bahamas, but were initiated by slight topographic relief caused by salt movement during middle Smackover time. The thin layer of superficial ooids and grapestones present above the ooid grainstone in some of the wells represents slightly lower energy deposits behind the main axis of the ooid shoal.

Landward of the ooid shoal are lagoonal deposits. In local areas at the top of these lagoonal deposits is an algal laminated mudstone which contains calcite and celestite pseudomorphs after gypsum; the gypsum disrupted the laminated sediment as it was precipitated. This facies is overlain by Buckner anhydrite and salt. This upper sequence is similar to intertidal and supratidal deposits in the Persian Gulf where intense evaporation causes the precipitation of gypsum and anhydrite landward of the algal mats (Kendall and Skipwith, 1969).

The upper Smackover represents a regressive upward-shoaling sequence (Fig. 13). The thickness of the coated-fossil and lagoonal deposits may have been caused by continued subsidence and vertical stacking of facies - a mechanism described by Malek-Aslani (1977) who categorized them as stretched-upward, shoaling cycles. In such situations an increase in sedimentation (or decrease in subsidence) results in rapid progradation of the

REGRESSIVE SEQUENCE

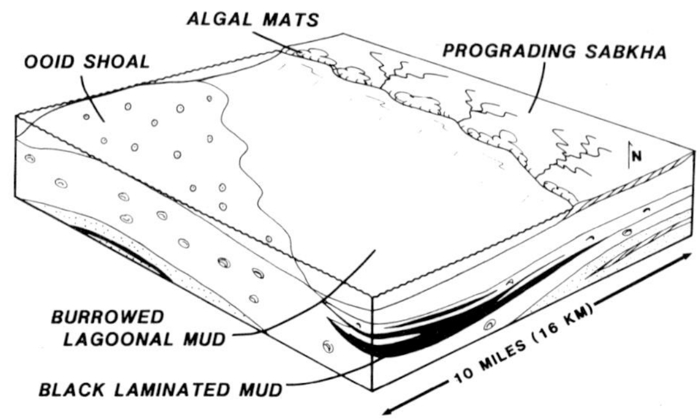

Figure 13. Depositional model of the regressive upward-shoaling sequence of the upper Smackover.

facies in an offshore direction. Progradation of the supratidal and intertidal environments over open-shelf facies occurs today in the Persian Gulf (Malek-Aslani, 1973; Patterson and Kinsman, 1981). Continued progradation will result in a situation similar to the Smackover in which anhydrite has prograded over the oolite shoal.

HYDROCARBON RESERVOIRS

The Smackover hydrocarbon reservoirs developed as a result of a combination of original depositional environment, subsequent dolomitization of select facies, and present-day structure. The Smackover carbonates have both primary intergranular porosity and secondary oomoldic and intercrystalline porosity which resulted from dolomitization of the upper Smackover grainstones and underlying mudstones. The porous grainstones provided a conduit for the dolomitizing fluids and only those mudstones associated with the grainstones were dolomitized (Figs. 9 and 14). In areas such as Flomaton Field, where no grainstones were deposited, the mudstones were not dolomitized (Figs. 9 and 15). Therefore, the original depositional environment provided an important control over later porosity-forming dolomitization. The overlying Buckner anhydrite provided a seal to this reservoir facies.

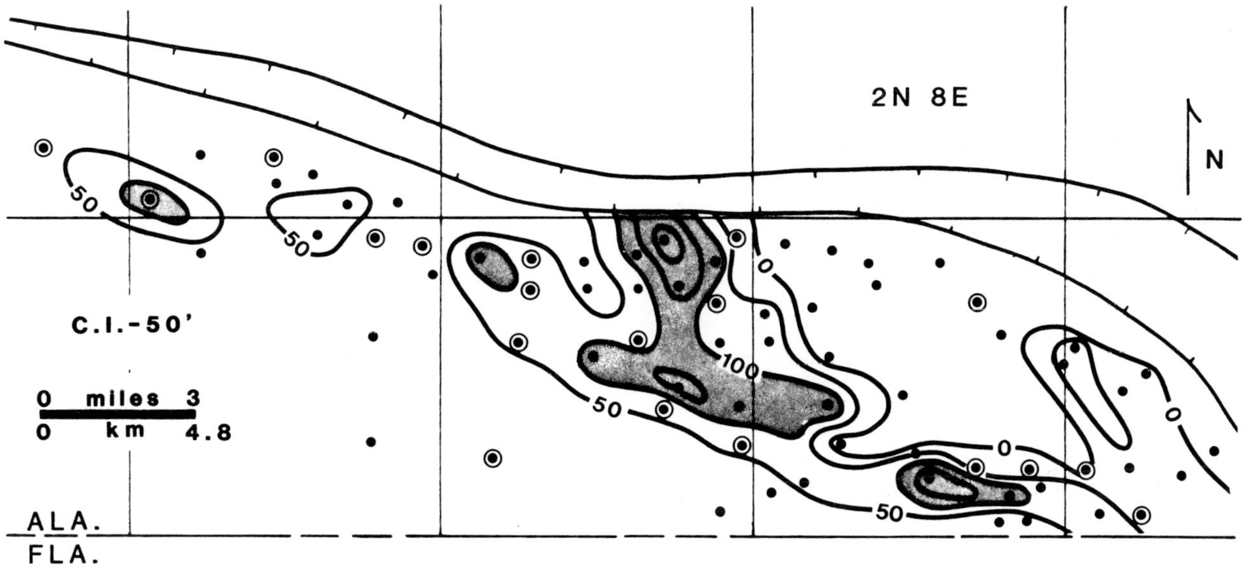

Figure 14. Isopach map of the dolomite present in the upper Smackover.

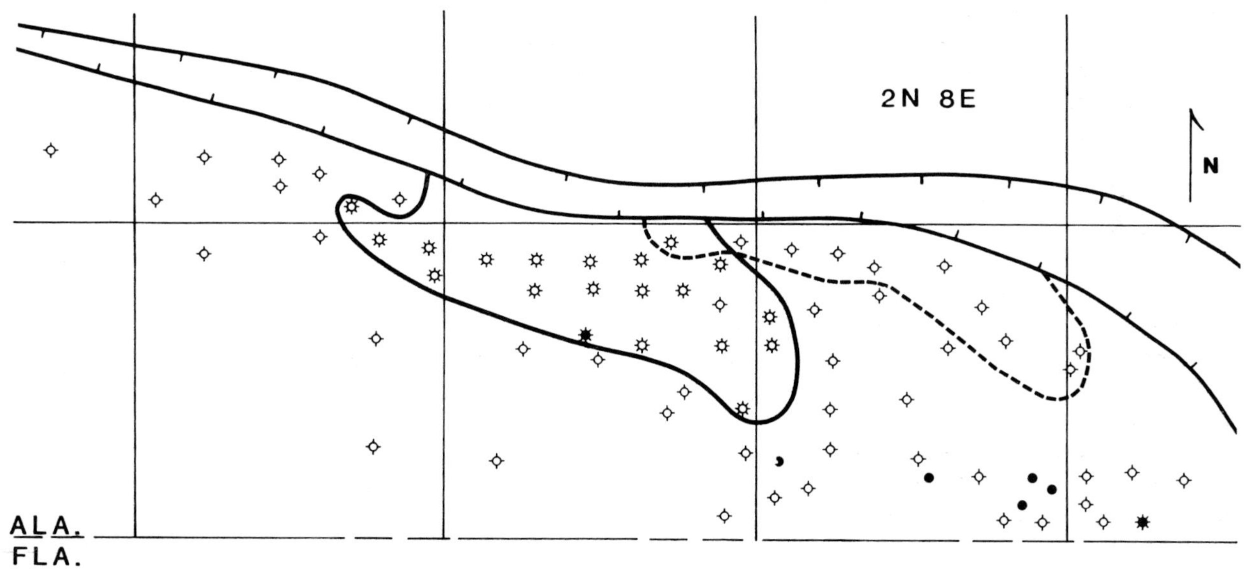

Figure 15. Extent of Big Escambia Creek (Smackover, solid line) and Flomaton (Norphlet, dashed line) Fields.

The original depositional environment was, in turn, controlled by topographic highs which were the result of salt movement. Salt movement during the Late Smackover was minor, but provided the slight differences in elevation needed for higher energy conditions to develop in the central and western part of the study area. A map of present-day structure (Fig. 16) shows that the undolomitized lagoonal mudstones are now structurally the highest in the area. These dense lime mudstones have no porosity and therefore contain no hydrocarbons. But hydrocarbons are present beneath this lime mudstone in the under-

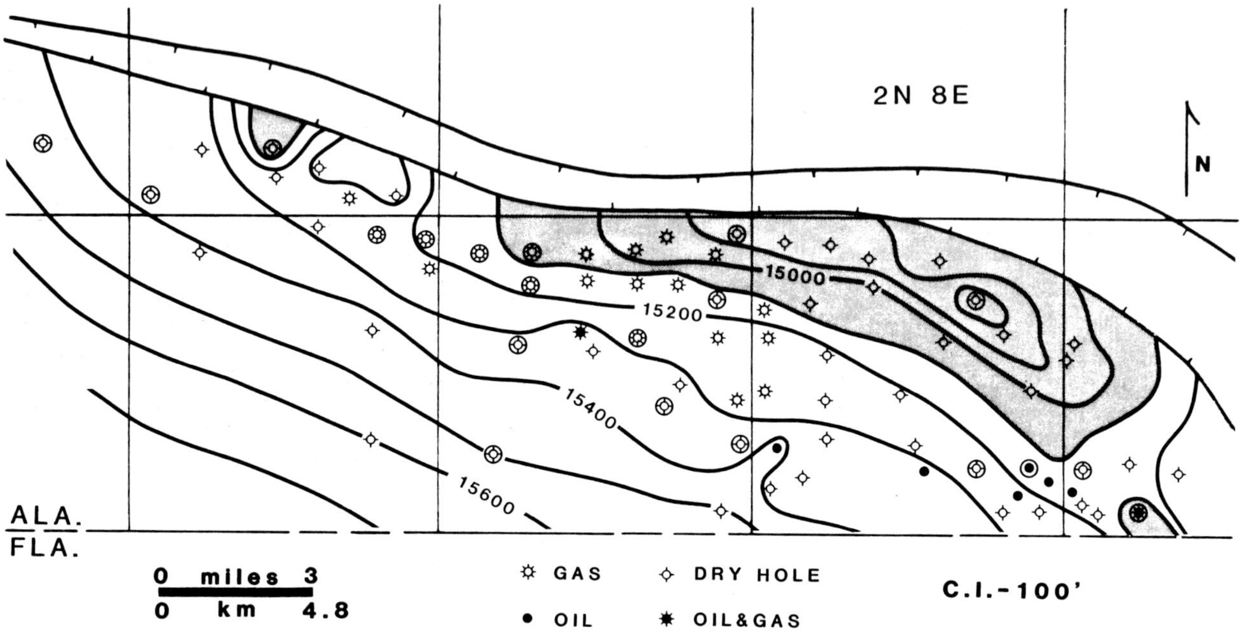

Figure 16. Structure map of the Smackover Formation with sea level as a datum.

lying Norphlet Sandstone in Flomaton Field (Fig. 15) where the dense mudstone forms a seal. In Big Escambia Creek Field, hydrocarbons occur in updip regions of the porous dolomitized grainstones and mudstones. The contact between dolomitized and nondolomitized rock provides the eastern boundary to the Smackover Field.

ACKNOWLEDGMENTS

This paper is a summary of part of a thesis study done at the University of Texas at Austin. I would like to thank Exxon Co. U.S.A. for providing me with the core material and Mobil Oil for providing the well logs used in this study.

REFERENCES

Ball, M. M., 1967, Carbonate sand bodies of Florida and the Bahamas: Jour. Sed. Petrology, v. 37, no. 2, p. 556-591.

Davies, G. R., 1970a, Carbonate bank sedimentation, Eastern Shark Bay, Western Australia: Am. Assoc. Petroleum Geologists Memoir, no. 13, p. 85-168.

__________, 1970b, Algal-laminated sediments, Gladstone Embayment, Shark Bay, Western Australia: Am. Assoc. Petroleum Geologists Memoir, no. 13, p. 169-206.

Demaison, G. J., and Moore, T., 1980, Anoxic environments and oil source bed genesis: Am. Assoc. Petroleum Geologists Bull., v. 64, no. 8, p. 1179-1209.

Hughes, D. J., 1968, Salt tectonics as related to several Smackover fields along the northeast rim of the Gulf of Mexico Basin: Gulf Coast Assoc. Geol. Socs. Trans., v. 18, p. 320-330.

Kendall, S. G. St. C., and Skipwith, P. A. d'E., 1969, Holocene shallow water carbonate and evaporite sediments of Khor Al Bazam, Abu Dhabi, southwest Persian Gulf: Am. Assoc. Petroleum Geologists Bull., v. 53, no. 4, p. 841-892.

Logan, B. W., Rezak, R., and Ginsburg, R. N., 1964, Classification and environmental significance of algal stromatolites: Jour. Geology, v. 72, p. 68-83.

Logan, B. W., and Cebulski, D. E., 1970, Sedimentary environments of Shark Bay Western Australia: Am. Assoc. Petroleum Geologists Memoir, no. 13, p. 1-38.

Loucks, R. G., and Budd, D. A., 1981, Diagenesis and reservoir potential of the Upper Jurassic Smackover Formation of South Texas: Gulf Coast Assoc. Geol. Socs. Trans.,

v. 31, p. 339-346.

Malek-Aslani, M., 1973, Environmental modeling: a useful exploration tool in carbonates: Gulf Cost Assoc. Geol. Socs. Trans., v. 23, p. 239-245.

__________, 1977, Plate tectonics and sedimentary cycles in carbonates: Gulf Coast Assoc. Geol. Socs. Trans., v. 27, p. 125-133.

Mancini, E. A., and Benson, D. J., 1980, Regional stratigraphy of Upper Jurassic Smackover carbonates of southwest Alabama: Gulf Coast Assoc. Geol. Socs. Trans., v. 30, p. 151-165.

Moore, C. H., and Druckman, Y., 1981, Burial diagenesis and porosity evolution, Upper Jurassic Smackover, Arkansas and Louisiana: Am. Assoc. Petroleum Geologists Bull., v. 65, no. 4, p. 597-628.

Newell, N. J., Purdy, E. G., and Imbrie, J., 1960, Bahamian oolitic sand: Jour. Geology, v. 68, no. 5, p. 481-497.

Ottman, R. D., Keyes, P. L., and Zeigler, M. A., 1973, Jay Field, a Jurassic stratigraphic trap: Gulf Coast Assoc. Geol. Socs. Trans., v. 23, p. 146-157.

Patterson, R. J., and Kinsman, D. J. J., 1981, Hydrologic framework of a sabkha along Arabian Gulf: Am. Assoc. Petroleum Geologists Bull., v. 65., p. 1457-1475.

Purser, B. H., and Evans, G., 1973, Regional sedimentation along the Trucial Coast, southeast Persian Gulf: *in* Purser, B. H. (ed.), The Persian Gulf-Holocene carbonate sedimentation and diagenesis in a shallow epicontinental sea, 471 p.

Sigsby, R. J., 1976, Paleoenvironmental analysis of the Big Escambia Creek-Jay-Blackjack Creek Field area: Gulf Coast Assoc. Geol. Socs. Trans., v. 26, p. 258-278.

Wilkerson, R. P., 1981, Depositional environments and regional stratigraphy of the Jurassic Norphlet Formation in south Alabama: Gulf Coast Assoc. Geol. Socs. Trans., v. 31, p. 417-419.

Wilson, J. L., 1975, Carbonate facies in geologic history: Springer-Verlag, New York, 471 p.

PALEOENVIRONMENT OF AN UPPER COTTON VALLEY (KNOWLES LIMESTONE) PATCH REEF, MILAM COUNTY, TEXAS

ALLEN K. CREGG

Shell Oil Company
P.O. Box 60775
New Orleans, Louisiana 70160

WAYNE M. AHR

Department of Geology
Texas A&M University
College Station, Texas 77843

ABSTRACT

The Knowles Limestone is the uppermost unit of the Cotton Valley Group, and in Milam County, Texas it is approximately 350 feet thick and consists of shales, terrigenous dolomitic limestones, grainy limestones, and algal boundstones with stromatoporoids and corals. The boundstones represent an elongate, wave-resistant, encrusted skeletal patch reef which probably developed on a subtle, salt-generated topographic high. The narrowest width of the reef is slightly more than one mile based on facies correlations of three cored wells in the study area.

The principal reef framebuilders included massive and columnar stromatoporoids, solitary and dendroid corals, and encrusting *Tubiphytes* and *Lithocodium* algae. Dendroid stromatoporoids, digitate corals, *Tubiphytes* oncoids, echinoids, and various mollusks were the principal contributors to the detrital infill of the reef.

Increased dip and dip reversal of draping beds suggest that the Knowles reef achieved relief above the sea floor. Dipmeter logs indicate that the beds dip 5° northwest on the updip reef flank and 10° southeast on the downdip reef flank.

Reef-core boundstones and reef talus were consistently present downdip, and lagoonal to tidal-flat facies were common updip throughout Knowles deposition. The reef organisms eventually became overwhelmed with terrigenous sediment transported downdip as the tidal-flat environment prograded over lagoon, reef-talus and reef-core facies at the end of Knowles deposition.

Early dolomitization of tidal-flat and lagoonal facies created local porous zones in some of these rocks. However, the reef, *per se,* is cemented by sparry calcite and is not a potential reservoir facies.

INTRODUCTION

The Knowles Limestone, the uppermost unit of the Cotton Valley Group (Fig. 1), was first described by Mann and Thomas (1964) from the subsurface of North Louisiana. The Knowles was deposited as a marine, lagoonal limestone and shale behind the western end of a barrier sand island (Thomas and Mann, 1966). However, it appears to be a more open-marine limestone in Milam County, Texas (Fig. 1). The Knowles Limestone is approximately 350 feet thick in the study area and consists of arenaceous shales, terrigenous dolomitic limestones, grainy limestones, and algal boundstones with stromatoporoids and corals, which represent a framebuilt patch reef. Movement on a salt structure, which underlies the area, could account for the formation of a subtle high upon which the Knowles reef developed.

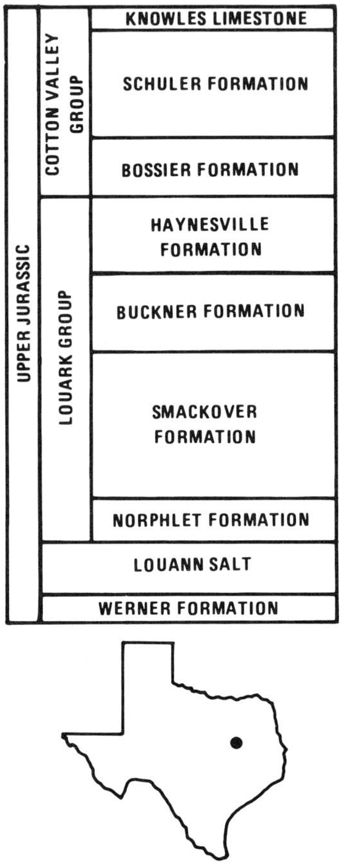

Figure 1. Stratigraphic column for the study area. After Philpott and Hazzard (1949), Shreveport Geological Society (1953 a,b) and Mann and Thomas (1964). Location of study area, Milam County, shown on inset.

Structural and Stratigraphic Setting

The Knowles Limestone is 300 to 400 feet thick and is present from the southern Arkansas-northern Louisiana area, around the Sabine uplift, to the southwestern edge of the East Texas Basin. The study area is located on the extreme southwestern portion of the East Texas Basin, which is a northwest trending reentrant of the Gulf Coast Basin (Fig. 2). The Mexia-Talco fault zone passes through Milam County, just updip from the study area. Andrews (1960) noted that the updip limit of salt and the Mexia-Talco, South Arkansas, and Pickens-Gilbertown fault systems are closely parallel, and that a cause-effect relationship appears to exist between them and the salt pinchout. The study area is located in an area of low-relief salt pillows (Fig. 3).

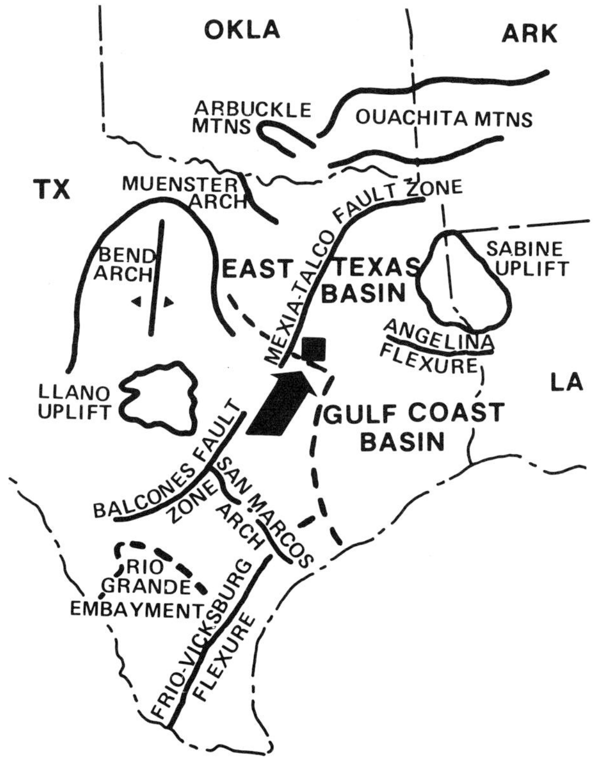

Figure 2. Regional structural features of the Gulf Coast. Arrow indicates location of study area. Modified from Bornhouser (1958).

Previous Work

Late Jurassic or Early Cretaceous reefs are not common in North America, and, therefore, little has been written about them. However, there is extensive literature on Jurassic reefs in Europe. Upper Jurassic European patch reefs formed in downslope positions and consisted of sponges, encrusting algae and foraminifers. These reefs existed below wave base and stood as much as 100 m above the basin floor. The buildups were followed by a period of regional shallowing, coral-sponge colonization and organic encrustation. Large mollusks became increasingly common as the buildups grew. Generally the buildups grew into wave base and extensive bioclastic debris accumulated on the reef crests. In places, the skeletal accumulations surrounded, protected, and helped to restrict basins deep enough for deposition of fine carbonate mud, which preserved a fauna derived from both the land and open sea.

Many of the European Jurassic reefs exhibit various high-diversity, climax communities that show considerable zonation, which depended on the bathymetry caused by growth of the reef crests. The Knowles reef exhibits a characteristically zoned pattern of different morphological types, which owe their form and distribution to depositional (biogenic) topography.

METHODS OF STUDY

Full diameter cores were obtained from three wells in the study area. The location of the three cored wells and Mobil Oil Company seismic line no. 11 are shown in Figure 4. A line of section for subsequent cross-sections through the study area is shown by the line connecting the updip Shell No. 1 Nell Ross (SR1), the downdip Shell No. 1 Adoue Estate (SA1) and the Shell No. 2 Adoue Estate (SA2) wells.

The cores from these wells were examined to determine rock texture, composition, and sedimentary structures in order to establish correlatable lithic units in the area. These units were correlated to borehole logs for further interpretation.

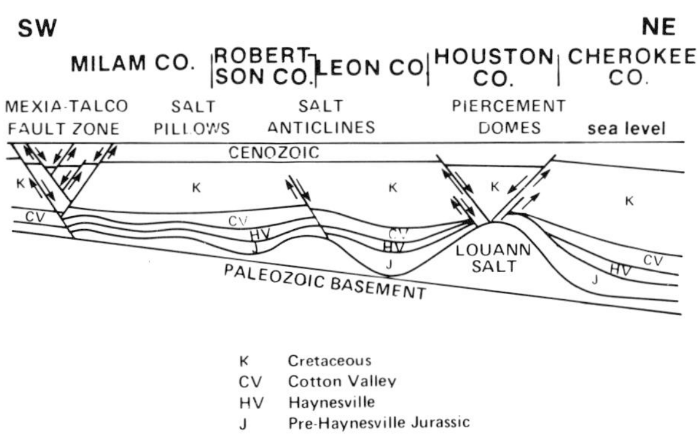

Figure 3. Salt tectonic features of the East Texas Basin. Modified from Hughes (1968).

A total of 109 samples from all major lithic units in the cores were thin-sectioned for petrographic determination of texture, composition, and diagenesis.

Percentages of carbonate particles, terrigenous grains, clay matrix, micrite matrix, and dolomitic matrix were calculated and normalized to 100 percent for the detrital and authigenic dolomitic matrix fraction of each sample. Cements were calculated and expressed as a percentage of the whole rock. Compositional and textural data were analyzed to determine microfacies, which in turn were compared to observed sedimentary structures to aid in the interpretation of depositional environments.

The diagenetic history of the rocks was determined by studying cement mineralogy, cement fabric, grain fabric, cement to grain relationships, and grain integrity.

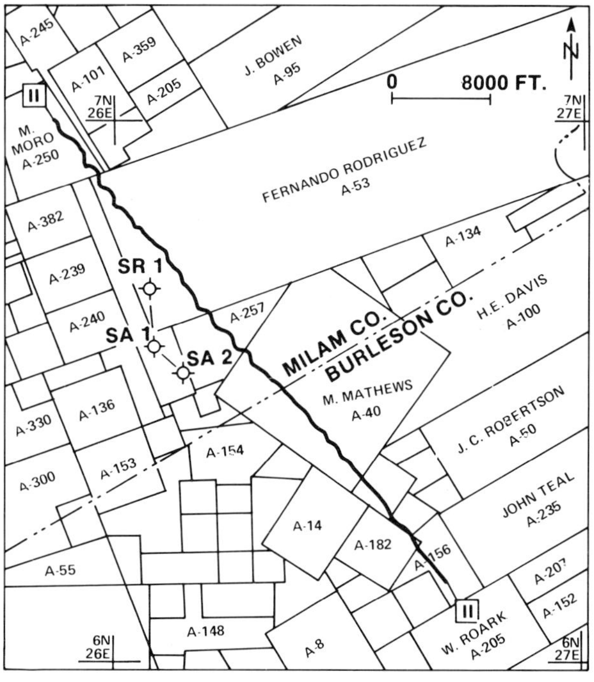

Figure 4. Index map of the study area showing the location of the three cored wells, and Mobil Producing Tx. and NM Inc.-United Geophysical Seismic Line No. 11 which were utilized in the study. The line of section for figures 5, 7, 8, 9, 12, and 13 is shown.

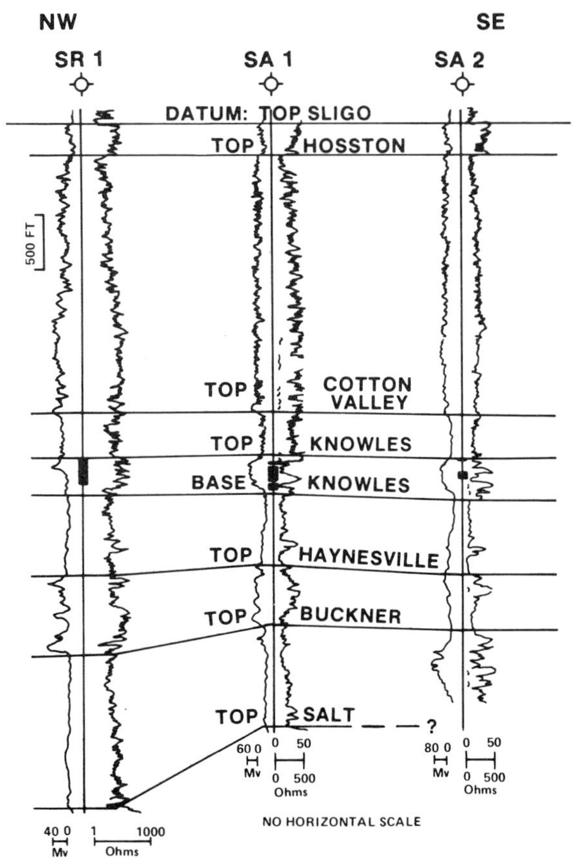

Figure 5. Stratigraphic cross section of the study area illustrating the relatively thin section below the Knowles in the SA 1 and SA 2 wells, which reflects the subtle salt-generated paleotopographic high upon which the Knowles reef developed. Cores are shown by heavy line in the three wells.

Spontaneous potential and resistivity logs from the three wells in the study area were used to establish stratigraphic correlations and to construct cross sections, which show the thickness and general configuration of the reef and surrounding facies. Dipmeter logs from the Shell No. 1 Nell Ross (SR1) and the Shell No. 2 Adoue Estate (SA2) wells were used to map the attitude of beds draping over the reef and reef flanks.

PALEOSTRUCTURE

The stratigraphic cross-section (Fig. 5), illustrates the paleotopography on which the Knowles Limestone was deposited. The SR 1 and SA 1 wells apparently penetrated salt near their total depths (the penetration of the salt in the SA 1 well is questionable). We interpret the top of the salt to be structurally higher in the SA 1 well than in the SR 1 well (Fig. 5). This salt structure is recognizable on the seismic section (Fig. 6). The top of the salt below the total depth of the SR 1 well was projected from the nearby seismic profile. The Knowles reef does not appear on this seismic section.

The stratigraphic cross-section shows that the interval between the Knowles Limestone and the salt is thin under the SA 1 and SA 2 wells and thickens under the SR 1 well. This variation in thickness represents early salt movement contemporaneous with deposition. The site of the SR 1 well is on the flank of the structure and must have received more sediment than the locations of the SA 1 and SA 2 wells. The Knowles Limestone reef apparently grew on the subtle, salt-generated paleo high. Salt movement declined until the end of the Hosston deposition.

KNOWLES ROCK PROPERTIES

Introduction

The SR 1 core consists mainly of arenaceous, argillaceous and dolomitic mudstones to wackestones, arenaceous shales, and intra clastic and skeletal wackestones to packstones with some terrigenous debris and calcerous red or green algal fragments. The two downdip cores consist mainly of coral, stromatoporoid and algal bindstones, framestones, bafflestones, and floatstones (terms from Embry and Klovan, 1971), with interbedded oncolitic, intraclastic, and skeletal wackestones to packstones, and arenaceous, argillaceous and dolomitic mudstones to wackestones.

Texture

Mean grain-size trends of carbonate grains and monocrystalline quartz grains in

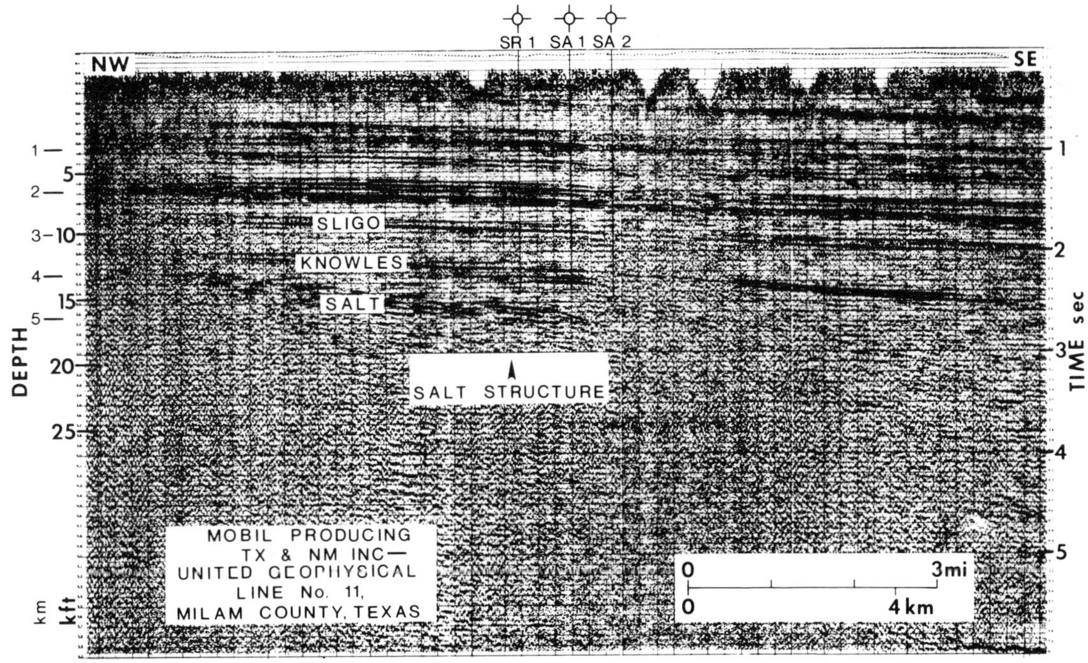

Figure 6. Seismic section through Milam and Burleson Counties. The projected location of the three wells in the study area is shown. Depth in kilometers (km) and thousands of feet (kft). Two-way time in seconds (sec.).

the three cores are shown in Figures 7-9.

There is a general coarsening-upward trend of carbonate and monocrystalline quartz grain sizes through the Knowles Limestone, which represents an overall regressive sequence. The percentages of carbonate grains, especially intraclasts, decrease as the rock types change upward from oncoidal wackestones and packstones, with mean sizes ranging from 0.56 mm to 2.37 mm, or coarse grained to granule size, to terrigenous, dolomitic mudstones and wackestones. These latter rock types contain smaller percentages of very fine sand to pebble size skeletal fragments (0.07 mm to 4.81 mm), and intraclasts of very fine sand to granule size (0.11 mm to 2.47 mm).

Sedimentary Structures

The sedimentary structures present in the Knowles Limestone are common to several different rock units. The principal sedimentary structures in the cored intervals are laminae, burrows, and contorted bedding.

Three types of laminae characterize the limestones and shale; they are, in order of decreasing abundance: (1) wavy, subparallel discontinuous laminae, (2) horizontal, parallel discontinuous laminae, and (3) horizontal, parallel continuous laminae. Wavy, subparallel discontinuous laminae are abundant in the terrigenous dolomitic mudstones and wackestones (Fig. 10a). Also, they are common in the intraclastic and algal wackestones to packstones, and they are present in the oncolitic, intraclastic, and skeletal wackestones to packstones and floatstones.

Horizontal, parallel discontinous laminae are most abundant in the terrigenous dolomitic mudstones and wackestones of all three cores (Figs. 7-9), but also occur in the intraclastic and skeletal wackestones and packstones with algal fragments of the SR 1 core (Fig. 7).

The horizontal, parallel continous laminae occur in the thinly interbedded arenaceous shales that are present in the updip,

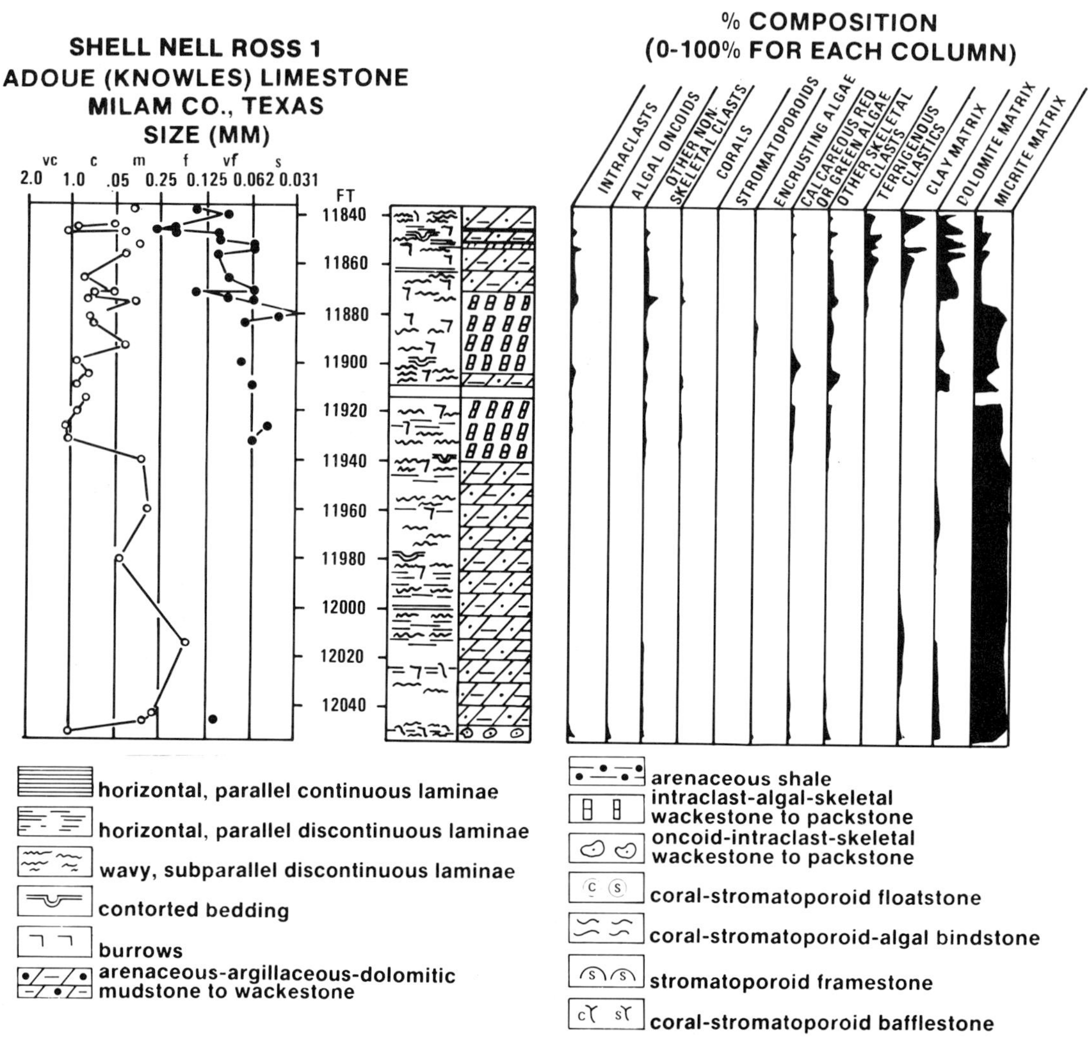

Figure 7. Rock properties and microfacies of the Knowles Limestone in the SR 1 core. The mean grain sizes for carbonate particles and quartz grains are marked by open and closed circles respectively. The normalized percentages of constituents are shown.

terrigenous dolomitic mudstones and wackestones of the SR 1 and SA 1 cores (Fig. 10c).

Bioturbation is common in the updip, terrigenous dolomitic mudstones and wackestones of the SR 1 and SA 1 cores (Fig. 10a), and in the intraclastic and skeletal wackestones and packstones which contain calcareous red or green algal fragments of the SR 1 core (Fig. 10d).

Contorted bedding is common in basal portions of updip, terrigenous dolomitic mudstones and wackestones of the SR 1 and SA 1 cores (Fig. 10b), and in the basal portions of the intraclastic and skeletal wackestones and packstones of the SR 1 core. Contorted bedding also occurs in oncolitic, intraclastic, and skeletal wackestones to packstones and coral-stromatoporoid floatstones of the SA 1 core (Fig. 10e, f). This feature was probably formed by rapid deposition of a relatively grainy facies into a softer, muddier facies.

All boundstone rock types are dominantly unstratified due to the organic growth which formed them (Fig. 11a, b, c). Occasional wavy, subparallel discontinuous laminations are present in the detrital infill of the boundstones.

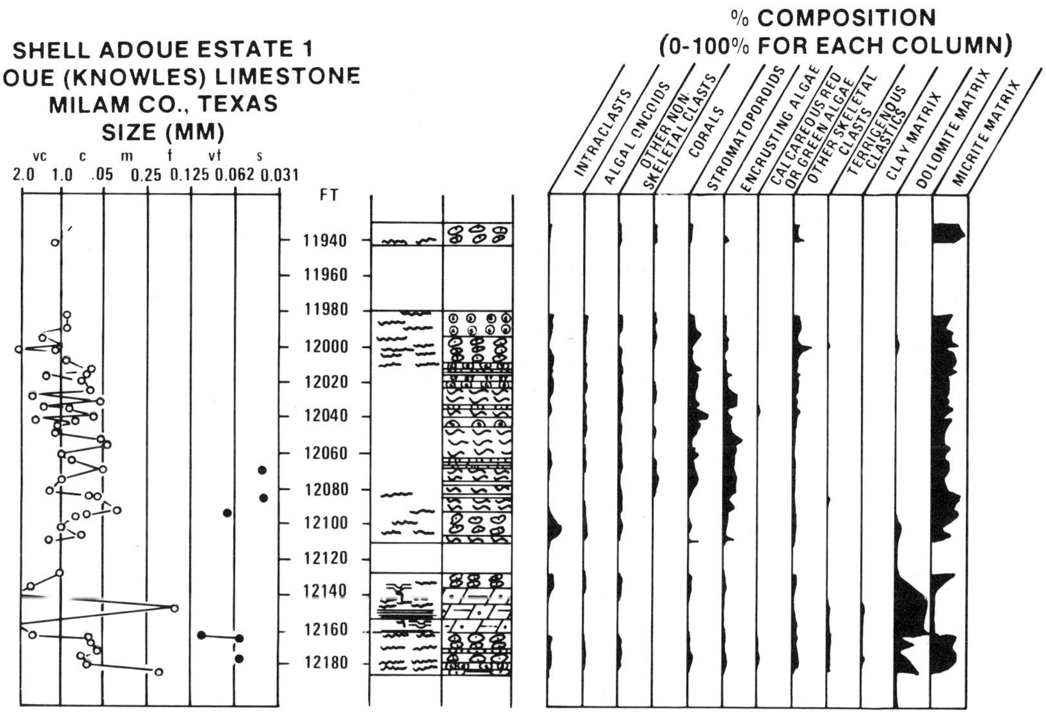

Figure 8. Rock properties and microfacies of the Knowles Limestone in the SA 1 core. The mean grain sizes for carbonate particles and quartz grains are marked by open and closed circles respectively. The normalized percentages of constituents are shown. See Figure 7 for symbols.

Composition

The normalized percentages of the important grain types are shown in Figures 7-9. Carbonate particles are more common downdip and terrigenous clastics are more common updip. The carbonate framework grains include both skeletal and non-skeletal types in nearly equal proportions. Non-skeletal grains are ooids, peloids, intraclasts, lithoclasts, grapestones and *Tubiphytes* oncoids. Skeletal grains include whole and broken scleractinian corals, stromatoporoids, pelecypods, gastropods, *Solenopora,* dasycladacean algae, codiacean algae, large mollusk fragments, echinoid fragments, and minor numbers of ostracods, benthic foraminifers, bryozoans, and serpulid worm tubes. Encrusting *Tubiphytes* and *Lithocodium* algae form much of the framework in the boundstone rock types.

Tubiphytes-algal oncoids are conspicuous as are encrusting algal intraclasts in the oncolitic, intraclastic, and skeletal wackestones to packstones, and coral, stromatoporoid floatstones. They are also abundant in boundstones. This form appears to be identical to the blue-green (cyanophycean) alga *Tubiphytes obscurus* Maslov, 1956 which Flugel (1981, p. 149-153) described from the Lower Permian of the Southern Alps in southern Austria and northern Italy. They have also been recently described from Upper Jurassic algal-sponge buildups of Germany (Flugel and Steiger, 1981).

Ooids and peloids are abundant updip in the intraclastic, and skeletal wackestones to packstones, but are also occasionally present in the terrigenous dolomitic mudstones to wackestones.

ORIGIN OF KNOWLES ROCK TYPES

Rock types were described within broadly defined microfacies types based primarily upon constituent compositions and depositional textures as seen in thin-sections or core slabs. The microfacies are defined by the

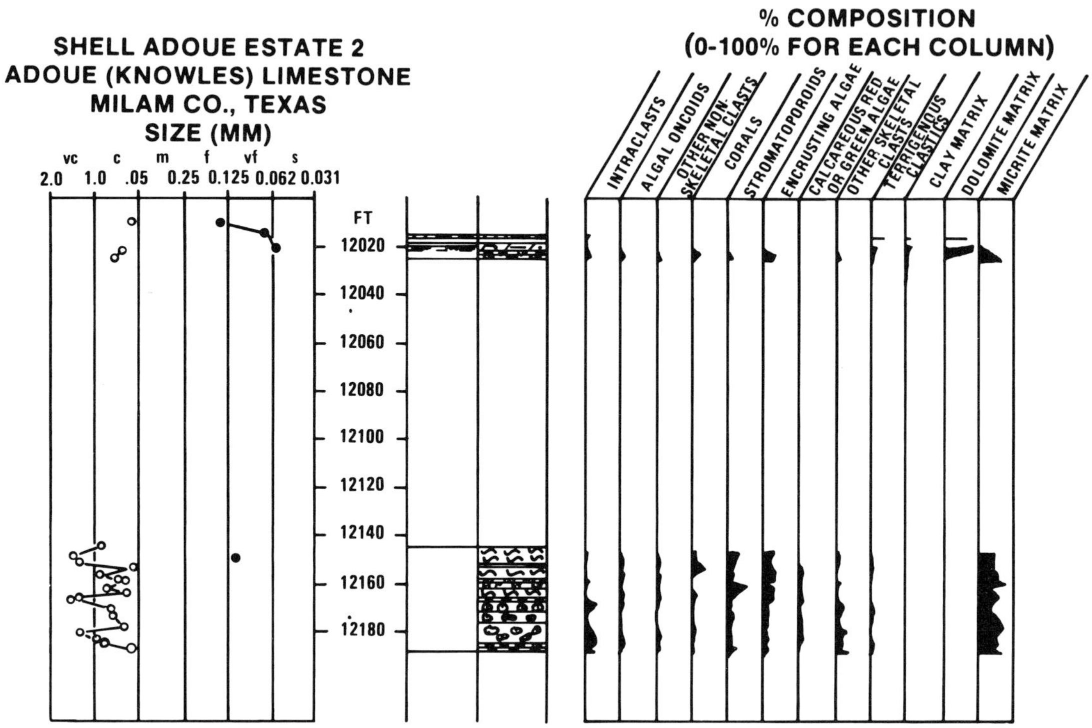

Figure 9. Rock properties and microfacies of the Knowles Limestone in the SA 2 core. The mean grain sizes for carbonate particles and quartz grains are marked by open and closed circles respectively. Twe normalized percentages of constituents are shown. See Figure 7 for symbols.

Figure 10. Photographs of core slabs showing basic fabrics and sedimentary structures in different rock types of the Knowles Limestone in the study area.

a. Wavy, subparallel discontinuous, horizontal to slightly inclined (<5°) laminae near top of photograph, which grades upward to faint horizontal, parallel discontinuous laminae to arenaceous-argillaceous-dolomitic mudstone to wackestone. Bottom portion is extensively bioturbated with some small burrows present. From SA 2 core at depth of 12,018.3 feet.

b. Contorted bedding due to loading of light gray arenaceous-argillaceous-dolomitic mudstone to wackestone above onto dark gray intraclast-algal-skeletal wackestone to packstone below. From SR 1 core depth of 11,851.0 feet.

c. Arenaceous shale diplaying fissility and very faint horizontal, parallel continuous laminae. From SA 1 core at depth of 12,149.7 feet.

d. Massive intraclast-algal-skeletal wackestone to packstone with large burrow. From SR 1 core at depth of 11,877.6 feet.

e. Dominantly massive oncoid-intraclast-skeletal wackestone to packstone with abundant fine, low amplitude stylolites. From SA 1 core at depth 12,130.1 feet.

f. Massive coral-stromatoporoid floatstone with abundant large coral and stromatoporoid fragments, and high amplitude stylolites present. From SA 1 core at depth of 11,981.3 feet.

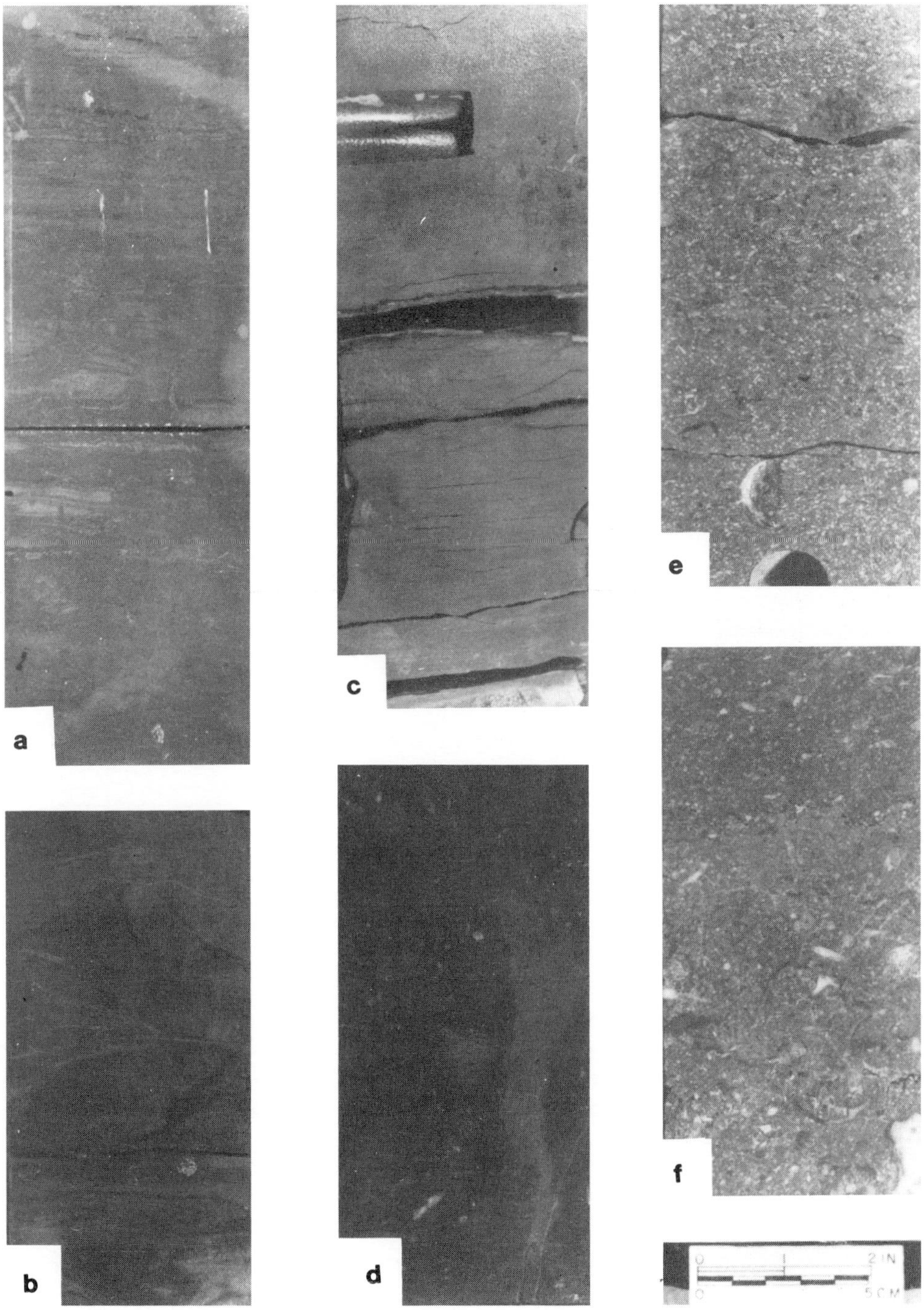

Figure 10

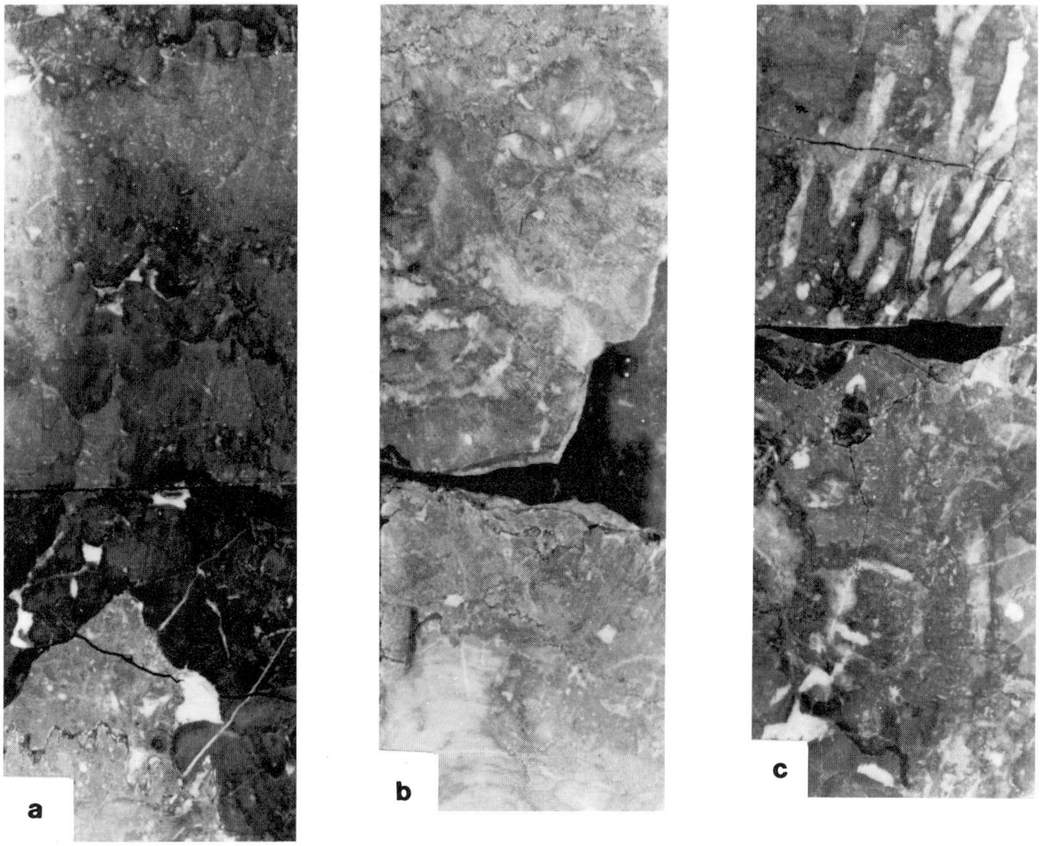

Figure 11. Photographs of core slabs showing basic fabrics and sedimentary structures in different boundstones of the Knowles Limestone in the study area.

a. Dark gray coral-stromatoporoid-algal bindstone above with light gray oncoid-intraclast-skeletal wackestone to packstone below. Note the algal constructed voids filled with calcite. From SA 2 core at depth of 12,157.6 feet.

b. Stromatoporoid framestone with massive columnar stromatoporoid at bottom left. From SA 1 core at depth of 12,038.3 feet.

c. Coral-stromatoporoid bafflestone with clustered, digitate corals and dendroid stromatoporoids near top, and coral-stromatoporoid-algal bindstone with encrusting *Tubiphytes*, tabular stromatoporoids, and a few solitary corals at bottom. Note the algal constructed voids filled with calcite in the bindstone. From SA 2 core at depth of 12,161.7 feet.

most frequent occurrence or the most conspicuous particle type, by a mud-to-particle ratio, and by the biota present. In addition, any widespread diagenetic feature is also taken into account, such as abundant dolomitized lime mud in the terrigenous dolomitic mudstones to wackestones. Eight different microfacies were identified.

1. Arenaceous-argillaceous-dolomitic mudstone to wackestone microfacies. This microfacies is present in all three cores, but it is most abundant in the SR 1 core. It consists of abundant quartz, terrigenous clay, lime mud, which is often dolomitized, and carbonate grains that include intraclasts, bivalves, gastropods, benthic foraminifers and calcareous algal fragments. Detrital grains are generally moderately sorted and are fine sand to granule in size. Wavy, subparallel, discontinuous laminae are dominant.

2. Arenaceous shale microfacies. A few thin, arenaceous shales are interbedded with-

in the previous microfacies. These shales contain quartz, other terrigenous clastics and abundant clay, and typically display horizontal, parallel, continuous laminae.

3. Intraclast-algal-skeletal wackestone to packstone microfacies. These rocks are present only in the upper half of the SR 1 core and are often interbedded with the arenaceous-argillaceous-dolomitic mudstone to wackestone microfacies. The rocks contain intraclasts, calcareous red or green algae, other skeletal grains such as whole or fragmented mollusks and echinoid fragments and lime mud; the lime mud may be dolomitized. Grains range from medium to very coarse sand in size. Bedding commonly has been homogenized by bioturbation, and large burrows are often present.

4. Oncoid-intraclast-skeletal wackestone to packstone microfacies. This facies is typical of the cores from downdip wells. Abundant *Tubiphytes* oncoids, intraclasts and skeletal grains that include whole or fragmented stromatoporoids, corals, various mollusks, echinoid fragments and serpulid worm tubes are characteristic of the microfacies. Grains are of coarse-sand to granule size and are generally very poorly sorted. Intraclasts consist of fragments of encrusting *Tubiphytes* or *Lithocodium* algae derived from the reef core. This microfacies is massively bedded or bioturbated, but also contains wavy, subparallel discontinuous laminae.

5. Coral-stromatoporoid floatstone microfacies. This microfacies is interbedded with the oncoid-intraclast-skeletal wackestone to packstone facies and the matrix is texturally and compositionally the same; but it is distinguished from the latter facies by having greater than 10% very coarse- to pebble-sized coral and stromatoporoid clasts floating in the wackestone to packstone matrix. The remaining three microfacies make up the unstratified reef-core boundstone rock types, which are only present downdip in cores from the SA 1 and SA 2 wells.

6. Coral-stromatoporoid-algal bindstone microfacies. This is the principal boundstone rock type. It consists of abundant dendroid or solitary corals and tabular stromatoporoids in growth position, encrusting *Tubiphytes* and *Lithocodium* algae, lime mud, and other grains including intraclasts, peloids, *Tubiphytes* oncoids, coral and stromatoporoid clasts, echinoids and mollusk fragments. Detrital grains are medium to very coarse sand size and generally are very poorly sorted.

7. Stromatoporoid framestone microfacies. This type dominantly consists of massive stromatoporoids in growth position which form the framework and make up much of the rock. The detrital fraction consists of lime mud and carbonate grains including coral and stromatoporoid clasts, echinoid and bivalve fragments, peloids, *Tubiphytes* oncoids, and intraclasts. The grains are coarse to very coarse sand size, and are poorly sorted.

8. Coral-stromatoporoid bafflestone microfacies. This microfacies contains *in situ* clusters of digitate corals, dendroid stromatoporoids, abundant lime mud and carbonate grains including intraclasts, peloids, and *Tubiphytes* oncoids, Grains are dominantly coarse sand size and are poorly sorted.

Environments of Deposition

There are four basic depositional environments, two of which have sub-environments, represented by the microfacies. They are:

1. Tidal flat, intertidal to supratidal.
   Tidal flat arenaceous-argillaceous-dolomitic mudstone to wackestone (minor arenaceous shale).
2. Lagoonal.
   Lagoonal intraclast-algal-skeletal wackestone to packstone.
3. Reef flank.
   A. Distal reef talus oncoid-intraclast skeletal wackestone to packstone.
   B. Proximal reef talus coral-stromatoporoid floatstone.
4. Reef core.
   A. High-energy zone stromatoporoid framestone.
   B. Intermediate-energy zone coral-stromatoporoid-algal bindstone.
   C. Low-energy zone coral-stromatoporoid bafflestone.

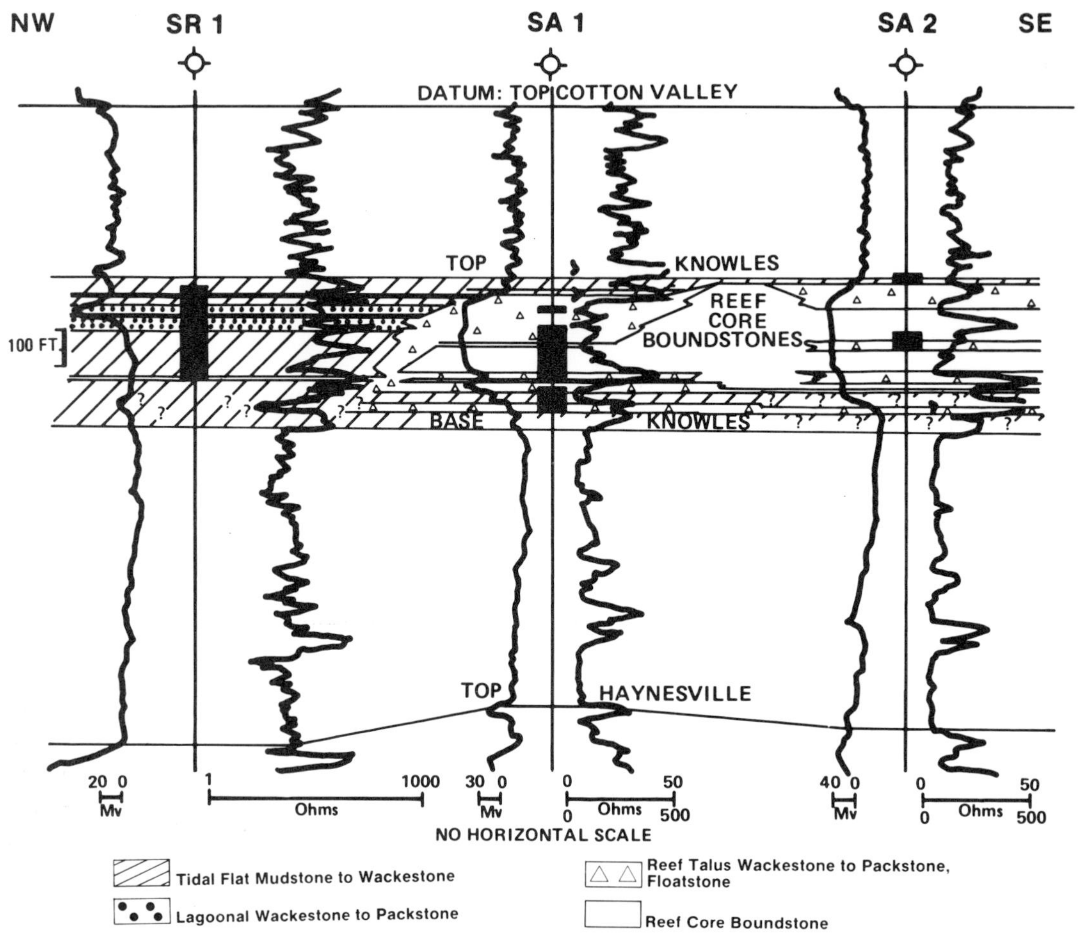

Figure 12. SP and resistivity log character and facies correlations shown in stratigraphic cross section through the study area. Core control is shown for each of the three wells. Note the difference in log character between the updip tidal-flat and lagoonal facies of the SR 1 well, and the reef-talus and reef-core boundstones of the SA 1 and SA 2 wells. The subtle salt-generated topographic high on which the Knowles reef developed is shown by the thin-section below the Knowles in the SA 1 and SA 2 wells (indicated by the top of the Haynesville Formation).

The microfacies have been grouped to form facies which characterize the basic depositional environments. The generalized facies (Fig. 12) are: tidal-flat mudstone to wackestone; lagoonal wackestone to packstone; reef-talus wackestone to packstone, floatstone; reef-core boundstone. The four facies have different SP and resistivity log characteristics. This difference enables identification of facies and environments of deposition to be made where core control is lacking (Fig. 13).

DIAGENESIS

Important types of diagenesis include early micritization of grains in the marine environment, and formation of fine-crystalline anhedral and limpid euhedral dolomite within tidal-flat and lagoonal mudstones to wackestones, during hypersaline and mixing-zone diagenesis, respectively. The formation of phreatic calcite cements, including equant calcite crusts, pore-filling calcite mosaic and coarse sparry calcite, occurred during

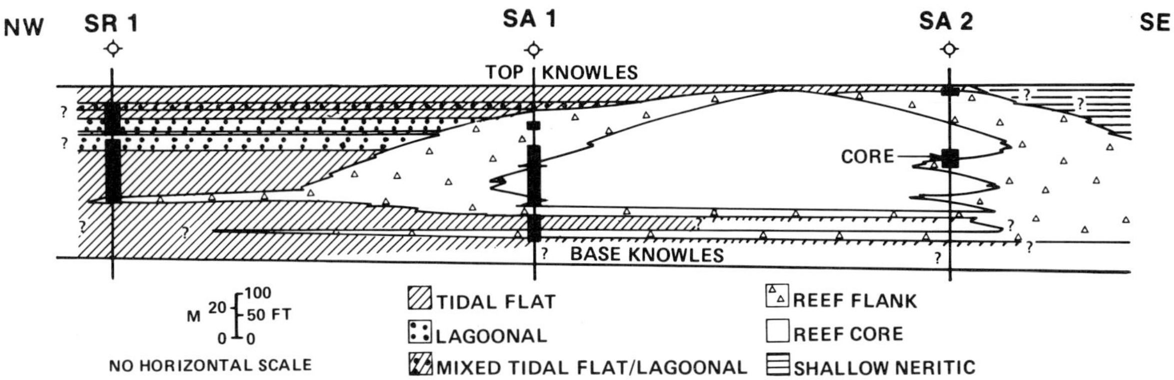

Figure 13. Stratigraphic cross section showing environments of deposition in the study area. Tidal flat and lagoonal environments are common updip to the northwest, reef-flank and reef-core environments are common downdip to the southeast. A shallow neritic depositional environment was probably present downdip from the reef core and reef flank to the southeast. Cores are shown by heavy line in the three wells.

meteoric/mixing zone and subsurface diagenesis. The dissolution of aragonite and inversion and recrystallization of skeletal grains and lime mud to microspar or calcite mosaic are common in most facies. In addition, stylolitization is common in all facies, but it is most prevalent in the reef talus, which also displays microstylolites and sutured grain contacts.

DEVELOPMENT OF REEF MORPHOLOGY

The Knowles reef is most like the Type 1, encrusted skeletal buildup, described by Heckel (1974). This patch reef exhibits *in situ* skeletal growth and encrustation. Massive columnar, subspherical, and tabular stromatoporoids and solitary or dendroid fasciculate corals functioned primarily as reef builders, and *Tubiphytes* and *Lithocodium* algae functioned primarily as encrusters that bound the framework and incorporated loose sediment into the reef mass. However, *Tubiphytes* and *Lithocodium* appear to have also functioned as primary framework organisms as suggested by constructional voids that are associated only with these algal forms.

The paleogeographic setting was most important in determining the gross morphology of the Knowles reef. Local bathymetric changes associated with a subtle, salt-generated paleotopographic high was most important in siting the reef. With only three cores on a dip line, it is impossible to determine the total extent of the Knowles reef, but based on observed facies distribution and the biota of the reef core and surrounding facies, it appears very likely that the Knowles patch reef grew as a lenticular to circular buildup.

The reef probably had a considerable amount of topographic relief above the surrounding sea floor. Although this relief cannot be measured directly in the subsurface, the study of dipmeter logs from the SR 1 and SA 2 wells has indicated the relative topographic relief on the Knowles reef (Fig. 14). Weighted, mean-dip azimuth values were calculated for thinly-bedded, tidal-flat mudstones and wackestones present in the SR 1 and SA 2 wells that drape over the top of the reef core and flanks (Figs. 12 and 13). There is a southeasterly structural dip of 7° in the area, which was removed from dipmeter values to obtain the stratigraphic dip values of the draping beds in the SR 1 and SA 2 wells. The beds dip generally northwest at 5° in the SR 1, and generally south-southeast at 10° in the SA 2 well (Fig. 14). This indicates that the reef crest is located between the SR 1 and SA 2 wells, and suggests that the Knowles reef had topographic relief above the sea floor, although differential

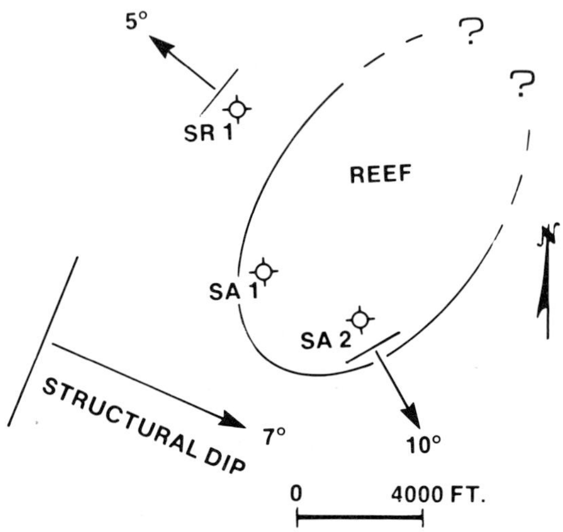

Figure 14. Map showing regional structural dip and stratigraphic dip of thinly-bedded, tidal-flat mudstones and wackestones present in the SR 1 and SA 2 wells which drape over the reef core and reef flanks. Dipmeter log was unavailable for the SA 1 well. Probable maximum lateral extent of the Knowles patch reef core is shown.

compaction probably enhanced the local depositional relief. Based on the similar facies distribution as in the SA 2 core, and the abundance of reef talus (Fig. 12), it is evident that the SA 1 well is situated on the updip reef flank, and the SA 2 well is on the downdip reef flank (Figs. 13 and 14).

STRATIGRAPHIC HISTORY

The Knowles Limestone in the study area was deposited in a shallow-water marine environment. Deposition began with tidal-flat mudstones and wackestones. Dolomitization of lime mud was widespread, and waves and tidal currents formed horizontal or wavy, discontinuous laminae. With increasing water depth, shoaling occurred on the subtle salt-generated high where fragments of mollusks and echinoids, ooids, and *Tubiphytes* oncoids accumulated. Sea level then dropped, causing a return to tidal-flat deposition across the area. Subsequently, shoaling on the high returned as water depth increased, and tidal flats existed updip. Eventually, delicate dendroid stromatoporoids and clustered digitate corals grew on the sediment pile, in a low-energy zone below average wave base. After reaching average wave base, this buildup was colonized by frame builders. Solitary and dendroid fasciculate corals, tabular stromatoporoids, and abundant encrusting *Tubiphytes* and *Lithocodium* grew in an intermediate-energy zone just below average wave base. Massive columnar or subspherical stromatoporoids grew in a high-energy zone on the reef crest. Large, reef-derived coral and stromatoporoid clasts, other bioclastic debris, and *Tubiphytes* oncoids were deposited as talus on the reef flanks.

After lateral and vertical reef growth during rising sea level, a regression began which continued throughout Knowles deposition. A shallow-water restricted lagoon formed behind the reef where intraclastic, algal, and skeletal wackestones and packstones were deposited. At this time, all the basic depositional environments characterized by various rock types seen in the cores existed simultaneously. The tidal-flat and lagoonal environments existed updip, and the reef flank and reef core, with its various facies zones and organism communities, existed downdip (Fig. 15). Crustaceans, mollusks and echinoids burrowed the tidal-flat and lagoonal sediments. The reef retreated seaward with continued regression, and tidal-flat and lagoonal environments advanced downdip. Eventually, the reef organisms were overwhelmed with terrigenous sediments brought in with the advancing tidal flat. Reef growth ended and the tidal flat advanced over the lagoonal, reef-flank, and reef-core facies, and Knowles deposition ended in the area.

POTENTIAL RESERVOIR PROPERTIES

Early dolomitization of the tidal-flat and lagoonal facies created local porous zones in some of these rocks. However, the reef, *per se*, is cemented by sparry calcite and is not a potential reservoir facies in this area.

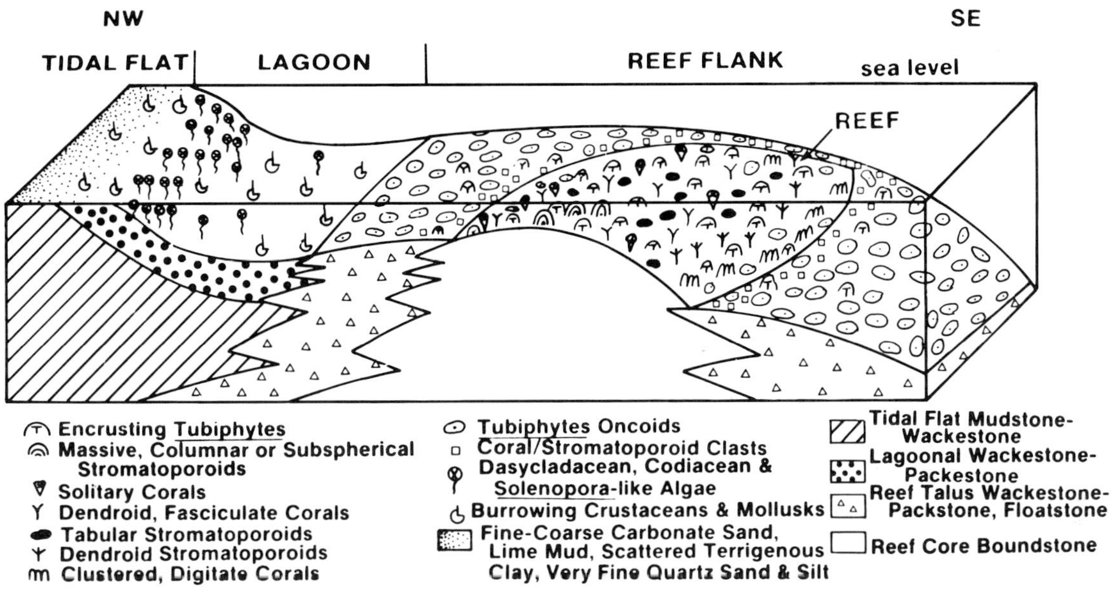

Figure 15. Interpretive reconstruction of the Knowles reef and associated facies.

CONCLUSIONS

1. The Knowles Limestone of southeastern Milam County, Texas, consists of reef-core boundstones, reef talus, lagoonal, and tidal-flat facies. Framestones, bindstones and bafflestones represent a wave resistant, encrusted skeletal patch reef that developed on a subtle, elongate, strike-trending salt high; reef talus was deposited on the flanks. Wackestones and packstones were deposited in a shallow lagoon formed behind the reef during a lowered sea-level stage. An intertidal to supratidal environment was present updip and, at least twice, spread downdip over the entire study area depositing mudstones to wackestones, and often abundant, terrigenous debris. Later the mudstones and wackestones were dolomitized.

2. A subtle paleotopographic high was generated by movement of the Louann Salt which continued through Knowles deposition and ended with Hosston deposition.

3. Massive, columnar and tabular stromatoporoids, solitary and dendroid corals, and encrusting *Tubiphytes* and *Lithocodium* algae were the principal contributors to the reef framework. Dendroid stromatoporoids, digitate corals, *Tubiphytes* oncoids, echinoids, pelecypods, and gastropods were the principal contributors to the detrital infill of the reef.

4. Reef organisms were overwhelmed by terrigenous debris carried downdip with the prograding tidal flat that advanced over the lagoonal, reef-flank and reef-core facies at the end of Knowles deposition.

5. Early dolomitization of tidal-flat and lagoonal facies has created local porous zones in some of these rocks. However, the reef, *per se*, is cemented by sparry calcite and is not a potential reservoir.

ACKNOWLEDGMENTS

This study was supported by grants from Tenneco Oil Exploration and Production Company, and Gulf Coast Association of Geological Societies. The seismic section utilized was provided by Mobil Producing, Texas and New Mexico Incorporated - United Geophysical. Special thanks are given to Shell Oil Company for supplying cores and drafting facilities during preparation of this paper.

REFERENCES

Andrews, D. I., 1960, The Louann Salt and its relationship to Gulf Coast salt domes: Gulf Coast Assoc. Geol. Soc. Trans., v. 10, p. 215-240.

Bornhauser, M., 1958, Gulf Coast tectonics: AAPG Bull., v. 42, p. 339-370.

Embry, A. F., and J. E. Klovan, 1971, A Late Devonian reef tract on northeastern Banks Island, Northwest Territories: Can. Petrol. Geology Bull., v. 19, p. 730-781.

Flugel, E., 1981, Lower Permian *Tubiphytes/-Archaeololithoporella* buildups in the Southern Alps (Austria and Italy), *in* D. F. Toomey, ed., European fossil reef models: SEPM Spec. Pub. 30, p. 143-160.

__________ and T. Steiger, 1981, An Upper Jurassic sponge algal buildup from the Northern Frankenalb, West Germany, *in* D. F. Toomey, ed., European fossil reef models: SEPM Spec. Pub. 30, p. 371-397.

Halbouty, M.T., 1966, Stratigraphic trap possibilities in Upper Jurassic rocks, San Marcos Arch, Texas: AAPG Bull., v. 50, p. 3-24.

Heckel, P. H., 1974, Carbonate buildups in the geologic record: A review, *in* L. F. Laporte, ed., Reefs in time and space--selected examples from the recent and ancient: SEPM Spec. Pub. 18, p. 90-154.

Hughes, D. J., 1968, Salt tectonics as related to several Smackover fields along the northeast rim of the Gulf of Mexico basin: Gulf Coast Assoc. Geol. Soc. Trans., v. 18, p. 320-330.

Mann, J.C., and W. A. Thomas, 1964, Cotton Valley Group (Jurassic) nomenclature, Louisiana and Arkansas: Gulf Coast Assoc. Geol. Soc. Trans., v. 14, p. 143-152.

Maslov, V. P., 1956, Iskopaemye izvestkovye vodorosli SSSR: Akad, Nauk SSSR, Trudy Instituta Geol. Nauk, v. 160, 301 p.

Philpott, T. H., and R. T. Hazzard, 1949, Preliminary correlation chart of upper Gulf Coast: Shreveport Geol. Soc. 17th Ann. Field Trip Guidebook (in pocket).

Shreveport Geological Society, 1953a, Correlation chart: *in* Shreveport Geol. Soc. 21st Ann. Field Trip Guidebook.

__________ 1953b, Reference report on certain oil and gas fields of north Louisiana, south Arkansas, Mississippi, and Alabama: Shreveport Geol. Soc., v. 3, 108 p.

Thomas W. A., and J. C. Mann, 1966, Late Jurassic depositional environments, Louisiana and Arkansas: AAPG Bull., v. 50, p. 178-182.

DEPOSITIONAL MODELS FOR JURASSIC REEFAL BUILDUPS

PAUL D. CREVELLO

Denver Research Center
Marathon Oil Company
P.O. Box 269
Littleton, Colorado 80160

PAUL M. HARRIS

Gulf Research and Development Company
P.O. Box 37048
Houston, Texas 77236

ABSTRACT

Studies of Jurassic carbonate reefal buildups from the surface and subsurface of the United States Gulf Coast (including the Smackover Formation), the eastern North American Margin, and the Tethyan reef belt of Europe and northwest Africa show that buildup development occurred in three depositional settings: (1) deep, quiet water below wave base; (2) gently sloping platforms (ramps) subjected to wave agitation; and (3) margins of steeply rimmed, wave-swept platforms. A classification developed for the buildups from these various depositional settings emphasizes the major reef-framebuilding assemblages which, if combined with studies of buildup-related facies, provide a basis for interpretation of depositional setting and paleoenvironmental reconstruction.

Buildups that developed in deep, quiet-water settings below wave base were constructed primarily by siliceous sponges and thrombolite/stromatolite algal(?) structures. Accessory reef-dwelling biotic communities and related lithofacies assemblages indicate that these sponge-algal buildups grew in normal marine waters either on gently sloping platforms, in settings downslope from platform margins, or on basin floor highs.

Buildups that developed on wave-agitated, gently sloping platforms show the greatest range in framebuilding assemblages and are the most common types in the subsurface Smackover Formation. In lower-energy settings, which could range in water depth depending on wave energy, these buildups were constructed by siliceous and calcareous sponges, chaetitids, corals, stromatolites/thrombolites, and calcareous algae. Such buildups apparently represent an intermediate assemblage between the sponge-algal deep-water mounds and the coral-dominated patch-reef mounds developed in more energetic platform settings. The coral-dominated patch reefs contain abundant calcareous biota such as skeletal algae, calcareous sponges, and stromatoporoids. Commonly, reworking of the reef core by waves produced piles of skeletal debris. In regions where salinities and temperatures of marine waters covering the platforms diverged from normal oceanic conditions, biotic-community assemblages were reduced, resulting in stromatolite algal(?) buildups.

Buildups replete with coral, stromatoporoid, calcareous sponge, and skeletal algae framebuilders developed on steepened platform margins facing open oceans. Stromatoporoid assemblages enabled these buildups to be differentiated into back-reef patch reefs and platform-margin reefs.

INTRODUCTION

Jurassic carbonate buildups have only recently been documented from the United

States Gulf Coast, whereas equivalent-age buildups in Europe and North Africa have been studied for over 100 years. This disparity is due primarily to the lack of surface exposures along the Gulf Coast, in contrast to the accessible buildups cropping out around the Tethyan reef belt. The outcrop studies define biotic assemblages, buildup geometries and trends, facies relationships, and paleogeographic settings of buildup development - all of which should be applied to investigations of subsurface Jurassic reefal buildups.

Baria *et al.* (1982) have recently discussed the occurrence of Oxfordian patch reefs that developed on a ramp shelf throughout the Gulf Coast during deposition of the Smackover Formation. Researchers have also reported that patch reefs (Cregg and Ahr, 1982) and a ramp-reef complex (Finneran *et al.* 1982; Scott, this volume) developed as a part of the Knowles Limestone in central East Texas. Exploration drilling and seismic studies provide the only means for gathering information on these buildups. Occurrences and general trends have been inferred from seismic and borehole log correlations, whereas facies trends and paleocommunity analyses require thorough examination of rock material with correlations to well-log signatures. Reefs, probably more than other carbon- ate facies, provide considerable insight in- to depositional setting and general paleooceanography because of their diverse biologic community assemblages. Modern reefs grow in a wide range of water depths, which can be reasonably interpreted from biotic assemblages. This paper shows that Jurassic (and possibly early Cretaceous) reef (or buildup) development also occurred in a variety of depositional settings which are also reflected by their biotic communities. In this paper, we develop a classification for Jurassic carbonate buildups based on the major framebuilding paleocommunity. This classification stresses an interpretation that can be made from limited subsurface rock samples which, through biotic community analyses, can be used to infer subsurface buildup geometries and trends, and paleoshelf setting.

Much of the information on buildup geometries and lithofacies relationships presented in this paper was taken from outcrop studies. The work of Baria *et al.* (1982) and additional microfacies analyses provided the framework for our classification of the Smackover buildups. A general classification for Jurassic buildups, which may also apply to early Cretaceous buildups, incorporates our Smackover model along with data from outcrop studies reported in the literature and our examinations of buildups that crop out in Europe and northwest Africa.

PREVIOUS STUDIES OF SUBSURFACE JURASSIC BUILDUPS FROM THE UNITED STATES GULF COAST

Prior to the work of Baria *et al.* (1982; Fig. 1), relatively little information on Smackover (Oxfordian) buildups appeared in the Gulf Coast literature. The first report on possible buildups in the upper Smackover Formation was provided by Imlay (1940) for widely distributed, coral-rich, fossiliferous beds in southern Arkansas. Similarities with coral-rich deposits from the Jurassic of England, described as fringing reefs by Arkell (1935), prompted Imlay (1940) to speculate that the Smackover shelf was also rimmed with fringing reefs. Wells (1942) discounted the fringing-reef hypothesis, however, based on his study of coral specimens provided by Imlay. Wells (1942) noted that the coral growth forms were more indicative of either a very shallow, quiet-water lagoon or deeper water undisturbed by surf, and that none of the corals described from the Smackover were similar to the massive forms described by Arkell (1935). Wells (1942) further noted that the corals probably grew in thin patches which were scattered across a gently sloping sea floor.

Not until Croft's (1980) study of two cores in the North Haynesville Field, northern Louisiana, was the first description and well-correlation study provided of reef and related facies in the Smackover. Croft (1980) described the reef interval as a coral and coralline-algal framework with multiple

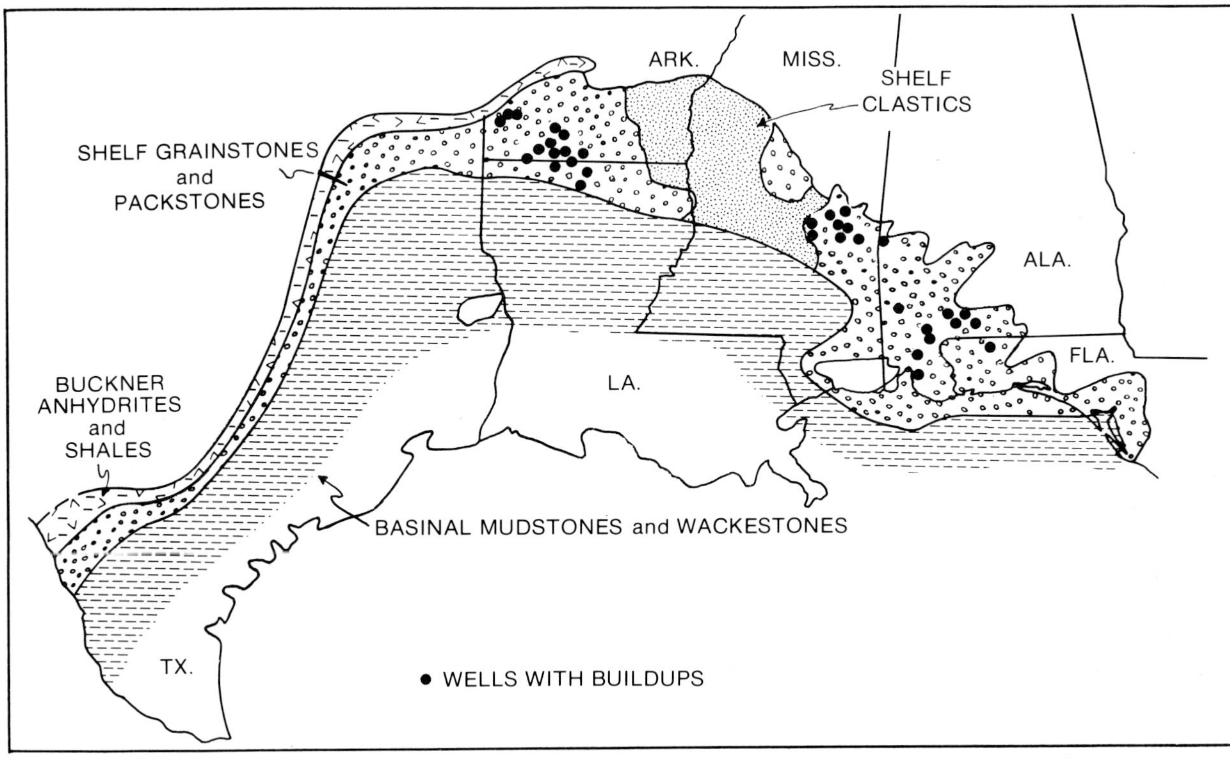

Figure 1. Generalized map of Smackover and Buckner lithologies along United States Gulf Coast (modified after Baria *et al.*, 1982; and Budd and Loucks, 1981). Black circles represent locations of subsurface Jurassic buildups described from conventional cores by Baria *et al.* (1982) and McGraw (1982).

generations of microborings and geopetal sediment; and he recognized that localization of the reef on the southern, seaward-facing edge of a horst was likely caused by penecontemporaneous faulting during deposition on the Smackover shelf. Croft's study, however, dealt primarily with local variations in the non-skeletal carbonate facies and did not address buildup paleocommunities.

Baria *et al.* (1982) concluded, from extensive core and borehole log studies, that reefs were common in the upper Smackover (Fig. 1) and grew on or near the seaward-facing crest of basement- or salt-cored paleostructures. Variations in the reef foundation (Fig. 2) through the trend, the stratigraphic level of reef development, and more importantly, the diversity of biotic assemblages led Baria *et al.* (1982) to suggest that circulation of normal marine waters must have varied along trend over the Smackover platform to account for the dramatic variation in the paleocommunities: more restricted marine conditions existed over the platform in Alabama, Mississippi, and Florida. Baria *et al.* (1982) recognized that similarities existed between the paleocommunities of the skeletal-rich Smackover buildups and the European examples, but did not compare the buildups of the two areas. Crevello and Harris (1982) refined the classification of Smackover buildups to facilitate comparisons with Tethyan buildups. Crevello *et al.* (in press) also examined the reservoir potential and diagenesis of several of these buildups, emphasizing that both depositional and diagenetic variations through the subsurface Smackover trend were important in developing reservoir-quality lithologies.

In buildups of the Knowles Limestone

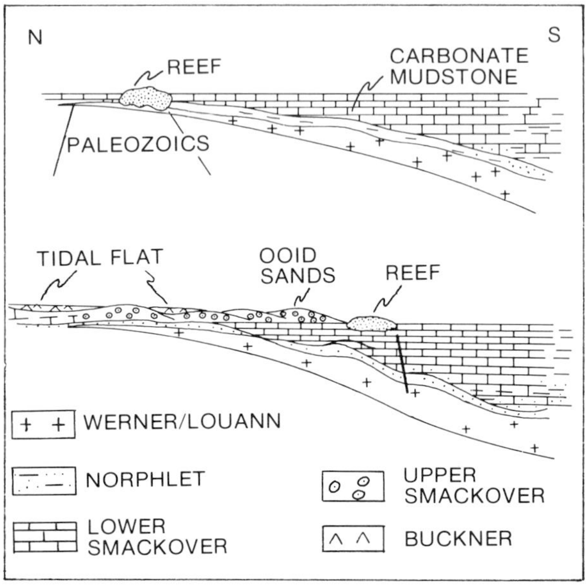

Figure 2. Schematic dip cross section illustrating development of late Jurassic reefs on basement blocks in the lower Smackover Formation of Alabama and Florida (top) and on salt swells in the upper Smackover Formation of Arkansas and Louisiana (bottom) (modified after Baria *et al.*, 1982).

(Finneran *et al.*, 1982; Cregg and Ahr, 1982), most recently determined to be Berriasian to Valanginian in age (Scott, this volume), corals along with stromatoporoids developed a more diverse and important role as reef builders. The Knowles Limestone also marks the development of a seismically and lithologically definable platform margin, in contrast to the gently sloping Smackover carbonate platform (or carbonate ramp).

CLASSIFICATION OF JURASSIC CARBONATE BUILDUPS

Our classification of Jurassic carbonate buildups is based on depositional setting and major fossil assemblages. Baria *et al.* (1982), although noting a variation in biotic constituents through the Smackover trend, broadly classified these facies as reefs because they contained attributes common to many modern and ancient reefs: framebuilding, binding, and encrusting organisms; marine cements; internal sediment; and a boring infauna. A refinement of the classification of subsurface buildups has enabled us to compare our findings with outcrop studies of other Jurassic buildups for which detailed facies relationships and paleogeographic reconstructions exist.

Wilson (1975) compiled several commonly used definitions for buildups based, in general, on their local configuration and composition. A recurring problem encountered in classifying subsurface carbonate buildups, however, is the inability to determine the buildup geometry and the relationship of the buildup to lateral and bounding facies; hence, thick biostromes may be indistinguishable from bioherms, and bioclastic beds may be indiscernible from bioclastic piles. In addition, borehole information is too widely spaced to resolve this problem, although, in some cases, where the buildups are thick and/or laterally extensive, seismic profiling may aid in the interpretation. Classification for subsurface buildups therefore must be based primarily on the biotic assemblages recognized in cores and cutting samples and on similarities with analogues from outcrop studies.

The outcrop and subsurface examples discussed here display some of the major differences in paleocommunities that exist in Jurassic reefal buildups. Our subsurface and outcrop studies as well as literature review suggest that the major framebuilding and binding organisms that contributed to reef development were environmentally sensitive to depositional settings and are useful for classifying the buildups. Accessory reef inhabitants, epifaunal and infaunal dwellers, and encrusting organisms were also important in reef building, but their distribution and recognition are much more complicated. Few studies have been published which adequately document entire Jurassic reef paleocommunities.

Jurassic carbonate buildups have been described from three depositional settings (Table 1): (1) deep, quiet water below wave base; (2) wave-agitated, gently sloping car-

Table 1. Classification, Depositional Setting, and Reef-Building Organisms of Jurassic Buildups

DEPOSITIONAL SETTINGS	DOMINANT REEF BUILDERS	ACCESSORY ORGANISMS	CLASSIFICATION
Deep, Quiet Water *fore-reef slope *deep-water lagoon *deep water of sloping carbonate platform (ramp) *basin floor	Siliceous Sponges (hexactinellids and lithistids) and stromatolites	Brachiopods, Echinoids and Ophioroids, Belemnites and Ammonites, Foraminifers (vagile-benthonic, sessile and planktic forms), Ostracods, Serpulids, Bryozoans, Pelecypods, Calcareous Sponges, *Tubiphytes* sp., Gastropods, Corals	Sponge-Algal Buildup Sponge Buildup Algal Buildup
Wave-Agitated Platform *lagoon behind reef belt *shallow water of gently sloping carbonate platform	Corals, Siliceous (hexactinellid and lithistid) and Calcareous Sponges, Stromatolites Calcareous Algae	Pelecypods, Chaetetids Foraminifers (vagile-benthic and sessile forms), Bryozoans, Gastropods, Brachiopods, Skeletal Algae, Echinoids, Stromatoporoids, Stromatolites, Corals	Coral-Dominated Patch Reef Coral-Stromatoporoid Patch Reef Sponge-Coral-Algal Buildup Stromatolitic Buildup Bioclastic Pile
Platform Margin *open ocean facing	Corals, Stromatoporoids, Calcareous Sponges	Chaetitids, Calcareous Sponges, Calcareous Algae, Pelecypods, Gastropods, *Tubiphytes* sp. Foraminifers (vagile-benthic and sessile forms), Echinoids, Bryozoans, Crinoids	Stromatoporoid, Coral, and/or Calcisponge Buildup

bonate platforms; and (3) platform margins. Buildups that developed on wave-agitated platforms display the broadest range in paleocommunity assemblages, some of which have been recognized in the subsurface Smackover Formation.

Outcrop Examples of Tethyan Buildups

Deep-Water Mounds

Sponge-Algal Mud Mounds.- During the late Jurassic, micrite-rich sponge-algal mounds developed across parts of Central Europe in relatively deep, quiet water. These mounds occur along a trend termed the Malm reef belt, which rimmed the northern margin of the Tethyan Sea (Wilson, 1975). In Germany, the well-documented mounds (late Oxfordian to early Tithonian in age) in the Swabian and Franconian Alb represent the "classic" examples of sponge-algal reefs (Fig. 3) (Wagenplast, 1972; Nitzopoulos, 1974; Zeiss *et al.* 1975; Gwinner, 1976; Flugel and Steiger, 1981). The cores of these mounds consist of a sponge-algal boundstone facies (Fig. 3) with siliceous sponges (Hexactinellidae and Lithistidae) in various growth forms (flat, dish-, tube-, or cone-shaped), algal crusts (thrombolitic forms and stromatolites), and an accessory population of tubular foraminifers, *Tubiphytes* sp., serpulids, brachiopods, bryozoans, and various molluscs (bivalves, ammonites, and belemnites)(Wagenplast, 1972; Gwinner, 1976; Flugel and Steiger, 1981). Volumetrically, the bulk of the buildup consists of sponges and cyanophycean algal crusts in a matrix of skeletal micrite and pellets.

Adjacent to the buildups are distinct, rhythmically bedded marls and biomicrites containing ammonites, foraminifers, thin-walled bivalves, brachiopods, and sponge spicules. Near the mounds, the marls and biomicrites grade laterally into biointramicrites containing oncolites and calcified fragments of algal crusts and sponges, referred to as tuberoids (Fritz, 1958, in Gwinner, 1976; Flugel and Steiger, 1981). A cross section through a sponge-algal mound (Fig. 3a) shows an abrupt lateral change from the core-mound facies to the tuberolitic debris facies. Apparently, the water overlying the mounds was relatively deep because wave turbulence or other currents did not produce significant accumulations of detritus.

Sponge-algal communities also produced biostromes, but more commonly formed buildups with several meters to 100 meters of seafloor relief (Gwinner, 1976). Gwinner (1976) could not envisage water depths greater than 100 meters for mound development because of the depth limitations placed on the photosynthetic, blue-green (cyanophycean), non-calcareous algae, providing, that is, these crusts and stromatolites were produced by algae. Ziegler (1967, in Gwinner, 1976) based water-depth zone interpretations on ammonite assemblages, placing sponge-algal mud-mound development in about 100 meters of water.

In the late Oxfordian and early Kimmeridgian of the Swabian Alb, the sponge-algal buildups occurred as isolated patches across a deep-water shelf. But during middle and late Kimmeridgian, bioherms often coalesced to form beds in excess of 250 meters thick, covering an area of several hundred kilometers (Gwinner, 1976). Nitzopoulos (1974) and Zeiss *et al.* (1975) suggested that submarine ridges provided the foundation on which many sponge-algal mounds developed. The sponge-algal mounds in this region commonly grade upward into coral-dominated patch reefs during the later Jurassic regression (see the discussion of buildups developed on wave-agitated platforms).

Sponge-algal mound development in downslope and basin-floor settings was also common in a number of other areas: in the Bajocian of Spain (Giner and Barnolas, 1979), in the Liassic of the Central High Atlas Mountains of Morocco, and offshore Nova Scotia in Jurassic and early Cretaceous carbonates (Eliuk, 1978; Jansa *et al.* 1982).

Along the axis of the Moroccan High Atlas Trough, sponge-algal mud mounds (Fig. 4) developed biotic assemblages similar to those of the sponge-algal mud mounds of the Swabian and Franconian Alb. The boundstone core facies of the Moroccan mud mounds (Fig. 4) is dominated by digitate- and branching-algal crusts with sponges, which produced a microframework infilled with pellets, skeletal debris, and micrite. The microframestone changes abruptly into bedded micrites and marls containing brachiopods, echinoid debris, and ammonites. The entire sequence is encased in marls and micrites that Evans and Kendall (1977) interpreted as deep-water deposits. They suggested that structural movement localized the mounds along upthrown ridges of tilted fault blocks.

Buildups on Gently Sloping Wave-Agitated Platforms

Coral-Dominated Patch Reefs.- Examples of late Jurassic patch reefs in outcrop that developed on shallow, wave-agitated platforms commonly contain coral-rich biotopes. Related facies include skeletal, oolitic, and oncolitic lime sands. The coral patch reefs developed on pre-existing seafloor relief that, in parts of the Schwabian and Franconian Alb, was created by the deeper water sponge-algal mounds discussed in the previous section (Gwinner, 1976). Examples cited from the Knoll reef ramps of Germany and the Oxfordian-aged patch reefs of the Paris Basin (Fig. 5; Wilson, 1975) show the change in ma-

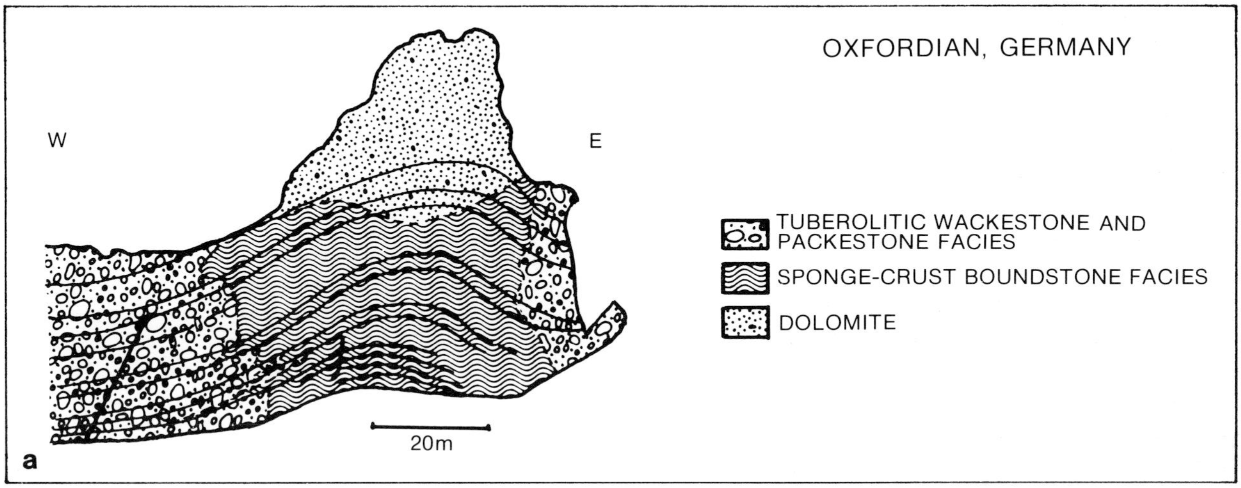

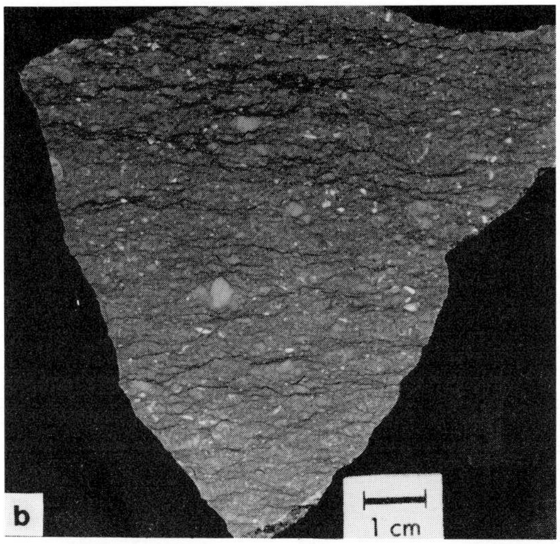

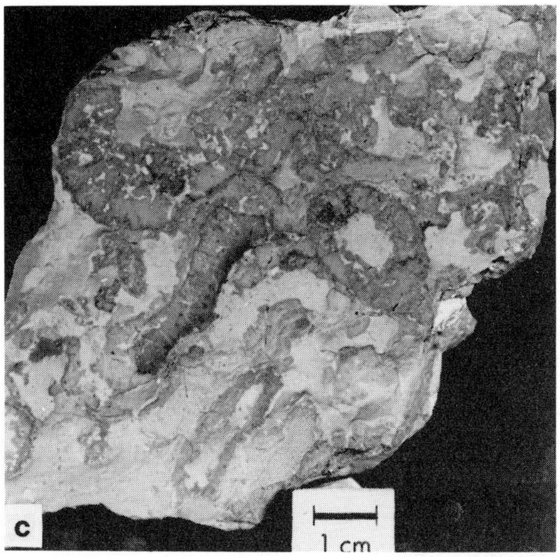

Figure 3. Line drawing of an outcropping sponge-algal mud mound and photographs of polished rock slabs.

a. Distribution of carbonate facies within the Middle-Upper Oxfordian Mullersfelen buildup near Streitberg, in northern Franconia, West Germany (modified after Flugel and Steiger, 1981).

b. and c. Polished slabs of outcrop samples from quarry near Urspring, in northern Franconia, West Germany, where sponge-algal buildup of approximately 80 to 100 meters is well exposed. b. Tuberolitic wackestone-packstone facies with rare sponge fragments from margin of buildup. c. Sponge-algal boundstone from core of buildup with stromatolite/thrombolite crusts and siliceous sponges. Lighter patches are pelleted and bioclastic packstones to mudstones filling between sponge-algal framestone (samples courtesy of E. Flugel and B. Lang).

jor reef-building paleocommunities and growth forms associated with the evolution from sponge-algal mounds to shallower water coral-patch reefs. The base of the patch reef is marked by lamellar (plate-shaped) and cup-shaped sponges, along with lamellar corals that occupied muddy, deeper water and lower energy settings of these buildups. With shoaling conditions, the change to coral-dominated buildups was accompanied by (1) a transition from thrombolitic, pelletal stromatolitic algal crusts to more laminated stromatolitic algal boundstones, and (2) a change from lamellar sponges, corals, and algal stromatolites to massive, fasciculate, dendroid, and encrusting corals, calci-sponges, skeletal algae, and chaetitids. The paucity of encrusting and binding red coralline algae and the apparent inability of non-skeletal blue-green algal stromatolites and crusts to colonize strongly wave-agitated environments made these reefs more susceptible to wave erosion. Bioclastic sand and rubble became volumetrically important in the reef matrix and formed extensive flanking beds. Barthel (1977) speculated that the development of a large spur-and-groove system in such an early Tithonian coral reef in Bavaria was enhanced by the lack of reef-binding organisms (discussed below). Wilson (1975) suggested that patch reefs of this type developed on ramps inclined only 1 to 2 degrees.

During Oxfordian time in the Swiss Jura Mountains (Bollinger and Burri, 1970; Fig. 6), a coral-rich horizon formed that was 10 kilometers wide and extended over 50 kilometers along depositional strike. It consisted of numerous patch reefs which were flanked by coalescing beds of bioclastic debris. To the southeast the patch reefs gave way to lower--energy, deeper-water limestones and marls. The carbonates contained an open-marine fauna, including ammonites. Updip from the reefs, oolite, oncolitic, and bioclastic lime sands formed in shoal environments that represented the crest of the ramp-shelf profile.

The cores of the coral patch reefs are, in general, rather small when compared to the extensive volume of the bioclastic limestones originating from this source. Steiger and Wurm (1980) described Oxfordian coral-stromatoporoid boundstone patch reefs from Austria, which they interpreted as having developed in a lagoon behind a barrier reef that was dominated by stromatoporoids rather than by corals. These patch reefs are up to 20 meters thick and 100 meters wide, and are surrounded by flanking bioclastic debris. These debris deposits extend for several hundred meters in all directions away from the reef core and show no obvious distinction into fore- or back-reef talus (Fig. 7).

The volumetric significance of bioclastic flanking beds compared to the size of the patch-reef core is also shown in the Bajocian patch reefs of the Central High Atlas Mountains, Morocco (Fig. 8). These patch reefs

Figure 4. Photographs of outcrops of Liassic sponge-algal mud mounds exposed at Jebels Amalou Tassacatt and Idignt, from Central High Atlas, Morocco.

a. Stacked series of sponge-algal mud mounds in well-bedded marls and bioclastic wackestones. Height of mound exposure approximately 50 meters.

b. Sponge-algal mud mound with abrupt contact to the left of buildup into well-bedded bioclastic mudstones to packstones. Person on right for scale.

c. Upper surface of a sponge-algal mound. Plan view illustrates algal crust heads surrounded by bioclastic wackestones and packstones. Probable spongiomorph (S-shape) occupies right portion of photograph.

d. Sectional view of sponge-algal mound illustrating laminar crusts (near base) and digitate, branching stromatolites/thrombolites (upper half). Crusts are dark gray versus lighter gray sediments. Note large sediment pocket in upper center of outcrop.

e. Bedding plane view of well-bedded bioclastic wackestone immediately adjacent to sponge-algal mud mound and of brachiopods, ammonite casts, and small thrombolite-crust head (to lower right of scale).

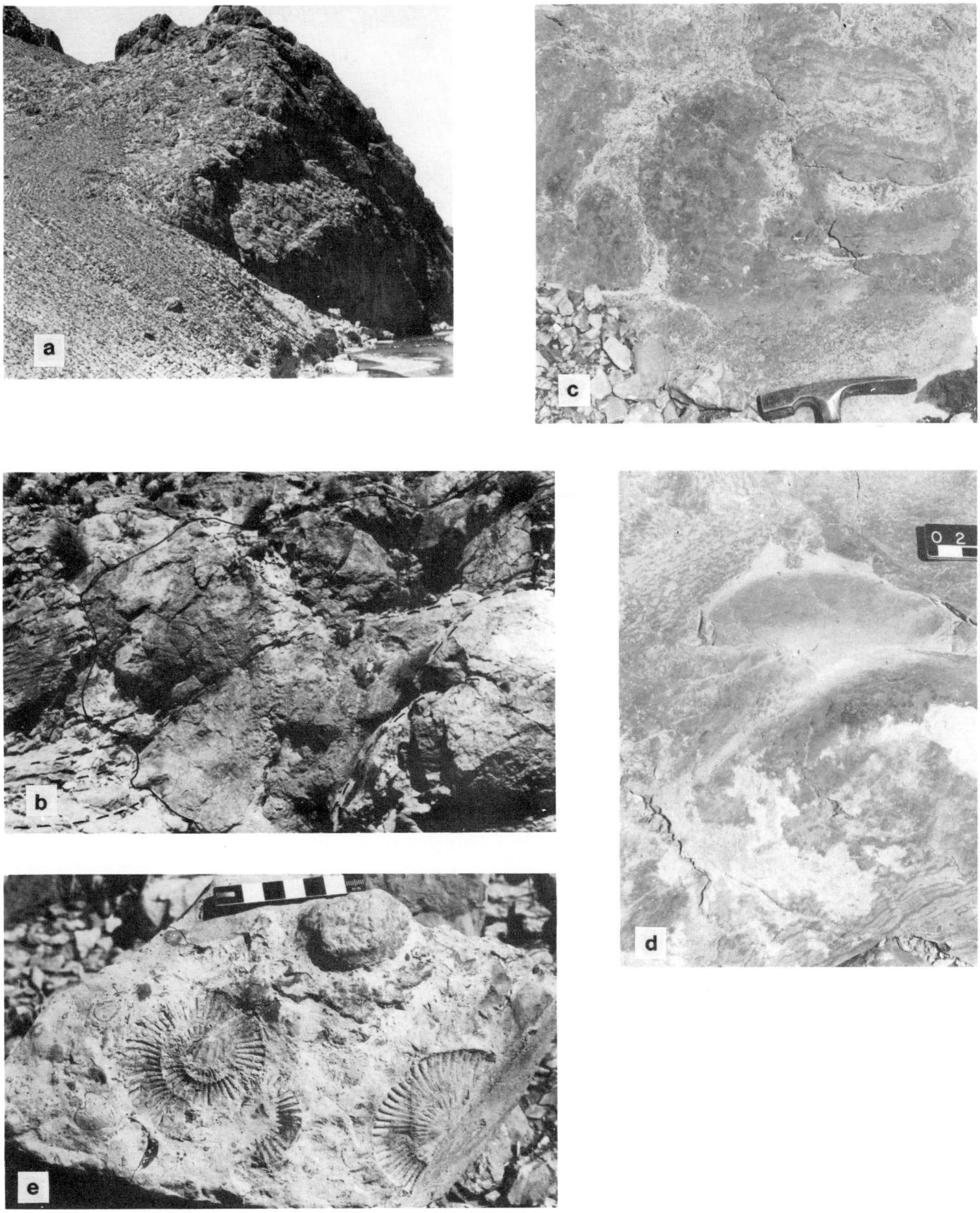

Figure 4

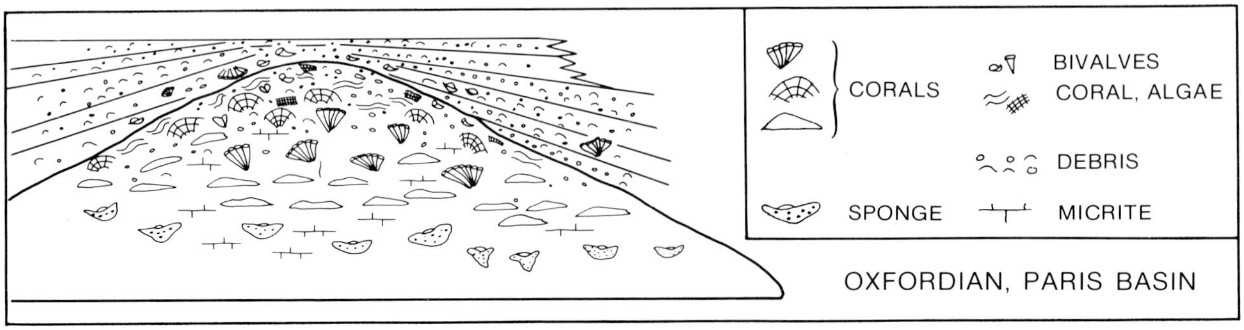

Figure 5. Idealized diagram of Oxfordian patch reef from Yonne Valley in southeastern Paris Basin illustrating changes in major reef-building paleocommunities and sediment type associated with evolution from sponge-algal mud mound into shallow-water coral patch reef (modified after Wilson, 1975).

and flanking beds are up to 40 meters thick: the central core appears to form only a third or less of the entire unit, while the flanking beds extend more than 100 meters away from the core and coalesce with debris from adjacent patch reefs. This horizon, rich in patch reefs, developed at the top of a shallowing sequence across a Jurassic atoll within the High Atlas Trough and covers an area of nearly 40 square kilometers (Warme *et al.* 1975; Stanley, 1981).

Most of the patch reefs that developed on wave-agitated platforms do not show preferred windward margins or orientations. However, Barthel (1977) described an early Tithonian coral patch reef from Bavaria, Southern Germany, that contains a spur-and-groove system which caps a later Kimmeridgian sponge-algal mound. The reef complex is exposed along quarry walls more than 1 kilometer in length and 400 meters in width. Barthel (1977) recognized a succession of reef-building organisms and growth forms as the patch reef grew up to sea level that is similar to that in the coral patch reefs in the Paris Basin (Wilson, 1975). The base of the coral patch reef consists of micritic sponge-algal mounds, overlain by platey corals that

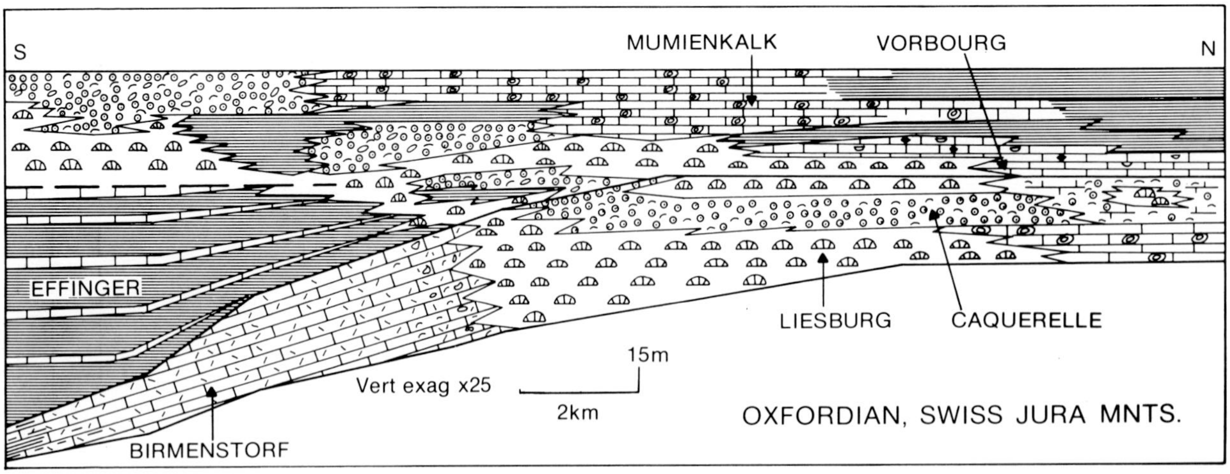

Figure 6. Diagrammatic cross section of Oxfordian strata illustrating facies transitions of a gently sloping, prograding carbonate platform (ramp) in the Swiss Jura Mountains near Basel to Gorges de Pichoux (modified after Bollinger and Burri, 1970; *in* Wilson, 1975).

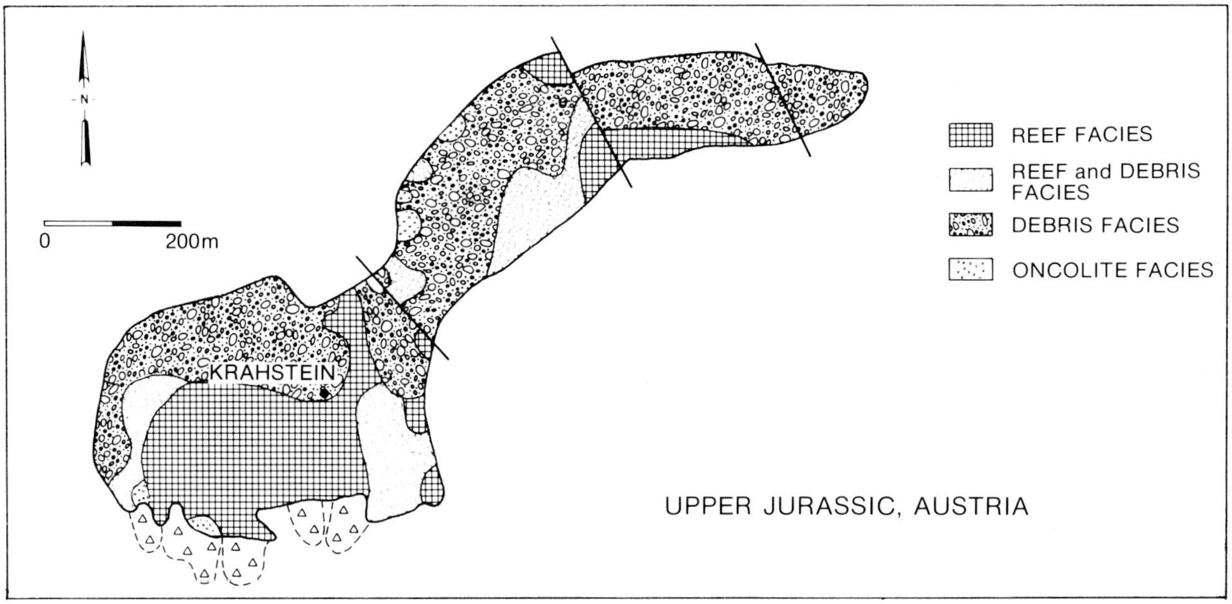

Figure 7. A plan-facies map of Upper Jurassic coral-stromatoporoid patch reefs and debris in the Plassen Limestone cropping out near Krahstein, Northern Alps of Austria (modified after Steiger and Wurm, 1980).

settled on, and bound, the talus deposits in the grooves. The reef-crest margins were colonized by large branching corals more than 1 meter in diameter. Barthel (1977) noted that stromatoporoids were extremely rare and speculated that the lack of binding organisms, such as corals and red algae, was probably responsible for the development of a large spur-and-groove system.

Coral patch reefs are commonly associated with regressive oolite shoals in the Iberian Ranges of Spain and in eastern Portugal (Giner and Barnolas, 1979; Wilson and Exton, 1979). Giner and Barnolas (1979) described Kimmeridgian coral patch reefs as coral boundstones, although their description of the reef biota suggests that the buildups were loosely bound, encrusted coral patch reefs. They described the growth of buildups in a shoaling-upward cycle. In this setting, sponge-rich mounds occurred in quieter water in downslope (or deeper outer platform or intrashelf basin) basinal marls and micrites. Whereas, the coral-rich facies developed in high-energy grainstone belts, usually seaward of oolite and oncolitic shoals (Fig. 9). The reefs range from 2 to 10 meters in height and from 5 meters to, more rarely, 30 meters in width. Although the biota of the reef are not well illustrated, Giner and Barnolas (1979) indicated a vertical zonation with a lower zone of encrusting sponges, few stromatoporoids, and small corals; and an upper zone dominated by coral colonies up to a meter in diameter. Biotic diversity is greater in the upper zone, which contains oysters, gastropods, brachiopods, byrozoans, echinoderms, encrusting solenoporacean algae, and stromatoporoids.

Wilson and Exton (1979) described coral patch reefs, which have a similar succession of facies to those descibed by Giner and Barnolas (1979), from the Lusitanian Basin in Portugal. The patch reefs recognized in Portugal reached a height of 10 meters and occur in large numbers along a common horizon, which is immediately overlain by thick (more than 60 meters) oolitic grainstones (Fig. 10). Patch reefs that commonly coalesced as intermound areas were filled with

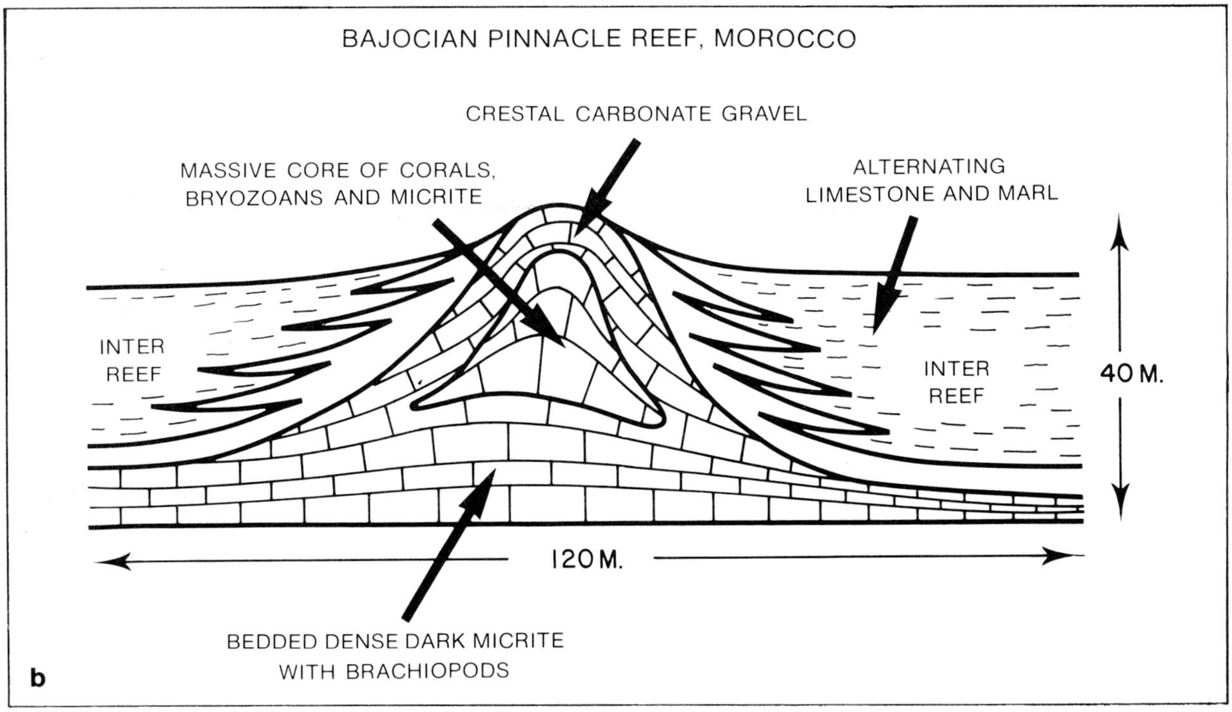

Figure 8. a. A view of a series of Bajocian pinnacle reefs exposed at Jebel Assameur n'Ait Fergane, Central High Atlas, Morocco. Reefs made of a resistant, massive core of scleractinian corals and skeletal wackestone while flanking beds, many which can be seen to coalesce between pinnacles, are mainly coral debris in bioclastic rudstones to wackestones. Well-bedded micrites and marls below the 40- to 70-meter-thick reef horizon represent a shallowing sequence from deep-water sedimentation below wave base to a moderately wave-agitated platform (see Stanley, 1981).

b. Line drawing of Bajocian pinnacle reef shown on Figure 8a (modified from Kendall and Evans, 1973; illustration courtesy of C.G.ST.C. Kendall).

bioclastic debris and patches of boundstone.

Reefs Developed on Platform Margins

Stromatoporoid-Coral Reefs.- Buildups described as barrier reefs occur in Austria (Steiger and Wurm, 1980), on the margin of the Dinaric Platform in Yugoslavia (Turnsek *et al*. 1981), and on the margin of the Panormide Platform of Sicily (Catalano and D'Argenio, 1981). The principal reef framebuilding organisms were stromatoporoids and corals. Two of the barrier complexes can be subdivided into platform edge and back reef based upon the distribution of two stromatoporoid faunal subzones: *Actinostromariid* and *Parastromatoporid,* respectively.

Catalano and D'Argenio (1981) described the Sicilian reef complex (Fig. 11) as an alternation of reef (patch-reef and inter-reef deposits) and fore-reef debris. In the Dinaric reef complex, stromatoporoids dominate and massive and phaceloid corals predominate over solitary ramose forms and chaetitids. However, non-skeletal algae, foraminifers, bryozoans, crinoids, echinoids, brachiopods, molluscs, and *Tubiphytes* sp. are of lesser importance (Turnsek *et al.* 1981).

Extensive linear reef trends are illustrated in both the Sicilian and Yugoslavian examples. The Dinaric complex is approximately 140 kilometers long and about 20 kilometers wide. Fore-reef and slope depositional settings are represented by calcareous breccias consisting of allochthonous blocks of reef limestone and bioclastic detritus. The presence of clean skeletal detritus in the reef areas, marginal to slope settings, suggests that high-wave energies impinged on the reef complex. A smaller reef complex mapped by Steiger and Wurm (1980), in Austria, is approximately 1.5 kilometers in length and is about 300 meters in width (Fig. 12).

Subsurface Examples of Smackover Buildups

Buildups on Gently Sloping Wave-Agitated Platforms

Sponge-Coral-Algal Mounds.- Subsurface examples of sponge-coral-algal buildups in the Smackover Formation (Fig. 13) are displayed best in cored intervals from Walker Creek Field in Columbia County, Arkansas (Arco No. 1 McFadden Well, Fig. 14; Appendix A; Baria *et al.* 1982; Harris and Crevello, 1983), and from North Haynesville Field in Claiborne County, Louisiana (Tenneco No. 1 Seegers-Waller Well, Fig. 15; Appendix B; Baria *et al.*, 1982).

The framework of these buildups was constructed mainly by non-calcifying, stromatolitic blue-green algae that apparently were rapidly lithified by micritic marine cements. These algae were volumetrically important constituents of the buildups. In addition, they served a dual role of binding reef sediment and producing a microframework for organisms to inhabit and sediment to infiltrate. The algae developed faint laminar crusts on sediments of encrusted skeletal debris, or developed delicate digitate branches and heads with only a few centimeters of re-

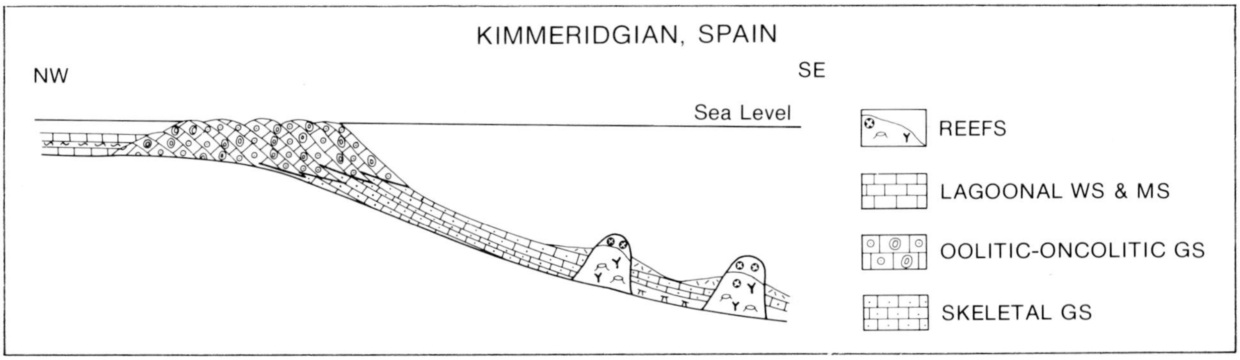

Figure 9. Depositional model for the outcropping lower Kimmeridgian of Spain (modified after Giner and Barnolas, 1979) illustrating a carbonate platform with moderate bottom agitation. Note that individual reefs and bioclastic debris occur on the shelf seaward of oolite shoals and lagoonal muds.

Figure 10. Photograph of outcrop and polished rock slabs illustrate buildup and associated lithologies from wall of working quarry, 2 kilometers south of Arruda, Portugal, exposing Kimmeridgian-age Corallien sequence of the Lusitanian Basin.

a. Reefal carbonates at base of quarry (upper boundary of the interval is highlighted) are overlain by well-bedded ooid grainstone which is regionally up to 60 meters thick. Individual mounds up to 10 meters thick in the quarry and up to 30 meters thick regionally developed over an area that covers 400 square kilometers.

b. Digitate algal-crust boundstone and wackestone. Boundstone (left half of slab) is stained orange-brown, appears lithified and fractured early, with fractures partially filled with wackestone and equant calcite spar; some borings. Wackestone is pelleted and thrombolitic in places, is oncolitic, and contains scattered clam shells.

c. Boundstone of algal-encrusted corals, red algae, and clams; internal cavities filled by wackestone with numerous fractures. Algal coatings are similar to *Lithocodium-Bacinella*. Corals and clams are replaced or filled with equant calcite spar, as is fracture-breccia fabric.

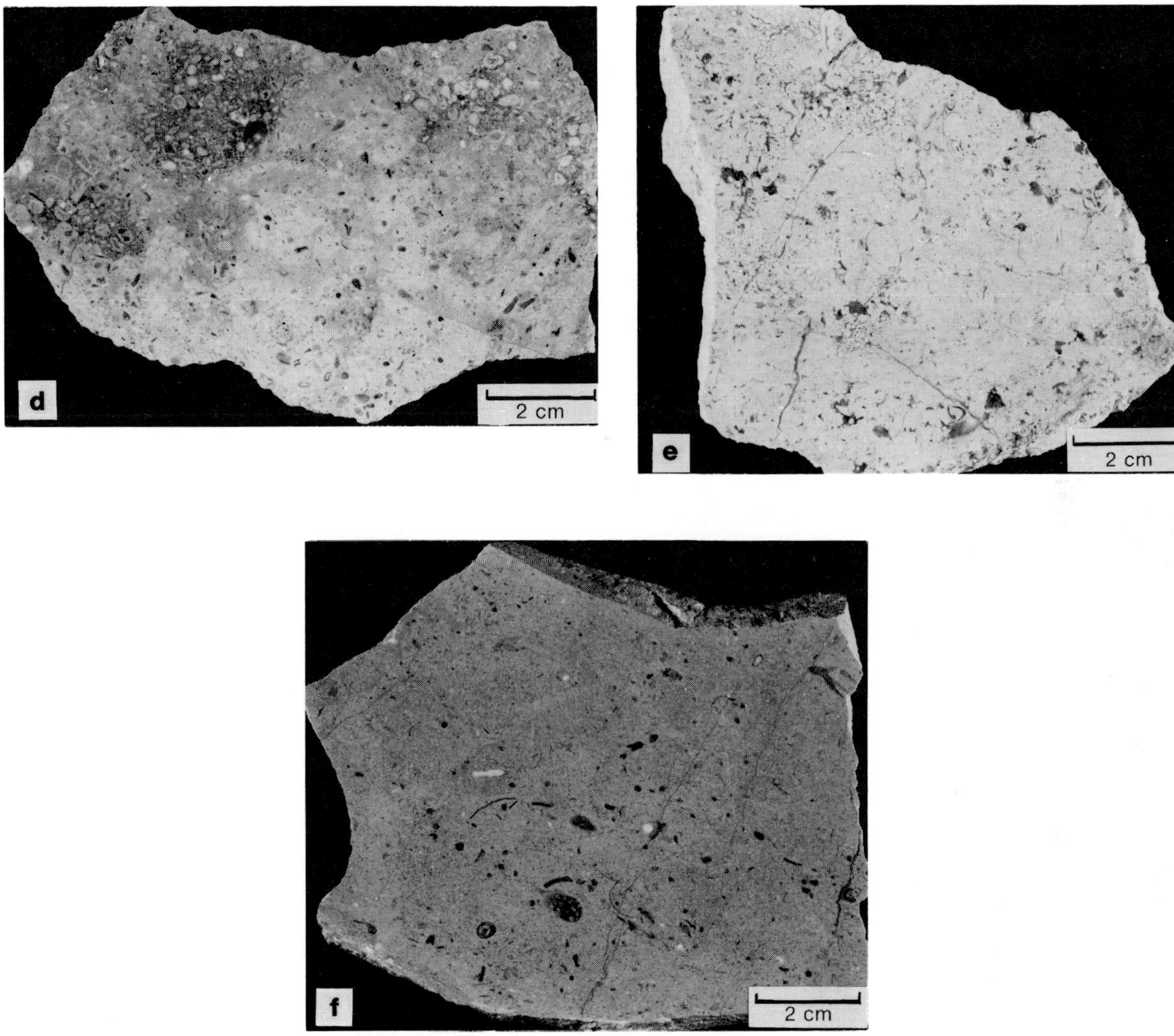

Figure 10. Continued.

d. Oncolite rudstone to packstone overlying buildup contains algal-coated echinoid plates and spines, gastropods, pelecypods, and corals, in addition to uncoated skeletal fragments. Mollusc and coral debris were replaced by blocky calcite spar.

e. Bored and encrusted cyanophycean algal boundstone (stromatolite/thrombolite) from crest of buildup. Framework and/or shelter pores are filled with layered grainstones and packstones of oncolites and bioclasts. Notice mollusc (borings?) scattered in boundstone fabric.

f. Intermound sediments of bioturbated fine packstone to wackestone of micropeloids and pellets with scattered pelecypods, gastropods, echinoderm debris, oncolites, and coral fragments.

lief (Fig. 16).

Commonly, pelleted or thrombolitic (unlaminated stromatolite) fabrics developed and were encrusted with *Tubiphytes* sp., agglutinated and calcareous worm tubes, bryozoans, foraminifers, and calcareous sponges (Fig. 17). All of these organisms seemingly required a substrate more stable than that of a

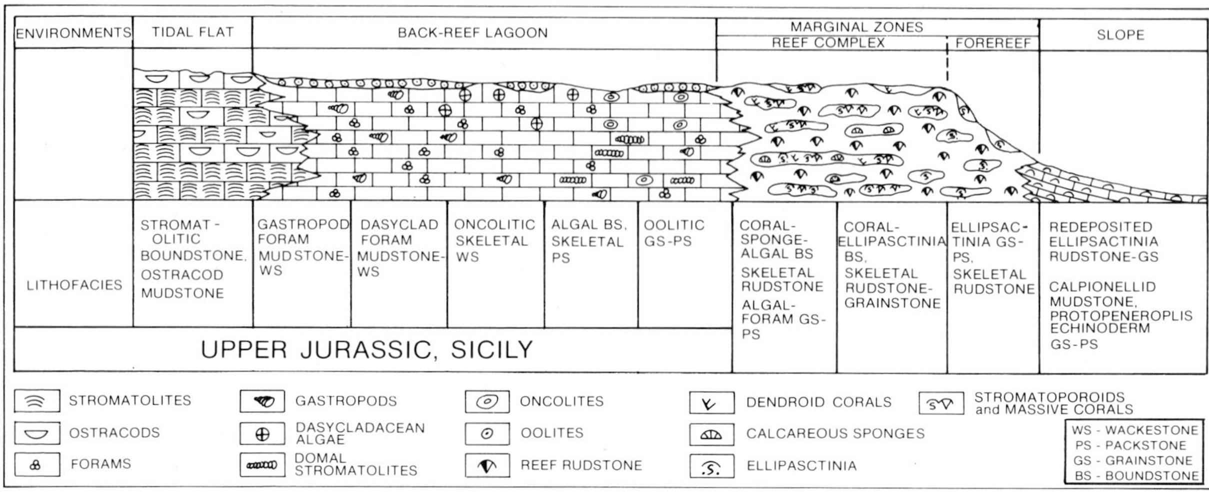

Figure 11. Diagrammatic cross section illustrating depositional settings and facies developed across the Panormide carbonate platform margin of western Sicily during Upper Jurassic (modified from Catalano and D'Argenio, 1981).

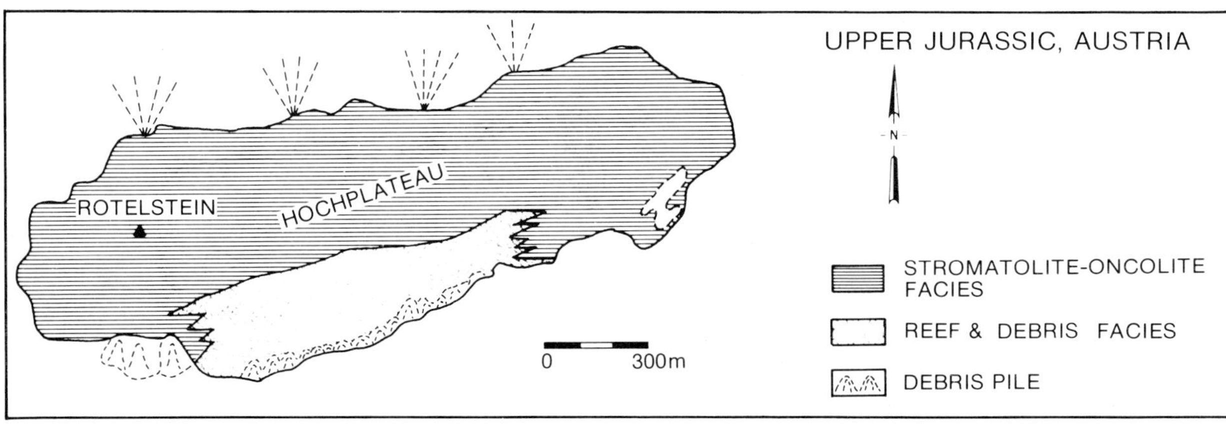

Figure 12. A plan-facies map of an Upper Jurassic stromatoporoid-coral reef margin complex in the Plassen Limestone in the Rotelstein area, northern Alps of Austria (modified from Steiger and Wurm, 1980).

weakly cemented mat. Although synsedimentary fracturing of the reef framework and boring of the algal fabric are rare, multiple cross-cutting relationships of clam and sponge borings into corals and crusts indicate marine lithification of the buildups (Figs.16h, 17a and 17b).

Sponges are also important in the construction of these buildups, particularly in the lower part of the buildup in the Arco #1 McFadden well. Here, laminar and finger-like lithistid sponges and hexactinellid sponges are common, and encrusting calcareous sponges and calcareous finger-like sponges are rare. Chaetitids are scattered through the buildups. A vertical variation in the paleocommunity of the buildup in the Arco No. 1 McFadden well occurs; the upward increase in size and number of corals suggests a shoaling-upward trend (Baria *et al.* 1982; Harris and Crevello, 1983). Calcareous red algae are abundant in the upper parts of the Tenneco No. 1 Seegers-Waller well buildup and are in the bioclastic pile in the Bass Smith

Appendix C), and their abundance suggests a paleocommunity evolution brought on by shallowing conditions.

Corals (primarily *Actinastrea* sp.), the problematical calcareous red alga *Parachaetetes* sp., and the blue-green alga *Cayeuxia* sp. served as the primary framebuilders of the reef. The only significant difference between the buildups described from the Walker Creek and North Haynesville fields is the greater abundance of red coralline algae that cap the buildup in the Tenneco No. 1 Seegers-Waller well (North Haynesville).

Imlay (1940) and Wells (1942) described at least six different genera of corals from the Smackover Formation, none of which was recognized in the core studies of Baria *et al.* (1982). Various types of small head, finger, and branching forms of corals were observed, but the preservation of the coral microstructure was generally poor and identifications could not be made. Partially dissolved and micritized calcareous algal structures also impeded identification of many of the green codiacean and blue-green algae. Echinoid and mollusc debris, including gastropods, cerithids and nerinids, along with whole and fragmented shells of the oyster *Lopha* sp., and unidentifiable bivalves is commonly associated with the buildups.

Immediately beneath the buildup in Walker Creek Field, poorly sorted biointraclastic and peloid packstones and oncolite rudstones are present. These lithologies may represent debris that was shed onto an unstabilized, deeper water sediment flat or channel adjacent to the buildup. Oncolites from this interval display a digitate growth fabric that resembles the framework developed in the buildups. Lithoclasts consist of pelleted, very fine-grained peloidal (micropeloid) packstones to mudstones resembling the blue-green algal fabric and sponge debris. Similar clasts observed in outcrop studies of flanking beds adjacent to sponge-algal mounds in Germany and Austria were termed tuberoids

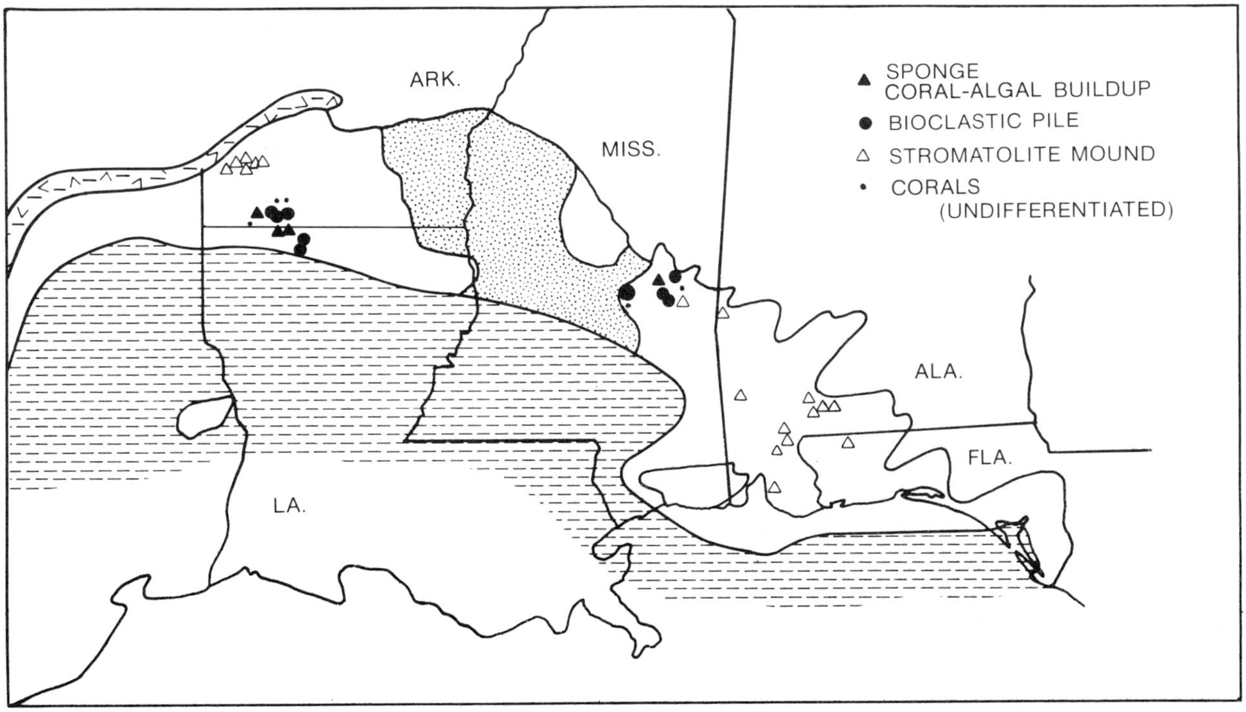

Figure 13. Types of Jurassic buildups recognized in conventional cores from the Smackover Formation; data from Baria *et al.*, 1982 and McGraw, 1982). Lithologies are identified in Figure 1. Corals (undifferentiated) designate isolated occurrences of corals or biostromes.

(Gwinner, 1976; Flugel and Steiger, 1981). In those studies, the tuberoids were interpreted as reworked calcified siliceous sponges and stromatolite crusts.

Underlying the biointraclastic and oncolitic intervals in the Arco No. 1 McFadden well buildup is a quartz sandstone and a poorly sorted ooid grainstone couplet (Appendix A). In contrast, below the buildup in the Tenneco #1 Seegers-Waller well, sediments grade downward to very fine-grained peloidal packstones and wackestone containing scattered oncolites, abundant echinoid debris, and brachiopod *Terebratellia* sp. (Appendix B).

Oncolite, peloidal, and intraclastic packstones and grainstones, containing abundant tuberolitic intraclasts, overlie both buildups. Somewhat greater water turbulence over the buildup in the Tenneco No. 1 Seegers-Waller well is suggested by the presence of a thick sequence of biointraclastic packstones and rudstones. In contrast, the buildup in the Arco No. 1 McFadden well is capped by less than 3 meters of fine- to medium-grained peloid packstones which are overlain by a coarse-grained ooid grainstone.

Bioclastic Piles.- The Bass No. 1-A Smith (Figs. 13 and 18; Appendix C) in Hico Knowles Field in Lincoln Parish, Louisiana, provides a subsurface example of skeletal rubble from the Smackover Formation from which hydrocarbons were produced (Baria *et al.*, 1982; Crevello *et al.*, in press). The skeletal rubble lacks a boundstone fabric and differs somewhat in texture and biotic constituents from the sponge-coral-algal mounds just described. Delicate or more fragile types of reefbuilding organisms, such as hexactinellid and lithistid sponges, are lacking. Also absent are encrusting organisms such as *Tubiphytes* sp., which commonly populated the stromatolite crust and framework. Corals, molluscs, echinoids, calcareous finger-like and en-

Figure 14. Photographs of thin sections in plane-polarized light illustrating microfacies from and associated with the sponge-coral-algal buildup in the Arco No. 1 McFadden well, Walker Creek Field, Arkansas.

a. Micropeloidal (very fine-grained sand- to silt-size peloids) packstone with coated echinoid and texturlarid foraminifer. 10,960 feet (3,340.6 m).

b. Reef-derived skeletal debris in an oolitic and skeletal-peloidal packstone: *Cayeuxia* sp. (lower right); *Ptychochaetetes* sp. (lower left); *Actinastrea* sp., bored and encrusted by calcisponge (upper right); and gastropod and pelecypod fragments. 10,966 feet (3,242.4 m).

c. Coral-bearing thrombolite/stromatolite-*Tubiphytes* boundstone with pelecypods, agglutinated worm tubes in a bioclastic-micropeloidal wackestone to packstone. 10,973 feet (3,344.9 m).

d. Coral-bearing thrombolite/stromatolite-*Tubiphytes* boundstone illustrating the intensely pelleted packstone framework filled in with bioclastic-micropeloid packstone to wackestone (lighter color). Note several generations of microborings and geopetal sediment in upper half of coral. 10,982 feet (3,347.3 m).

e. Boundstone of laminar sponge (lower half) encrusted by calcisponge and bryozoan. *Lopha* sp. (oyster) on left is bored. 10,985 feet (3,348.2 m).

f. Biointraclastic packstone of echinoid debris, intraclasts of micropeloid*iphytes* packstone to wackestone (tuberoids) and oyster debris in dolomicrospar matrix. 10,990 feet (3,349.7 m).

g. Biointraclastic oncolite packstone to rudstone with fragments of echinoids, pelecypods (*Lopha* sp.), calcisponges, chaetitids, reworked thalli of *Cayeuxia* sp., and intraclasts of micropeloid (tuberoids)-*Tubiphytes* packstones. 10,994 feet (3,351 m).

h. Quartzose, intraclastic and oolitic oncolite rudstone, poorly sorted, with perched micrite, grapestones, and oolite intraclasts. Large oncolite (right) is encrusted with serpulid worm tubes and foraminifers. 10,997 feet (3,351.9 m).

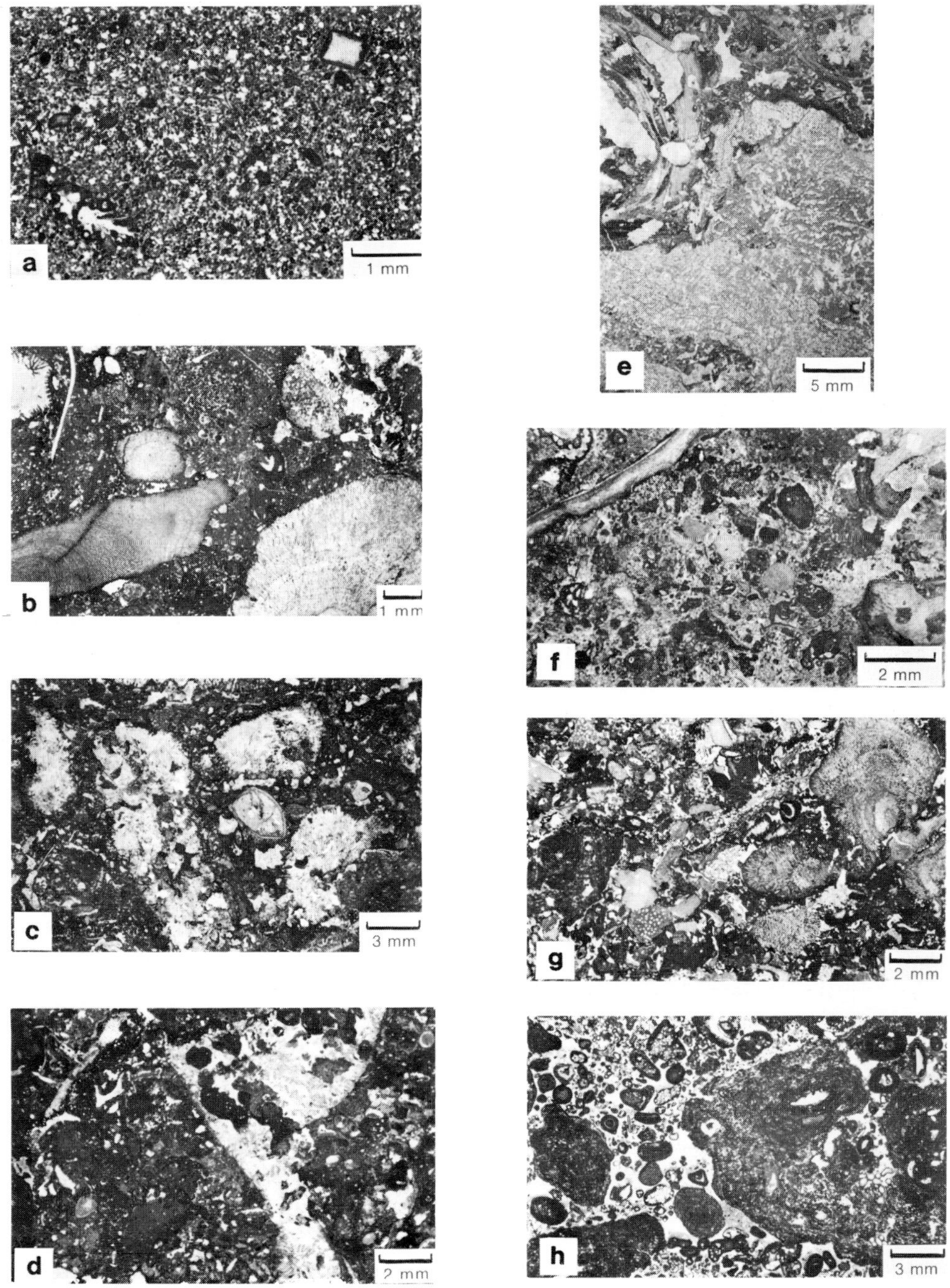

Figure **14**

crusting sponges, chaetitids, and various calcareous algae (*Cayeuxia* sp., *Parachaetetes* sp.) and abundant red algae are the dominant constituents of the bioclastic pile.

The combination of (1) fragmented, gravel-supported skeletal debris with bioclastic grainstones and packstones between the clasts (Fig. 18c); (2) the variable depositional sequence of bioclastic rudstones, grainstones, packstones, and terrigeneous silt/sandstones; (3) the lack of reef-building organisms in growth position; and (4) extensive marine cementation - all suggest that the skeletal interval represents locally reworked detritus, possibly flanking beds of a nearby patch reef (Crevello *et al.*, in press). Based on the biotic assemblage and thickness of these deposits, as well as outcrop analogues, we believe this facies forms a pile, perhaps with a nearby reef core, rather than an extensive sheet deposit.

Stromatolite Mounds.- Stromatolite mounds (Figs. 13 and 19) are the only types of buildups encountered, to date, in the Alabama and Florida part of the subsurface Smackover trend (Baria *et al.*, 1982). These buildups are dominated by laminated stromatolites, but pelleted thrombolite growth forms are also common (Fig. 19). The framebuilding and encrusting organisms, common in the other two types of buildups recognized in the Smackover Formation, are minor constituents in the stromatolite mounds. Corals, skeletal algae, echinoderms, oysters, and other organisms in growth position are rare, although their presence verifies that these mounds developed in a subtidal environment. These buildups also developed near ooid shoals as they are covered by peloid and ooid sands. The stromatolite mounds are up to 40 meters thick, which is thicker than the other buildups recognized in the Smackover. Baria *et al.* (1982) suggested that the development of these mounds on subsiding basement fault blocks caused them to be thicker than the buildups in Louisiana and Mississippi. The foundation for the latter buildups was salt-cored structures, which were capped by buildups with thicknesses of less than 20 meters.

Stromatolite mounds were recently reported from Paup Spur Field in Southern Arkansas (McGraw, 1982), about 50 kilometers northwest of Walker Creek Field (Fig. 13).

Figure 15. Photographs of thin sections in plane-polarized light illustrating microfacies from and associated with the sponge-coral-algal buildup in the Tenneco No. 1 Seegers-Waller well, North Haynesville Field, Louisiana.

a. Fine- to medium-grained dolomitic oopeloid packstone to poorly washed and bioturbated grainstone. 10,056 feet (3,065.1 m)

b. Intraclastic rhodolite-oopeloid rudstone to packstone. 10,058 feet (3,065.7 m).

c. Biointraclastic rhodolite rudstone to grainstone with alga thalli of *Cayeuxia* sp. and red alga thalli of probable *Parachaetetes* sp. that is partially micritized and replaced by anhydrite. 10,066 feet (3,068.1 m).

d. Intraclastic floatstone in dolomicrospar matrix. Intraclasts are reworked boundstone (left center) and bioclastic micropeloid (tuberoid) packstones to wackestones (center and right). 10,089 feet (3,075.1 m).

e. Thrombolite/stromatolite boundstone to framestone (lower half) overlain by lamellar sponge with large vertical fractures and hexactinellid sponge (top). Boundstone to framestone crusts consist of faintly laminated, pelleted, or micropeloid texture with rare agglutinated worm tubes. 10,093 feet (3,076.3 m).

f. Coral-bearing *Tubiphytes*-thrombolite boundstone illustrating abundant *Tubiphytes* encrusting boundstone which grew out from coral. 10,095 feet (3,077 m).

g. Biointraclastic and oncolite floatstone in a bioclastic peloid packstone with a dolomicrospar matrix. Bioclasts are calcareous thalli of *Cayeuxia* sp., pelecypods, and *Tubiphytes*. Intraclasts are micropeloid (tuberoid) packstones and wackestones. 10,109 feet (3,3081.2 m).

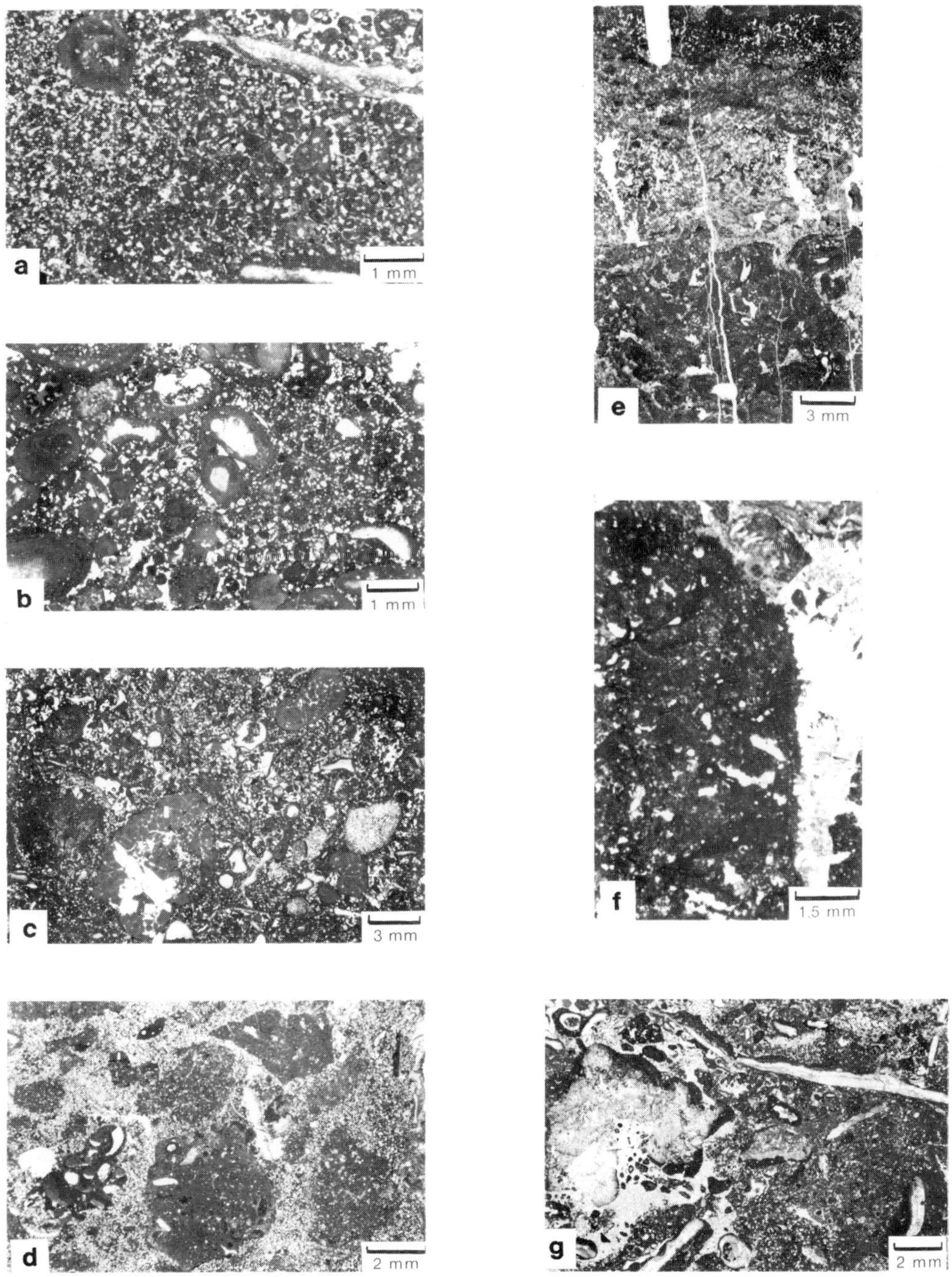

Figure 15

McGraw's (1982) work provides the first documentation of stromatolite mound development in the Smackover of southern Arkansas. In Paup Spur Field, the algal stromatolites are associated with ooid, oncolite, and pellet packstone-grainstone facies, similar to those of the stromatolite mounds reported by Baria *et al.* (1982) from Alabama to Florida. Using borehole logs and cores, McGraw (1982) correlated the algal stromatolite unit, which is commonly less than 8.5 meters thick, for nearly 5 kilometers along depositional strike and 1.6 kilometers in a shelfward direction. The algal facies is overlain by a blanket of ooid grainstones, suggesting mound development immediately seaward of ooid shoals.

DISCUSSION

Platform Setting for Carbonate Reefal Buildups

Platform Profiles

Two figures illustrate the various positions that reefal buildups occupied on carbonate platforms during the Jurassic (and early Cretaceous)(Figs. 20 and 21). The reconstruction of the Smackover platform (Fig. 20) shows the considerable variation in types of buildups along the depositional trend (see also Fig. 13). Whereas, the platform profiles (Fig. 21) show only the variation in types of buildups across the depositional dip: (A) a gently sloping platform (commonly referred to as a carbonate ramp (Ahr, 1973) or knoll reef ramp (Wilson, 1975) as the Smackover platform, (B) a steepened platform margin as presented by Catalano and D'Argenio (1981; Fig. 11) and Eliuk (1978), and (C) a steepened or rimmed platform margin with an intrashelf basin or deep lagoon that passes laterally updip into a gently sloping carbonate platform as profile A. Profiles A and B (Fig. 21) represent platforms several tens of kilometers across: (A) such as a transect from Tongue of the Ocean onto the steepened margin of the Great Bahama Platform or across the subsurface Jurassic Abenaki platform margin offshore from Nova Scotia (Eliuk, 1978); and (B) a transect across the gently sloping Trucial Coast of the Persian Gulf (*in* Budd and Loucks, 1981) or the Jurassic Smackover platform. Profile C (Fig. 21) may be several times larger than A and B (Fig. 21), on the order of ten's to hundred's of kilometers across. Analogues similar in scale and profile may be comparable to a transect from the Atlantic margin of the Great Bahama Bank,

Figure 16. Photographs of thin sections in plane-polarized light of stromatolite and thrombolite boundstone to framestone fabrics.

A.M. (Arco No. 1 McFadden well), S.W. (Tenneco No. 1 Seegers-Waller well), B.S. (Bass No. 1-A Smith well).

a, b, c, and d. Laminated and pelleted stromatolites illustrating boundstone to framestone fabric (a, b, and d) and isolated, upward-expanding crust (c). a. A.M., 10,988 feet (3,349.1 m); b. S.W., 10,092 feet (3,076 m); c. A.M., 10,988 feet (3,349.1 m); d. A.M., 10,985 feet (3,348.3 m).

e. Micropeloidal framestone encrusted with *Tubiphytes* and probable nubeculinellid foraminifer. Note geopetal sediment and shelter porosity developed in the thrombolite framestone. S.W., 10,095 feet (3,077 m).

f. Thrombolite framestone encrusted by serpulid worm and agglutinated worms with sponge spicules in geopetal sediment; micritic marine cement in serpulid and framestone, and late-stage ferroan dolomite pore-filling cement. A.M., 10,985 feet (3,348.2 m).

g. Photograph of close-up view of encrusted stromatolite framestones with algal(?) cysts (middle- and lower-center). A.M., 10,985 feet (3,348.2 m).

h. Microboring of micropeloidal intraclasts and pelecypod in oncolite floatstone. A.M., 10,961 feet (3,340.9 m).

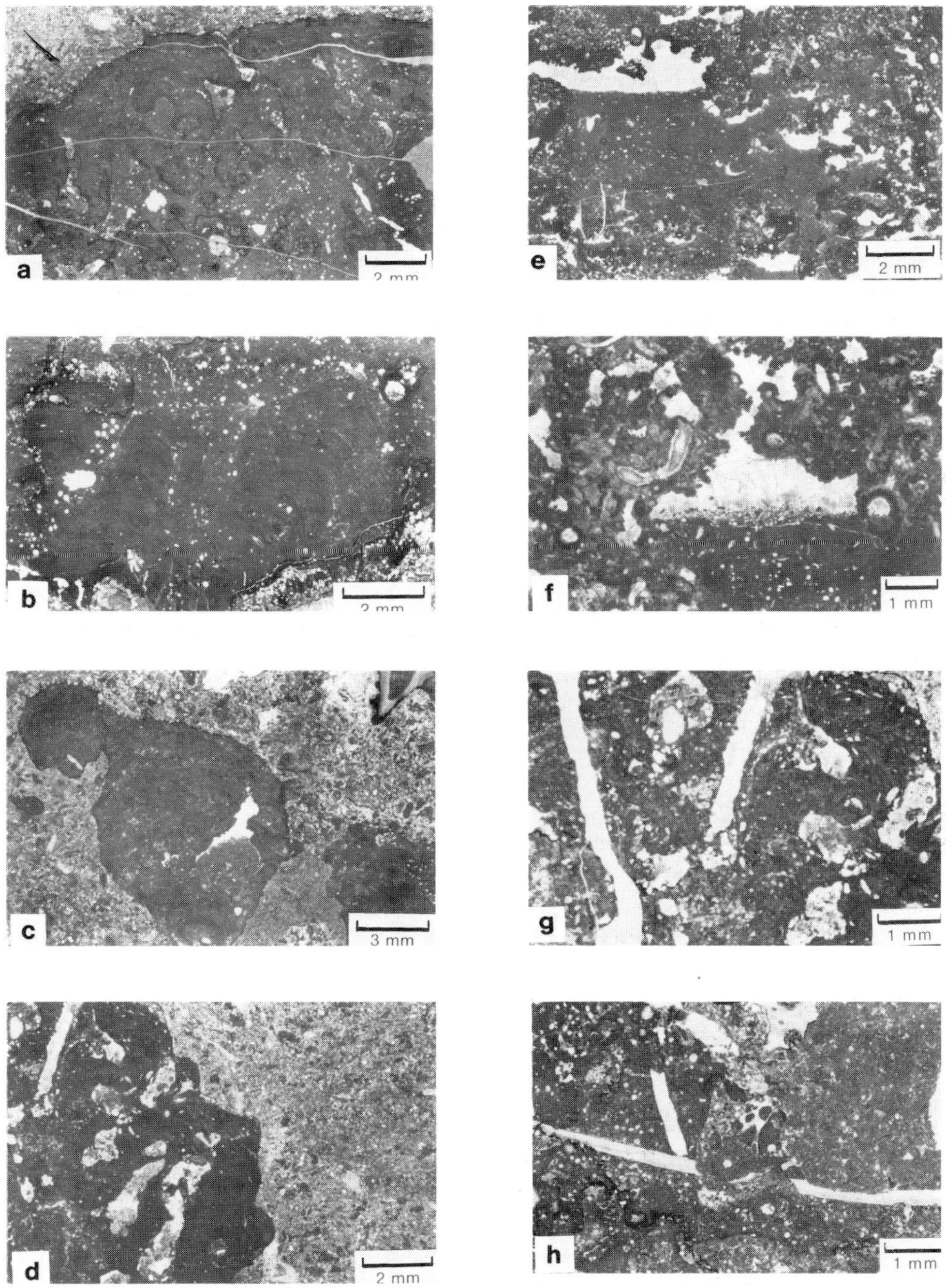

Figure 16

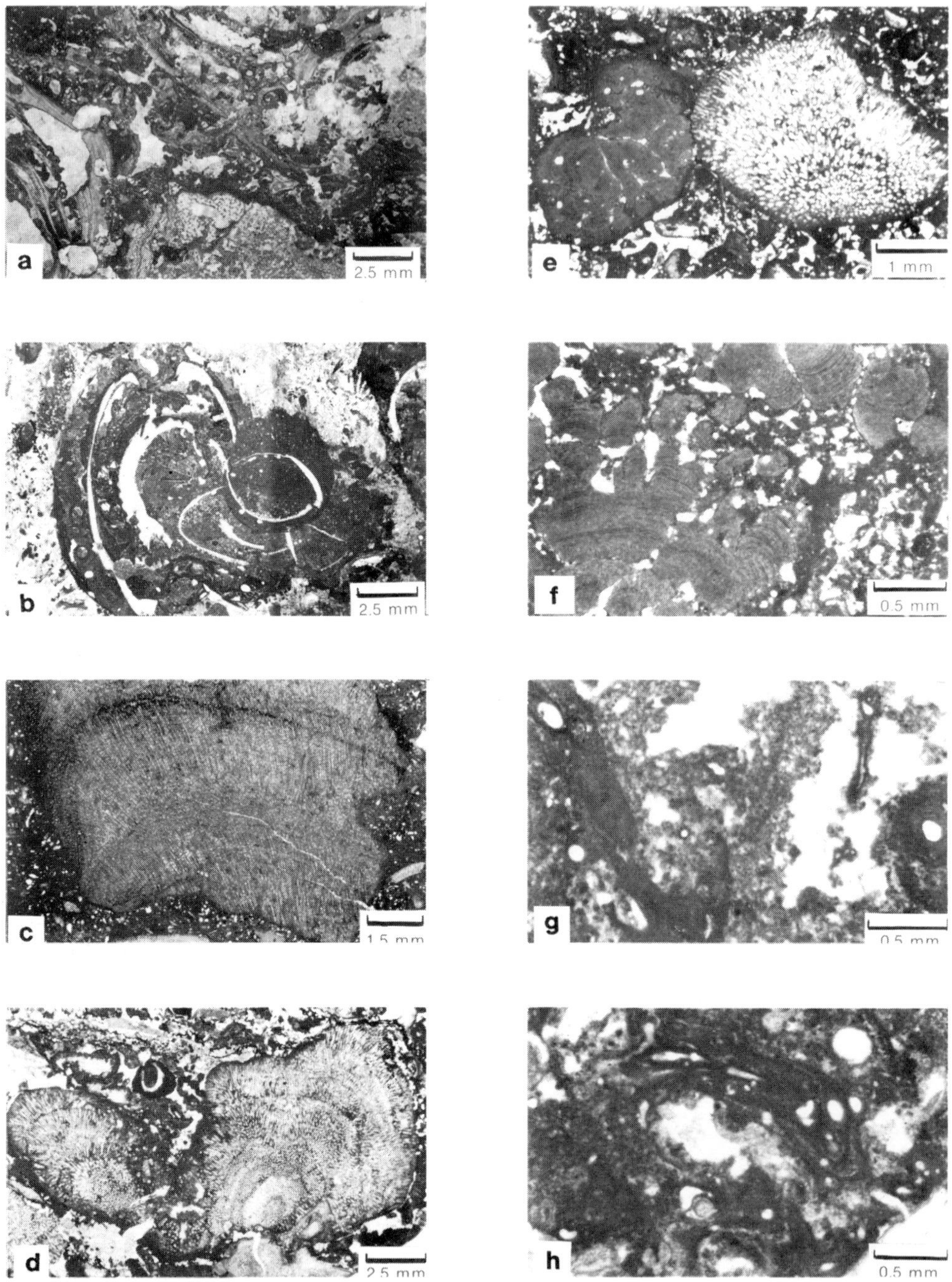

Figure 17

across the intraplatform basins and platforms, and onto the Florida peninsula; or across the Jurassic Friuli Platform to the Lombard Basin of the southern Alps (Winterer and Bosselina, 1981).

Buildup Type and Platform Position

Sponge-Algal Mounds (Type 1).- These mounds are developed downslope from gently dipping platforms and steepened margins (Fig. 21). These mounds occupied areas of lower sediment input, below the fore-reef talus slopes, favoring intergulley or seafloor highs. They have not been documented from shallow, wave-agitated settings, yet the algal frameformers presumably required light for photosynthesis, thus limiting their development to within the photic zone. Such deep-water mud mounds have not yet been reported from the subsurface of the Smackover Formation. However, Jansa *et al.* (1982) interpreted algal-stromatolite buildups occurring in the subsurface Abenaki Formation as indicative of a deeper water neritic environment.

Coral-Bearing Reefs (Types 2a, b, and 3; Fig. 21).- These reefs show marked variations in

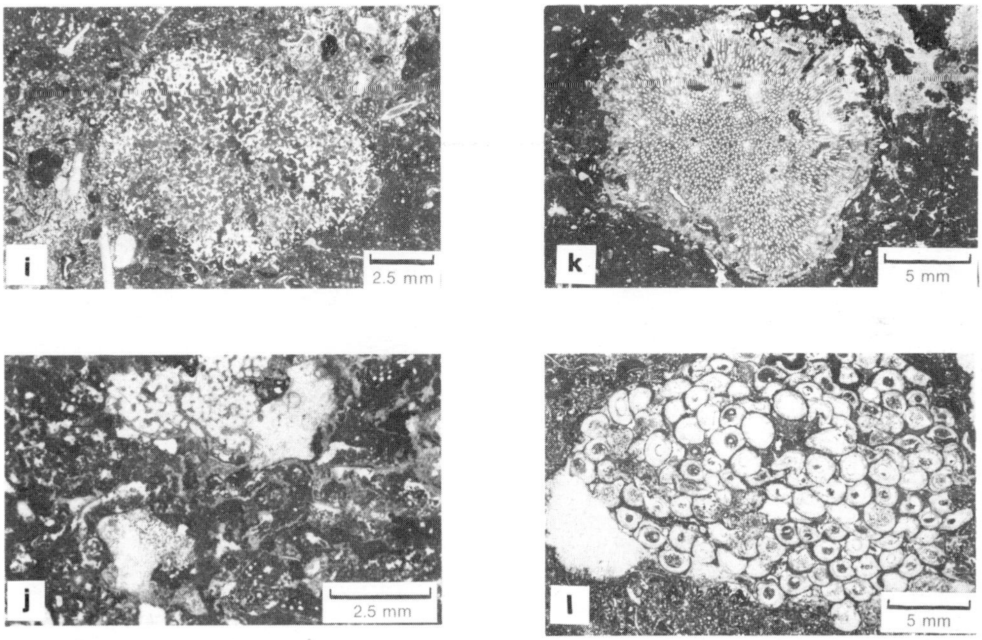

Figure 17. Photographs of thin sections of reef framebuilders and encrusters, and microborings from Smackover buildups.

a and b. Boring in pelecypod and boundstone (a) of sponge-coral-algal boundstone and multiple borings (b) in coral. a. A.M., 10,985 feet (3,348.2 m); b. S.W., 10,106 feet (3,080.3 m). Thallus of calcareous algae: c. *Parachaetetes* sp., B.S., 10,755 feet (3,278.1 m); d. *Cayeuxia* sp., A.M., 10,944 feet (3,350.9 m); e. *Cayeuxia* sp. or *Ortonella* sp. (right) and red alga of probable *Parachaetetes* sp. (left), S.W., 10,066 feet (3,068.1 m); f. Red alga *Parachaetetes* sp. or micritized coralline alga; g. *Tubiphytes morronensis* Crescenti, S.W., 10,095 feet (3,077 m); h. *Tubiphytes* sp., A.M., 10,985 feet (3,348 m); i. Cross section of a lithistid sponge, S.W., 10,106 feet (3,080.3 m); j. Calcisponge and leached codiacean alga thalli, A.M., 10,976 feet (3,345.5 m); k. Chaetitid (*Acanthochaetetes* or *Pseudochaetetes*), S.W., 10,105 feet (3,080 m); l. Annelid worm tubes encrusted by chaetitid, S.W., 10,116 feet (3,083.4 m).

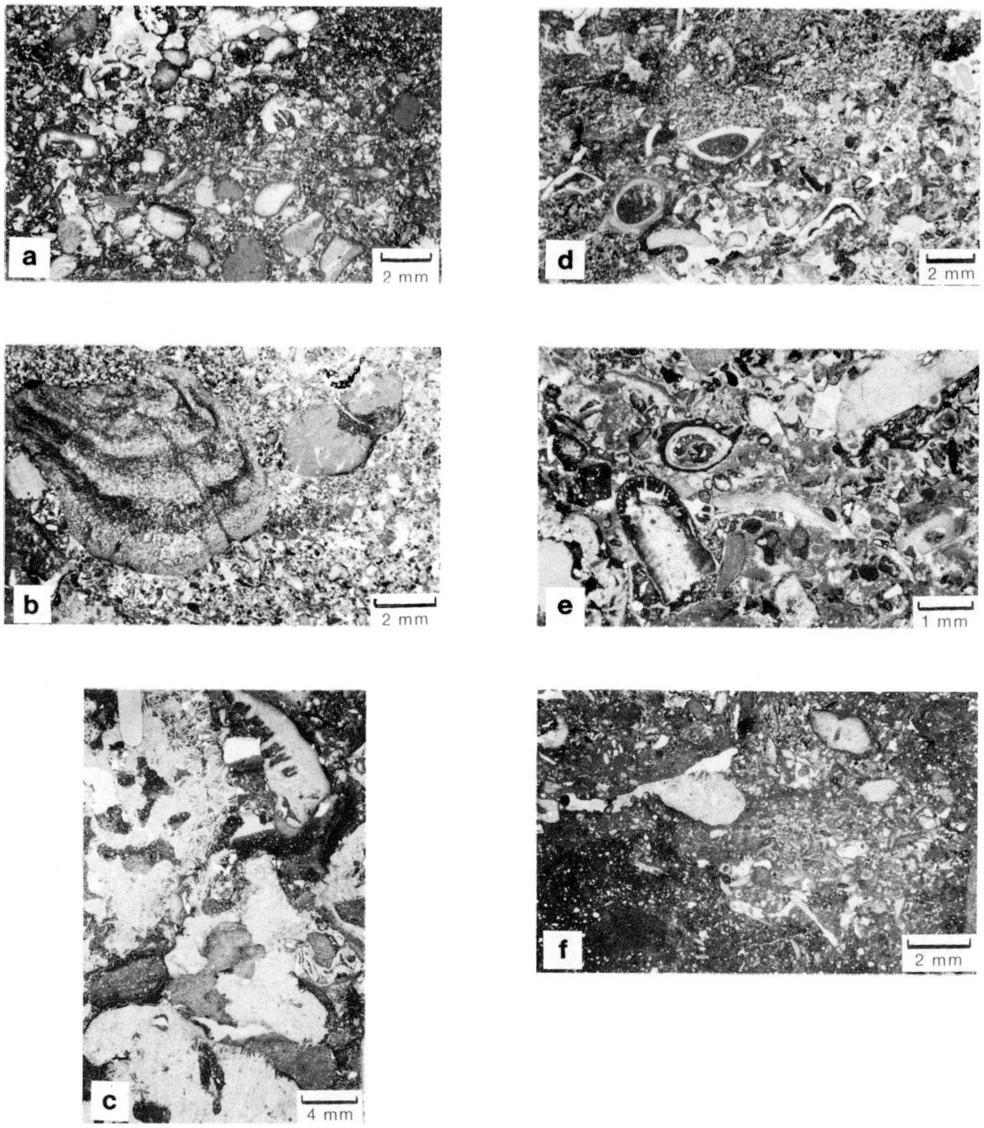

Figure 18. Photographs of thin sections in plane-polarized light illustrating microfacies in the bioclastic pile (reef rubble) from the Bass No. 1-A Smith, Hico Knowles Field, Louisiana.

a. Peloidal bioclastic packstone to rudstone illustrating fragmented and abraded debris of corals, red algae, pelecypods, and echinoids with preserved primary interparticle porosity. 10,733 feet (3,271.4 m).

b. Coral and red algal floatstone in bioclastic grainstone matrix with preserved interparticle and skeletal moldic porosity. Reworked thalli of calcareous red alga (*Parachaetetes* sp.), or possibly micritized coralline alga, on right. 10,739 feet (3,273.2 m).

c. Coral rudstone with multiple generations of sediment flooring shelter cavities; preserved shelter porosity. 10,744 feet (3,274.8 m).

d, e, and f. Peloidal bioclastic packstones and grainstones to floatstones (in bioclastic packstone-grainstone). Skeletal debris consists of annelid worm tubes, echinoid spines and plates, corals, pelecypods, red algae, and codiacean green algae. Note the leached coral (lower right) and the reworked codiacean alga thalli in e. d. 10,748 feet (3,276 m); e. 10,754 feet (3,277.8 m); f. 10,755 feet (3,278.1 m) their overall biotic assemblages, coral growth forms, and species diversity. Beauvais' (1973) review of Upper Jurassic hermatypic coral deposits listed three principal types of coral formations: bioherms or patch reefs (types 2a, b, and 3) that produced deposits which ranged from a few to 30 meters thick; thin beds or biostromes, which also have organisms in growth position (not discussed in this paper); and bioclastic deposits (type 2c), in which reworked coral and other skeletal debris are often associated with sandy or oolitic sediment, a lack of binding organisms, and poor preservation of the original reef core. In these coral-bearing bioherms (Fig. 21), at least three framebuilding associations, with minor variations, dominate: sponge-coral-algal (type 2a) crust buildups developed on low-energy, less frequently wave-agitated platforms; coral-dominated patch reefs (type 2b) and associated bioclastic piles (type 2c) developed on highly wave-agitated platforms; and stromatoporoid-coral buildups (type 3) developed on steepened platform margins.

Coraliferous buildups that occupied deeper water (type 2a) were micrite-rich and contained lamellar corals, siliceous sponges, and stromatolite boundstones. Their assemblage and laterally equivalent muddy, pelleted, moderately fossiliferous, and oncolite-bearing sediments suggest lower energy settings, probably near the lowermost reaches of storm wave base. In contrast, coral-dominated bioherms (type 2b) with robust and encrusting growth forms of corals, calcareous algae, and calcisponges reflect bioherm development in shallow-water wave-agitated conditions. Associated bioclastic, oolitic, and oncolitic grainstone and packstone facies support the interpretation of a high-energy depositional setting.

The sponge-coral-algal mounds (Fig. 20; type 2a in Fig. 21) of the subsurface Smackover Formation (Walker Creek and North Haynesville Fields) contain nearly equal amounts of siliceous sponges, corals, and algal crusts, in addition to calcareous algae. Insignificant beds of skeletal debris flank these buildups, although the association with bioturbated, poorly sorted, oncolite-peloid packstones and fine-peloid and ooid packstones indicates a moderately wave-agitated shelf setting. In the buildup from Walker Creek Field, cross-bedded oolite grainstones overlying the peloidal packstones formed in highly wave-agitated shoal conditions. The red algal-rich rubble zone in the upper part of the buildup in Haynesville Field suggests, perhaps, that more normal marine conditions existed in a seaward direction from Walker Creek Field, where water depth or wave energy could also have been greater. The thicker peloidal packstones overlying the buildup in Haynesville Field suggest greater distances to the crestal ooid shoals, though crossbedded sand and crestal or flanking gravels indicate high wave (storm generated or normal) energies.

In even higher energy, wave-dominated environments, such as promontories exposed to ocean or deep-shelf wave swells, coral-bearing patch reefs are nearly completely reworked to form extensive and often coalescing piles of bioclastic debris (Fig. 20; type 2c in Fig. 21), in which little *in situ* framework is preserved. This type of buildup appears to be representative of the subsurface Smackover example from Hico Knowles Field, and the reef core diminution displayed by the Bajocian patch reefs of Morocco and the Oxfordian coral-stromatoporoid patch reefs from Austria (Steiger and Wurm, 1980).

Stromatoporoid-Coral Buildups (Type 3).- These buildups dominated the wave-swept platform margins facing open oceans. Reef talus of bioclastic detritus and allochthonous blocks were deposited in basins containing micrites with normal marine biota. Turnsek *et al*. (1981) suggested this association as evidence for a shelf-margin location of the buildups and paleorelief at the margin. A seismic line across the reef complex of the Knowles Limestone in East Texas (Finneran *et al*.,1982) shows a morphologic differentiation into platform and gently dipping slope. The description of carbonate facies and paleocommunities by Finneran *et al*. (1982) suggests closer affinities to the stromatoporoid-coral

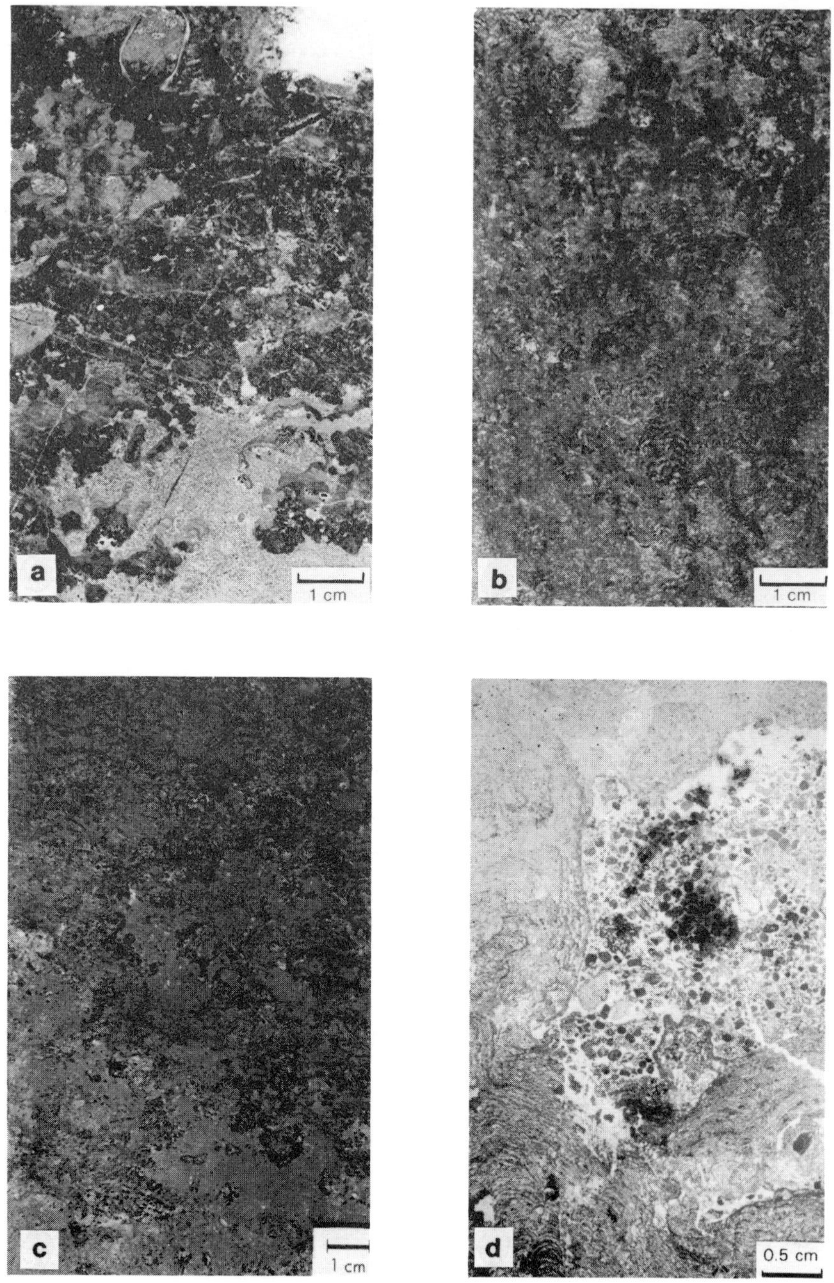

Figure 19

reefs of Turnsek *et al.*(1981) and Steiger and Wurm (1980) than to the types of carbonate reefal buildups recognized in the Smackover Formation. Eliuk (1978) and Jansa *et al.* (1982) also document a Kimmeridgian-Tithonian stromatoporoid-coral dominated shelf margin occurring in the subsurface along the eastern North American margin offshore from Nova Scotia. Seismic profiling revealed a relatively steep carbonate margin with several hundreds of meters of depositional relief facing the Jurassic Atlantic Ocean. Eliuk (1978) and Jansa *et al.* (1982) also recognized that algal-stromatolite buildups (type 1) and calcisponge buildups had occupied deeper water environments along this margin.

Stromatolitic Algal-Mounds (Type 4).- Development of this type of mound has only been reported from the subsurface Smackover wave-agitated platform (Figs. 20 and 21). These buildups may be a variation of the sponge-coral-algal mounds, where the sponges and corals were not very prolific, possibly because of elevated salinities (Baria *et al.* 1982); however, other normal marine, stenohaline organisms, particularly red algae and echinoids, are present. Such mounds have not been reported from studies of Jurassic outcrops, although similarities exist with deepwater mounds.

The development of a subtidal stromatolitic sequence in Paup Spur Field north of the skeletal-rich and biotically diverse buildups of Walker Creek, Hico Knowles, and Haynesville fields was not recognized by Baria *et al.* (1982). Its occurrence north of the skeletal-rich belt, in a more shelfward position, suggests that stromatolite buildups occur some distance up on the shelf, whereas the skeletal-rich buildups are restricted to the margin of the gently sloping platform. This distribution may explain why stromatolite mounds are more common in the eastern part of the Smackover trend where the broader shelf in this region would be less conducive to the development of biotically diverse mounds.

The association of ooid shoals and stromatolitic algal mounds in the Smackover Formation may have a modern counterpart in the Bahamas. Dravis (1983) speculated that the development of columnar stromatolite heads in water depths of 1 to 5 meters is promoted by high physical stress caused by migrating ooid bars, which restricted the activities of grazing organisms. Another modern example of stromatolite development was reported by James and Ginsburg (1979; their Figs. 16-17a and b) from the Belize barrier reef. Here, columnar-, domal-, and digitate-laminated mudstones lithified by aragonite marine cements occur in the reefs. Both of these examples are important modern analogues from which to infer similar associations for thrombolitic or stromatolitic structures found in ancient buildups.

Buildup Geometries and Trends

Subsurface studies of the buildups in the Smackover Formation indicate that they attain thicknesses of 3 to 40 meters. Out-

Figure 19. Polished core slabs and photographs of thin sections illustrating stromatolite mound boundstones to framestones.
a. Thrombolite boundstone (dark color) illustrating boundstone to framestone fabric. Oyster *Lopha* sp. (upper center) and anhydrite-filled coral (upper right); dolomitized and fractured.
b. Stromatolite boundstone illustrating strongly laminated and bifurcating fabric.
c. Dolomitized and oil-stained thrombolite boundstone. a, b, and c are polished core slabs from depths between 15,500 and 16,200 feet (4,724-4,938 m), Baldwin County, Alabama.
d. Photograph of thin section of algal stromatolite illustrating layered columnar and sheet-growth fabric with marine-cemented *Favreina* pellet-peloid grainstone filling constructional reef cavity. Tesoro No. 1 Land Brothers well, Melvin Field, Alabama. 11,252 feet (3,429.6 m).

Table 2. Geometries and Trends of Some Jurassic Buildups

BUILDUP CLASSIFICATION*	THICKNESS (m) individual (stacked)	WIDTH (individual)	TREND (width x length)
Tethyan Buildups			
Sponge-Algal Mounds (1,2) (type 1)	2-100 m (250 m)		
Coral-Dominated Patch Reefs (3,4,5,6) (type 2b)	2-40 m (100 m)	105 m-1 km	10 km x 50 km
Coral-Stromatoporoid (7,8) Patch Reefs (type 3)	2-20 m	0.1-0.3 km	1-5 km x 60-140 km
Stromatoporoid-Coral (7,8) Buildup Margin (type 3)	5-30 m (200-600 m)	0.2-1.5 km	6-10 km x 60-140 km
*U. S. Gulf Coast Subsurface***			
Sponge-Coral-Algal Mounds (9) (type 2a)	3-10 m	0.3-1.5 km	1-2 km x 5 x 10 km
Bioclastic Piles (9) (type 2c)	7-21 m		
Stromatolite Mounds (9,10) (type 4)	4-27 m	0.3-1.5 km	1.6-5 km
Coral-Stromatoporoid Patch Reefs (11) (type 3)	3-25 m		
Stromatoporoid-Coral Margin (12) (type 3)	16 (150 m)		5-11 km x 35 km

* Numbers in parentheses refer to sources as follows: (1) Flugel and Steiger (1981); (2) Gwinner (1976); (3) Barthel (1977); (4) Bollinger and Burri (1970); (5) Giner and Barnolas (1979); (6) Wilson and Exton (1979); (7) Steiger and Wurm (1980); (8) Turnsek *et al.* (1981); (9) Baria *et al.* (1982); (10) McGraw (1982); (11) Cregg and Ahr (1982); (12) Finneran *et al.* (1982).

**Widths and trends of subsurface buildups are based on borehole log correlations and may not represent individual buildups.

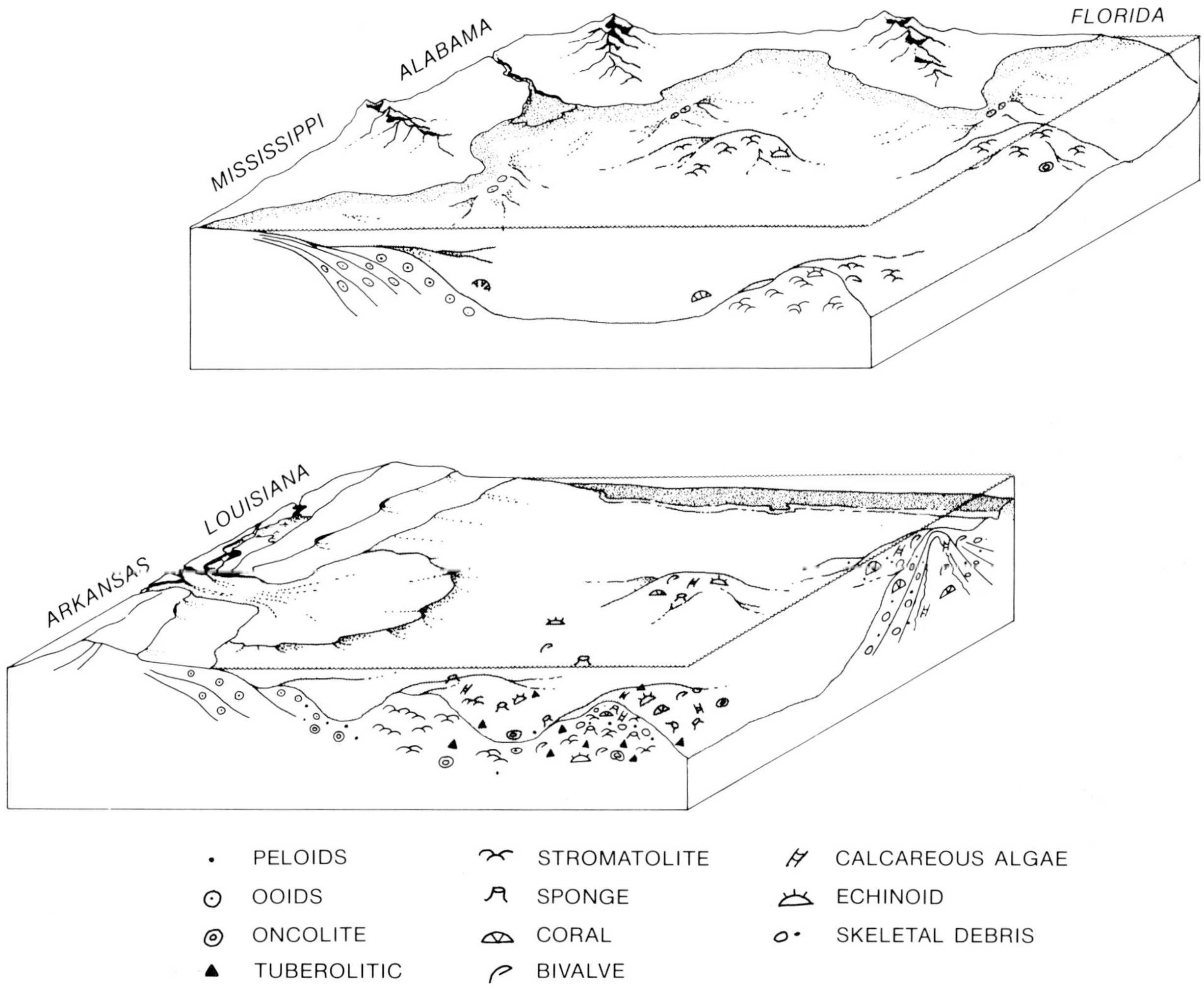

Figure 20. An idealized diagram illustrating the position of buildups developed across the Oxfordian Smackover platform from Arkansas to Florida. Stromatolite mounds occur over the board platform from Mississippi to Florida and up on the Arkansas segment of the platform. Skeletal-rich sponge-coral-algal buildups and bioclastic piles developed on the platform in open-ocean conditions from southern Arkansas to northern Louisiana.

crop studies show similar thicknesses for buildups (Table 2), with those which are 5 to 30 meters thick being the most common. Outcrop studies also show that the areal and lateral extent of individual buildups are generally limited: reef cores range in size from commonly less than 100 meters to, rarely, 1000 meters in diameter.

Complexes made up of mound-rich intervals commonly developed during coalescing mound growth. Coral limestones of the early Kimmeridgian of the Swiss Jura Mountains developed a trend 10 kilometers wide and over 50 kilometers long. Wilson (1975) suggested that the widespread patch-reef trend in the Jura Mountains was developed in several stages during a regressive phase and does not represent simultaneous reef development over the entire area.

The stromatoporoid-coral reef margins range from small developments with dimensions of only 300 meters by 1.5 kilometers (Steiger

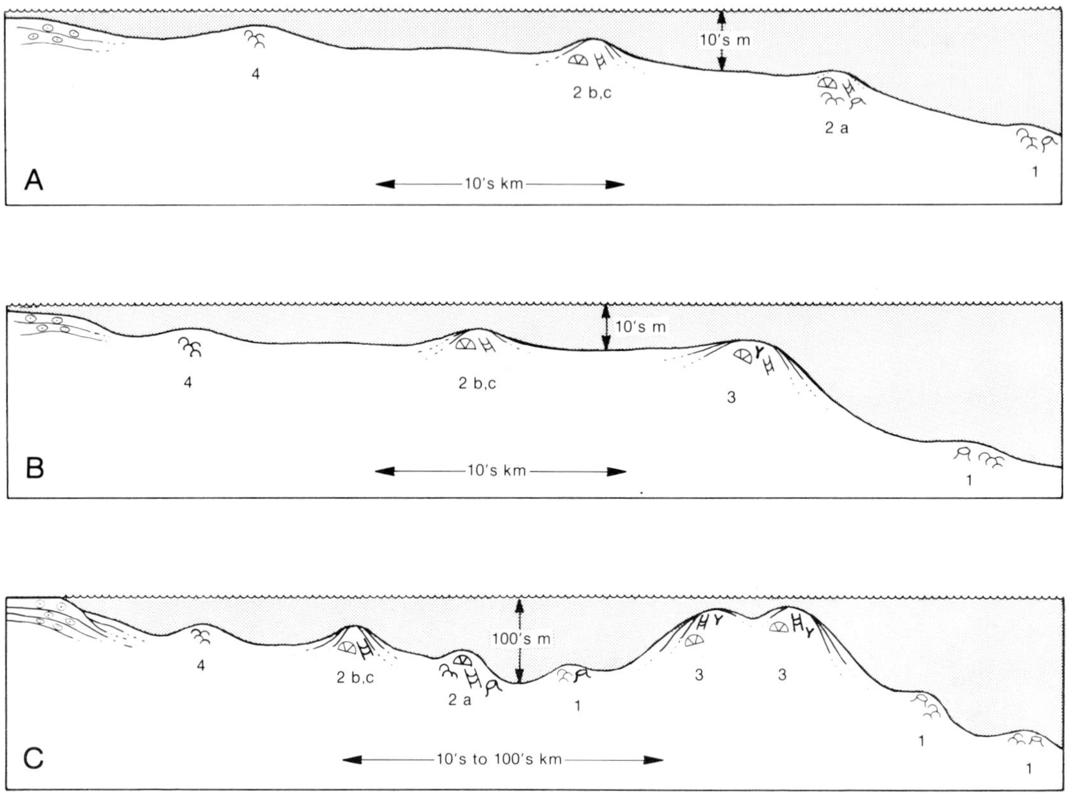

Figure 21. Idealized cross sections illustrating the location of Jurassic carbonate reefal buildups across a gently sloping platform (A), a steepened platform margin (B), and a steepened or rimmed platform margin facing an open ocean with an intrashelf basin or deep-water lagoon (C). Distribution of major reef framebuilders appears to be related to physical energy (wave related) and local environmental stresses (temperature, salinity, nutrients, or turbidity). (type 1) sponge and sponge-algal mounds, (type 2a) sponge-coral-algal buildups, (type 2b and c) coral-dominated patch reefs and bioclastic piles, (type 3) stromatoporoid-coral platformmargin buildups and near back-reef coral-stromatoporoid patch reefs, and (type 4) stromatolite buildups.

and Wurm, 1980) to linear trends several tens to hundreds of kilometers in length (Dinaric Platform; Turnsek *et al.*, 1981). Using borehole logs and seismic data, Finneran *et al.* (1982) estimated that a stromatoporoid-coral boundstone reef complex in the subsurface Knowles Limestone of Texas to be 5 to 11 kilometers wide and to extend a distance of nearly 35 kilometers along trend. This reef complex is similar to the outcropping stromatoporoid-coral platform margins in that the complex consists of interbedded boundstone core facies and bioclastic sediments. Total thickness of the Knowles Limestone complex is about 150 meters.

From well-log correlations, Baria *et al.* (1982) estimated that the Smackover buildups extend for 1 to 2 kilometers along paleotopographic highs and more than one kilometer bankward. Outcrop studies of the sizes, trends, and biotic assemblages of these buildups suggest that the well-log correlations represent entire mound-rich intervals, as commonly observed in outcrop, rather than individual buildups. Preliminary well- log correlations in the Walker Creek Field suggest that the mound-rich interval extends over an area of 50 square kilometers and may be present throughout the entire field (Baria *et al.*, 1982; Harris and Crevello, 1983).

CONCLUSIONS

Modern carbonate reef-building organisms occur in a variety of associations that reflect organism adaptation to specific physical, chemical, and biological conditions. Similarly, Jurassic reef-building paleocommunities adapted to specific depositional settings. Generally, the dominant biotic association enables one to distinguish between deep-water mounds, wave-agitated platforms, and platform-margin reef complexes. Outcrop studies of Jurassic mound geometries, biotic assemblages, and related facies confirm that their counterparts in the Smackover Formation formed on a wave-agitated platform. In addition, these studies suggest that subsurface well-log correlations of Smackover buildups may actually represent mound-rich horizons rather than an individual continuous mound or reef trend. As more buildups are recognized in the subsurface and are mapped seismically, a better understanding of reef-building communities of the Jurassic of the United States Gulf Coast will undoubtedly evolve.

ACKNOWLEDGMENTS

We thank Marathon Oil Company and Gulf Research and Development Company for the opportunity to undertake this worldwide study of Jurassic carbonate buildups and for permission to publish this study. Reviews of the manuscript by Wolfgang Schlager, Philip W. Choquette, and Noel P. James are greatly acknowledged. Helen Hodgson provided editing suggestions that improved the manuscript.

REFERENCES

Ahr, W.M., 1973, The carbonate ramp - an alternative to the shelf model: Gulf Coast Assoc. Geol. Soc. Trans., V. 23, p. 221-225.

Arkell, W.J., 1935, On the nature, origin, and climatic significance of the coral reefs in the vicinity of Oxford: Geol. Soc. London Quart. Jour., V. 91, p. 77-110.

Baria, L.R., D.L. Stoudt, P.M. Harris and P.D. Crevello, 1982, Upper Jurassic reefs of the Smackover Formation, U.S. Gulf Coast: Am. Assoc. Petrol. Geol. Bull., V. 66, p. 1449-1482.

Barthel, K.W., 1977, A spur and groove system in Upper Jurassic coral reefs of southern Germany, *in* Taylor, D.L. (ed.), Proc. Third Intern. Coral Reef Symposium: Miami, Florida, p. 201-208.

Beauvais, L., 1973, Upper Jurassic hermatypic corals, *in* Hallam, A. (ed.), Atlas of Paleobiogeography: Elsevier, p. 317-328.

Bollinger, W. and P. Burri, 1970, Sedimentologie von schelf - carbonaten und beckenablagerungen im Oxfordien des zentralen Schweizer Jura. Mit Betiragen zu Stratigraphie und Okologie: Beitr. zur Geol. Karte der Schweiz, N.F. 140.

Budd, D.A. and R.G. Loucks, 1981, Smackover and Lower Buckner formations, South Texas: Depositional systems on a Jurassic carbonate ramp: Univ. of Texas, Bur. Econ. Geol., Rept. Inf. 112, 38 p.

Catalano, R. and A. D'Argenio, 1981, Paleogeographic evaluation of a continental margin in Sicily: Guidebook, Penrose Conference, Controls of Carbonate Platform Evolution, 142 p.

Cregg, A.K. and W.M. Ahr, 1982, Paleoenvironment of an upper Cotton Valley (Knowles Limestone) patch reef, Milam County, Texas, *in* The Jurassic of the Gulf Rim: Gulf Coast Section, Soc. Econ. Paleontologists and Mineralogists, 3rd Annual Research Symposium, Program and Abs., p. 21-24.

Crevello, P.D. and P.M. Harris, 1982, Depositional models and reef-building organisms, Upper Jurassic reefs of the Smackover Formation, *in* The Jurassic of the Gulf Rim: Gulf Coast Section, Soc. Econ. Paleontologists and Mineralogists, 3rd Annual Research Symposium, Program and Abs., p. 25-28.

______________, P.M. Harris, D.L. Stoudt and L.R. Baria, in press, Porosity evolution of Upper Jurassic Smackover reef rubble; Hico Knowles Field, *in* Roehl, P.O. and P.W. Choquette (eds.), Carbonate Petroleum Reservoirs - A Casebook: Springer-Verlag.

Croft, W.S., 1980, Deposition and diagenesis of the Jurassic Upper Smackover Formation, North Haynesville Field, northern Louisiana: unpublished Masters thesis, Louisiana State Univ., Baton Rouge, 96 p.

Dravis, J.J., 1983, Hardened subtidal stromatolites, Bahamas: Science, v. 219, p. 385-386.

Eliuk, L.S., 1978, The Abenaki Formation, Nova Scotia Shelf, Canada - A depositional and diagenetic model for a Mesozoic carbonate platform: Bull. Canadian Petrol. Geol., V. 26, p. 424-514.

Evans, I. and C.G.St.C. Kendall, 1977, An interpretation of the depositional setting of some deep-water Jurassic carbonates of the Central High Atlas Mountains, Morocco; *in* Cook, H.F. and P. Enos (eds.), Deep-water carbonate environments: Soc. Econ. Paleontologists and Mineralogists Spec. Publ. No. 25, p. 249-261.

Finneran, J.M., R.W. Scott, G.A. Taylor and G.H. Anderson, 1982, Late Jurassic ramp reefs: "Knowles Limestone", southwest flank of the East Texas Basin, *in* The Jurassic of the Gulf Rim, Gulf Coast Section, Soc. Econ. Paleontologists and Mineralogists, 3rd Annual Research Symposium, Program and Abs., p. 36-44.

Fritz, G., 1958, Schwammstotzen, Tuberolithe und Schuttbreccien im Weiben Jura der Schwabischen Alb, Eine vergleichende petrogenetische Untersuchung: Abr. Geol.-Palaont. Inst. Techn. Hochsch. Stuttgart, V. 13, 119 p.

Flugel, E. and T. Steiger, 1981, An Upper Jurassic sponge-algal buildup from the Northern Frankenalb, West Germany, *in* Toomey, D.F. (ed.), European fossil reef models: Soc. Econ. Paleontologists and Mineralogists Spec. Publ. No. 30, p. 371-397.

Giner, J. and A. Barnolas, 1979, Las Construcciones Arrecifales del Jurasico Superior de la Sierra de Albarracin (Teruel): Cuad. Geol., V. 10, p. 73-82.

Gwinner, M.P., 1976, Origin of the Upper Jurassic Limestones of the Swabian Alb (Southwest Germany): Contributions to Sedimentology, V. 3, 75 p.

Harris, P.M. and P.D. Crevello, 1983, Upper Jurassic Smackover Reefs - an example from Walker Creek Field, Arkansas, *in* Harris, P.M. (ed.), Carbonate buildups - a core workshop: Soc. Econ. Paleontologists and Mineralogists, Core Workshop No. 4, p. 356-380.

Imlay, R.W., 1940, Lower Cretaceous and Jurassic formations of southern Arkansas and their oil and gas possibilities: Arkansas Geol. Serv. Information Circular 12, 4 p.

James, N.P. and R.N. Ginsburg, 1979, The seaward margin of Belize Barrier and Atoll Reefs: Inter. Assoc. Sedimentologists, Spec. Publ. No. 3, 191 p.

Jansa, L.F., 1981, Mesozoic carbonate platforms and banks of the eastern North American margin: Mar. Geol., V. 44, P. 97-117.

__________, G. Termier and H. Termier, 1982, Les biohermes a algues, spongiaires et coraux des series carbonatees de la flexure bordiere du paleoshelf au large du Canada Oriental: Revue de Micropaleontologie, V. 25, p. 181-219.

Kendall, C.G.St.C. and I. Evans, 1973, Coral-reef and bioherm morphology - criteria for interpretation of evolution of Jurassic basin of Central High Atlas Mountains, Morocco: Am. Assoc. Petrol. Geol. Bull., abs. V. 57, p. 788.

McGraw, M.M., 1982, Carbonate facies of the Upper Smackover Formation, Paup Spur Field, Miller Co., Arkansas, *in* the Jurassic of the Gulf Rim: Gulf Coast Section, Soc. Econ. Paleontologists and Mineralogists, 3rd Annual Research Symposium Program and Abs., p. 71-77.

Nitzopoulos, G., 1974, Faunistish-okologische, stratigraphische und sedimentologische Untersuchungen am Schwammstotzen-Komplex bei Spielberg am Hahnenkamm (Ob. Oxfordien, Sudliche Frankenalb): Stuttgarter Beitr. Naturkde., Ser. B, 16, 143 p.

Stanley, R.G., 1981, Middle Jurassic shoaling of the Central High Atlas Sea near Rich, Morocco: Jour. Sed. Petrol., V. 51, p. 895-907.

Steiger, T. and D. Wurm, 1980, Facies pattern of Upper Jurassic platform carbonates (Plassen Limestone, Northern Alps, Styria/Austria): Facies, V. 2, p. 241-284.

Turnsek, D., S. Buser and B. Ogorelec, 1981, An Upper Jurassic reef complex from Slovenia, Yugoslavia, *in* Toomey, D.F. (ed.), European Fossil Reef Models: Soc. Econ. Paleontologists and Mineralogists, Spec. Publ. No. 30, p. 371-397.

Wagenplast, P.O., 1972, Okologische Untersuchung der Fauna aus Bankund Schwammfazies des Weiben Jura der Schwabischen Alb: Abr. Geol. Palaont. Inst. Univ. Stuttgart, n. F. 67, 99 p.

Warme, J.E., R.G. Stanley, J.L. Wilson and A.A. Ekdale, 1975, Middle Jurassic Reef tract, Central High Atlas, Morocco: Ninth Internl. Cong. of Sedimentology, Nice, 11 p.

Wells, J.W., 1942, Jurassic corals from the Smackover Limestone, Arkansas: Jour. Paleontology, V. 16, p. 126-129.

Wilson, J.L., 1975, Carbonate Facies in Geologic History, Springer-Verlag, New York, 471 p.

Wilson, R.C.L. and J. Exton, 1979, Excursion to the Mesozoic of the Lusitanian Basin, West Central Portugal: Petroleum Society of Great Britain Field Guidebook, 80 p.

Winterer, E.L. and A. Bosselini, 1981, Subsidence and sedimentation on Jurassic Passive Continental Margin, Southern Alps, Italy: Am. Assoc. Petrol. Geol. Bull., V. 65, p. 394-421.

Zeiss, A., H. Keupp, R. Meyer, E. Flugel and J.T. Groiss, 1975, Upper Jurassic of the Southern Frankenalb (Algal-sponge reefs, coral reefs, Solnhofen lithographic limestones); *in* Flugel, E. (ed.), Guidebook, Internl. Symp. on Fossil Algae, Erlangen. p. 147-228.

Appendix A. Photograph of continuous core from the ARCO No. 1 McFadden well, Walker Creek Field, Columbia County, Arkansas. 3-inch-diameter core.

10,945-10,957 ft. (3,336-3,339.6m)
Coarse ooid grainstone, cross laminated.

10,957-10,964 ft. (3,339.6-3,341.8m)
Peloid packstone with oncolites, intraclasts of reef framework (tuberoids or micropeloid clasts), corals, brachiopods, echinoderms, and pelecypods; coarser toward base.

10,964-10,976 ft. (3,341.8-3,345.5m)
Numerous large corals of *Actinastrea* sp. in a reef frame work with digitate stromatolites and thrombolites (very gray), various encrusters, and pelecypods; internal reef sediment (light gray) is fine bioclastic packstone to mudstone. Fractures developed between corals through the reef framework are the result of compaction-fracturing of coral molds and reef matrix.

10,976-10,988 ft. (3,345.5-3,349.1m)
Reef framework as above with fewer large corals but numerous finger-like corals, encrusters, siliceous sponges, echinoids, and pelecypods.

10,988-10,992 ft. (3,349.1-3,350.4m)
Biointraclastic packstone and patches of framework with oysters, bivalves, calcisponges, echinoids, stick-like corals, and intraclasts of micropeloid packstones to wackestones.

10,992-10,998 ft. (3,350.4-3,352.2m)
Poorly sorted biointraclastic and peloid packstones and oncolite rudstones; many oncolites have a digitate growth fabric.

10,998-11,003 ft. (3,352.2-3,353.7m)
Quartz sandstone and poorly-sorted ooid grainstone.

Appendix A - Plate 1

Appendix A - Plate 2

Appendix A - Plate 3

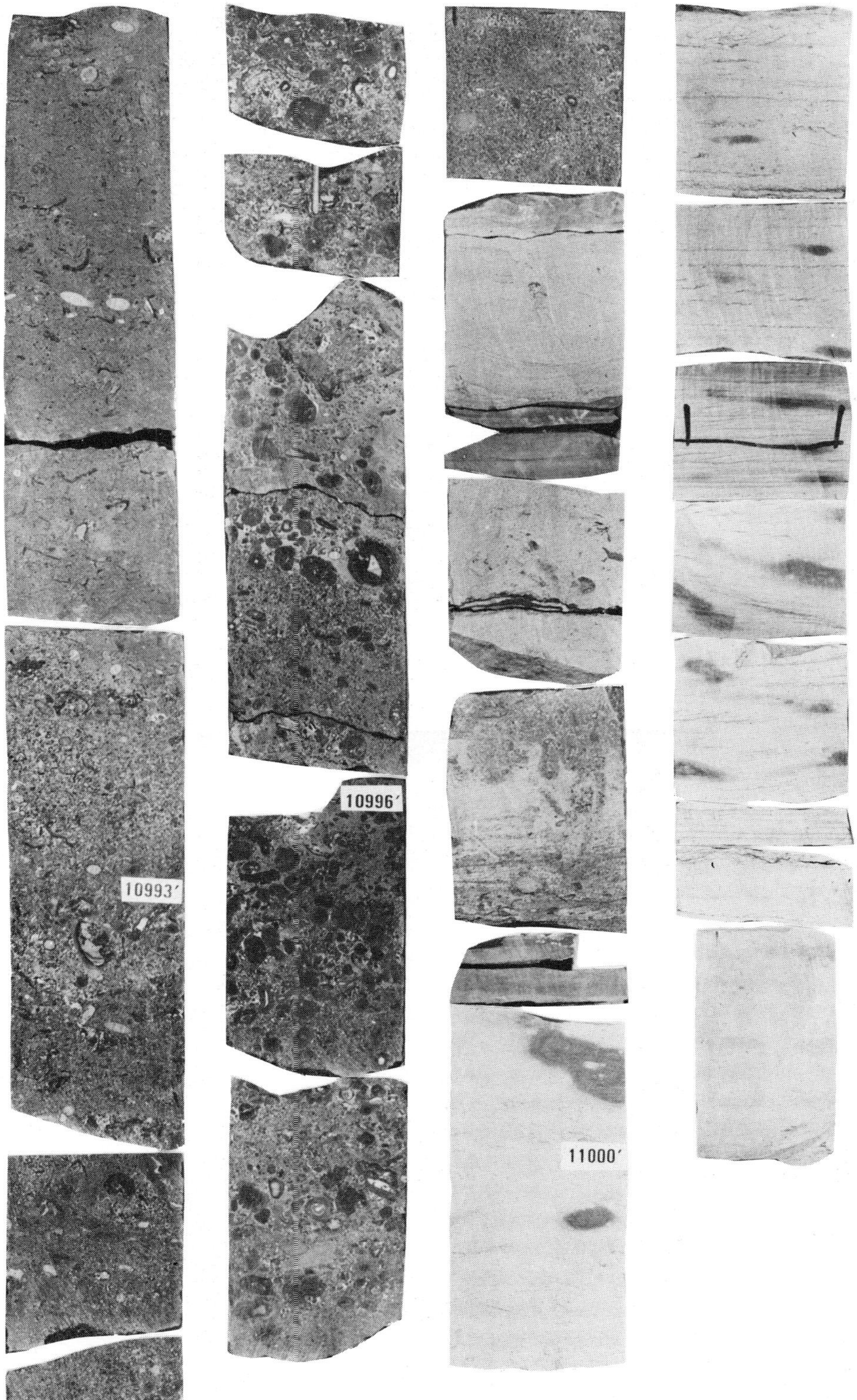

Appendix A - Plate 4

Appendix B. Photograph of continuous core from the Tenneco No. 1 Seegers-Waller well, North Haynesville Field, Claiborne Parish, Louisiana. 3-inch-diameter core.

10,051-10,066 ft. (3,063.5-3,068.1m)
Fine-grained peloid packstone to grainstone with echinoids, thin-walled pelecypods and brachiopod, scattered oncolites and oolites.

10,066-10,076 ft. (3.068.1-3,071.1m)
Missing Core.

10,076-10,090 ft. (3,071.1-3,075.4m)
Biointraclastic rubble to packstone locally cross-stratified, and bioclastic wackestone interbedded with digitate stromatolite/thrombolite sponge-coral-algal boundstone containing red algae and various encrusting organisms, pelecypods, and echinoids; numerous stylolites and microfractures; corals filled with anhydrite.

10,090-10,106 ft. (3,075.4-3,080.3m)
Sponge-coral-algal boundstone as above.

10,106-10,109 ft. (3,080.3-3,081.2m)
Biointraclastic and oncolite rudstones to packstones with pelecypods (oysters), echinoids, and intraclasts of micropeloid (tuberoid) packstones to wackestone.

10,109-10,116 ft. (3,081.2-3,083.4m)
Very fine peloidal packstone and wackestone with oncolites, echinoid debris, and brachiopods.

Appendix B - Plate 1

Appendix B - Plate 2

Appendix B - Plate 3

Appendix B - Plate 4

Appendix C. Photograph of continuous core from the Bass #1-A Smith well, Hico Knowles Field, Lincoln Parish, Louisiana. 2-inch-diameter core.

10,732-10,738 ft. (3,271.1-3,272.9m)
Skeletal peloid packstone to grainstone with coral debris, molluscs, and echinoids.

10,738-10,763 ft. (3,272.9-3,280.6m)
Interbedded and crudely layered, possibly graded, sequences of bioclastic rudstones, grainstones, and packstones of corals, calcareous algae, calcisponges, echinoids, molluscs, and chaetitids; punctuated by siltstone at 10,745 ft.(3,275.1m) and 10,747 ft. (3,275.7m), and by silty mudstone at 10,756 ft.(3,278.4m).

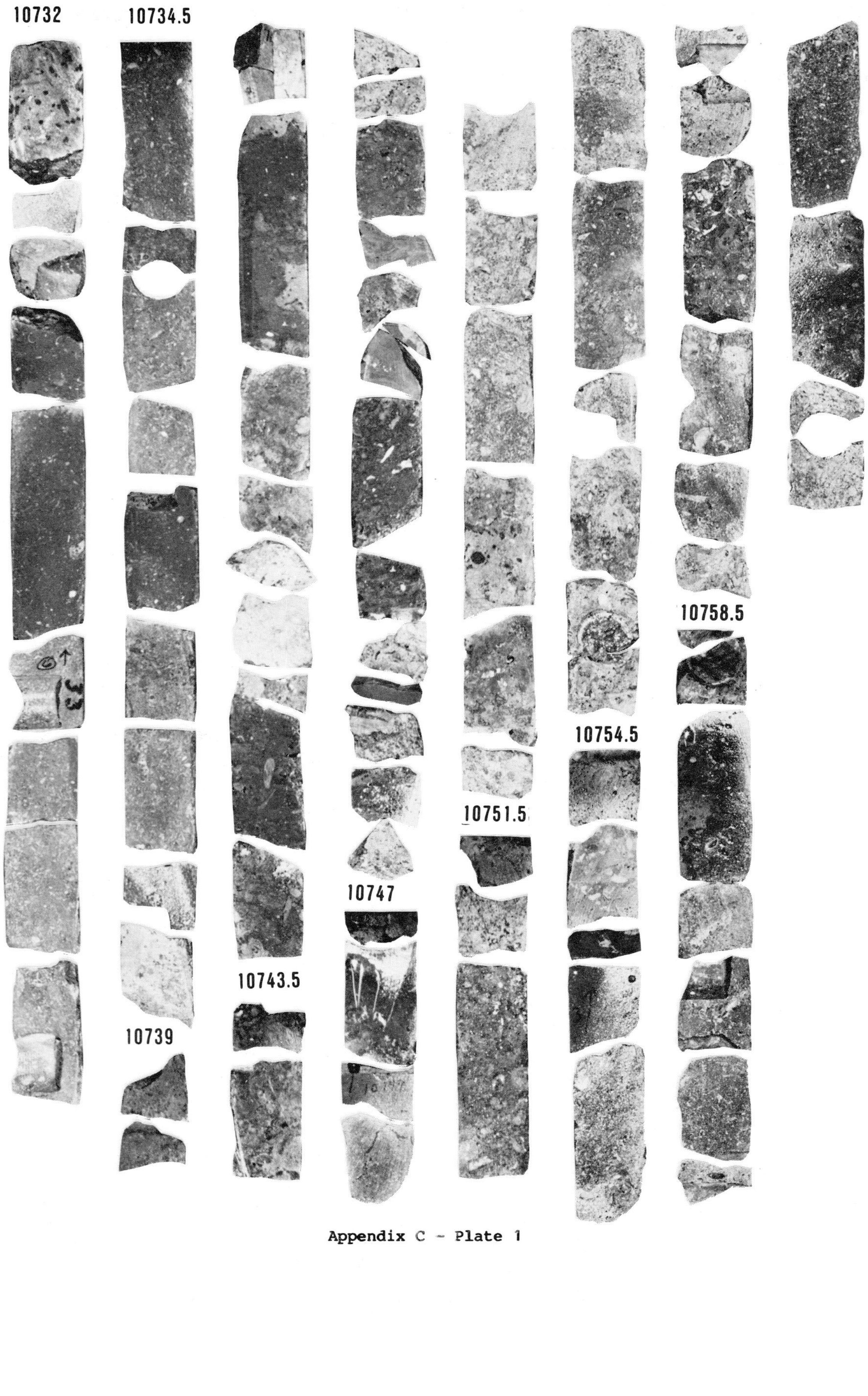

Appendix C - Plate 1

DEPOSITIONAL AND DIAGENETIC HISTORY OF UPPER JURASSIC HAYNESVILLE FORMATION, TEAGUE TOWNSITE FIELD, FREESTONE COUNTY, TEXAS

ROBERT C. FAUCETTE

Getty Oil Company
5177 Richmond, Suite 800
Houston, Texas 77056

WAYNE M. AHR

Department of Geology
Texas A&M University
College Station, Texas 77843

ABSTRACT

The Upper Jurassic Haynesville Formation in Teague Townsite Field, Freestone County, Texas, consists of grainy limestones, muddy limestones, shales, and quartz conglomerates. Constituent grains in the limestones are primarily non-skeletal with ooids being the most common grain type.

Composition and textural data from petrographic studies were grouped into seven microfacies. These microfacies represent three depositional environments: (1) tidal bar, (2) open marine, and (3) restricted marine. The distribution of these environments was controlled by contemporaneous salt movement.

A major shoaling upwards sequence is present in the Haynesville Formation. Nine smaller, local cycles are also present; these local cycles were probably caused by intermittent salt movement during deposition.

Four major diagenetic environments are recognized: (1) marine phreatic, (2) mixed or brackish phreatic, (3) meteoric phreatic, and (4) subsurface environments. Dissolution of ooids in the meteoric phreatic environment created very fine, intragranular porosity in oolite grainstones. Leached zones occur at the crests of the upper shoaling cycles. Virtually all intergranular porosity has been filled by subsurface coarse, mosaic spar cement. Production is from the intragranular porosity created by early meteoric leaching.

INTRODUCTION

The Upper Jurassic Haynesville Formation in the east Texas area was depositied on a shallow-water, gently dipping ramp (Ahr, 1973). Regionally, the Haynesville consists predominately of lime mudstones with grainy facies restricted to the upper 300 feet (91m) of the formation (Collins, 1980). The ramp model of Ahr (1973) predicts that grainy reservoir facies will be found in updip positions. Some Haynesville reservoirs, similar to Teague Townsite Field, occur downdip associated with paleohighs. Some of these reservoirs are in areas previously considered to be too far downdip for grainy facies to be developed (Collins, 1980).

Teague Townsite Field is located in eastern Freestone County, Texas (Fig. 1). Here the Haynesville is a massive, regressive carbonate unit that overlies the Smackover and Buckner Formations. It is overlain by the Bossier Formation (Fig. 2). The Haynesville grades into continental sandstones updip and massive, lime mudstones and black shales downdip. Figure 3 illustrates the relative stratigraphic position of Teague

Townsite Field.

The primary objectives of this study have been to: (1) determine the environments in which the reservoir rock was deposited and diagenetically altered, and (2) determine the depositional, diagenetic and tectonic processes that affected the porosity in this reservoir.

HAYNESVILLE ROCKS

Cores and borehole logs from three wells along with borehole logs from eighteen field wells were studied to determine depositional and diagenetic history. The cored intervals are from the upper 200 feet (60m) of the Haynesville, and consist of grainy limestones, muddy limestones, shales and quartz pebble conglomerates.

Constituent grains in the limestones are primarily non-skeletal with ooids being the most common type. Other non-skeletal particles include grapestones, pisolites, intraclasts, pellets, peloids, and oncoids. Skeletal grains are less common and include molluscs, echinoderms, algae, foraminifera, scleractinian corals, stomatoporoids, and bryozoans. The distribution of the important grain types in the studied cores is shown in Figures 4-6.

Petrographic study of texture and composition enabled the following microfacies to be identified:

1. Pisolite-grapestone-ooid grainstone facies.
2. Oncoid-grapestone-ooid wackestone to packstone facies.

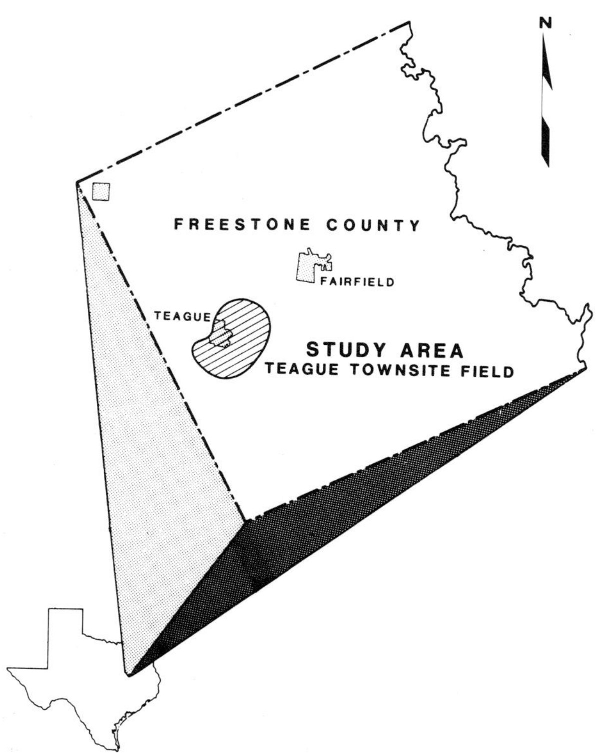

Figure 1. Index map showing location of the Teague Townsite Field, Freestone County.

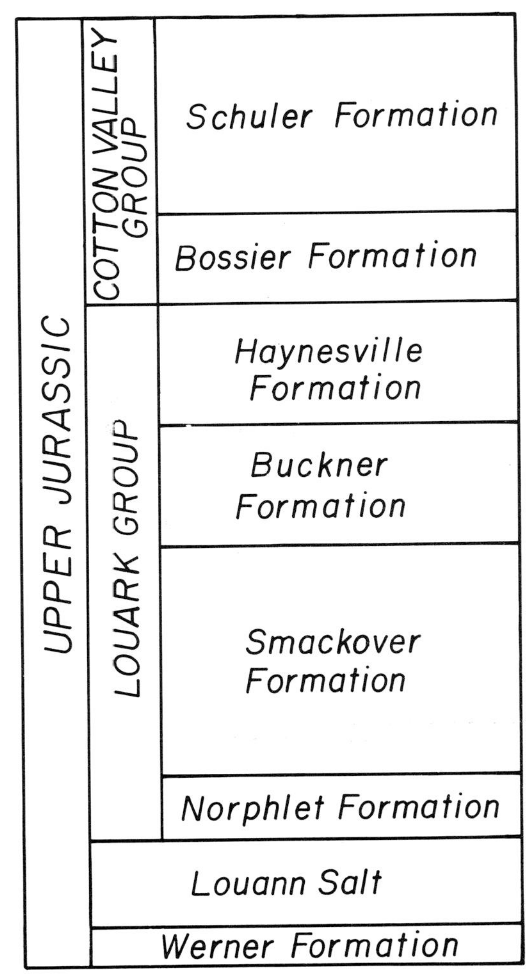

Figure 2. General stratigraphic column of the Upper Jurassic for the Gulf Coast subsurface.

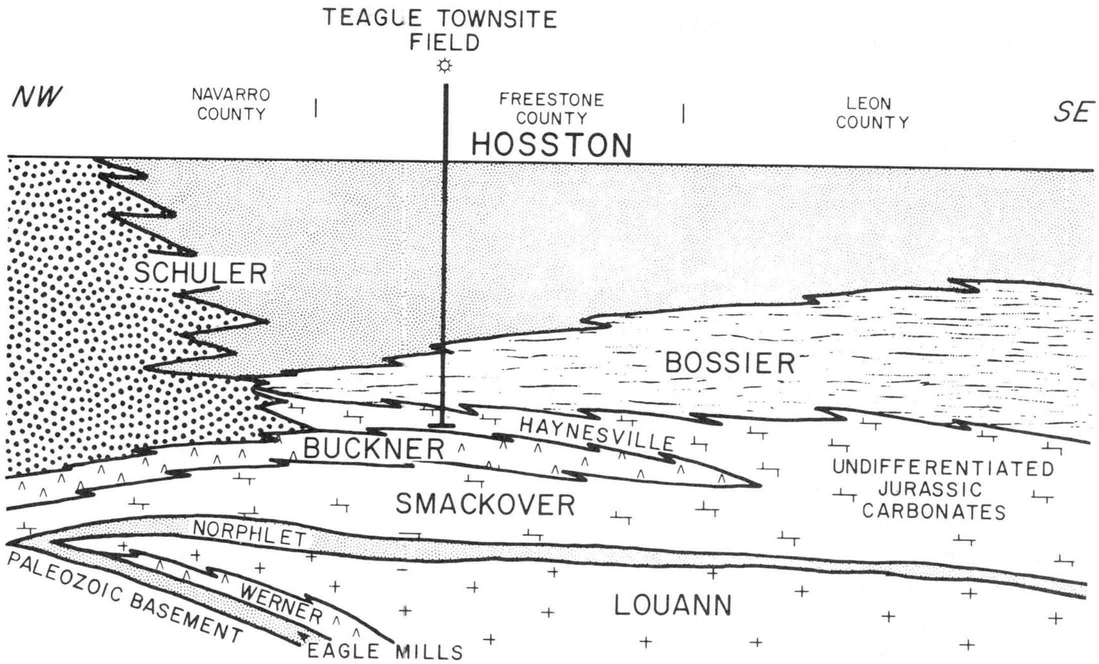

Figure 3. Diagrammatic dip section showing the stratigraphic relationships of the Jurassic and Triassic stata (Modified from Collins, 1980).

3. Intraclast-ooid packstone to grainstone facies.
4. Skeletal-ooid mudstone to wackestone facies.
5. Skeletal-peloid-pellet wackestone to to packstone facies.
6. Ooid-peloid-pellet grainstone facies.
7. Shale-conglomerate facies.

Pisolite-Grapestone-Ooid Grainstone Microfacies

This microfacies contains 30-90% ooids and pisolites and 0-45% grapestones. The average coated-grain content is 76%. This unit contains less than 10% lime mud matrix; no whole skeletal grains are present.

These units are usually massive, although horizontal to slightly inclined laminae and festoon crossbeds are present. Depositional textures are all grain supported. These intervals show either no grain-size trend or a coarsening upwards trend, although some thin, repetitive fining upwards units are also present. Bed thicknesses range from 1 foot (0.3m) to 17.5 feet (5.3m) and average 5.5 feet (1.7m). These units are usually dark gray to black, although a light tan to brown color is also present.

Oncoid-Grapestone-Ooid Wackestone To Packstone Microfacies

This microfacies contains 30-80% ooids and oncoids and 1-20% grapestones. The average coated-grain content is 50%. These units contain greater than 10% lime mud matrix, but whole skeletal grains are rare.

These units are generally massive and bioturbated. Depositional textures are mainly grain supported, although mud-supported textures are present. Bed thicknesses range from 2 feet (0.6m) to 5 feet (1.3m) and average 3 feet (0.9m). The color is light gray to black and brown.

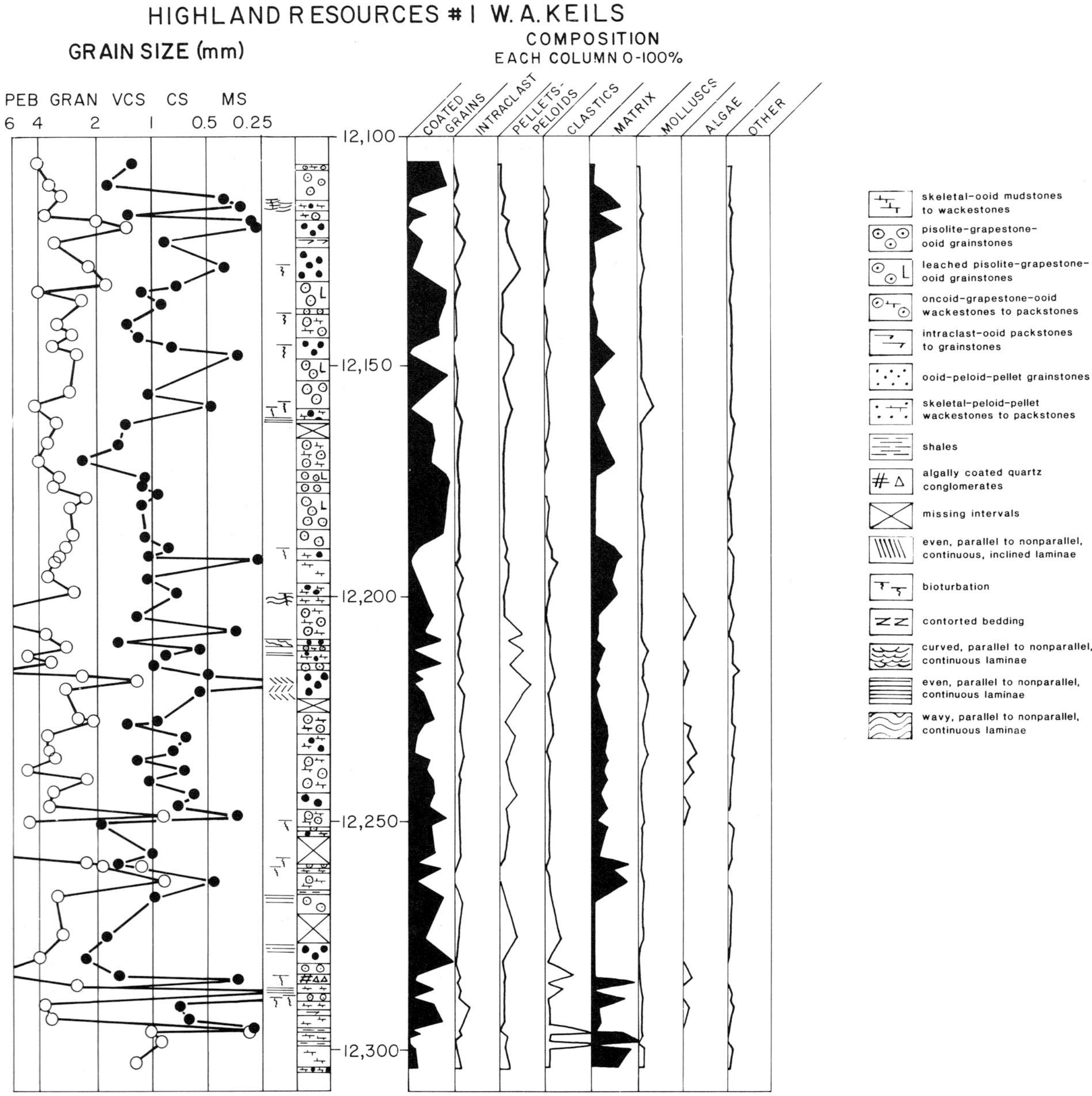

Figure 4. Texture, composition, and sedimentary structures and their relationships to microfacies in the Highland Resources 1 W.A. Keils core. Closed circles in grain-size column are mean carbonate grain sizes, open circles are maximum carbonate grain sizes.

Skeletal-Peloid-Pellet Wackestone To Packstone Microfacies

This microfacies contains 15-65% *Favreina* pellets and peloids and 5-25% skeletal grains. Whole and broken mollusc remains comprise 75% or more of the skeletal fraction; lime mud content is greater than 10%. This unit contains more whole mollusc remains and fewer coated grains than the grainy facies.

Wavy to even horizontal laminae are present as they are in the other muddy facies. Depositional textures are primarily

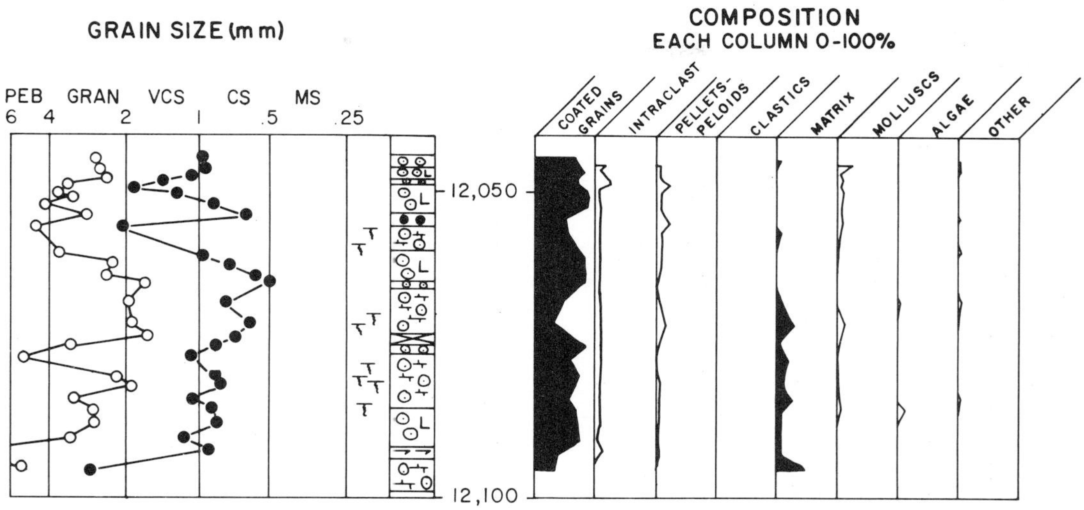

Figure 5. Texture, composition, and sedimentary structures and their relationships to microfacies in the Highland Resources 2 Roper G.U.2 core. Closed circles in grain size column are mean carbonate grain sizes, open circles are maximum carbonate grain sizes. See Figure 4 for symbols.

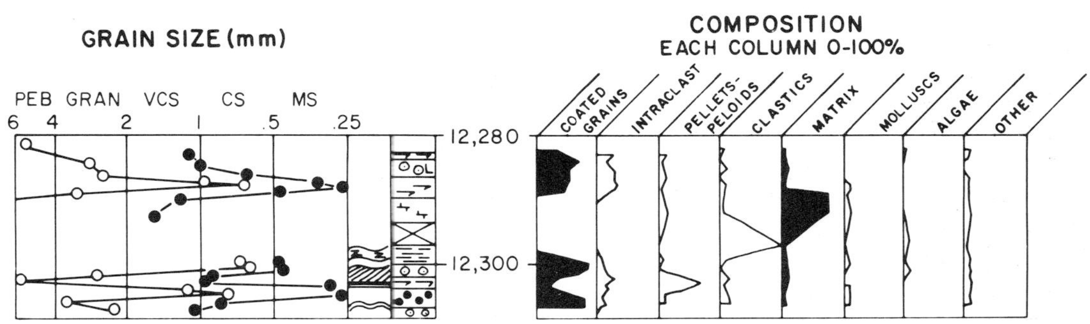

Figure 6. Texture, composition, and sedimentary structures and their relationships to microfacies in the Highland Resources 1 Zeno King G.U.2 core. Closed circles in grain size column are mean carbonate grain sizes, open circles are maximum carbonate grain sizes. See Figure 4 for symbols.

mud supported. Bed thicknesses range from 2 feet (0.6m) to 4.5 feet (1.4m) and average 3 feet (0.9m). The color is light gray to black to brown.

Ooid-Peloid-Pellet Grainstone Microfacies

This microfacies contains 25-55% peloids and pellets and 25-50% ooids. Lime mud content is less than 10%. This unit contains fewer coated grains than the other grainstone facies and no whole skeletal grains.

Horizontal to inclined laminae and festoon crossbeds are present. Depositional textures are grain supported. Bed thicknesses range from 1 foot (0.6m) to 7.5 feet (2.3m) and average 4 feet (1.2m). The color is light gray to black.

Shale-Conglomerate Microfacies

This microfacies is composed of 75-100% terrigenous clastics. The shales are composed of clay and silt-sized quartz and feldspar. Lime mudstone clasts make up less than 5% of the detrital fraction of the shale units. The shale beds are wavy and horizontal, with even laminae and sheared and contorted laminae.

The conglomerate unit from 12,284 to 12,286.5 feet in the Highland Resources No. 1 W.A. Keils well core (Fig. 4) is composed of very coarse and sand-sized and larger algal-coated, polycrystalline quartz grains. Bed thicknesses for both types of units range from 1 foot (0.3m) to 3.5 feet (1.1m) and average 2 feet (0.6m). The color is light gray to black.

DEPOSITIONAL HISTORY

Knowledge of the hydrological and sedimentological origins of the various components in each of the microfacies enables an interpretation to be made of the depositional environments that existed in the study area. The seven microfacies represent three depositional environments: (1) tidal bar, (2) open marine, and (3) restricted marine. The distribution of these environments in the studied cores is illustrated in Figure 7.

Haynesville sedimentation was controlled by contemporaneous salt movement. A Lower Bossier isopach map (Fig. 8) shows that the Lower Bossier thins from the northeast to the southwest. The Upper Haynesville shows similar thinning (T. Thompson, *pers. comm.*, 1981). Thinning on both the Haynesville and Bossier indicates the presence of an actively growing salt structure during deposition of these units. A seismic line across Teague Townsite Field (Fig. 9) shows the salt structure and a thinning of the Jurassic units across the crest of the structure. The Continental No. 1 Brown well (CB1) penetrated the top of the Louann Salt at a depth of 14,027 feet (4,275m) after drilling through a complete Upper Jurassic section, indicating the lack of piercement structures in the study area. A comparison of the present Haynesville structure (Fig. 10) with the Lower Bossier isopach thin (Fig. 8) shows that the crest of the structure has migrated updip. This updip migration is probably due to a regional basinward tilting of the section.

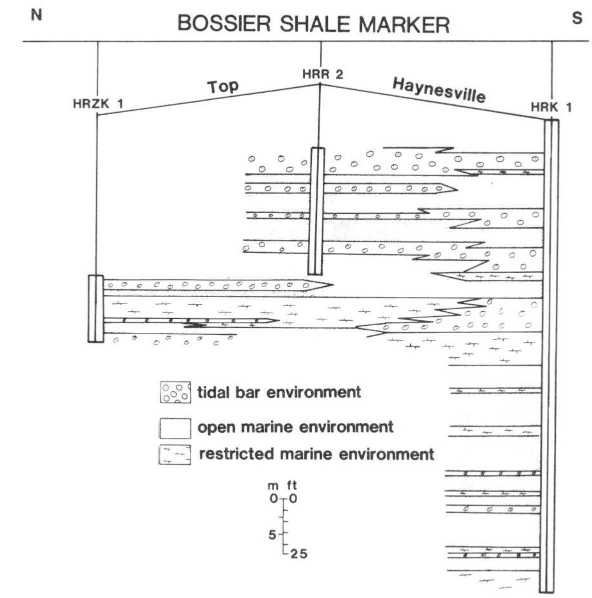

Figure 7. Stratigraphic cross section showing the distribution of depositional environments in the study area. The section is oriented approximately along depositional strike.

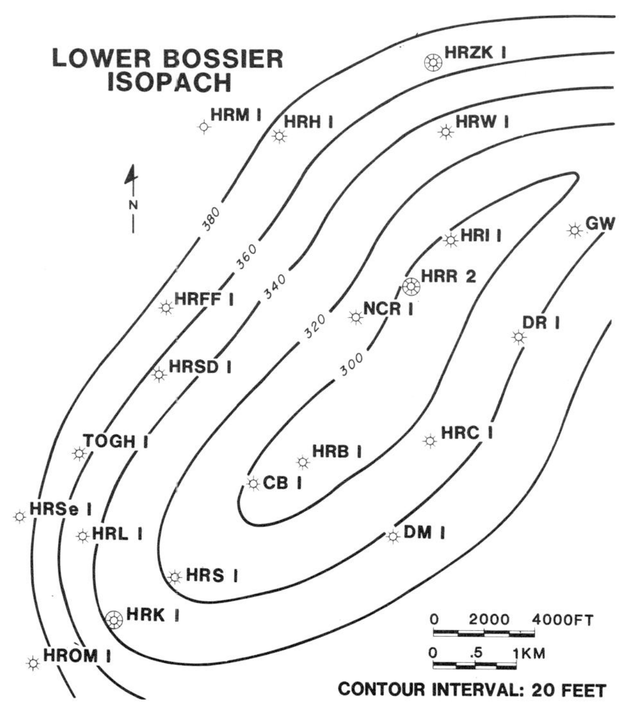

Figure 8. Isopach map of the Lower Bossier Formation. Cored wells are circled.

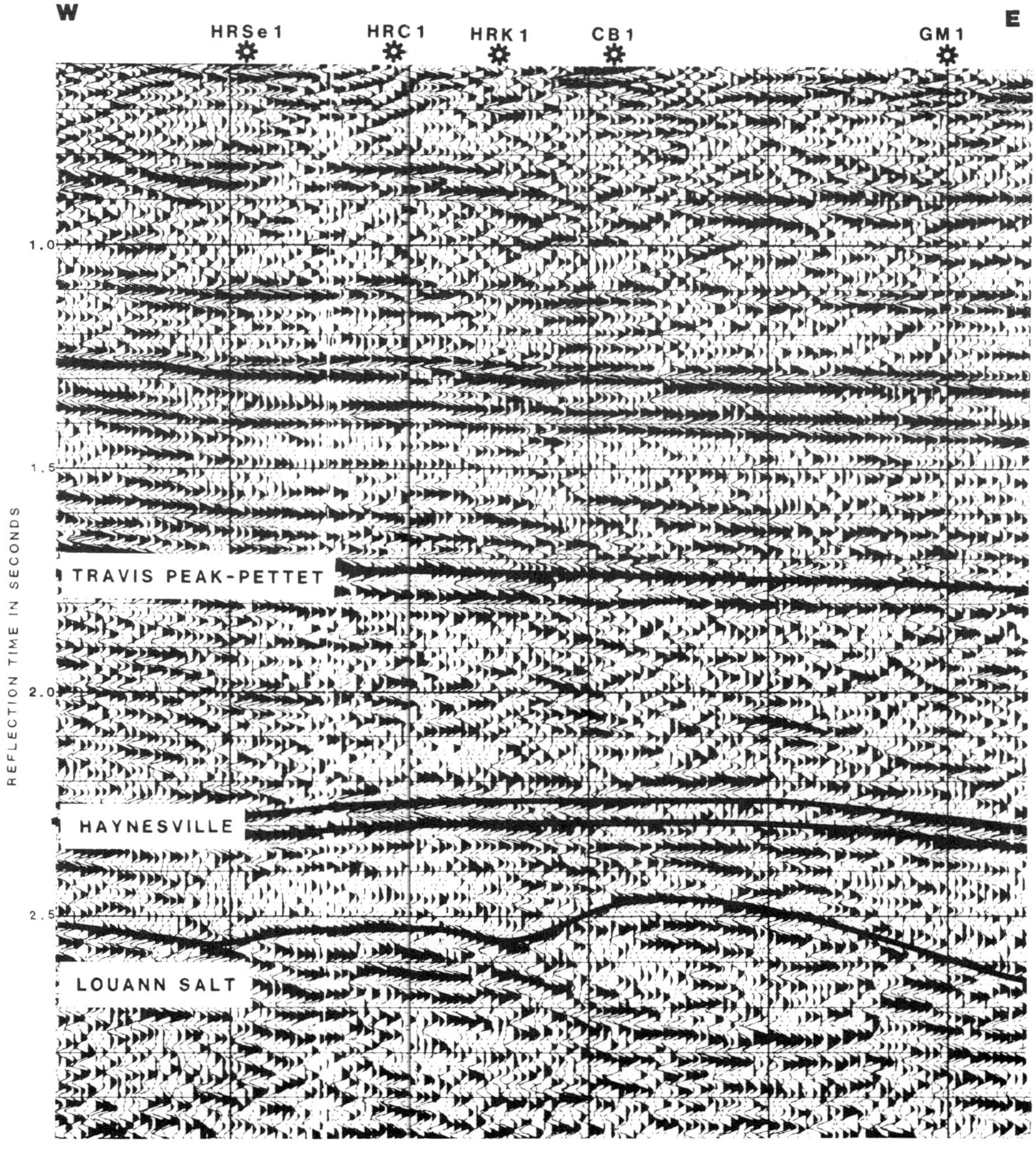

Figure 9. Seismic section across Teague Townsite Field showing the salt structure and depositional thinning of the Jurassic carbonate section across the crests of the structures.

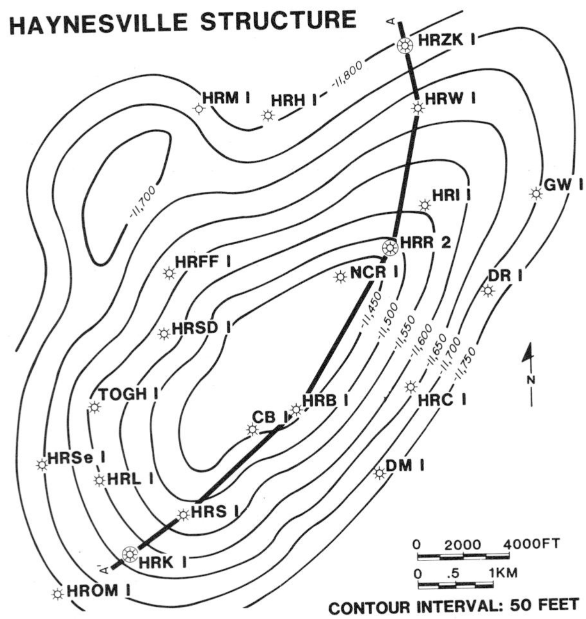

Figure 10. Structural map of the Haynesville Formation. Cored wells are circled.

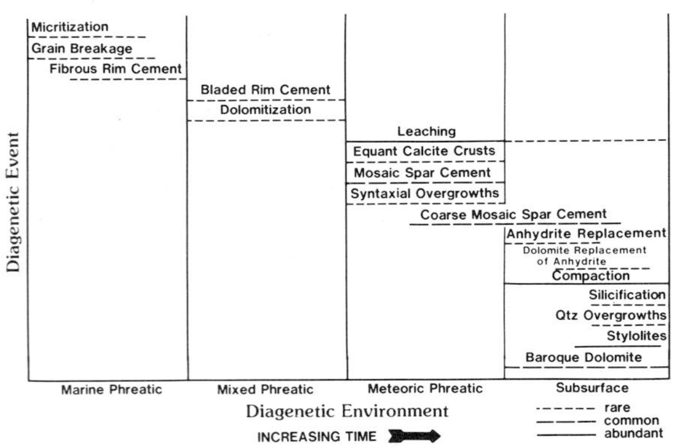

Figure 11. Relative severity and sequence of the diagenetic events in the field and the diagenetic environments in which each occurred (Modified from Loucks, 1977).

Tidal-Bar Environment

The tidal-bar environment formed on the crests of salt-related highs. It is represented by the pisolite-grapestone-ooid grainstone microfacies. This microfacies was deposited in a marine environment, in shallow, highly agitated water. Borehole log correlations indicate that the paleohighs were covered by a series of strike-trending oolite bars. The lower parts of some of these oolite grainstone intervals exhibit repetitive fining upwards sequences suggesting that they were deposited by intermittent currents. The festoon crossbeds also found in these intervals are indicative of migrating dunes formed by unidirectional current flow (Harms, *et al.*, 1975). These sedimentary structures suggest a tide-dominated, shallow-water marine origin for these bars. The lack of bedding in many of the intervals may be due to the presence of a narrow range of grain sizes or to the destruction of original bedding by diagenesis.

Open-Marine Environment

The open-marine environment occurred in the shallow-marine waters surrounding the bathymetric highs. It is represented by the ooid-intraclast packstone to grainstone microfacies, the oncoid-ooid wackestone to packstone microfacies, the ooid-peloid-pellet grainstone microfacies and the conglomerate microfacies. These microfacies were deposited in a marine environment in shallow, quiet to moderately agitated water. They were subsequently covered by oolite grainstones. Many of the ooids and pisolites were transported off the oolite shoals into this environment. Figures 4-6 show that, in general, the percentage of coated grains decreases with distance away from the oolite bar intervals. An increase in lime mud indicates a more stable substrate than that found in the oolite-bar environment.

Sedimentary structures indicate a moderately agitated-water environment. The wavy laminae present in the muddier intervals may be indicative of low-velocity ripples (Harms, *et al*, 1975). The crossbeds found in the grainstones represent a bedform produced by a higher flow velocity which was sufficient to winnow out the matrix, but not high enough to transport large grains. These grainstone intervals are finer grained than the surrounding intervals. This environment contains more uncoated skeletal remains than the oolitic shoal environment.

Restricted-Marine Environment

The restricted-marine environment existed in quiet water "ponded" updip and behind the strike-trending oolite shoals. It is represented by the skeletal-peloid-pellet wackestones to packstones, skeletal-ooid mudstones to wackestones, and shale microfacies. Restricted conditions are indicated by the sedimentary structures, the low faunal diversity, and the microfacies. Ooids and other coated grains are least abundant, and lime mud matrix is most abundant in this environment. Skeletal remains are common in the restricted-marine environment and a large number of whole fossils are present.

Low-velocity currents are suggested by the presence of wavy laminae. The abundance of preserved organic material and the presence of hydrogen sulfide are indicative of reducing conditions.

DEPOSITIONAL CYCLES

An increase in coated-grain content upwards with a corresponding decrease in lime mud content (Fig. 4) indicates a major shoaling upwards sequence. In addition to this major cycle, several smaller, local shoaling upwards cycles are also present. An ideal cycle would be composed of (in descending order):

- I. Tidal-bar environment.
  - A. Highly inclined (25°) and festoon crossbedded oolitic grainstone.
  - B. Horizontal to slightly inclined, graded beds of oolitic grainstone.
- II. Open-marine environment.
  - A. Wavy to even, inclined, laminated oolite-oncoid wackestones to packstones
  - B. Evenly laminated to crossbedded peloid-pellet grainstones.
- III. Restricted-marine environment.
  - A. Wavy to evenly laminated, bioturbated pellet-peloid wackestones to packstones.
  - B. Basal wavy to evenly laminated, bioturbated skeletal-ooid mudstones to wackestones and/or shales.

Figure 7 shows that a minimum of nine complete and incomplete shoaling cycles are present in the studied cores. These local shoaling cycles were probably caused by intermittent salt movement. The effect of the major regressive cycle on the local cycles can be seen in Figure 7. As regional sea level fell, the upper shoaling cycles were exposed to highly agitated shoaling conditions more frequently and for longer periods of time than the lower cycles. As a result, the upper cycles contain thicker ooid shoal sequences.

DIAGENETIC HISTORY

The diagenetic history of Teague Townsite Field provides the key to understanding its porosity distribution. Petrographic relationships indicate that four major diagenetic environments were present in the Haynesville. In order of occurrence, they are: (1) the marine phreatic, (2) mixed or brackish phreatic, (3) meteoric phreatic, and (4) subsurface environment. The relationships between observed diagenetic alterations and the diagenetic environments, the relative severity of the diagenetic types, and their order of occurrence is shown on Figure 11.

Marine Phreatic Environment

Three types of diagenesis occurred in this environment (in chronological order): (1) grain breakage by biological activity, (2) micritization by endolithic algae and fungi, and (3) isopachous, fibrous rim cementation. Grain breakage by biologic activity was the earliest diagenetic event. Micrite rims, which were the second diagenetic event, may cover the broken ends of grains. Isopachous, fibrous rim cementation was the last stage of marine diagenesis (Fig. 12a). These rim cements were observed covering broken grains and micrite rims. Occasionally, these rim cements are micritized. All of these diagenetic types are rare in the studied cores.

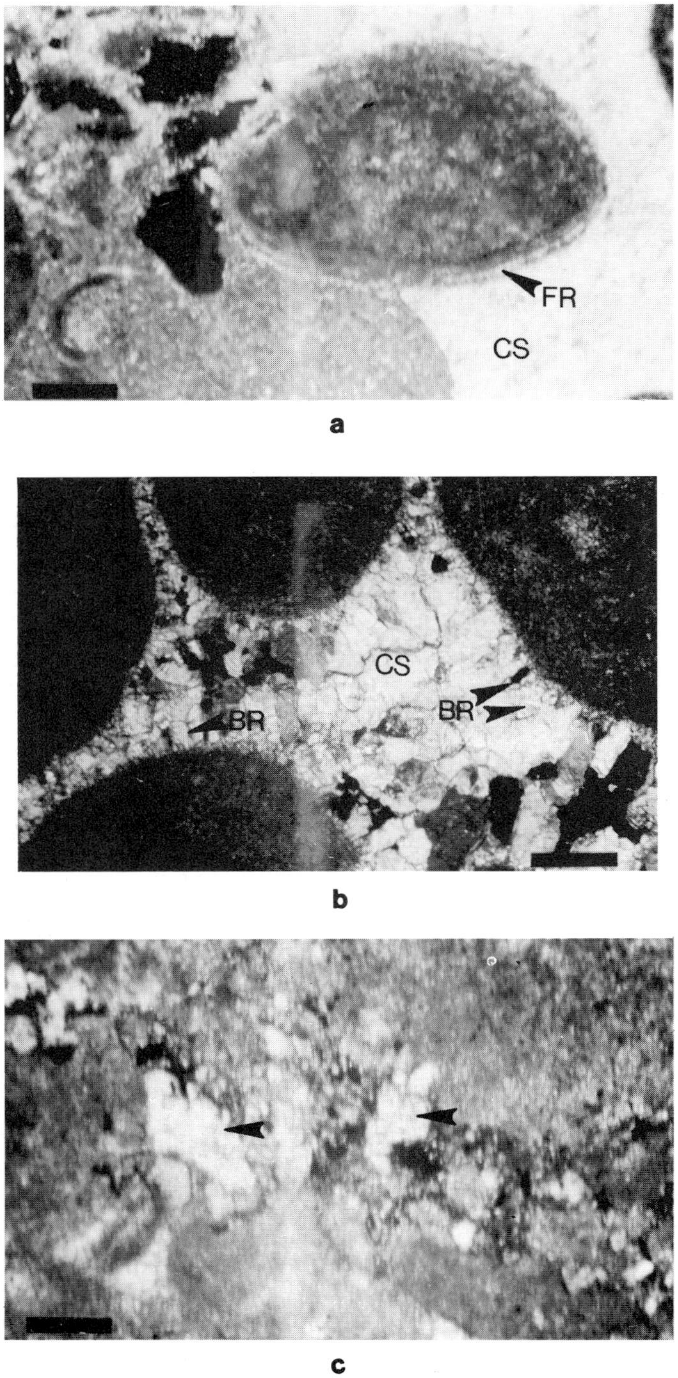

Figure 12. Photomicrographs of diagenetic types found in Teague Townsite Field.

a. Fibrous rim cement (FR) surrounding the ooid is in sharp contact with coarse mosaic spar cement (CS). Bar scale is 0.01 mm.

b. Bladed rim cement (BR) surrounds ooids and is in sharp contact with coarse mosaic spar cement (CS). Bar scale is 0.01 mm.

c. Dolomite rhombs (arrows) replacing lime mud. Bar scale is 0.01 mm.

Mixed Or Brackish Phreatic Environment

Two types of diagenesis occurred in the mixed or brackish phreatic environment: (1) isopachous, bladed rim cementation, (Fig. 12b) and (2) dolomitization (Fig. 12c). Isopachous bladed rim cements overlie the marine fibrous rim cements, showing them to be a result of later and distinct stages. Finely crystalline dolomite rhombs were observed to replace lime mud in only a few thin sections. Both of these diagenetic types are rare in the studied cores.

Meteoric Phreatic Environment

Four important types of diagenesis occurred in the meteoric phreatic environment (in chronological order): (1) leaching and dissolution of ooids, (2) formation of equant calcite crusts, (3) cementation by fine equant spar, and (4) cementation by coarse, mosaic spar. Dissolution of ooids created very fine intragranular porosity in oolite grainstones. This dissolution gives the ooids a "chalky" or "pithy" appearance and destroyed most of the microstructure in the affected ooids (Fig. 13a,b,c). The pores that were created are very small and cannot be seen with a petrographic microscope. Oomoldic porosity may have been created if the process had been allowed to continue. This diagenetic type is abundant in the upper ooid grainstone intervals.

The formation of equant calcite crusts was the next diagenetic event. These cements are similar to the equant crusts of Becher and Moore (1976), except that the crusts in the study area are not concentrated in smaller pores or near grain contacts (Fig. 14a). The crusts are similar to the equant calcite crusts of Longman (1980), which he interpreted to have formed in the meteoric phreatic environment. This diagenetic type is rare.

Cementation by fine, equant spar cements occurred next. This cement is found on the outside of large pores and often completely fills small pores (Fig. 14a). Fine equant spar occurs in sharp contact with equant crusts, and is common in the study area.

Cementation by coarse mosaic, pore-filling spar was the next diagenetic event. Coarse mosaic spar overlies and is in sharp contact with all of the previously mentioned cements and is believed to have formed in this environment only where there is little or no evidence of compaction. It represents a small percentage of the coarse mosaic spar in the studied cores.

Subsurface Environment

Seven types of diagenesis occurred in this environment: (1) compaction, (2) replacement by anhydrite, (3) dissolution of anhydrite and subsequent replacement by dolomite, (4) cementation by coarse, mosaic pore-filling spar and baroque dolomite, (5) leaching of grains and cements, (6) replacement by silica, and (7) pressure solution.

Compaction was the first subsurface diagenetic event. Grain deformation, grain breakage, formation of sutured contacts and fracturing are the compactional effects observed in the study area.

Replacement of grains and cements by anhydrite is also present. The anhydrite was subsequently removed and the resultant molds were filled by dolomite (Fig. 14b). Petrographic relationships indicate that anhydrite replacement was an early subsurface event. Anhydrite was observed to replace grains and cements.

Most of the coarse mosaic, pore-filling spar cement along with baroque dolomite cement was formed in the subsurface. These cements fill all intergranular pore space, overlie all other cement types, and fill crushed grains. They are interpreted to have been deposited primarily in the subsurface since they are often found in thin sections from highly compacted intervals and postdate many subsurface compactional features (Fig. 15a,b).

Leaching of grain edges, rim cements and pore-filling spar cements was the next diagenetic event in the subsurface environment (Fig. 16a). The resultant intergranular porosity is filled with organic matter, which is

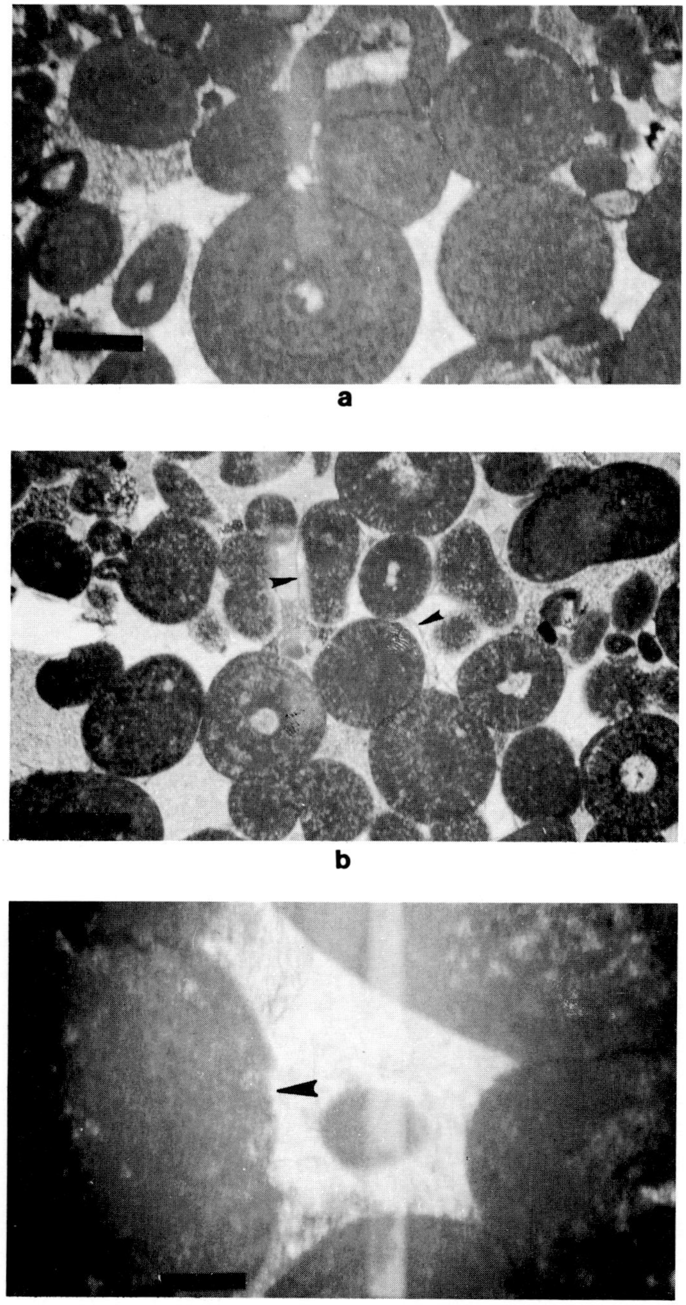

Figure 13. Photomicrographs of diagenetic types found in Teague Townsite Field.

a. Oolite grainstone illustrating the "chalky" or "pithy" appearance created by meteoric leaching. Bar scale is 0.5 mm.

b. Oolite grainstone illustrating spalled ooid rims (arrows), which have been filled in by coarse mosaic spar, and good preservation of ooid microstructure.

c. Corrosion of the edge of an ooid (arrow) created by leaching in the meteoric phreatic environment. Bar scale is 0.01 mm.

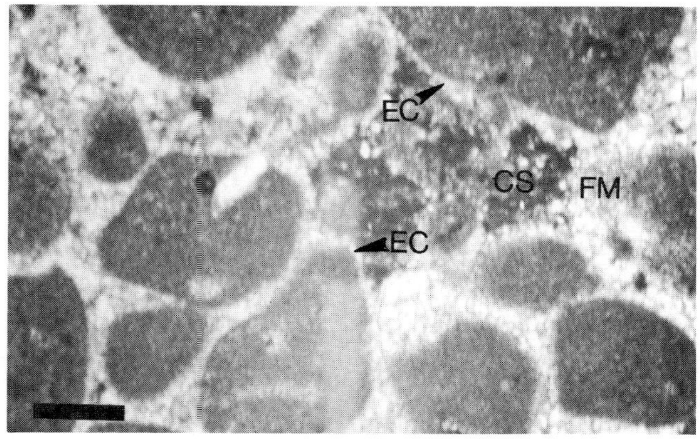

a

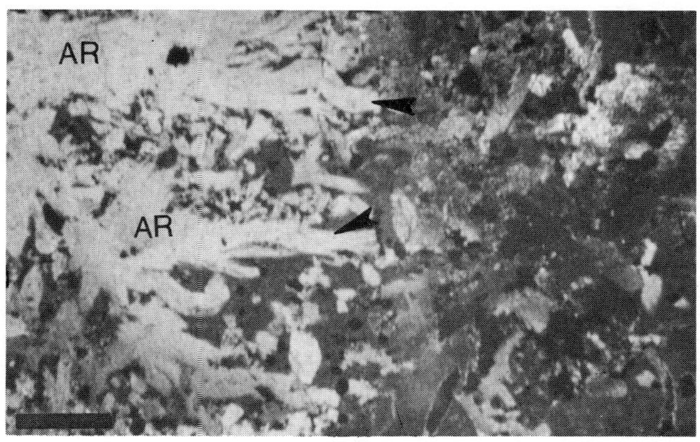

b

Figure 14. Photomicrographs of diagenetic types found in Teague Townsite Field.
a. Equant calcite crusts (EC) surround grains and are in sharp contact with fine mosaic spar cement (FM) and coarse mosaic spar cement (CS). Bar scale is 0.5 mm.
b. Dolomite pseudomorphs after anhydrite (AR) which replaced grains and cements (arrows). Bar scale is 0.5 mm.

probably dead oil (Fig. 16a). This leaching stage is similar to the late, subsurface leaching described by Moore and Druckman (1981) and Mazzullo (1981). The leached zones are thin (1-2 feet or less) and isolated, and represent a relatively rare diagenetic event in the studied cores.

Silica replacement occurred primarily as overgrowths on quartz in ooid nuclei, on silt-sized detrital quartz, and on polycrystalline quartz grains (Fig. 16b). Quartz overgrowths were observed to replace early cements, subsurface spar cements, and sutured contacts and fractures.

Formation of stylolites was the final diagenetic event in the studied thin sections.

DISCUSSION

Figures 4-6 show that the zones leached by fresh-water lenses are present at the crests of the upper shoaling cycles. The thicker oolite grainstone intervals in the upper cycles were exposed more commonly than those in the lower cycles because the upper shoals had the highest depositional topogra-

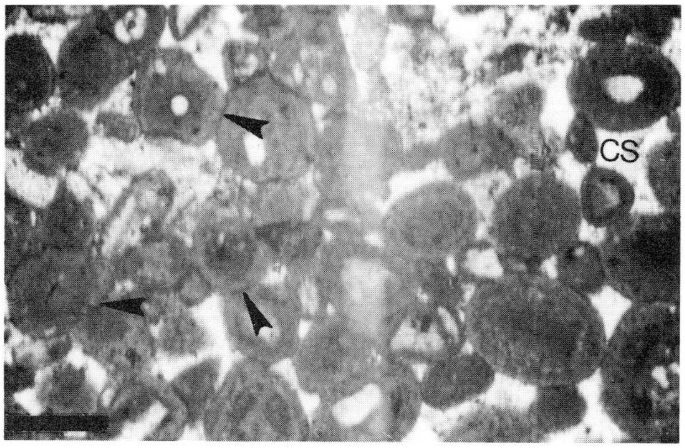

a

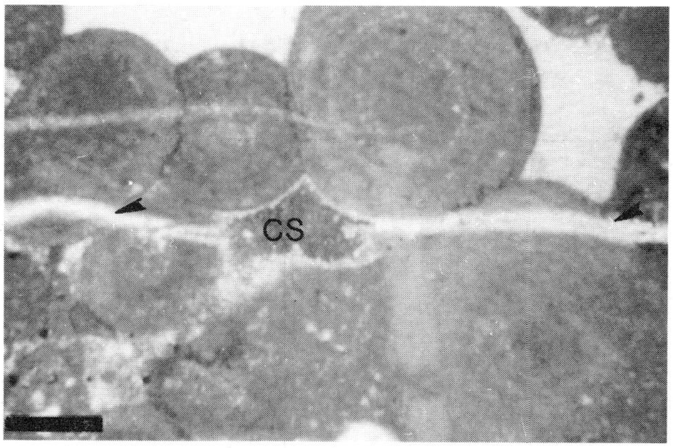

b

Figure 15. Photomicrographs of diagenetic types found in Teague Townsite Field.
a. Highly compacted oolite grainstone in which all pore space has been filled with coarse mosaic spar (CS). Arrows point to sutured contacts between grains. Bar scale is 0.5 mm.
b. Fracture which cuts grains and sutured contacts between grains (arrows) is filled with coarse mosaic spar (CS). Bar scale is 0.5 mm.

phy in the section. The leached zones were identified on the basis of the poorest grain preservation and most severe grain-boundary corrosion observed both in the cores and in thin section.

Figure 17 illustrates the sequence of events which formed these leached intervals. Initially, ooid shoals formed on the crests of salt related highs. Subsequently, salt movement caused parts of these shoals to be subaerially exposed. Invasion by undersaturated, meteoric water created "pithy", intragranular porosity in the upper part of the exposed shoals. Evaporite pans formed in the low areas between and behind these shoals. A relative rise in sea level caused another set of ooid shoals to be deposited on the first set. Further upward salt movement subaerially exposed the next set of shoals, thus repeating the cycle. The early nearsurface origin of this leaching is based on four main lines of evidence: (1) The leached intervals are facies selective. (2) The leached intervals are at the tops of the upper shoaling cycles. (3) Late, pore-filling cements are not leached. (4) Anhydrite zones, representing evaporite pans, are stratigraphically equivalent to the leached zones and occur updip and between the leached zones.

The porosity in Teague Townsite Field

was created by leaching in an undersaturated meteoric environment. Figure 18 shows the relationship between porous zones identified from core analysis and density logs, and the intervals leached in fresh-water lenses from one of the cored wells. Virtually all of the intergranular pore space in these intervals has been filled by cement. A stratigraphic cross-section (Fig. 19) shows the distribution of the leached porous zones and the stratigraphic relationship between the porous and anhydrite zones. Porosity thicks on a net porosity isopach map for Teague Townsite Field (Fig. 20) represent the crests of the early salt highs which were exposed during Haynesville deposition.

The three Haynesville crests lie approximately on the crest of the Bossier anticline (Fig. 21). The shape of the salt structure changed from these small anticlines on top of a strike-trending salt swell during Haynesville deposition to a long, linear, strike-trending anticline during lower Bossier deposition.

CONCLUSIONS

1. Oolitic grainstone bars were deposited on the crests of salt-related paleotopogra-

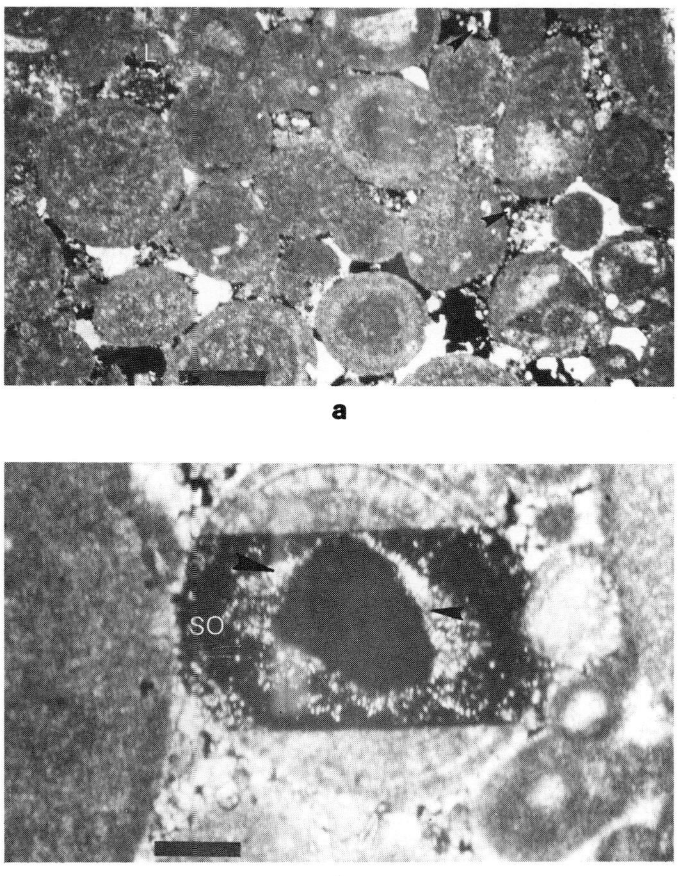

Figure 16. Photomicrographs of diagenetic types found in Teague Townsite Field.
a. Oolite grainstone showing leaching of grains and cements which occurred in the subsurface. Arrow points to an isolated patch of spar cement which is in optical continuity with the cement patches around it. Bar scale is 0.5 mm.
b. Silica overgrowth (SO) replacing an ooid. Arrows point to calcite islands in silica. Bar scale is 0.01 mm.

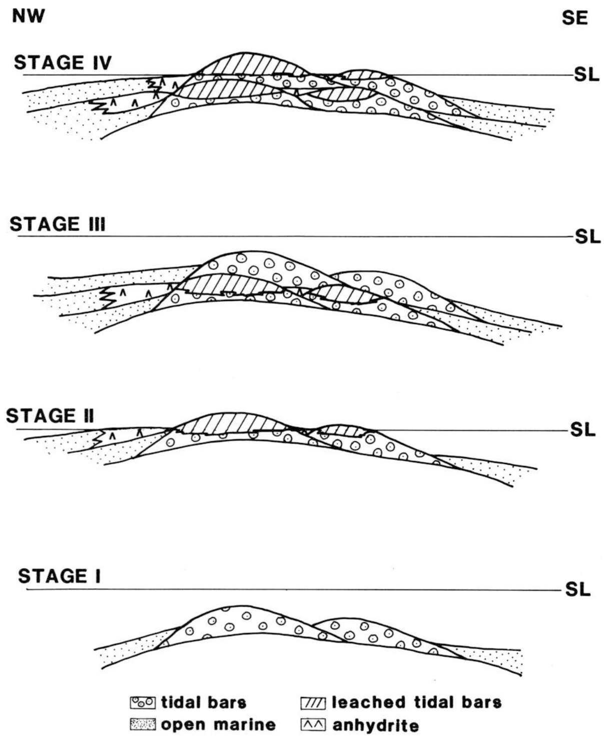

Figure 17. Diagrammatic illustration of depositional and early diagenetic cycles in the Haynesville in the study area.

phic highs, and have a northeast-southwest trend. Lower energy facies were deposited in the areas surrounding these highs.

2. The ooid shoals formed on paleotopography created by early vertical movement of the underlying Louann Salt. The shape of the salt structure changed through time. During Haynesville deposition, three small, distinct salt anticlines were present, which later merged into one long, northeast-southwest trending salt anticline by the time of lower Bossier deposition. The present structure is similar in shape to the Bossier structure, but the crest has migrated updip due to a regional basinward tilting of the section.
3. A minimum of nine shoaling cycles are probably related to vertical salt movement. A regional fall in sea level caused the uppermost shoaling cycles to be exposed to highly agitated shoaling

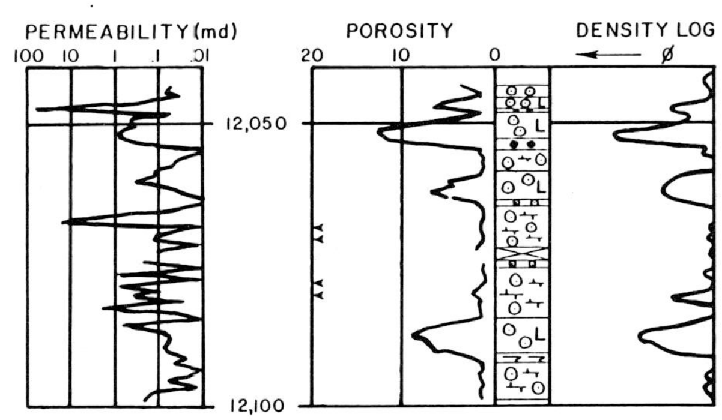

Figure 18. Relationship between leached zones and porous zones from core analysis and density logs in the Highland Resources 2 Roper G.U.2 core. Distribution of permeability in this core and its relationship to porous zones and depositional microfacies is also shown.

conditions more frequently and for longer periods of time than the lower shoaling cycles.

4. The ooid shoals were subaerially exposed at the ends of the upper shoaling cycles.
5. The ooid shoals were leached during exposure by undersaturated meteoric waters creating the very fine, intragranular ("pithy") porosity from which this reservoir produces. The intergranular pore spaces have been filled by cements.

ACKNOWLEDGMENTS

We thank Tenneco and Universal Resources for their financial support of this work. Highland Resources provided the cores, core analyses, logs, and other data. Many of the logs used in the study were provided by Universal Resources. Lengthy discussions with Mr. Tommy Thompson, Highland Resources, and Mr. Joe Kast, Universal Resources, provided much valuable information which was used in this study. Getty Oil Company provided

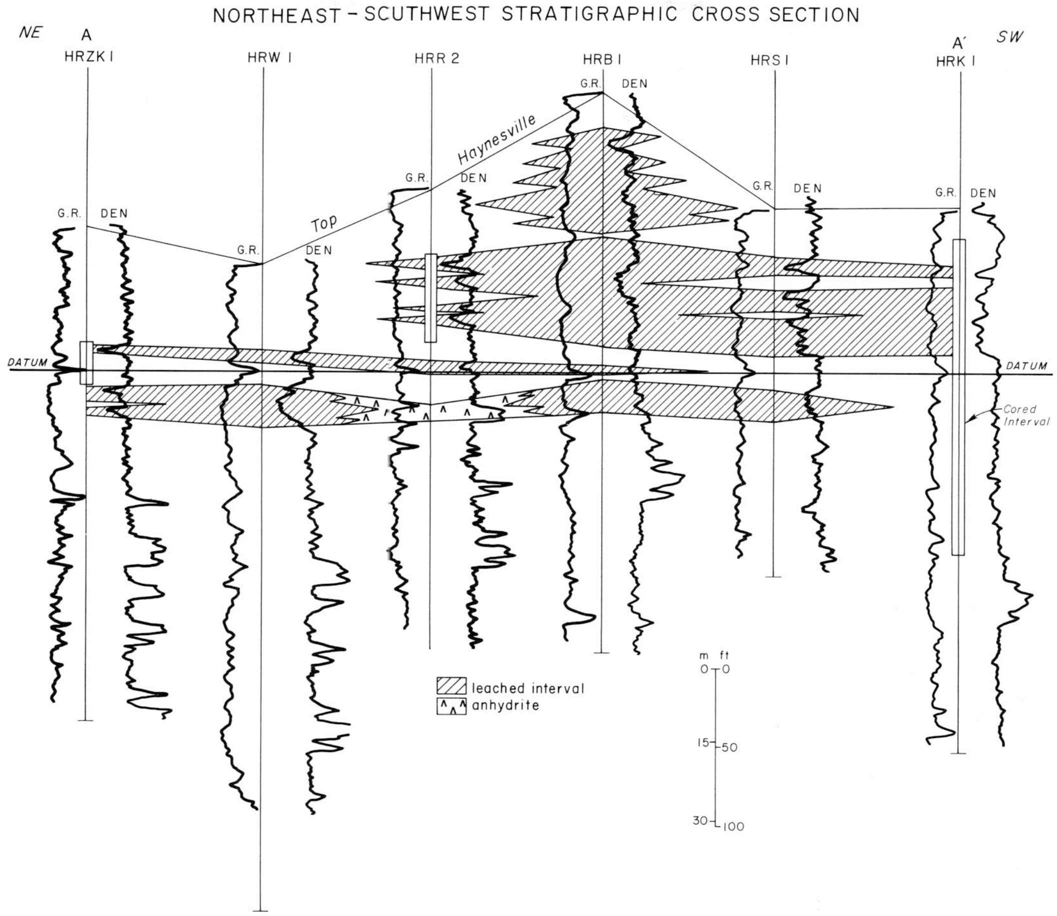

Figure 19. Stratigraphic cross section A-A' showing the distribution of porous leached zones in the study area and the relationship of the leached zones to anhydrite zones. The leached zones finger out away from the crests of salt paleohighs. Line of section is shown on Figure 20.

drafting support and seismic data. Joan Taylor typed the manuscript. R. B. Gayle and R. J. Le Blanc, Jr. improved the paper with their criticism.

REFERENCES

Ahr, W. M., 1973, The carbonate ramp: an alternative to the shelf model: Gulf Coast Assoc. Geol. Soc. Transactions, v. 23, p. 221-225.

Becher, J.W., and C. H. Moore, 1976, the Walker Creek Field: a Smackover diagenetic trap: Gulf Coast Assoc. Geol. Soc. Transactions, v. 26, p. 34-56.

Collins, S. E., 1980, Jurassic Cotton Valley and Smackover reservoir trends east Texas north Louisiana and south Arkansas: Am Assoc. Petroleum Geologists Bull., v. 64, p. 1004-1013.

Harms, J. C., J. B. Southard, D. R. Spearing, and R. G. Walker, 1975, Depositional environments as interpreted from primary sedimentary structures and stratification sequences: Soc. Econ. Paleontologists and Mineralogists Short Course No. 2, 161 p.

Longman, M. W., 1980, Carbonate diagenetic

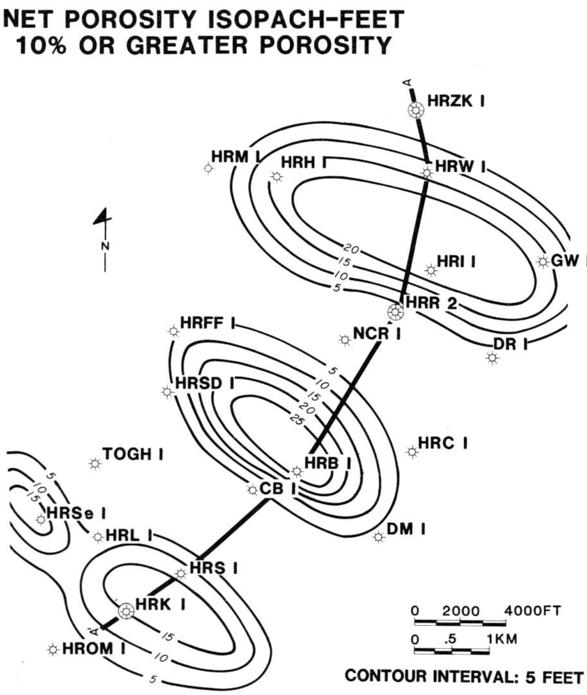

Figure 20. Isopach map of net feet of 10% or greater porosity in the study area. Porosity thicks indicate the positions of salt structures which were active during Haynesville deposition. Line of section A-A' is shown. Cored wells are circled.

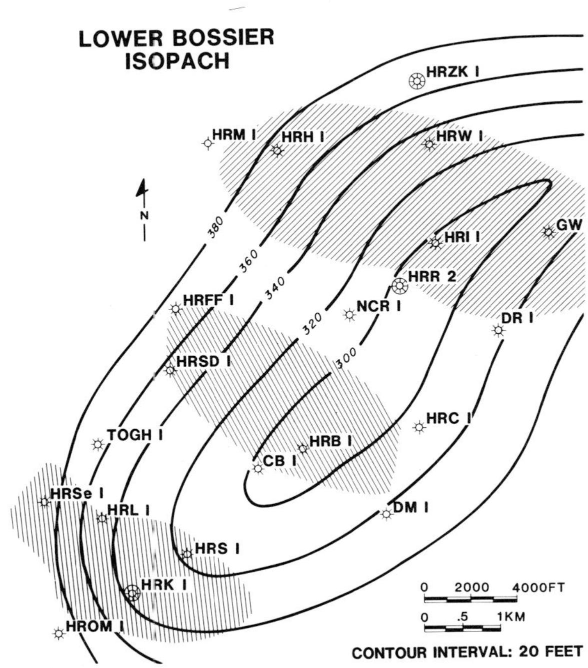

Figure 21. Lower Bossier isopach map with positions of Haynesville salt crests (cross-hatched areas) superimposed.

textures from nearsurface diagenetic environments: Am. Assoc. Petroleum Geologists Bull., V. 64, p. 461-487.

Loucks, R. G., 1977, Porosity development and distribution in shoal-water carbonate complexes--subsurface Pearsall Formation (Lower Cretaceous) South Texas: Texas Un. Bu. Eco. Geol. Rpt. Inv. 89, p. 97-126.

Mazzulo, S. J. 1981, Facies and burial diagenesis of a carbonate reservoir: Champman Deep (Atoka) Field, Delaware Basin, Texas: Am. Assoc. Petroleum Geologists Bull., v. 65, p. 850-865.

Moore, C. H., and Y. Druckman, 1981, Burial diagenesis and porosity evolution, Upper Jurassic Smackover, Arkansas and Louisiana: Am. Assoc. Petroleum Geologists Bull., v. 65, p. 597-628.

BAYOU MIDDLE FORK (SMACKOVER) FIELD, CLAIBORNE PARISH, LOUISIANA A CASE HISTORY: DISCOVERY TO WATERFLOOD

C. KIPP FERNS

ANR Production Company
111 Capital Building, Suite 400
Jackson, Mississippi 39201

MARK E. YORK

Inexco Oil Company
633 North State Street, Suite 500
Jackson, Mississippi 39205

Bayou Middle Fork field, Claiborne Parish, Louisiana is located in north-central Louisiana near the Arkansas-Louisiana border in an area known as the "State Line Trend" of the Upper Jurassic Smackover Formation. Smackover production in the area is associated with a complex fault system masked by approximately 10,000 feet of younger sediments. A geophysical program, combined with geological studies, indicated a faulted, deep-seated, east-west trending anticline (Fig. 1). Based on this evidence, Cities Service Company drilled a test well on the structure resulting in the discovery of the Smackover reservoirs at Bayou Middle Fork field in March, 1975.

Core and sample studies made as the field developed showed the Smackover Formation at Bayou Middle Fork field to be a limestone composed mainly of oolites, hardened pellets, pisolites, oncolites, and micrite. This limestone has been divided into three

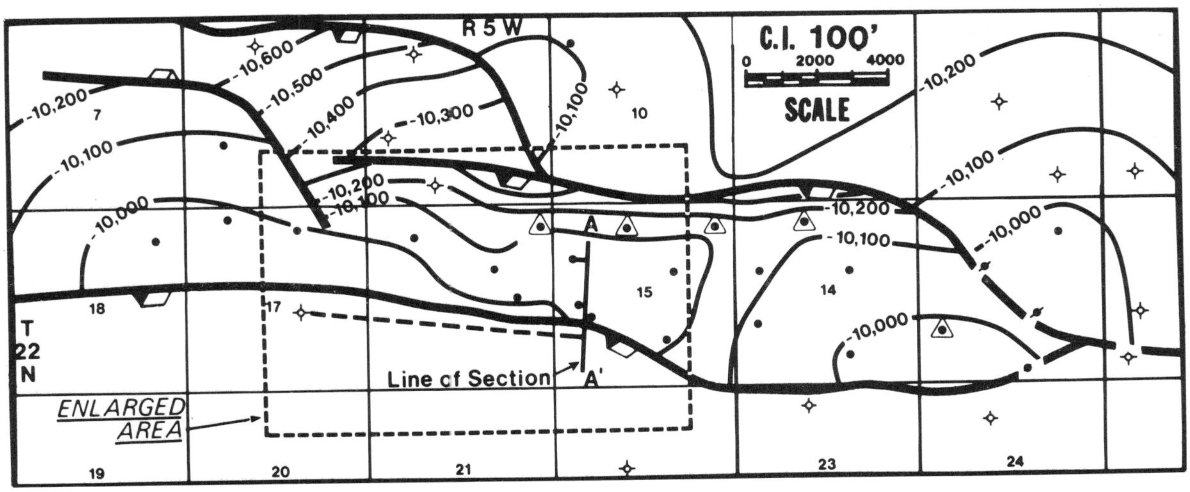

Figure 1. Structure map of the top of the Smackover, Bayou Middle Fork Field, Claiborne Parish, Louisiana.

units, designated as the Smackover "A", "B", and "C" (Fig. 2). The general environmental setting that produced these sediments varied from low to high energy conditions over a broad, shallow, gradually south-sloping marine shelf. This environmental setting underwent continual minor sea-level fluctuations and structural changes which produced an interfingering and mixing of the various carbonate sediments. One major change occurred as sea level completely receded. The shelf was exposed to supratidal conditions which resulted in the deposition of evaporitic and continental sediments. This regression ended Smackover "C" deposition and produced the Buckner "B" member of the stratigraphic section. This was followed by partial transgression and subsequent regression resulting in the Smackover "B" and "A" being deposited in an off-lap sequence. This second withdrawal of the sea ended Smackover deposition and again produced conditions for accumulation of supratidal and continental sediments.

Porosity preserved within these oolitic rocks is primary intergranular, which has

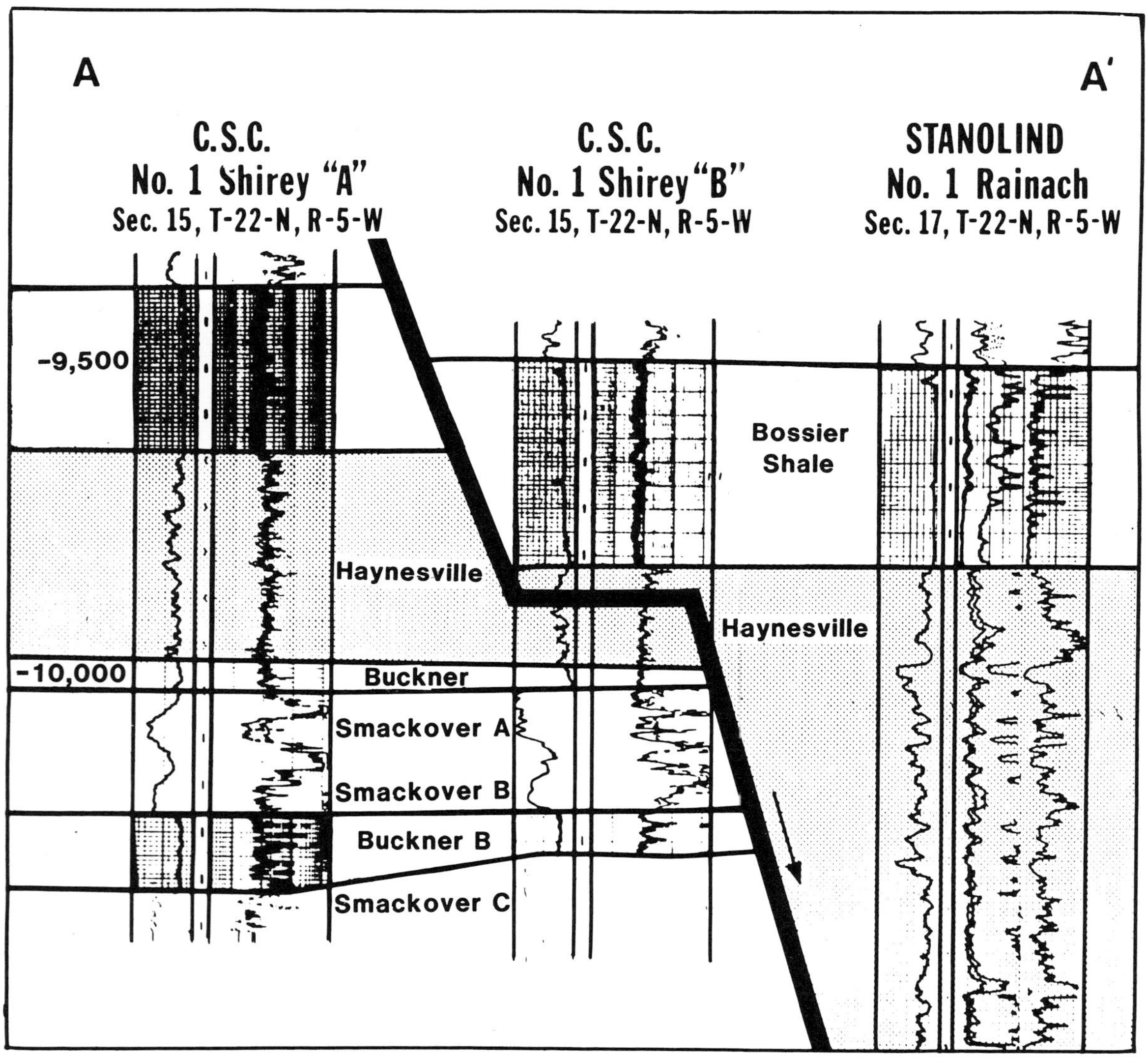

Figure 2. Bayou Middle Fork Field, Claiborne Parish, Louisiana.

been enhanced by leaching of the oolites. Effective porosity varies from a low of 8% to a high of 23.7%, while permeability ranges from less than 1 md to 270 md. During early development of the field, the porosity and water values from log analysis indicated the possibility of substantial water production; however, as wells were completed no water was produced. This can be explained by the presence of microporosity containing irreducible water, as noted by scanning electron microscopy and petrographic examination.

Smackover production at Bayou Middle Fork field is from three separate reservoirs, the Smackover "C", and the upper and lower "A". The lower Smackover "A" reservoir is the largest of the three. This reservoir contains volatile oil, and the mechanism for primary production would be solution-gas drive, recovering only 20% of the oil in place. It was determined that a pressure maintenance system would be necessary to recover the maximum amount of hydrocarbons. A water-drive system was chosen as the most favorable means to accomplish pressure maintenance. It is estimated that an additional 20% of the oil in place will be produced by this water-flood program.

LOWERMOST CRETACEOUS RAMP REEFS: KNOWLES LIMESTONE, SOUTHWEST FLANK OF THE EAST TEXAS BASIN

J.M. FINNERAN

Amoco Production Company
P.O. Box 3092
Houston, Texas 77001

R.W. SCOTT

Amoco Production Company
P.O. Box 591
Tulsa, Oklahoma 74102

G.A. TAYLOR

Amoco Production Company
P.O. Box 3092
Houston, Texas 77001

G.H. ANDERSON

Ballard Exploration Company
1706 First City National Bank Building
Houston, Texas 77002

INTRODUCTION

The Late Jurassic-Early Cretaceous of the Gulf Coast is primarily a terrigenous clastic sequence, which has been termed the Cotton Valley Group (Swain, 1944). Clastic deposition of the upper Cotton Valley Group was interrupted by a transgressive marine limestone unit, the Knowles Limestone (Mann and Thomas, 1964).

Along the southwest flank of the East Texas Basin the Knowles Limestone was deposited on a short-lived, carbonate ramp. The objective of this paper is to describe facies distribution on this ramp, concentrati cong on reefal buildups of the outer ramp.

Upper Jurassic and Lowermost Cretaceous reefs of the Gulf Coast have only recently been documented. Their exploration potential, facies patterns and paleocommunities are not well understood. Oxfordian patch reefs developed as a part of the Smackover (Baria *et al.*, 1981), and Berriasian-Valanginain ramp reefs developed in the Knowles Limestone. During Aptian through Cenomanian time, large-scale ramp and rimmedshelf reefs encircled the Gulf of Mexico. Carbonate ramps develop on gently sloping surfaces and evolve into a rimmed-shelf reef as the result of high carbonate production, and sediment starvation of outer-shelf areas (Reed,1982b). The reefs of the Knowles Limestone formed on a ramp which never developed into a rimmed-shelf type due to clastic influx that produced the overlying units.

Stratigraphic control is based on the study of over 100 electric logs from wells in the area which penetrated the Knowles Limestone. The wells from which lithologic data (cores or sample logs) have been studied are spotted on the location map (Fig. 1).

The authors wish to thank Amoco Production Company for permission to publish this paper and Seiscom Delta for release of the

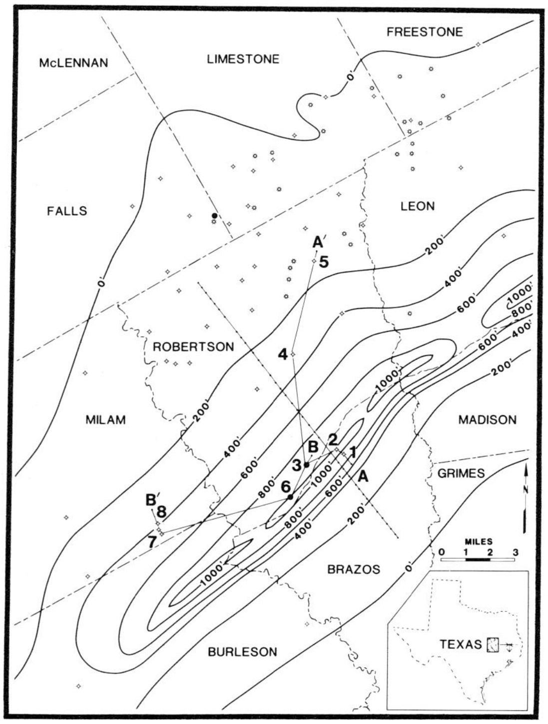

Figure 1. Isopachous map of Knowles Limestone. Numbered wells correspond to wells on cross sections, Figures 3a and 4. Dotted line is approximate location of seismic record section, Figure 3b.

seismic record section. We also gratefully acknowledge the work and support of our colleages in Houston and Tulsa.

STRATIGRAPHY

The Cotton Valley Group is an entirely subsurface stratigraphic unit which underlies much of the upper Gulf Coastal Plain. Numerous authors have published regional stratigraphic frameworks for the Cotton Valley Group. Most notable of these are: Hazzard (1939), Swain (1944), Mann and Thomas (1964), Collins (1980), and Coleman and Coleman (1981). The stratigraphic framework of Mann and Thomas (1964) emphasizes the distinct lithic units of the Cotton Valley Group, and is utilized in this report. Mann and Thomas (1964) recognized four formations of the Cotton Valley Group, which in ascending order are the Bossier Shale, the Terryville Sand, the Hico Shale and the Knowles Limestone and they form a genetically related depositional sequence. The Bossier was deposited as marine mud offshore to the Terryville barrier-sand complex; the Hico lagoonal mud was deposited landward of the barrier island (Thomas and Mann, 1966). The Knowles Limestone is a regionally transgressive limestone. Updip, these formations grade into the non-marine redbeds of the Schuler Formation (Mann and Thomas, 1964).

Incorporating new downdip data from Winn Parish, Louisiana, Coleman and Coleman (1981) have documented that: (1) the Terryville sands pinch out into the Bossier shale, (2) the Knowles Limestone thickens to over 1,200 feet and (3) there is an additional lime-sand couplet (Winn Limestone and Calvin Sand) above the Knowles Limestone and below the Hosston Formation.

The stratigraphic framework developed in north Louisiana is applicable to the southwest flank of the East Texas Basin except that the Hico Shale is not recognized and the Terryville Sand is informally referred to as the "Cotton Valley Sand". Also, a significant limestone unit is clearly recognizable in this area from samples and electric logs. A typical electric log from central Robertson County is illustrated by Figure 2, and demonstrates that the Knowles Limestone is easily identified by the general lack of spontaneous potential (s.p.) and the unusually high resistivity. This electric log response is very similar to the electric log response from the type section of the Knowles Limestone (Mann and Thomas, 1964).

The Knowles Limestone is bounded by two clastic sections; the Cotton Valley Sand/Bossier Shale below and the Travis Peak Formation above (Fig. 3a and 3b). These stratigraphic relationships are similar to the stratigraphic relationships of the type Knowles of North Louisiana (Coleman and Coleman, 1981).

The age of the lower part of the Knowles Limestone is upper Berriasian-Valanginian and the upper part of the Knowles Limestone is Valanginian based on the tintinnids and for-

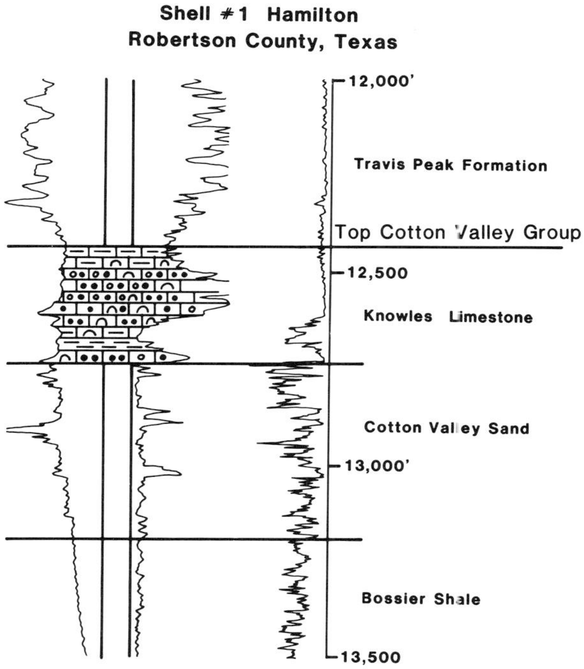

Figure 2. Electric log response of Knowles Limestone from an updip well. Lithology symbols from Figure 6.

aminifers. Documentation for these age determinations and references are provided in the companion paper by Scott in this volume.

Although we can document similar age, electric log response and statigraphic position of the Knowles Limestone of the study area to the Knowles Limestone of north Louisiana, it is difficult to correlate physically over this great distance.

THE KNOWLES LIMESTONE

The Knowles Limestone reaches its maximum thickness at the Robertson-Brazos County line (Fig. 1). To the north, the formation gradually thins to its updip limit, and to the south it also thins toward the downdip limit of well control.

Both the electric log response and the thickness of the Knowles Limestone are variable throughout the study area. A typical electric log from an updip well is illustrated by Figure 2. Farther updip in Limestone County the Knowles Limestone thins and becomes more argillaceous, so that it becomes difficult to identify the formation without sample logs. To the south the Knowles Limestone thickens and its electric log characteristics change.

Near its maximum thickness (Inexco No.1 Seale, Fig. 4 or Fig. 5), the Knowles Limestone can be subdivided into two units based on electric and gamma ray log responses. The upper, massive limestone unit is characterized by s.p. development and consistently high resistivity, while the lower, thin-bedded limestone unit generally lacks s.p. development and the resistivity is very erratic. Electric logs from intermediate wells show that the Knowles Limestone thins first by loss of the lower unit and finally by a character change in the log response of the upper unit.

Based on these electric log and isopachous variations, the Knowles Limestone is divided into inner- and outer-ramp deposits. Inner-ramp deposits have an electric log response similar to that shown in Figure 2 and are restricted to the area between the 400 foot isopachous contour and the updip limit of the Knowles Limestone. Outer-ramp deposits have an electric log response with the characteristics of either the upper, massive limestone unit or the lower, thin-bedded limestone unit, or both. The outer-ramp deposits are restricted to the area between the 400 foot isopachous contour and the southern end of the study area.

LITHOFACIES DESCRIPTION

The inner ramp deposits are divided into two facies based on thin sections of drill cuttings. These are a fossiliferous wackestone and a pelletal packstone facies.

The fossiliferous lime wackestone facies, penetrated by numerous wells in northern Robertson County, consists of sandy fossiliferous wackestone, impure dolomite and dark-gray shale. Common fossil types are benthic foraminifers, gastropods, and abundant oyster-like pelecypods. The matrix is lime mud and clay mud and the impure dolomite

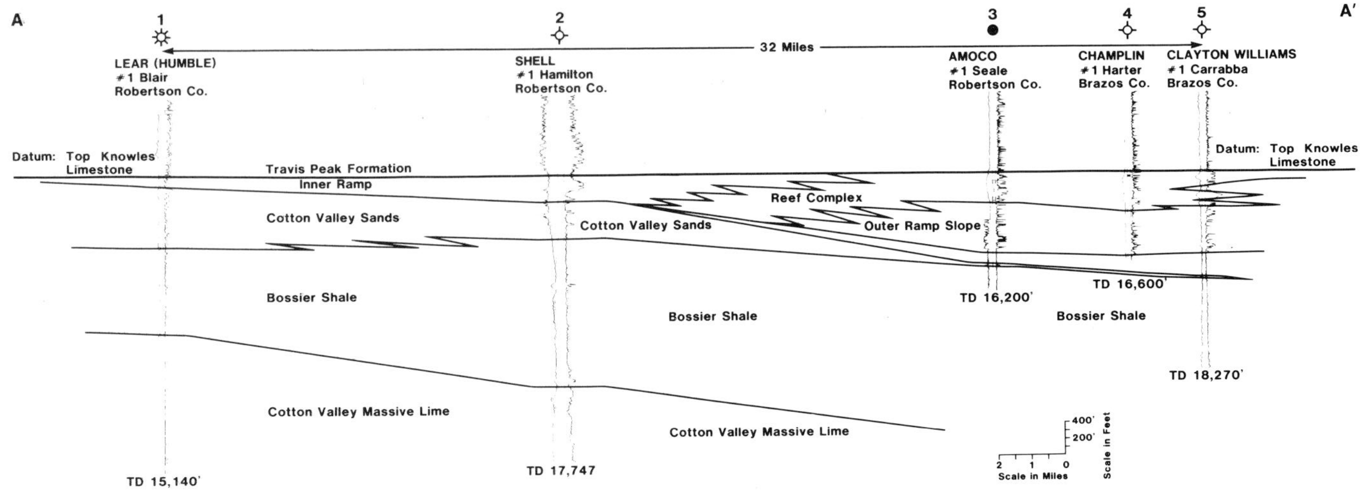

Figure 3a. Stratigraphic cross section (A - A') hung on the top of the Knowles Limestone. For location refer to Figure 1.

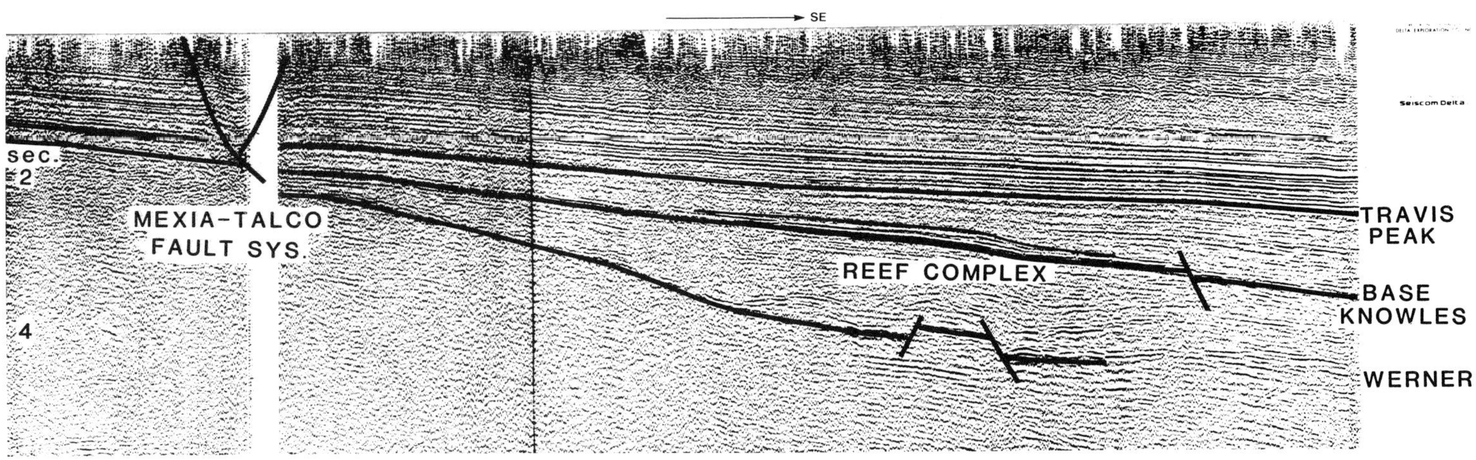

Figure 3b. Seismic record section illustrating reef buildup in the Knowles Limestone. Approximate location is indicated by dotted line on Figure 1.

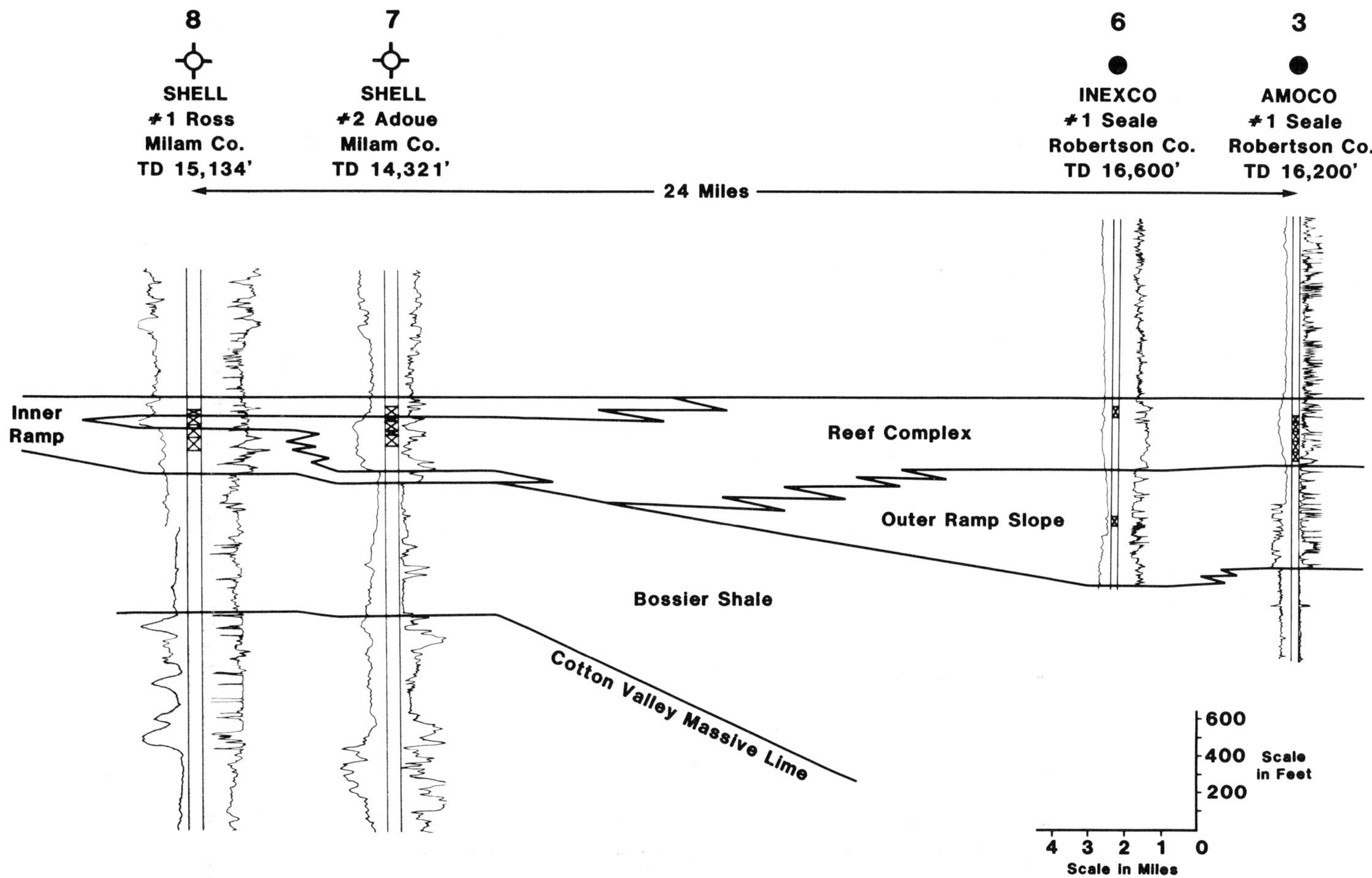

Figure 4. Stratigraphic cross section (B - B') hung on the top of the Knowles Limestone. Core control indicated by X symbol in well bore. For location refer to Figure 1.

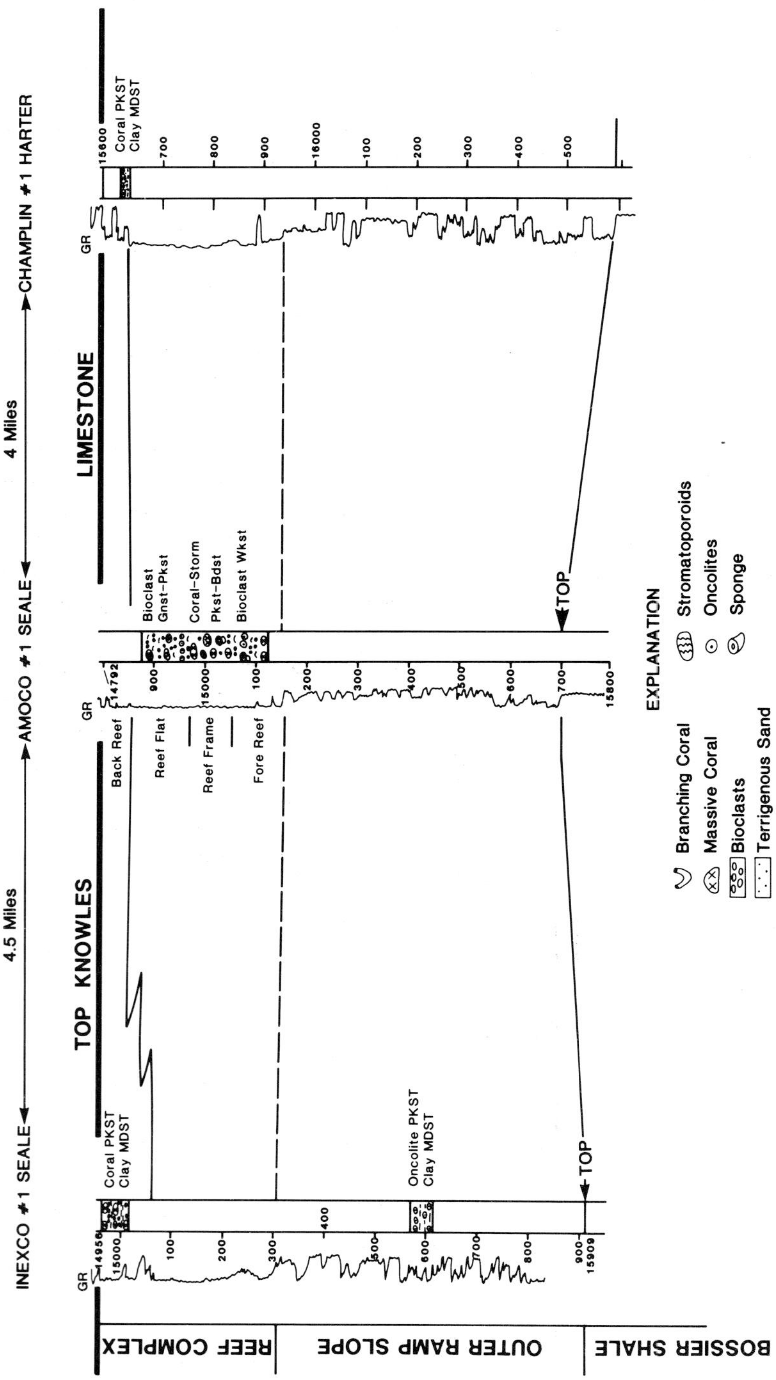

Figure 5. Detailed cross section from cores illustrating outer ramp facies and inferred depositional environment.

consists of subhedral dolomite rhombs in a clay and lime-mud matrix.

The pelletal packstone facies, typified by samples from the Shell No. 1 Hamilton well from 12,500 to 12,600 feet, is moderately well-sorted, pelletal oolite grainstone and packstone. Pellets or foraminifers commonly form the nucleus of the oolites. Ooid coatings are generally thin or surficial, and detrital quartz is common. Bivalves, gastropods, echinoderm plates and spines, dasyclad algae, codiaceae algae *(Cayeuxia)* and benthic foraminifers are common fossil types. The matrix is lime mud and dark-gray shale interbeds are common.

The outer-ramp deposits consist of a lower, thin-bedded limestone and an upper, massive limestone. From core and sample control, the lower unit is determined to be an oncolite-wackestone facies and the upper unit is a group of facies, which is referred to as the reef complex.

The oncolite-wackestone facies, represented by the deeper core (15,557 to 15,599 feet) in the Inexco No. 1 Seale (Fig. 5), is comprised of interbedded packstone and terrigenous mudstone with a matrix of lime mud and clay mud. Larger fossils such as brachiopods, corals and sponges are rare. Local clumps of serpulid tubes supporting large spiney brachiopods, both in life position, are present and pelagic microfossils, tintinnids and calcispheres are common.

The reef complex is divided into four facies. These are, in ascending order; bioclastic wackestones, coral-stromatoporoid boundstone, bioclastic grainstone and bioclastic packstone facies. The first three facies are present in the Amoco No. 1 Seale cores (14,873 to 15,120 feet) (Fig. 5). The bioclastic packstone facies is present in cores from the Inexco No. 1 Seale (14,962 to 15,017 feet) and the Champlin No. 1 Harter (15,615 to 15,630 feet) (Fig. 5).

Common fossils of the bioclastic wackestone facies are sponges, branching corals, echinoderms and oncolites in a lime-mud matrix.

The coral-stromatoporoid boundstone facies contains large colonies of coral (four taxa), branching and encrusting stromatoporoids (three taxa), sponges, and algae (six taxa) which either encrust one another or are encased in a matrix of mud or fine-grained debris. A succession of coral-colony morphology is developed. In the upper part of the facies, large, massive corals are common; below this, laminar corals are thickly encrusted by algae; below this, vase-shaped corals are usually algal coated.

The bioclastic-grainstone facies is composed of moderately to well-sorted skeletal fragments. Large bioclasts of stromatoporoids, bivalves including dicerid rudists, and gastropods are common. *Lithocodium,* a green alga, coats many grains; and dasyclad algae, along with a diverse assemblage of foraminifers associated with occasional beds of oncolites and oolites, are present.

The bioclastic packstone facies in the upper core from the Inexco No. 1 Seale well and the core from the Champlin No. 1 Harter well contains fragments of corals, echinoderms, stromatoporoids and gastropods. Thin-bedded, well-sorted skeletal grainstones are interbedded with dark-gray shale and rippled-laminated sandstone. This facies may represent a gradation from the underlying bioclastic grainstone facies to the overlying terrigenous clastic rocks.

FACIES DISTRIBUTION

Facies tracts on a carbonate ramp follow paleobathymetry (Ahr, 1973). Facies tracts of the Knowles Limestone parallel the trend of isopachous contours suggesting a southward paleoslope. The outer ramp, oncolite wackestone facies pinches-out updip near the 500 foot isopachous contour, and the reef complex progrades south-eastward across depositional stike (Fig. 3a and 4). The reef complex extends no further north than the 400 foot isopachous contour based on core control in the Shell No. 2 Adoue (Fig. 4). At its basinward extent (well No. 1, Fig. 3a), the reef complex interfingers with the oncolite wackestone facies. On a carbonate ramp continuous reef tracts are not usually developed (Ahr,

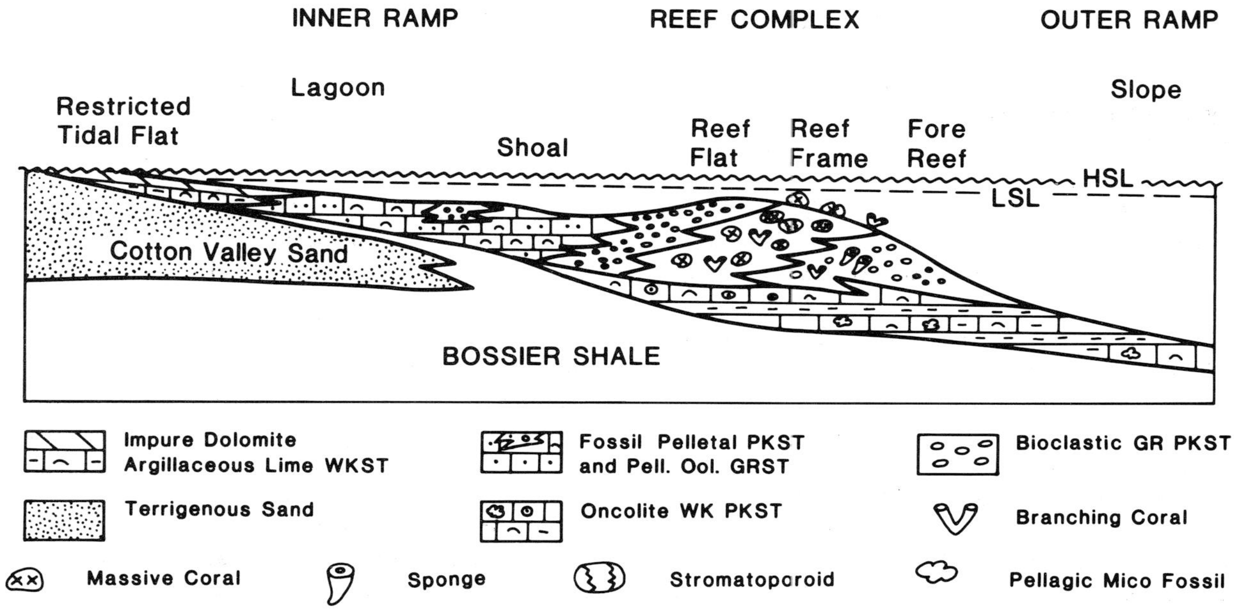

Figure 6. Environmental model of Knowles Limestone illustrating distribution of facies. Figure not to scale.

1973); however, Reed (1982a) has documented ramp buildups in the Ordovician of Virginia, which have a definite trend parallel to the strike of the ramp. The southern limit of the reef complex may be controlled by the large, down-to-the-basin fault scarp to the south or by an underlying basement horst block (Fig. 3b).

The inner-ramp, fossiliferous-lime wackestone facies extends from the updip limit of the Knowles Limestone to about the 200 foot isopachous contour. The pelletal packstone facies of the inner ramp is present between the 100 foot and the 400 foot isopachous contours.

DEPOSITIONAL MODEL

The facies development of the Knowles Limestone was in response to initial flooding of the ramp followed by progradation of facies basinward and finally by progradation of terrigenous clastics over the platform.

The inner-ramp facies are interpreted as having been deposited in a lagoonal environment (Fig. 6). The fossiliferous-lime wackestone facies was deposited in the shallowest part of the lagoon. Clastic influence was continuous, but periodic events deposited discrete shale beds. Oyster-like bivalves and gastropods are abundant. The pelletal packstone facies was deposited seaward from the fossiliferous-lime wackestone facies. This environment supported a moderately well-developed assemblage of bivalves, gastropods, foraminifers, echinoderms and calcareous algae. Locally, oolite shoals may have developed.

The outer-ramp reef complex is depicted as a prograding system (Fig. 6), as suggested by the shoaling-upward sequence of facies (Fig. 5) and by its regional distribution. The coral-stromatoporoid boundstone facies is interpreted as the reef frame. Vase-shaped coral, stromatoporoids, encrusting algae and sponges were pioneer communities of the reef frame. As the reef grew upwards towards wave base, the coral colony morphology evolved to laminar corals and finally to large massive corals, which is a succession similar to coral zones in modern reefs. Algae and stromatoporoids encrusted and bound the organisms and matrix into a wave resistant structure.

The bioclastic wackestone facies was deposited in a fore-reef environment, where bioclastic debris and pioneer reef communities were deposited below normal wave base. The bioclastic grainstone facies was deposited within wave base directly shoreward of the reef and is an accumulation of debris from the reef. The bioclastic packstone facies was deposited in a similar environment with the addition of terrigenous clastics, which signaled the end of reef development.

The oncolite wackestone facies was deposited below wave base, but within the photic zone on the gentle slope of the outer ramp basinward from the reef.

CONCLUSIONS

Earliest Cretaceous ramp reefs of the Knowles Limestone developed along the southwest flank of the East Texas Embayment. Our data indicate that the lower, outer-ramp slope deposits of the Knowles Limestone wedgeout updip on the carbonate ramp and the upper reef complex progrades seaward. Seismic data and regional isopachous maps indicate a westsouthwest to east-northeast trend to the reef complex. The continuity of the reef complex is uncertain and must be defined by further drilling.

REFERENCES

Ahr, W.M., 1973, The Carbonate Ramp: an alternative to the shelf model. Trans. Gulf Coast Association of Geol. Soc., 23rd Annu. Conv., pp. 221-225.

Baria, L.R., Stoudt, D.L., Harris, P.M., and Crevello, P.D., 1981, Late Jurassic Reefs of the Smackover Formation - a preliminary report. Trans. Gulf Coast Associaton of Geol. Soc., 31st Annu. Conv., p. 18.

Coleman, J.L., and Coleman, C.J., 1981, Stratigraphic, sedimentologic and diagenetic framework for the Jurassic Cotton Valley Terryville Massive Sandstone complex, north Louisiana. Trans. Gulf Coast Association of Geol. Soc., 31st Annu. Conv., pp. 71-79.

Collins, S.E., 1980, Jurassic Cotton Valley and Smackover Reservoir tends, east Texas, north Louisiana and South Arkansas. AAPG Bull., V.64, No. 7. pp. 1004-1013.

Hazzard, R.T., 1939, Notes on the Comanche and pre-Comanche? Mesozoic Formations of the Ark-La-Tex area. Shreveport Geol. Soc. 14th Annu. Field Trip Guidebook, pp. 155-178.

Mann, D.J., and Thomas, W.A., 1964, Cotton Valley Group (Jurassic) nomenclature Louisiana and Arkansas. Trans. Gulf Coast Association of Geol. Soc., 14th Annu. Conv., pp. 143-152.

Read, J.F., 1982a, Geometry, facies and development of middle Ordovician carbonate buildups, Virginia Appalachians. AAPG Bull., V.66, No. 2, pp. 189-210.

______________, 1982b, Carbonate platforms of passive (extensional) continental margins: types, characteristics and evolution. In: A. L. Hales (editor), Geodynamics Final Symposium. Tectonophysics, 81: 195-212.

Swain, F.M., 1944, Stratigraphy of Cotton Valley beds of northern Gulf Coastal Plain. AAPG Bull. V.28, No. 5, pp. 577-614.

Thomas, W.A., and Mann, C.J., 1966, Late Jurassic depositional environments, Louisiana and Arkansas. AAPG Bull. V.50, No. 1, pp. 178-182.

SMACKOVER DEPOSITION AND DIAGENESIS AND STRUCTURAL HISTORY OF THE BRYAN'S MILL AREA, CASS AND BOWIE COUNTIES, TEXAS

GILL HARWOOD

Department of Geology
University of Newcastle-Upon-Tyne
Newcastle-Upon-Tyne NE17RU
United Kingdom

CHRIS FONTANA

Phillips Petroleum Company
3743 Pitkin Circle
Aurora, Colorado 80013

ABSTRACT

Production from the Bryan's Mill, Frost, and Carbondale Fields in the Bryan's Mill area is from Jurassic Upper Smackover dolomitized ooid grainstones. Within the Upper Smackover, three coarsening-upwards sequences culminating in ooid grainstones extend across the area. Shortly after deposition, leaching by meteoric fluids resulted in development of oomoldic porosity in many of the ooid grainstones; other carbonate facies were little affected. During early diagenesis, dolomitization enhanced permeability by preserving existing porosities and by generating an effective intercrystalline porosity. After dolomitization, brittle compaction during burial further increased permeability by interconnecting oomolds.

The center of the Bryan's Mill area was a positive feature during late Smackover deposition. Faulting during Buckner deposition resulted in a series of southeasterly tilted fault blocks, perhaps due to regeneration of basement structures. Later, post-Hosston faulting had an east-west trend and was possibly associated with doming of the sediments over two highs. There is little evidence of Jurassic or Early Cretaceous movement on the east-west trending faults.

Hydrocarbon production in the area depends on a combination of structural, stratigraphic, and diagenetic factors. Although the grainstones are continuous across the northernmost structural high, production is limited to the dolomitized areas. Thus, diagenesis, and in particular dolomitization, is the ultimate control on production in the area.

INTRODUCTION AND STRATIGRAPHIC FRAMEWORK

The Bryan's Mill area is located in northern Cass and southern Bowie Counties, East Texas (Fig. 1). The area includes the Carbondale, Frost, and Bryan's Mill Fields. Since the discovery of hydrocarbons in the early 1960's, some 95 wells, which penetrated the Upper Jurassic Smackover Formation, have been drilled in the area. Production is gas and liquid hydrocarbons from dolomitized ooid grainstones. This research was based on the study of 40 electric logs and twelve Smackover continuous cores of which two, the Shell No.2-B MacDonnell and the Shell No. 1 W. H. Morgan, are used as examples. This paper briefly reviews the sedimentology, diagenesis, and structural setting of the area.

Jurassic stratigraphic nomenclature (Fig. 2) used in the area is modifed from

Dickinson (1968). The Smackover carbonates are overlain by the Buckner Members of the Haynesville Formation (Fig. 3). No long sections of the Buckner have been cored in the area although small intervals of Buckner sediments were cored above the Smackover in both the Shell No. 1 W.H. Morgan and the Shell No. 2-B MacDonnell wells (Fig. 4). The basal Buckner sediments comprise anhydrite interbedded with thin dolomite mudstones and some red clays. The anhydrite shows displacive textures (*i.e.*, it formed while the surrounding sediments were still soft). Anhydrite dominates the lower 100 to 200 feet of the Buckner Members (Figs. 2 and 3). The overlying salt is recognized from electric log characteristics and sample logs from wells elsewhere in East Texas. The salt may not be pure halite; more likely it is a mixture of halite and anhydrite or of halite, anhydrite, and red mudstones. Evaporites are less common in the overlying section, which is dominated by fine clastic sequences interbedded with several thin (20 feet or less) carbonate beds (Fig. 2). These carbonates provide good

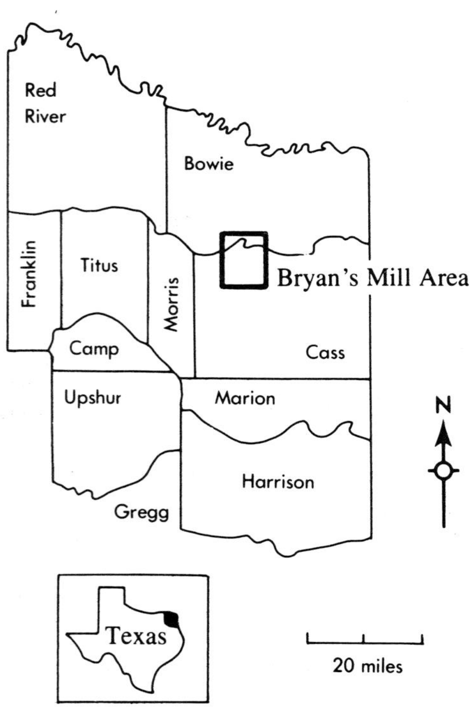

Figure 1. Location of the Bryan's Mill area in East Texas.

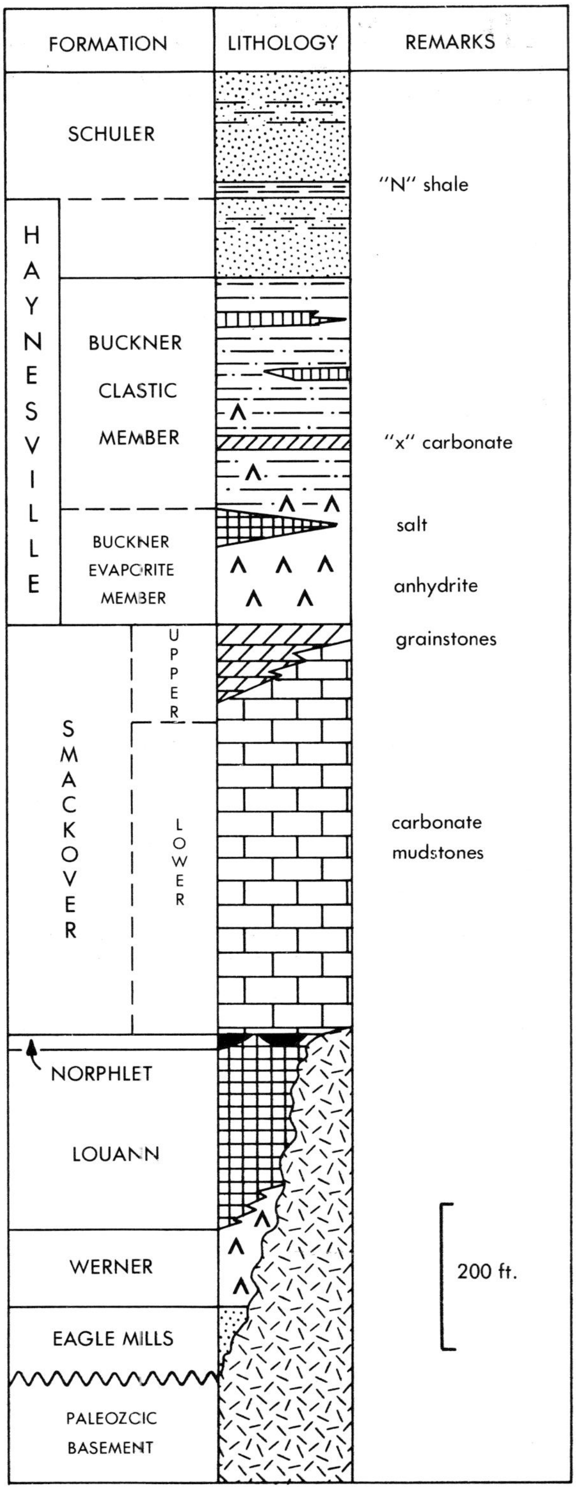

Figure 2. Generalized stratigraphic profile through the Jurassic and older sediments in the Bryan's Mill area.

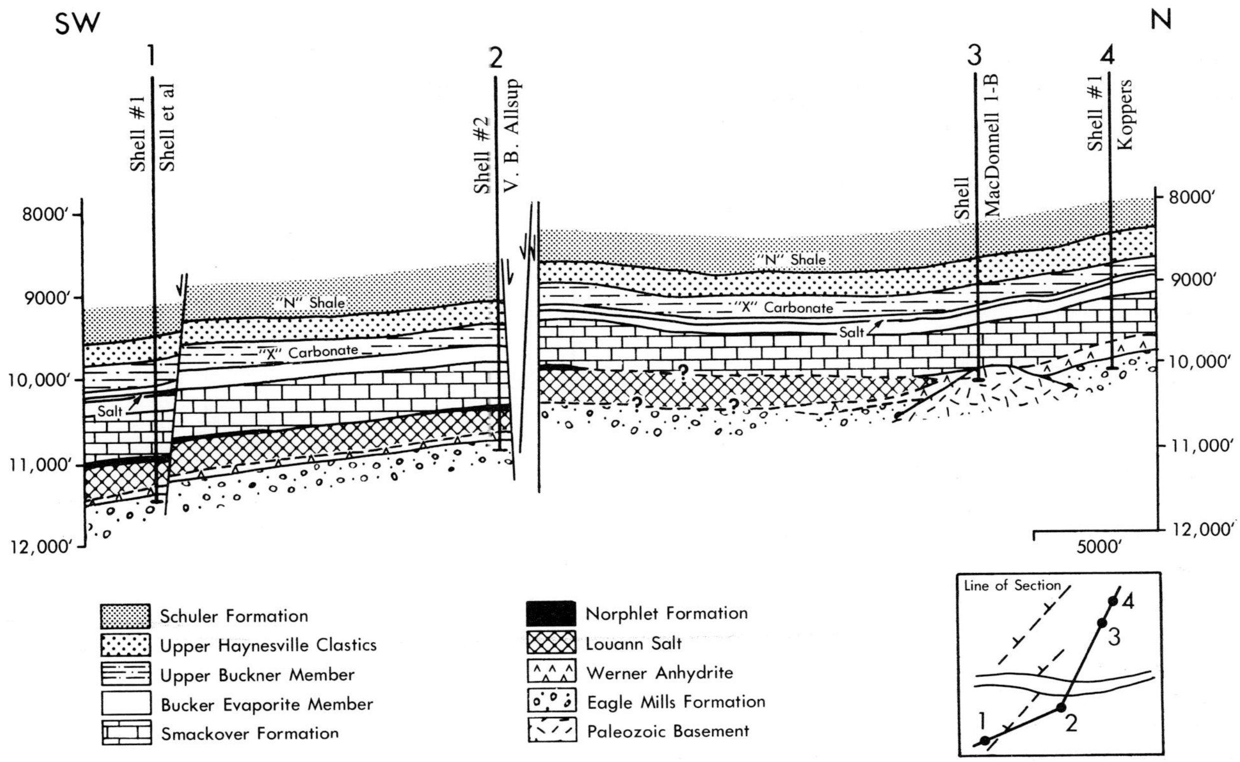

Figure 3. Cross section through the pre-Cretaceous sediments from southwest to north across the Bryan's Mill area.

marker horizons; the "X" carbonate, for example, can be correlated across the Bryan's Mill area (Fig. 3).

The Buckner is overlain by a major clastic sequence which includes the upper part of the Haynesville and Schuler Formations. The top of the Buckner Formation in the area was formerly correlated with the base of the Schuler Formation. However, more regional correlations show that a persistent shale, the "N" shale (Figs. 2 and 3), can be correlated laterally with the Bossier Formation (K. A. McGillis, pers. comm.), although the "N" shale apparently does not have typical Bossier lithology within the Bryan's Mill area. Thus, the clastic sequence below the "N" shale is equated with the clastic upper member of the Haynesville Formation (Moore, this volume).

Palezoic basement rocks were penetrated directly under the Smackover in the Shell No. 1-B MacDonnell well (Fig. 3); they have not been encountered elsewhere in the area, although their presence is suggested by a positive gravity anomaly. The oldest sedimentary rocks present, the Eagle Mills Formation, were found south (Shell No. 1 Shell *et.al.*) and north (Shell No. 1 Koppers) of the area. The Werner anhydrite is present beneath the Louann Salt in the Bryan's Mill Field, although to the north, in the Carbondale Field, it becomes the updip equivalent of the Louann (Fig. 3). The Norphlet sandstones are thin (less than 40 feet) and discontinuous throughout the area (Figs. 2 and 3).

SMACKOVER CARBONATES

All the cores examined in this study were from near the top of the Smackover Formation and no core exceeded 90 feet in length; thus detailed conclusions can only be made for the upper Smackover carbonates.

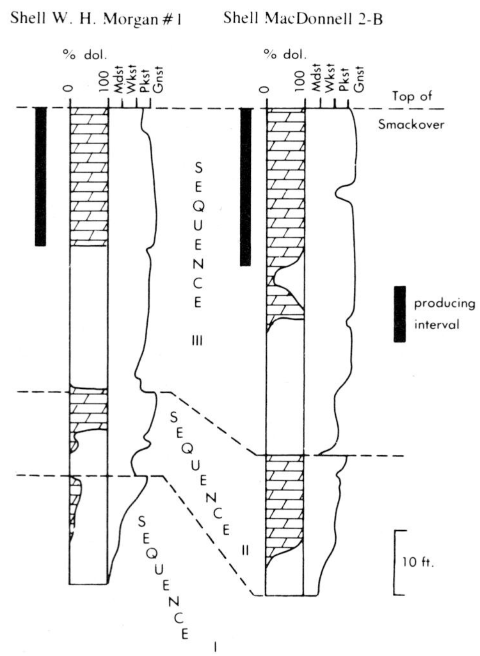

Figure 4. Carbonate facies and percentage of dolomitization in Upper Smackover cores from the Shell No. 1 W. H. Morgan and Shell No. 2-B MacDonnell wells. Detailed locations of these wells are shown on Figure 8.

Three major coarsening-upwards sequences (Fig. 4) are known to extend across the area but are not present in each core. The sequences show gradation upwards from carbonate mudstones through pellet wackestones, pellet-*Favreina* pellet packstones and pellet-*Favreina* pellet grainstones, and ooid-*Favreina* pellet grainstones to ooid grainstones. Intraclast and rhodolite horizons within grainstones and packstones cannot be correlated between wells. In most cores, true mudstones are only present in the lower sequence (sequence I, Fig. 4); electric log characteristics suggest that much of the underlying Lower Smackover is also composed of mudstones. Thick grainstones (up to 100 feet thick) are only present in the upper sequence; grainstones at the top of the two lower sequences are thinner, rarely reaching 40 feet in thickness (Fig. 4). A gradation from grainstones to packstones occurs near the top of the upper sequence in two wells (Shell No. @-A MacDonnell and Shell No.1 W. H. Morgan). Details of constituent carbonate grains are discussed by Fontana (1983).

Sedimentary structures within the grainstones comprise parallel and cross laminations together with bioturbation (Fig. 5a and b). Grainstone sedimentary structures are best exhibited where grains are poorly or moderately sorted; sedimentary structures are not obvious where the ooid grainstones are extremely well sorted (Fig. 5c, d and e). The contact between grainstones and packstones in most sequences is gradational, perhaps due in part to homogenization by bioturbation. Packstones rarely exhibit parallel or cross laminations; bioturbation, however, is common. In many sections, the packstones are intimately interbedded with grainstones on a fine (inches) scale (Fig. 5c). The rare intraclasts within the packstones are derived from grainstones. Packstones form the base of the upper sequence (sequence III, Fig. 4) in cores from the center of the Bryan's Mill area, whereas away from the central region (for example in the Shell No. 1 Koppers and

Figure 5. Photographs of core from the Shell No. 2-B MacDonnell well. Bars represent 1 cm.

a. Cross-laminated dolomitized ooid grainstone, top of sequence II, 9,790'.

b. Interbedded parallel laminations of grainstones (lighter colored laminae) and packstones, sequence I, 9,779'.

c. Well-sorted dolomitized ooid grainstone. Sedimentary structures are only apparent where zone of enhanced porosity (arrowed) has followed cross laminations, sequence I, 9,758'.

d. Lack of sedimentary structure in well-sorted dolomitized ooid grainstone with intergranular porosity, sequence I, 9,758'.

e. Well-sorted dolomitized ooid grainstone with oomoldic porosity, sequence I, 9,765'. Note the lack of sedimentary structures.

f. Wispy lamination and isolated burrows (arrows) in wackestones, sequence II, 9,807'.

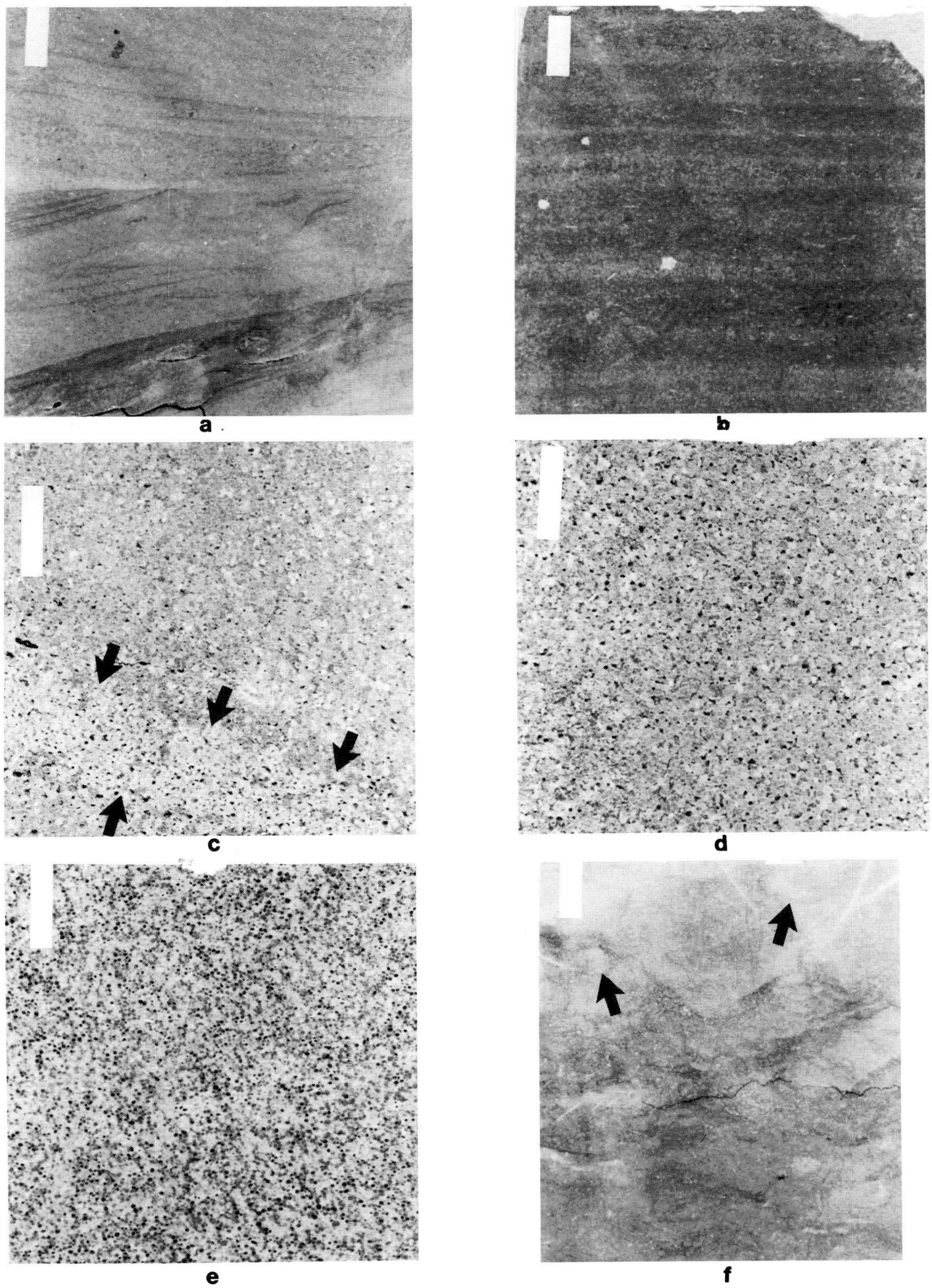

Figure 5

Shell No.1 MacDonnell cores) sequence III shows an uninterrupted gradation from mudstones and wackestones through packstones to grainstones stones. Intervals of wackestones and pack- stones within the grainstones represent a periodic decrease in energy during deposition.

The wackestones and mudstones are disrupted by bioturbation in many sections of the core; other sections show wispy laminations resulting from a slight winnowing of the sediments by low-energy bottom currents and/or from surface grazing organisms (Fig. 5f). Fossils are rare throughout all of the facies; however, gastropods, echinoid fragments and spines, and some pelecypods occur within the grainstones but are less abundant in the lower energy, more mud-rich facies. Bioturbation was most likely caused by crustaceans and soft-bodied fauna.

Each individual coarsening-upwards sequence represents a progressive increase in wave or current energy. Sedimentary structures within the grainstones and the abundance of ooids indicate continuous agitation in shallow-water marine tidal bars, tidal deltas, and mobile sand shoals. In Arkansas and Louisiana, the upper Smackover ooid grainstones have been compared with the modern ooid sandshoal blanket of the inner Bahamian Platform described by Ball (1967) (Moore and Druckman, 1981). This analogy is also appropriate for the Smackover ooid grainstones of north-east Texas. Regional facies studies show that these grainstones were desposited on the gentle northern ramp of the East Texas Salt Basin. These studies help to clarify this depositional model.

Although the Smackover blanket ooid sands continue across most of northeast Texas, within the Bryan's Mill area some variations in depositional environment can be recognized. The coarsening-upwards sequences indicate three periods of shallowing when sedimentation exceeded wave base across the area, terminating in ooid shoals. Continued subsidence and possible lateral migration of the locus of sedimentation out of the area led to deeper water and finer-grained sedimentary depositional environments at the commencement of the next sequence. The packstone intervals at the top of the Smackover (Shell No.1 W. H. Morgan, sequence III, Fig. 4) are thought to have formed in the lee of the shoal crests.

In the upper sequence (sequence III, Fig. 4), there are no mudstones or wackestones in the center of the Bryan's Mill area. This suggests that depositional energy remained high there, probably in response to a positive feature on the Smackover seafloor. This positive feature remained as a bottom high, promoting high-energy environments and ooid generation. This feature must, however, have had a small amplitude as there is no noticeable thinning of either sequence II or III across the center of the area, and no sequences have been found to onlap the high.

DIAGENESIS AND POROSITY-PERMEABILITY RELATIONSHIPS

Only a brief summary of diagenesis as it relates to porosity and permeability is discussed in this paper; a more detailed analysis of diagenesis in the Bryan's Mill area will be presented elsewhere. Two major diagenetic events are relevant here - the formation of oomoldic porosity and dolomitization.

During and immediately after deposition of the Smackover ooid sands, the ooids were coated with an isopachous calcite circumgranular cement (Fig. 6a). Before the aragonite ooids were converted to calcite, the area was subjected to flushing by meteroic fluids. These fluids selectively leached aragonite from the ooids, resulting in a well-developed oomoldic porosity throughout many grainstone sections (Fig. 6b). Calcite was also precipitated, and occluded some of the remaining intergranular porosity. Meteoric leaching rarely penetrated the packstone intervals and did not noticeably affect the mudstones or wackestones.

Dolomitization, the next major diagenetic event, followed early in the diagenetic history. Dolomitization was facies selective, being more efficient in the grainstones than the other carbonate facies (Fig. 4).

Dolomitization did not, however, only take place in grainstones with pre-existing oomoldic porosity (Fig. 5d and e), and not all oomoldic grainstones were dolomitized. Therefore, it is concluded that dolomitization was a separate event from the formation of oomoldic porosity and that it took place under a different hydrologic system, possibly linked to compaction of the overlying Buckner evaporites and faulting during Buckner sedimentation.

Although meteoric leaching produced limestones with porosities as high as 18 percent (core analysis), permeability in the limestones is generally low as the oomolds are not interconnected. A porosity-permeability plot for limestone and dolomite grainstones (Fig. 7) demonstrates that effective permeability is only present in the dolomites. This difference in permeabilities is because dolomitization not only preserved the existing oomoldic and intergranular porosities in the grainstones (Fig. 6c), but it also produced an effective intercrystalline porosity (Fig. 6c) which generates the high permeability values.

Burial compaction after dolomitization also increases the permeability of some oomoldic grainstones, while at the same time it decreases the porosity. This compaction is brittle, with fracture and collapse of the oomolds (Fig. 6d) and results in a chain of interconnected oomolds (Fig. 6e) with permeability along the fracture. One extreme display of this texture is recognized in Bryan's Mill Field where the dolomitized grainstone now only consists of broken shards of intergranular cement (Fig. 6f) with porosity and permeability around the individual fractured shards.

Hydrocarbon production in the Bryan's Mill area is primarily from the dolomitized grainstone at the top of sequence III (Fig. 4). There is no production where this grainstone has not been dolomitized. The distribution of dolomite is, therefore, one of the key factors in determining the amount and precise location of hydrocarbon production in the Bryan's Mill area, although the overall location of the fields is due to the structural development of the region.

STRUCTURE OF THE BRYAN'S MILL AREA

Structural contours of the top of the Smackover Formation show a major structural high in the center of the area, bisected by an east-west graben (Fig. 8). A smaller structural high is present in the northern part of the region. Production is concentrated on the southern high (Bryan's Mill and Frost Fields), whereas production is only from the flanks of the northern high as dolomitization is restricted to these flanks (Carbondale Fields). Both structures are coincident with the site of higher energy deposition (grainstone and packstones) in the upper coarsening-upwards sequence (Fig. 4); this area was a positive feature during late Smackover deposition.

The existence of structural highs immediately after Smackover deposition is demonstrated by the thinning of Buckner sediments over the positive features (Fig. 9). Thus, later movements appear to have amplified the Smackover paleohighs, which may, in turn, reflect basement structure (Collins, 1980). Doming of the overlying sediments over these highs produced a classic structural trap for hydrocarbons.

Two sets of faults cross the area. The older, northeast-southwest trending faults are cut by a major east-west graben with overall southerly throw (Fig. 8). The northeast-southwest faults can be demonstrated to have been active during Buckner time. The thickness of Buckner sediments (Fig. 9) was affected by these faults. Salt within the Buckner is thicker on the downthrown sides of these faults, whereas little or no salt is present on top of the structural highs on the upthrown sides of the faults (Fig. 9). We conclude that movement on these faults during Buckner time resulted in a tilting of the interfault blocks to the southeast. This tilting may have resulted from reactivation of basement structure as this trend transacts the later East Texas structural trends shown by Collins (1980) and Wood and Guevera (1981). Isopachs of Buckner sediments above the "X" carbonate (Fig. 2) shows no thickening on the downthrown side of the northeast-southwest trending faults. Therefore, we conclude that most of the movement on these faults was prior to the deposition of the "X" carbonate and that the lack of salt over the structural highs was possibly due to syn-Buckner erosion after faulting. Little detectable movement has taken place on these faults since Buckner time, although detailed correlation using electric logs in the overlying Schuler and Hosston Formations is difficult because of the rapid facies changes across the area. The development within the Smackover of a separate hydrologic system, in contrast to that which formed the oomoldic porosity, may well have occurred during this interval of fault movement and possible later erosion; thus dolomitization may have taken place during Bucker sedimentation.

The faults that form the east-west graben (Fig. 8) have been considered to be the eastern extremity of the Mexia-Talco fault system which borders the East Texas Salt Basin (Wood and Guevera, 1981). In the Bryan's Mill area, an overall downthrow to the south of nearly 600 feet is produced by these faults; the fault at the southern side of the graben is probably an antithetic fault to the major basin-rim fault. There is little evidence of early movement along this fault trend in the Bryan's Mill area, although the trend has been considered by many workers to be a reactivated basement structure. In the upper Smackover, the three coarsening-upwards sequences can be traced both north and south of the fault graben and contain no sedimentologic evidence of fault movement during their deposition. Both the total Buckner and Buckner salt are continuous across the graben (Fig. 9); therefore, syn-Buckner movement on the east-west faults is unlikely. Similarly, the Schuler and Hosston Formations cross the graben with no indication of fault movement

Figure 6. Photomicrographs of upper Smackover carbonates from the Bryan's Mill area.

a. Ooid grainstone; Shell No. 2 Koppers well, 9,601'. Micritized ooids show retention of concentric laminae; in other ooids micrite aggraded to microspar and destroyed original fabric. Circumgranular cement growth continued in intergranular areas, occluding porosity (arrowed). Resultant limestone has low porosity and permeability.

b. Oomoldic porosity in dolomitized ooid grainstone; Shell No. 2-C MacDonnell well, 9,862'. Oomolds contain little dolomite and are not interconnected. Dolomitization postdated formation of oomoldic porosity and created an effective intercrystalline permeability with retention of oomoldic porosity.

c. Oomoldic and intergranular porosity in dolomitized ooid grainstone; Shell No. 1 MacDonnell well, 9,678'. Dolomitization preserved isopachous circumgranular cement. Oomoldic porosity not fully developed, resulting in partial dolomite, and hydrocarbon (black), fill, with geopetal nature in places. Good intergranular porosity (arrowed). Combined with intercrystalline permeability, this fabric results in excellent porosities and permeabilities.

d. Brittle compaction of oomoldic grainstone; Shell No. 2-C MacDonnell well, 9,670'. Collapse of intergranular cement with subsequent interconnection of oomolds (arrowed). Intergranular cement is calcite with low permeability. Although oomoldic porosity was *reduced* by compaction, permeability was enhanced by fracturing. Isolated dolomite rhombs present on margins of oomolds post date oomolds' formation.

e. Brittle compaction of ooid grainstone; Shell No. 2-C MacDonnell well, 9,670'. Collapse of intergranular cement formed chain of interconnecting oomolds and caused reduction of oomoldic porosity but enhancement of permeability.

f. Brittle compaction of dolomitized ooid grainstone, Shell Elna Teague well, 10,529'. Broken shards of intergranular cement (arrowed) in matrix of dolomite rhombs originally within oomolds. Black is hydrocarbon stain. Excellent permeability although actual porosity is lower than 6b-6e.

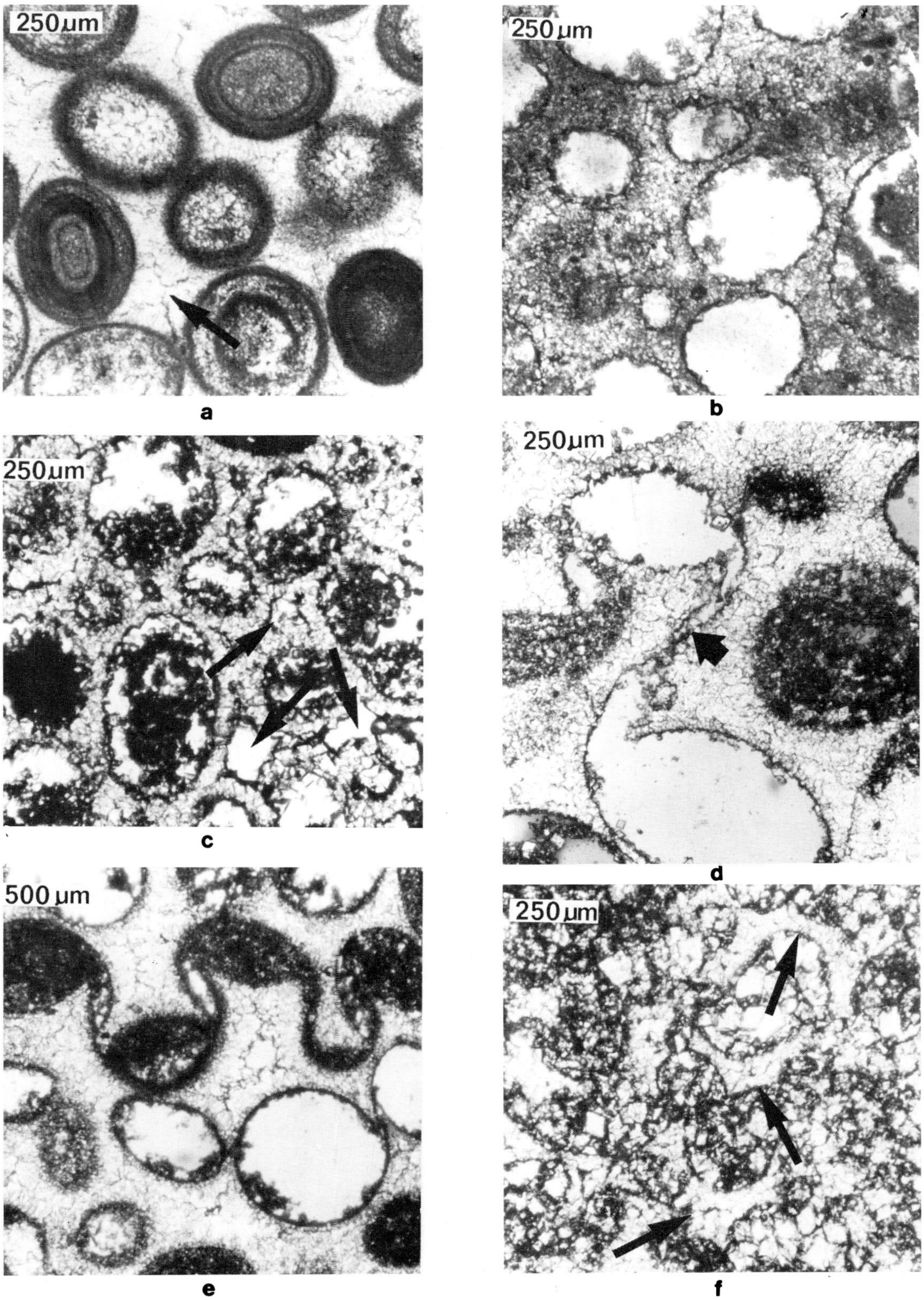

Figure 6

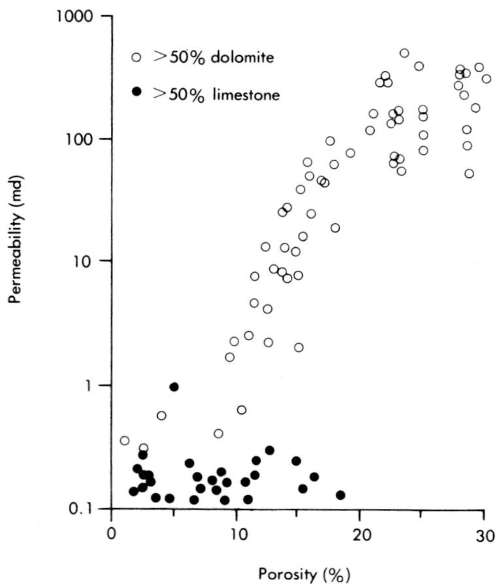

Figure 7. Porosity-permeability plot for upper Smackover limestones (from Shell No. 1 Koppers and No. 2 Koppers cores) and dolomites (Shell No. 1 W. H. Morgan and Shell No. 1 Allsup cores). Data for plots was determined by Core Lab from core plugs.

during their deposition. Major movement on the east-west fault system displaces the northeast-southwest trending faults and thus is younger than lower Buckner deposition. As we can detect no evidence of early movement on the east-west trend in the sedimentologic record, we conclude that graben development and the associated southwards downthrow were post-Hosston events.

There is one piece of evidence against this hypothesis. The electric log of the Phillips Frost No. 1 well which drilled through the graben, shows an expanded Buckner section but with faulted-out intervals in both the Schuler and Hosston Formations. The top of the Buckner in this well is structurally higher than the top of the Buckner on either side of the graben, whereas the top of the Smackover is at 9,730 feet. In the State Line Trend of the Louisiana-Arkansas border, a thickened Buckner sequence within the graben indicates syn-Buckner faulting (G. Herman, pers. comm.). There, however, the top of the Buckner is at approximately the same structural level as on the north side of the graben. In the Bryan's Mill area, the elevated top of the Buckner represents a slight compressional movement associated with faultslice movement within the graben, although the overall movement is extensional. Minor compressional movement could, however, explain both the elevated top of the Buckner and the anomalous Buckner thickness, by repetition of the lowermost Buckner sediments (also suggested from electric logs).

The east-west faults form a barrier to hydrocarbon migration because the Bryan's Mill Field, which is a gas field, is structurally lower than the Frost Field, which produces liquid hydrocarbons (Fig. 8). Liquid-hydrocarbon migration into the higher structural trap (Fig. 8) must, therefore, have preceded both east-west fault movement and gas migration (C.H. Moore, pers. comm.).

SUMMARY AND CONCLUSIONS

Within the Bryan's Mill area hydrocarbon production is related to carbonate facies, carbonate diagenesis, and structural development of the area. The interpreted sequence of events follows:

1. During the deposition of the upper Smackover, three coarsening-upwards sequences terminated in the deposition of ooid grainstones over most of the area. Grainstones were thickest in the upper sequence.
2. Most of the high-energy sediments were deposited over the center of the area, at least during the upper sequence, indicating the presence of a low-amplitude positive feature on the sea floor.
3. Shortly after deposition of the high-energy sediments, meteoric water leached the aragonite of the ooids, creating oomoldic porosity within many of the grainstones.
4. Dolomitization during early diagenesis preserved existing grainstone porosities and also generated an effective, intercrystalline porosity and high permeabilities.

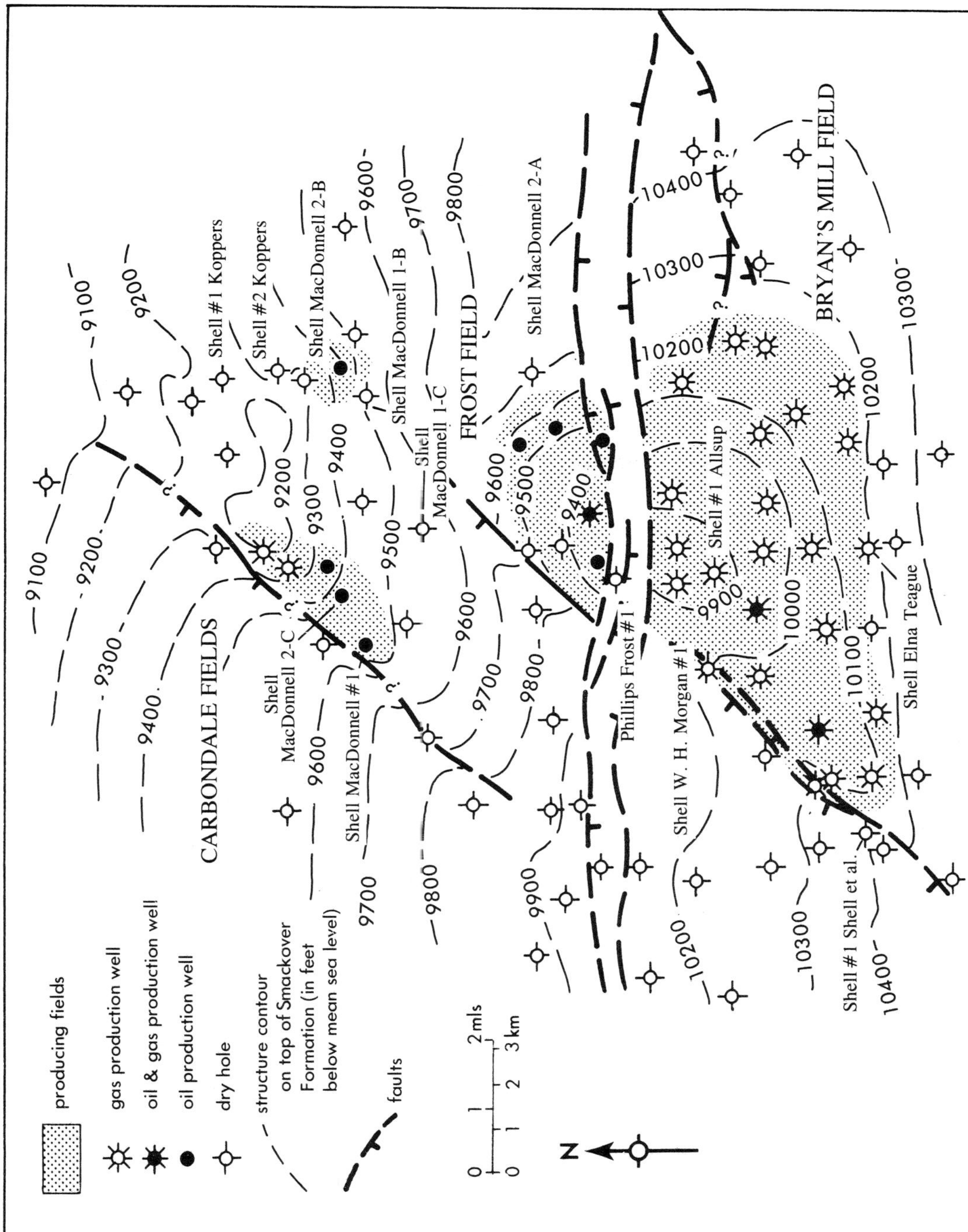

Figure 8. Structure map of the top of the Smackover Formation, faulting, and production in the Bryan's Mill area.

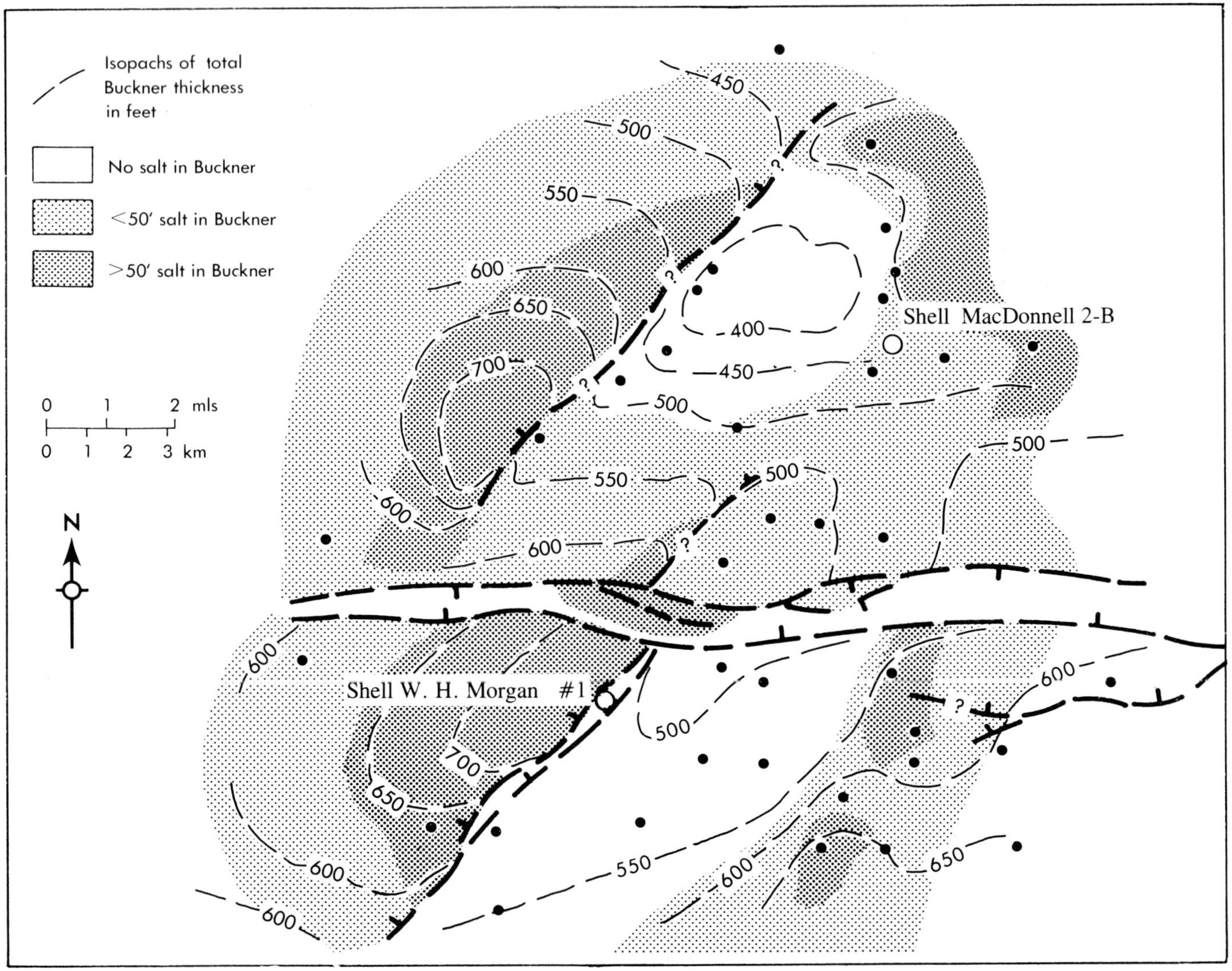

Figure 9. Map of the total thickness of Buckner sediments. Also shown are approximate thickness and lateral extent of salt near the top of the Buckner evaporite member. Salt thickness is interpreted from electric log (resistivity and S.P.) characteristics.

5. Dolomitization took place in a separate hydrologic regime to that which generated the oomoldic porosity. This may have been related to early diagenesis of the Buckner evaporites and to faulting during Buckner sedimentation.
6. Faulting and southeast tilting of fault blocks took place during Buckner deposition, with possible erosion of the uptilted structural highs.
7. During deeper burial of the Smackover carbonates, brittle compaction further enhanced permeability in the dolomites.
8. Doming of the sediments over the two structural highs and faulting along an east-west trend were post-Hosston in age. There is little evidence of early initiation of the major fault graben.

Hydrocarbon production in the Bryan's Mill area is, therefore, controlled by a complex integration of many factors. The majority of production is from the large structural dome where dolomitization pervades the grainstones of the upper sequence. In the smaller, northern structure, dolomitization is restricted to the flanks and it is only from these areas that production is viable. Therefore, although structure and facies are important factors, the distribution of dolomitization ultimately controls production in the Bryan's Mill area. Petrographic studies of the Smackover carbonates and associated geochemical investigations now in progress will contribute to determining some of the factors that control dolomite distribution in the Bryan's Mill area and surrounding portions of East Texas.

ACKNOWLEDGMENTS

We would like to thank George Herman of Shell Oil Company, New Orleans, for making the cores available for study and contributing to the discussion of our interpretation of the area. Clyde H. Moore helped with core and petrological studies and constructively criticized the preliminary draft of this manuscript. We also acknowledge the help and criticism of other colleagues at the Applied Carbonate Research Program (A.C.R.P.) at L.S.U., in particular Barbara Delaville for typing the manuscript and for doing all the additional jobs she did not realize would be part of her job with A.C.R.P.

REFERENCES

Ball, M.M., 1967, Carbonate sand bodies of Florida and the Bahamas: Jour. Sed. Petrology, v. 43, p. 965-984.

Collins, S.E., 1980, Jurassic Cotton Valley and Smackover Reservoir Trends, East Texas, North Louisiana and South Arkansas: A.A.P.G. Bull., v. 64, p. 1004-10 3.

Dickinson, K.A., 1968, Upper Jurassic stratigraphy of some adjacent parts of Texas, Louisiana and Arkansas: U.S. Geol. Survey Prof. Paper, 594E, 25 p.

Fontana, C., 1983, Deposition and diagenesis of Upper Smackover Carbonates, Bryan's Mill Field, East Texas: unpubl. Masters Thesis, A.C.R.P., Louisiana State University, Baton Rouge, Louisiana

Moore, C.H., and Y. Druckman, 1981, Burial diagenesis and porosity evolution, Upper Jurassic Smackover, Arkansas and Louisiana: A.A.P.G. Bull., v. 65, p. 597-628.

__________, (this volume), Regional patterns of diagenesis, porosity evolution and hydrocarbon production, Upper Smackover of the Gulf Rim.

Wood, D. H., and E. H. Guevera, 1981, Regional structural cross sections and general stratigraphy, East Texas Basin: Bureau of Economic Geology, The University of Texas at Austin, 21 p.

UPPER JURASSIC AMMONITES FROM THE SUBSURFACE OF TEXAS, LOUISIANA, AND MISSISSIPPI

RALPH W. IMLAY

U.S. Geological Survey
E-501 U.S. National Museum
Washington, DC 20560

GEORGE HERMAN

Shell Oil Company
P.O. Box 60775
New Orleans, Louisiana 70160

ABSTRACT

Ammonites ranging in age from late middle Oxfordian to latest Tithonian have been found in cores obtained from nine wells in Louisiana, one well in Mississippi, six wells in eastern Texas, and one well in southern Texas. The ammonites found in the middle and upper parts of the Smackover Formation are of late middle to late Oxfordian age based on the presence of *Ochetoceras, Proscaphites, Dichotomosphinctes, Discosphinctes, Orthosphinctes, Idoceras,* and *Euaspidoceras*. The lower part is probably also entirely of that age based on the presence of *Discosphinctes* at the top of the lower fifth of the Smackover Formation in the Carter-Hope Fee No. 1 well. No ammonites have been found in the red beds typical of the Buckner Formation, but the overlying Gilmer Limestone has furnished a few specimens of *Idoceras* of late Oxfordian, or early Kimmeridgian age. The basal beds of the younger Bossier Formation have yielded the early Kimmeridgian ammonites *Nebrodites* and *Ataxioceras* in association with *Idoceras*. Above these strata are beds characterized by the Kimmeridgian genus *Procraspedites* and also by *Glochiceras* and *Taramelliceras*, which range as high as the basal Tithonian. The upper and middle parts of the Bossier Formation are characterized by a few ammonite genera of Tithonian age including *Virgatosphinctes, Salinites, Durangites?,* and *Substeueroceras*.

STRATIGRAPHIC SUMMARY

The formational nomenclature that has been applied to the subsurface Jurassic sedimentary rocks of the southeastern states has been modified many times since the late thirties (Weeks, 1938, p. 953-983; Hazzard, 1939, p. 155-178) and the early forties (Imlay, 1940, 1943; Swain, 1944) as more wells were drilled over a large area. Much additional data and analyses concerning the Cotton Valley Group were published by Forgotson (1954) and Mann and Thomas (1964). These data and analyses of the Jurassic formations were eventually summarized for Texas, Louisiana, and Arkansas by Dickinson (1968) and for Mississippi and Alabama by Oxley, Minihan, and Ridgeway (1968a; 1968b). More recently, additional discussion of the other stratigraphy, lithology, and diagenesis interpretations of the Jurassic formations in those states were published by Forgotson and Forgotson (1976), Badon (1975), and Moore and Druckman (1982).

The lithologic changes described by these geologists reflect deposition of the sediments at various distances from the shoreline or from sources of sediment. Thus, the upper and middle members of the Smackover Formation interfinger southward in northern Louisiana with the lower part of the Bossier Formation. Similarly, the Gilmer Limestone, which occurs in eastern Texas directly above red beds typical of the Buckner Formation, grades northward and eastward into red to white sandstone and shale that have been referred to the Haynesville Formation (Forgotson and Forgotson, 1976). Both the Gilmer Limestone and Haynesville Formation have been considered by some geologists to be distinct formations and by other geologists to be upper members of the Buckner Formation. On the whole, the nomenclature proposed by Dickinson (1968) is appropriate except possibly for the term Haynesville, which in some areas is applied to beds that consist primarily of limestone and should be referred, therefore, to the Gilmer Limestone.

BIOLOGIC ANALYSIS

The ammonites described herein (Table 1) were obtained from 17 wells of which 1 is located in southern Texas, 6 in eastern Texas, 9 in

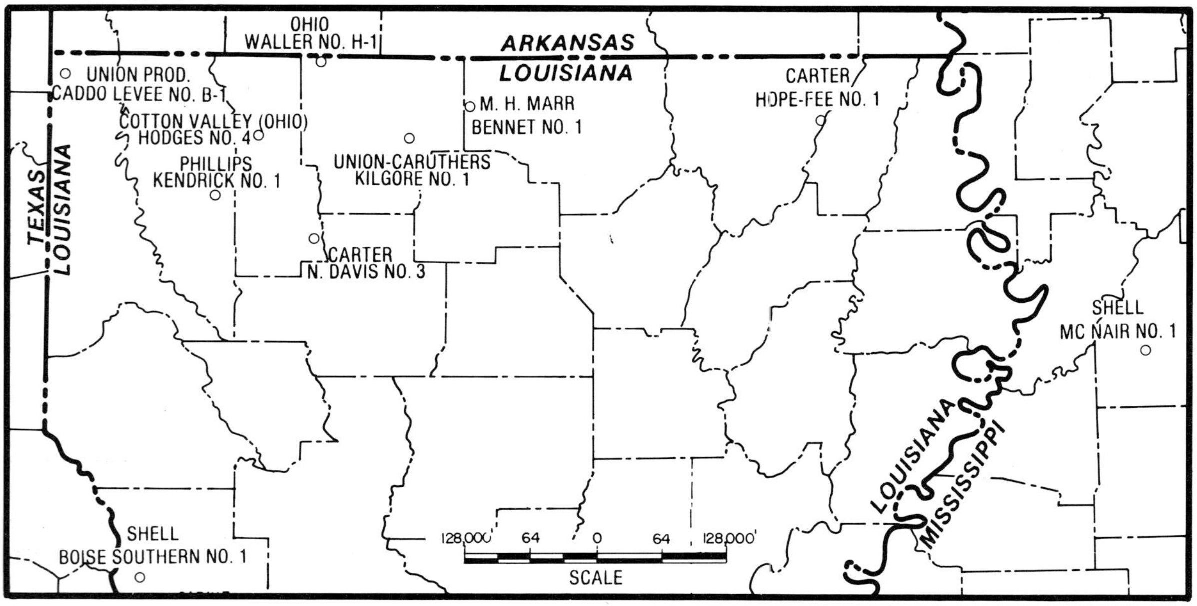

Figure 1. Index map of northern Louisiana and southwest Mississippi showing locations of wells from which Upper Jurassic ammonites have been obtained.

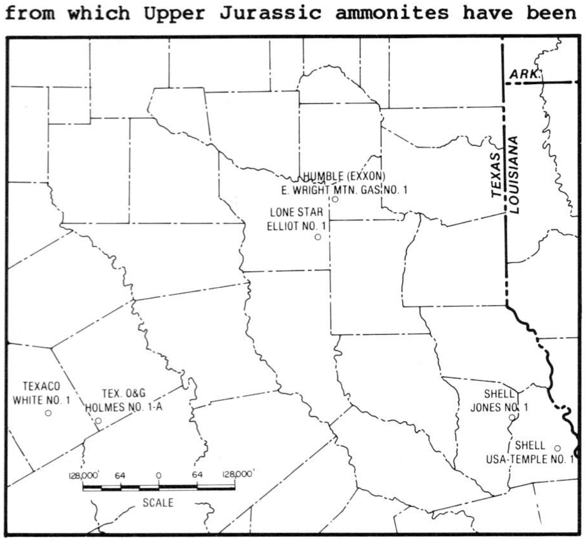

Figure 2. Index map of part of eastern Texas showing locations of wells from which Upper Jurassic ammonites have been obtained.

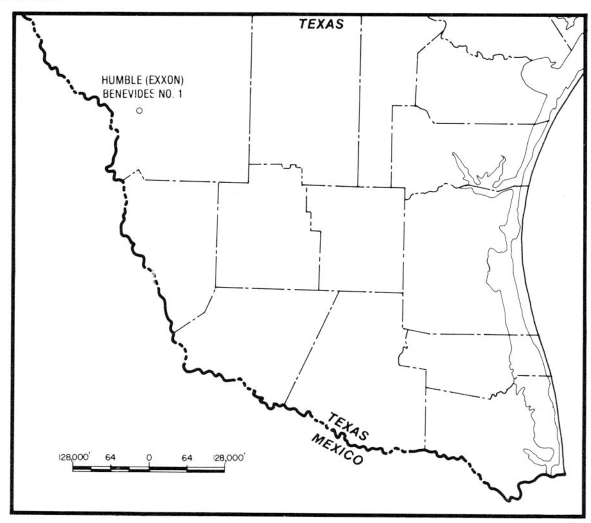

Figure 3. Index map of southernmost Texas showing location of a well from which Upper Jurassic ammonites have been obtained.

Table 1. Ammonite genera and subgenera of Late Jurassic age in the southeastern United States

Family	Subfamily	Genus and subgenus	Number of specimens
Lytoceratidae		*Lytoceras*	1
Haploceratidae		*Haploceras*	4
		Glochiceras	1
		Salinites	29
Oppeliidae	Ochetoceratinae	*Ochetoceras*	3
	Tarameliceratinae	*Taramelliceras*	2
		T. (Metahaploceras)	2
		T. (Proscaphites)	2
Perisphinctidae	Perisphinctinae	*Perisphinctes (Dichotomorsphinctes)*	1
		P. (Discosphinctes)	7
		P. (Orthosphinctes)	1
Ataxioceratidae	Ataxioceratinae	*Idoceras*	21
		Procraspedites	1
		Ataxioceras	2
	Virgatosphinctinae	*Virgatosphinctes*	1
Aspidoceratidae		*Euaspidoceras*	1
		Nebrodites	2
Olcostephanidae		*Proniceras?*	1
Berriasellidae		*Substeueroceras*	2
		Durangites?	5

northern Louisiana, and 1 in southwestern Mississippi. The locations of these wells within counties are shown herein on Figures 1 to 3 and are described in detail on Tables 2 and 3.

AGES AND CORRELATIONS

The Smackover Formation is dated as middle to late Oxfordian based on the presence of the ammonite taxa *Ochetoceras, Proscaphites, Dichotomosphinctes, Discosphinctes, Orthosphinctes, Idoceras,* and *Euaspidoceras* (Fig. 4). Most of these genera (Table 4) were found near or a little below the middle of the formation, but *Discosphinctes* also occurs only 107 feet (33 m) below the top of the formation in the No. 1 Marr-Bennett Est. well and only 156-162 feet (48-50 m) above the base of the formation in the No. 1 Carter Hope Fee well. Its presence at those depths is evidence of an age not older than later middle Oxfordian, or younger than late Oxfordian for most of the Smackover Formation.

The Gilmer Limestone overlying the Buckner Formation has furnished a few ammonites of late Oxfordian or early Kimmeridgian age. *Idoceras cajaense* Burchkardt and *I. zacatecanum* Burckhardt were obtained from cores in the Texas Oil and Gas-Holmes No. 1 well at depths of 13,789-13,802 feet, which are 14-27 feet below the top of the Gilmer Limestone and about 1,440 feet below the top of the Bossier Formation.

The Bossier Formation is dated as early Kimmeridgian to late Tithonian. An early Kimmeridgian age is shown by the presence of the ammonite *Nebrodites* and *Ataxioceras* (Fig. 4) in the basal beds and by their association with *Idoceras* in the lower 100 feet of the formation. An early to late Kimmeridgian age for slightly higher beds is shown by the presence of *Glochiceras* and *Procraspedites* in association with early forms of *Haploceras* and *Metahaploceras* and with late forms of *Taramelliceras*. A Tithonian age for the upper part of the Bossier Formation is based on a few ammonites. These include *Virgatosphinctes* which ranges throughout the Tithon-

Table 2. Wells From Which Upper Jurassic Ammonites Have Been Obtained in the Southeastern United States

Field	Company	Well	Location	County or Parish and State	Elev. D.F. in Feet	Top of Schuler Formation	Top of Bossier Formation	Top of Buckner Formation	Top of Smackover Formation	Top of Norphlet Formation	Total Depth in Feet
	Shell Oil	Jones No. 1	1604' FNL & 467' FWL of 74.35 AC tract in the William Roberts Survey A-39	San Augustine County, Texas	350	11880	12810	13515 Top of Gilmer Limestone	Equivalent Beds are Probably Not Present	No Norphlet / SS. at 13618 / Paleoz. Top at 13817 Feet	13942
	Shell Oil	USA-Temple No. 1	1556' FWL & 1903' FSL of the T&NO RR Survey No. 15 A212	Sabine County, Texas	267	13310	14013 (Base 14428)	Not Present	Not Present	Not Present Paleozoic Top at 14428	15885
	Humble Oil & Gas	Benevides No. 1	2640' FNL & 660' FWL GC&SF RR No. 1601 Survey	Webb County, Texas	630	17166	17578	18365 Top of Gilmer Ls. Buckner Facies Equiv. in Carbonate Seq.	Present in Unbroken Carbonate Seq.	Not Penetrated	19852
Oletha	Texaco	White No. 1	J. L. Chavert Survey	Limestone County, Texas	388	9560?	10620	11840 Top of Gilmer Ls. 12733 Top of Buckner Facies	Faulted Out	Norphlet Faulted Out Louann Salt at 13582 Ft.	13694
Chapel Hill	Lone Star (Sun)	Elliot No. 1	660' FSL & 1240' FWL Thomas P. Payne Survey	Smith County, Texas	521	9720	11355	11930 Top of Gilmer Ls. Buckner Facies Replaced by Carbonates	Not Differentiated	Norphlet 13415 / Louann Salt 13531	13623
Bear Creek	Texas Oil and Gas	Holmes No. 1-A	7100' FNEL, 3800' FNWL Isaac Connelly Survey	Freestone County, Texas	490	11535	12753?	13775 Top of Gilmer Ls. 14410? Buckner Equivalent	Smackover Equivalent Beds at 14860?	Not Penetrated	14966
East Texas	Exxon (Humble)	East Wright Mtn. Gas No. 1	Wm. W. Friar Survey A-68	Gregg County, Texas	395	9405?	10963	12540 Top of Gilmer Limestone	Not Differentiated	Norphlet? 12633? / Louann 12695	12700
	Shell Oil	Boise Southern No. 1	Sec. 7, T. 7 N., R. 13 W.	Sabine Parish, Louisiana	334	10610	11422	11910 Top of Gilmer Ls. and Other Carbonates	Not Differentiated	No Norphlet / 12415 Sandstone / 12475 Top of Paleozoic	13317
West Lisbon	Union Prod. and Caruthers et al	Kilgore No. 1	Sec. 20, T. 21 N., R. 5 W.	Claiborne Parish, Louisiana	285	7767	9396	10243 Sandstone & Carbonates Interbedded	10577	Not Penetrated	11329
Haynesville	Ohio Oil	Waller Est. No. H-1	Sec. 3, T. 23 N., R. 8 W.	Claiborne Parish, Louisiana	326	7305	Top 9434	9710 Sandstone 9923 Red Beds 10247 "A" Carbonates	10330 Smackover	Norphlet 11315 / Louann 11360	11376
Rodessa	Union Prod	Caddo Levee Board D-1	Sec. 28, T. 23 N., R. 16 W.	Caddo Parish, Louisiana	200	8023 (By Core)	9935	10250 Buckner "A" Carbonates	10580? Smackover Equivalent	Norphlet 11431 / Louann 11444	11450
Bellvue	Phillips	Kendrick No. 1	Sec. 22, T. 19 N., R. 11 W.	Bossier Parish, Louisiana	334	4950	6523	8230? or 8392? Basinal Facies of Bossier Fm.	8405 Top Ls. in Ls. and Shale Sequence Smackover? Fm.	Norphlet 9045 / Louann 9055	9070
Ada	Carter	Nora Davis No. 3	Sec. 34, T. 18 N., R. 8 W.	Webster Parish, Louisiana	213	9497	11330	Not Penetrated	Not Penetrated	Not Penetrated	14146
Cotton Valley	Cotton Valley Operators Comm.	Ohio-Hodges No. 4	Sec. 23, T. 21 N., R. 10 W.	Webster Parish, Louisiana	227	7653	9032 Top 9970 Base	Not Differentiated	10050? Probably Equivalent to Haynesville or Buckner	Not Penetrated	10856
	M. H. Marr	Bennett Est. 1	Sec. 30, T. 22 N., R. 3 W.	Union Parish, Louisiana	219	7150	9190	10110 Lower Cotton Valley Sandstone	11016 Sandstone 11123 Limestone	Not Penetrated	11620
	Carter (Exxon)	Hope Fee No. 1	Sec. 7, T. 21 N., R. 9 E.	Morehouse Parish, Louisiana	104	2680?	Facies Not Present	5670	5830	Norphlet 6732 / Louann 6740	6751
	Shell Oil & C. J. Calhoun	McNair No. 1	Sec. 13, T. 4 N., R. 4 W.	Hinds County, Mississippi	269	16220	21860? Facies Not Present	22385? Red Beds	Not Penetrated	Not Penetrated	22736

Table 3. Distribution of Upper Jurassic Ammonites in the Subsurface of Louisiana, Texas, and Mississippi

Species Numbers show depth in feet. All specimens are from cores.	Shell-Jones No. 1	Shell-Temple No. 1	Humble-Benevides No. 1	Texaco-White No. 1	Lone Star (Sun)-Elliot No. 1	Texas O&G-Holmes No. 1A	Exxon (Humble) E. Wright Mtn. Gas Unit 1	Shell-Boise Southern No. 1	Union-Carrathers-Kilgore No. 1	Ohio-No. H-1, Waller Est.	Union Prod.-Caddo Levee Board B-1	Phillips-Kendrick No. 1	Carter-Nora Davis No. 3	Cotton Valley-Ohio Hodges No. 4	Marr-Bennett Est. No. 1	Carter (Exxon)-Hope Fee No. 1	Shell Oil & C. J. Calhoun McNair No. 1
Lytoceras				11928													
Haploceras costatum Burckhardt		14380, 14419															
sp.	13501-503																
Glochiceras Fialar (Oppel) of Burckhardt												4048-63					
Salinites grossicostatum (Imlay)								11638-65									
sp.				11727, 11928													
Ochetoceras cf. O. mexicanum Burckhardt									11195-208 11210-224								
Taramelliceras sp.	13502-503																
(Metahaploceras) cf. T. (M.) nereus (Fontannes)												7654-64					
(M.) sp.		14380															
(Proscaphites) sp.														10620			
Perisphinctes (Dichotomosphinctes) sp.									11234								
(Discophinctes) cf. P. (D.) carribeanus Jaworski							12432		11210						11123		
(D.) sp.					12755-61							8741-56				6570-76	
(Orthosphinctes) cf. P. (O.) rutteni Jaworski										10748							
Idoceras cajense Burckhardt						13789-802							12770				
zacatecanum Burckhardt						13760-763 13788-800											
densicostatum Imlay						13799											
cf. I. durangense Burckhardt						13755-788						8048-63					22363
santarosanum Burckhardt											10150-170						
cf. I. santarosanum Burckhardt				11547-647													
Procraspedites praecursor (Burckhardt)				11547-647													
Ataxioceras sp.												8279-94 8377-92					
Virgatosphinctes cf. V. aquilari (Burckhardt)				11547-647													
Euaspidoceras cf. E. occonnelli Sanchez Roig										10777							
Nebrodites cf. N. flexuosus Burckhardt		14361-391															
Proniceras ? sp.								11638-650									
Substeueroceras cf. S. alticostatum Imlay			17803-817														
Durangites? cf. D. vulgaris Burckhardt			17803														

Table 4. Stratigraphic Occurrences of Jurassic Ammonites in the Gulf Region of Louisiana, Texas, and Mississippi

Genera and Subgenera	Wells	General Locations	Formational Occurrences	Depths in Feet of Ammonite Occurrences
Lytoceras	Texaco-White No. 1	Limestone County, Texas	Gilmer Ls. 88 Ft. Below Top	11928
Haploceras	Shell-Temple No. 1	Sabine County, Texas	Bossier, 19-48 Ft. Above Base	14380, 14419
Glochiceras	Phillips-Kendrick No. 1	Bossier Parish, Louisiana	Bossier, 182 Ft. Above Base	8048-8063
Salinites	Texaco-White No. 1	Limestone County, Texas	Bossier, 107 Ft. Above Base	11727-11728
	Shell-Boise Southern No. 1	Sabine Parish, Louisiana	Bossier, Near Middle	11638-11665
Ochetoceras	Union Carruthers-Kilgore No. 1	Claiborne Parish, Louisiana	Smackover, Upper, 618-647 Ft.	11195-11224
Taramelliceras	Shell-Jones No. 1	San Augustine County, Texas	Bossier, 13 Ft. Above Base	13502-13503
Metahaploceras	Phillips-Kendrick No. 1	Bossier Parish, Louisiana	Bossier, Near Middle	7564-7666
	Shell-Temple No. 1	Sabine County, Texas	Bossier, 48 Ft. Above Base	14380
Proscaphites	Cotton Valley-(Ohio)Hodges No.4	Webster Parish, Louisiana	Smackover, 618-647 Ft. Below Top	10668-10693
Dichotomosphinctes?	Union Carruthers-Kilgore No. 1	Claiborne Parish, Louisiana	Smackover, 657 Ft. Below Top	11234
Discosphinctes	Exxon-Wright Mtn., Gas No. 1	Gregg County, Texas	Bossier, 108 Ft. Above Base	12432
	Union Carruthers-Kilgore No. 1	Claiborne Parish, Louisiana	Smackover, 633 Ft. Below Top	11210
	Marr-Bennett Est. No. 1	Union Parish, Louisiana	Smackover, 107 Ft. Below Top	11123
	Texaco-White No. 1	Limestone County, Texas	Bossier, 200-300 Ft. Above Base	11547-11647
	Lone Star-Elliot No. 1	Smith County, Texas	825 Ft. Below Top of Gilmer Ls.	12755-12761
	Phillips-Kendrick No. 1	Bossier Parish, Louisiana	Smackover, 336-351 Below Top	8741-8756
	Carter-Hope Fee No. 1	Morehouse Parish, Louisiana	Smackover, 156-162 Ft. Above Base	6570-6576
Orthosphinctes	Ohio-Waller Est. No. H-1	Claiborne Parish, Louisiana	Smackover, 418 Ft. Below Top	10748
Idoceras	Shell-McNair No. 1	Hinds County, Mississippi	"Bossier", 22 Ft. Above Base	22363
	Texas Oil-Holmes No. 1A	Freestone County, Texas	Gilmer Ls. 14-27 Ft. Below Top	13788-13800
	Texas Oil-Holmes No. 1A	Freestone County, Texas	Bossier, Basal Few Feet	13760-13763
	Carter-Nora Davis No. 3	Webster Parish, Louisiana	Bossier?, 1440 Ft. Below Top	12770
	Phillips-Kendrick No. 1	Bossier Parish, Louisiana	"Bossier", Basinal Facies?	8279-94, 8377-92
	Union B-1, Caddo Levee Bd. B-1	Caddo Parish, Louisiana	Bossier, Lower 80-100 Ft.	10150-10170
	Texaco-White No. 1	Limestone County, Texas	Bossier, Lower 200-300 Ft.	11547-11647
Procraspedites	Texaco-White No. 1	Limestone County, Texas	Bossier, Lower 200-300 Ft.	11547-11647
Ataxioceras	Phillips-Kendrick No. 1	Bossier Parish, Louisiana	"Bossier", Basinal Facies?	8279-94, 8377-92
Virgatosphinctes	Texaco-White No. 1	Limestone County, Texas	Bossier, Lower 200-300 Ft.	11547-11647
Euaspidoceras	Ohio-Waller Est. No. H-1	Claiborne Parish, Louisiana	Smackover, Base of Upper Third	10777
Nebrodites	Shell-Temple No. 1	Sabine County, Texas	Bossier, Near Base	14361-14391
Proniceras?	Shell Boise-Southern No. 1	Sabine Parish, Louisiana	Bossier, Near Middle	11638-11650
Substeuroceras	Humble-Benevides No. 1	Webb County, Texas	Bossier, Base of Upper Third	17803-17817
Durangites?	Humble-Benevides No. 1	Webb County, Texas	Bossier, Base of Upper Third	17803

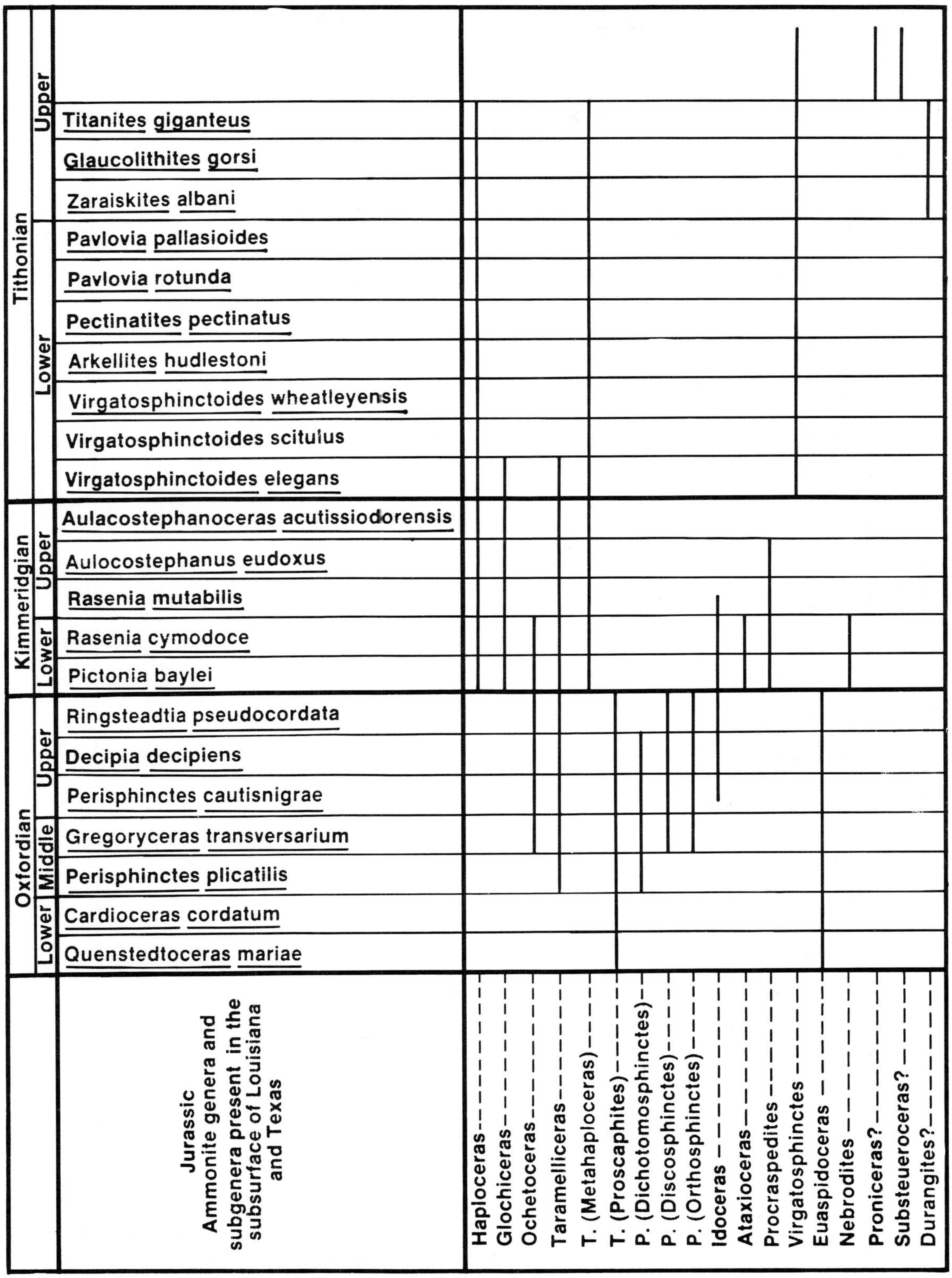

Figure 4. European ranges of Upper Jurassic ammonite genera, or subgenera, present in the Gulf Region of the United States.

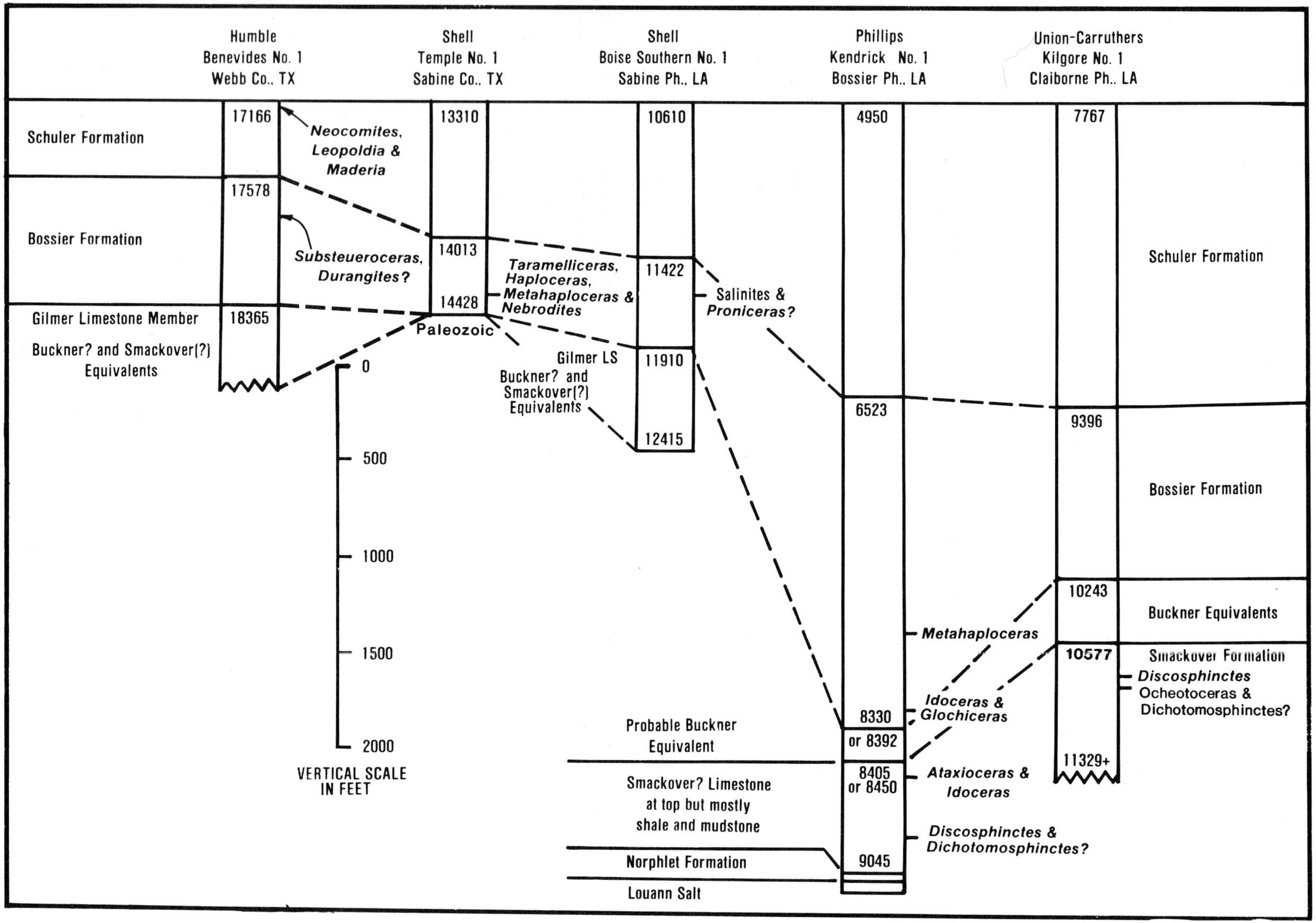

Figure 5. Stratigraphic occurrences of certain Oxfordian, Kimmeridgian, and Tithonian ammonite genra in the Smackover and Bossier Formations of Texas and Louisiana.

ian; *Salinites* which ranges through the Late Tithonian; *Durangites*(?) which is fairly common in the early Late Tithonian and *Proniceras*(?) and *Substeueroceras* which characterize the latest Late Tithonian. In addition, the genera *Haploceras* and *Metahaploceras* range from Early Kimmeridgian to the early Late Tithonian (Fig. 4). The stratigraphic occurrences of all these genera relative to the upper and lower boundaries of the Bossier Formation are shown on Table 4.

The Schuler Formation until recently has been dated as Late Jurassic on the basis of its stratigraphic position above the Bossier Formation and on the presence of some bivalves that are identical with species that occur in outcrops of Upper Jurassic beds in Mexico and in western Texas (Imlay, 1945, p. 261-269; Swain, 1944, p. 599). That dating needs additional confirmation, however, considering that (1) Jurassic ammonites have not yet been found in the Schuler Formation; (2) ammonites of latest Tithonian age, such as *Substeueroceras*, and *Durangites(?)*, have been found in the upper part of the Bossier Formation (Table 4); and (3) middle Neocomian ammonites, such as *Neocomites*, *Maderia* and *Leopoldia*, have been found in the uppermost beds of the Schuler Formation in the Humble-Benevides No. 1 well in Webb County, Texas.

Systematic Descriptions

Family LYTOCERATIDAE Neumayr 1875
Lytoceras sp.
Plate 1, Figure 1

The genus *Lytoceras* is probably represented by one laterally crushed specimen that is characterized by highly evolute coiling; and fairly fine, straight, closely spaced, simple ribs that trend nearly radial and do not bifurcate.

Figured specimen: USNM 340016.

Occurrence: Gilmer Limestone, 88 feet below top. Texaco Oil Co. No. 1 White well at a depth of 11,928 feet, Limestone Co. Texas.

Family HAPLOCERATIDAE Zittel, 1884
Genus HAPLOCERAS Zittel, 1870
Haploceras costatum Burckhardt
Plate 1, Figure 2

Haploceras costatum Burckhardt, 1906, Inst. Geol. Mexico Bol. 23, p. 96, pl. 25, figs. 1-10.

Two laterally crushed ammonites resemble the type specimens of *H. costatum* Burckhardt by having fairly strong falcate ribs on the upper part of the outermost septate whorl, whereas the adjoining body chamber bears much weaker ribs.

Hypotype: USNM 340017.

Occurrence: Bossier Formation, near base. Shell Oil Company No. 1 Temple well at depths of 14,380 and 14,419 feet, Sabine County, Texas.

Haploceras? sp.
Plate 1, Figure 3

The genus *Haploceras* is probably represented by many laterally crushed specimens that have a small umbilicus and nearly smooth surfaces. A few specimens bear faint falcate striae such as have been illustrated by Burckhardt (1906, pls. 21-24) under several different specific names.

Figured specimens: USNM 340018.

Occurrences: Bossier Formation, near base. Shell Oil Company No. 1 Jones well at depths of 13,501-13,503 feet, San Augustine County, Texas.

Genus GLOCHICERAS Spath, 1924
Glochiceras fialar (Oppel) of Burckhardt
Plate 1, Figure 4

Haploceras fialar Oppel. Burckhardt, 1906, Inst. Geol. Mexico Bol. 23, p. 77, pl. 19, figs. 1-19; pl. 20, figs. 1-6, 12, 14, 15.

Glochiceras fialar Burckhardt (not Oppel).

PLATE 1

(Figures natural size unless otherwise indicated)

Figure 1. *Lytoceras* sp. (p.157). Specimen, USNM 340016, Gilmer Limestone at depth of 11,928 feet, Texaco No. 1 White well.

Figure 2. *Haploceras costatum* Burckhardt (p. 157). Hypotype, USNM 340017, Gilmer Limestone at depth of 14,419 feet, Shell No. 1 Temple well.

Figure 3. *Haploceras?* sp. (p.157). Specimen, USNM 340018; Gilmer Limestone at depth of 13,503 feet, Shell No. 1 Jones well.

Figure 4. *Glochiceras fialar* Oppel (p.157). Hypotype, USNM 340019, Bossier Formation, basal part, at depth of 8,048-8063 feet, Phillips No. 1 Kendrick well.

Figures 5-6. *Ochetoceras* cf. *O. mexicanum* Burckhardt (p.161). 5, specimen, USNM 340024. Smackover Formation at depth of 11,208 feet, Union Prod. Co. and Carruthers *et al.*-Kilgore No. well; 6, specimen, USNM 340025. Smackover Formation at depth of 11,195 feet, Union Prod. Co.-Carruthers *et al.* No. 1 Kilgore well.

Figure 7. *Taramelliceras (Proscaphites)* sp. (p. 162). Specimen (X 2), USNM 340028, Smackover Formation at depth of 10,620 feet, Cotton Valley Operators (Ohio) No. 4 Hodges well.

Figure 8. *Salinites* sp. (p. 160). Specimen, USNM 340023. Bossier Formation at depths of 11,727-11,728 feet, Texaco No. 1 White well.

Figures 9-11. *Salinites grossicostatum* (Imlay) (p.160). 9, hypotype, USNM 340020; 10, hypotype, USNM 340021; 11, hypotype, USNM 340022. All three hypotypes are from the Bossier Formation at depths of 11,635-11,665 feet, Shell No. 1 Boise Southern well.

Figure 12. *Taramelliceras* sp. (p.161). Specimen, USNM 340026. Bossier Formation at depth of 13,503 feet, Shell No. 1 Jones well.

Figure 13. *Taramelliceras (Metahaploceras)* sp. (p.161). Specimen, USNM 103661. Bossier Formation at depth of 7,564-7,566 feet, Phillips No. 1 Kendrick well.

Figures 14-17. *Perisphinctes (Discosphinctes)* cf. *P. (D). carribeanus* Jaworski (p. 164). 14, specimen, USNM 340030. Bossier Formation at depth of 11,223 feet, No. 1 Marr Bennett Est. well; 15, specimen, USNM 340031. Smackover Formation at depth of 12,210 feet, Union Prod. Co.-Carruthers *et al.* No. 1 Kilgore 1 well; 16, 17, two specimens, USNM 340032. Bossier Formation at depth of 12,432 feet, Humble-East Wright Mtn. Gas Co. Unit 1.

Figure 18. *Ataxioceras* sp. (p. 167). Specimen (X 2), USNM 103667. "Bossier" Formation at depth of 8,279-8,294 feet, Phillips No. 1 Kendrick well.

Figure 19. *Perisphinctes (Orthosphinctes)* cf. *P. rutteni* Jaworski (p. 165). Specimen 340033. Smackover Formation at depth of 10,748 feet, Ohio Oil Co. No. H-1 Waller well.

Figure 20. *Proniceras?* sp. (p. 168). Specimen, USNM 340051. Bossier Formation at depth of 11,638-11,650 feet, Shell No. 1 Boise Southern well.

Figure 21. *Euaspidocras* cf. *E. o'connelli* (Sanches Roig) (p. 168). Specimen, USNM 340048. Smackover Formation depth of 10,777 feet, Ohio Oil Co. No. H-1 Waller Est. well.

Figure 22. *Procraspedites* cf. *P. praecursor* Burckhardt (p. 167). Specimen, USNM 340046. Bossier Formation at depth of 11,547-11,647 feet, Texaco No. 1 White well.

PLATE 1

Imlay, 1939, Geol. Soc. America Bull., v. 50, p. 26.

This species is characterized by the presence of a strong spiral furrow and by very fine falciform striae and folds that are very weak on the internal whorls and strongest on the outer whorl above the spiral furrow.

Figured specimen: USNM 340019.

Occurrence: Bossier Formation, basal part. Phillips Petroleum Company No. 1 Kendrick well at depths of 8,048-63 feet, sec. 22, T. 19 N., R. 11 W., Bossier Parish, Louisiana.

Genus SALINITES Cantu, 1968
Salinites grossicostatum (Imlay)
Plate 1, Figures 9-11

Hildoglochiceras grossicostatum Imlay, n.sp., 1939, Geol. Soc. America Bull., v. 50, p. 27, pl. 2, figs. 5-11, pl. 3, figs. 1-7, 9-11.

Salinites grossicostatum (Imlay). Cantu Chapa, 1968, Instituto Mexicano del Petroleo, Seccion Geologia, Monografia No. 2, p. 20, 21, pl. 4, figs. 1, 4, 7, 9; pl. 5, figs. 3, 8, 10; 1982, p. 315-317, pl. 1, fig. 7a.

This species is represented in Louisiana by 27 crushed specimens that bear all the features shown on the type specimens from Coahuila in northern Mexico. It is characterized by smooth inner whorls, a narrow umbilicus, a narrowly rounded keel, sickle-shaped ribs, striae that during growth become stronger ventrally, and a weak to fairly weak spiral groove that becomes broader and shallower during growth.

This species was made the type of new genus *Salinites* by A. Cantu Chapa (1968, p. 20, 21) who notes that it differs from *Hildoglochiceras* (1) by having a fairly narrow instead of wide umbilicus, (2) sickle-shaped ribs that curve forward on its venter instead of inclining backward, and (3) a distinct slightly elevated keel on the outer whorl of adult specimens instead of having a merely sharpened keel.

Salinites grossicostatum (Imlay) in northern Mexico occurs in the same beds as *Durangites*, *Parodontoceras*, *Proniceras*, and *Kossmatia* of late Tithonian age (Cantu Chapa, 1982, p. 315-317). In Louisiana it was found with an ammonite that probably represents *Proniceras*.

Hypotypes: USNM 340020-340022.

Occurrence: Bossier Formation, near middle, Shell Oil Company No. 1 Boise Southern well in Core 3 at depths of 11,638-11,665 feet in sec. 7, T. 7 N., R. 13 W., Sabine Parish, La.

Salinites sp.
Plate 1, Figure 8

Fragmentary specimens of *Salinites* have been obtained at two levels in the Texaco No. 1 White well, Limestone County, Texas. The higher level is at the depth of 11,727-28 feet near the base of the Bossier Formation. The lowest level is at the depth of 11,928 feet in the Gilmer Limestone. The specimen from the depth of 11,727-28 feet differs from *S. grossicostatum* (Imlay) by having a more distinct spiral groove.

Figured specimen: USNM 340023.

PLATE 1 - continued

Figure 23. ***Perisphinctes (Dichotomosphinctes?)* sp. (p. 164). Specimen, USNM 340029. Smackover Formation at depth of 11,234 feet in Union Prod. Co.-Carruthers *et al.* No. 1 Kilgore well.**

Figure 24. ***Durangites?* cf. *D. vulgaris* Burckhardt (p. 169). Rubber imprint of specimen (X 2), USNM 340053. Bossier Formation at depth of 17,803 feet in Humble No. 1 Benavides well.**

Occurrence: Bossier Formation, just below middle. Texaco No. 1 White well at depths of 11,727-28 feet, Limestone Co., Tx.

Genus OCHETOCERAS Haug, 1885

Ochetoceras cf. *O. mexicanum* Burckhardt

Plate 1, Figures 5,6

Three external molds of *Ochetoceras* bear ornamentation that is identical with that present at a comparable size on the holotype of *Ochetoceras mexicanum* (Burckhardt) (1912, p. 7, pl. 1, figs. 8-12; O'Connell, 1920, p. 686, pl. 38. figs. 1-3). Their venters bear a sharp median keel that is separated from the flanks by grooves. Their flanks bear a rather deep median lateral groove. The upper parts of their flanks bear strong falcoid ribs. The lower parts of the flanks bear much weaker ribs that trend adorally from the umbilicus and become slightly stronger near the lateral groove. The specimens differ from *O. pedroanum* Burckhardt (1912, p.9, pl. 1, figs. 13-17) by having a much sharper venter. They have stronger ribbing and deeper lateral grooves than the specimens that were referred to *O. canaliculatum* d'Orbigny by Burckhardt (1912, pl. 1, figs. 1-7).

Figured specimens: USNM 340024-25.

Occurrence: Smackover Formation 618-647 feet below top, probably in middle third. Union Prod. and Carruthers *et al.* No. 1 Kilgore well at depths 11,195 and 11,208 feet, sec. 20, T. 21 N., R. 5 W., Claiborne Parish, Louisiana.

Genus TARAMELLICERAS Del Campana, 1904

Taramelliceras sp.

Plate 1, Figure 12

Two small laterally crushed specimens bear falciform ribs that are similar to those on *Taramelliceras* cf. *T. trachynotum* Fontannes in Burckhardt (1906, p. 72, pl. 17, fig. 5). They differ by lacking tubercles on their ventral margins and having less prominent ribs. Their ribbing is similar to that on immature forms of *Salinites grossicostatum* (Imlay) (1939, p. 27, pl. 3, figs. 1-7, 9-11), but they differ by lacking a keel.

Figured specimen: USNM 390026.

Occurrence: Bossier Formation, 12 feet above base. Shell Oil Company No. 1 Jones well at depth 13,503 feet, San Augustine County, Texas.

Subgenus METAHAPLOCERAS Spath, 1925

Taramelliceras (Metahaploceras) cf.

T. (M.) nereus (Fontannes)

Metahaploceras cf. *M. nereus* (Fontannes). Imlay, 1945, Jour. Paleontology, v. 19, no. 3, p. 271, pl. 41, fig. 19.

This species is characterized by falciform striae on the lower two-thirds of its flanks and by striated, variably developed, forwardly inclined folds on the upper third of its flanks. These folds are most swollen on the ventro-lateral margins and arch forward gently on the venter.

Specimen: USNM 103661.

Occurrence: Bossier Formation near base of upper third. Phillips Petroleum Company No. 1 Kendrick well at depth of 7,564-66 feet, sec. 22, T. 19 N., R. 11 W., Bossier Parish, Louisiana.

Taramelliceras (Metahaploceras) sp.

Plate 1, Figure 13

One laterally crushed internal mold is characterized by very involute coiling, mostly smooth flanks, and closely spaced ventro-lateral swellings that curve forward and terminate in tubercles. These swellings occur on the upper third of half a whorl but are absent on the adoral fifth of the outermost preserved whorl.

This specimen shows some resemblance to *Metahaploceras affine* Spath (Arkell, *et al.* 1957, p. 281, fig. 332) but has more closely spaced tubercles. It also resembles *Metahaploceras* aff. *M. strombecki* (Oppel) as figured by Burckhardt (1906, p. 74, pl. 18, figs. 4-7) but has much stronger tubercles.

Figured specimen: USNM 340027.

Occurrence: Bossier Formation, near base. Shell Oil Company No. 1 Temple well at depth of 14,380 feet, Sabine Co., Texas.

Subgenus PROSCAPHITES Rollier, 1909
Taramelliceras (Proscaphites) sp.
Plate 1, Figure 7

One fragmentary ammonite bears spirally elongate nodes (serrations) along the midline of the venter. The upper part of its flanks bears forwardly inclined ribs that become stronger ventrally on the flanks but weaken considerably on the margin of the venter, except for two ribs at the adoral end of the fragment. Each ventral serration joins laterally with a single rib except for one that joins with two ribs.

This specimen in ventral view resembles *T. (Proscaphites) pichleriformis* (Burckhardt) (1912, p. 14, pl. 2, figs. 9-12) and T. (P.) *neohispanica* Burckhardt (1912, p. 12, pl. 2, figs. 5-8) from late Oxfordian beds near San Pedro del Gallo, Mexico. It differs by having slightly coarser ribbing.

PLATE 2

(All figures are natural size.)

Figures 1-3. *Idoceras zacatecanum* **Burckhardt (p. 166). 1, hypotype, USNM 340043. Gilmer Limestone at depths of 13,787.5-13,800 feet, Texas O&G No. 1-A Holmes well; 2, hypotype, USNM 340044 in same well and limestone at depth of 13,796 feet; 3, hypotype, USNM 340045 in same well and limestone at depth of 13,800 feet.**

Figures 4-8. *Idoceras cajense* **Burckhardt (p. 165). 4, 5, hypotype, USNM 340046. Bossier Formation at depth of 12,770 feet, Carter No. 3 Nora Davis well; 6, hypotype, USNM 340047. Gilmer Limestone at depth of 13,761 feet, Texas O&G No. 1-A Holmes well; at depth type, USNM 340048. Gilmer Limestone at depth of 13,789 feet, Texas O&G No. 1-A Holmes well; 8, hypotype, USNM 340049. Gilmer Limestone at depth of 13,798.5 feet, Texas O&G Holmes No. 1-A well.**

Figures 9, 11, 12. *Idoceras* **cf.** *I. durangense* **Burckhardt (p. 165). 9, specimen, USNM 340039. Bossier Formation at depth of 22,363 feet, Shell No. 1 McNair well; 11, specimen, USNM 340040. Gilmer Limestone at depth of 13,763 feet, Texas O&G No. 1-A Holmes well; 12, specimen, USNM 340041. Gilmer Limestone at depth of 13,785.5 feet, Texas O&G No. 1-A Holmes well.**

Figure 10. *Idoceras densicostatum* **Imlay (p. 165). Hypotype, USNM 340038. Gilmer Limestone at depth of 13,799 feet, Texas O&G No. 1-A Holmes well.**

Figure 13. *Idoceras santarosanum* **Burckhardt (p. 166). Hypotype, USNM 340042. Bossier Formation at depth of 10,150-10,170 feet, Union Prod. Co. No. B-1 Caddo Levee well.**

Figure 14. *Virgatosphinctes* **cf.** *V. aquilari* **(Burckhardt) (p. 167). Rubber specimen, USNM 340047. Bossier Formation at depth of 11,547-11,647 feet, Texaco No. 1 White well.**

Figure 15. *Substeueroceras* **cf.** *S. alticostatum* **Imlay (p. 169). Rubber specimen, USNM 340052. Bossier Formation at depth of 17,817 feet, Humble No. 1 Benavides well.**

Figures 16, 17. *Nebrodites* **cf.** *N. flexuosus* **Burckhardt (p. 168). 16, specimen, USNM 340049. Bossier Formation at depth of 14,391 feet, Shell No. 1 Temple well; 17, specimen, USNM 340050. Bossier Formation at depth of 14,361 feet, Shell No. 1 Temple well.**

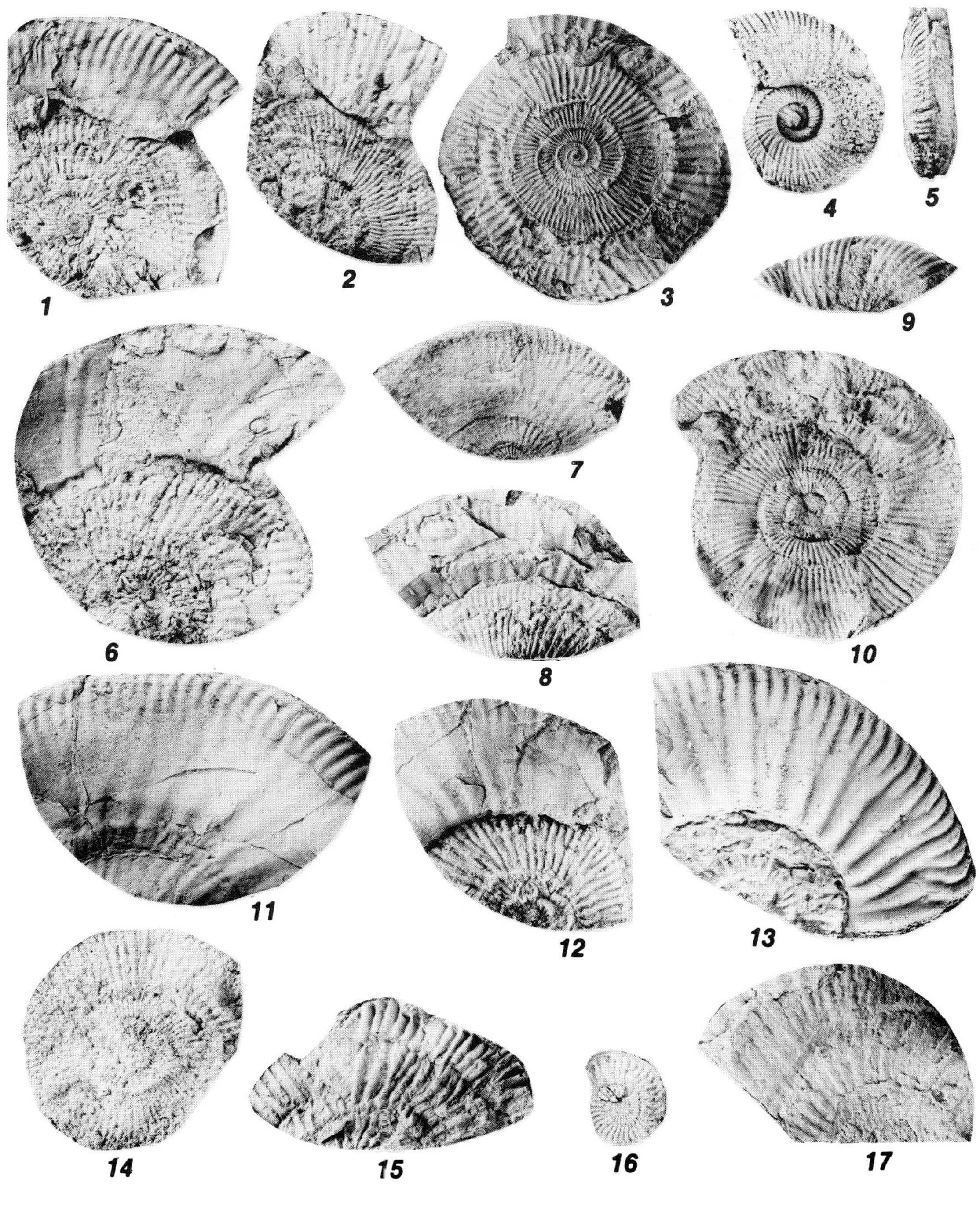

PLATE 2

Figured specimen: USNM 340028.

Occurrence: Smackover Formation, 618-647 feet below probable top of formation at depth of 10,050 feet. Cotton Valley Operators Commission (Ohio) No. 4 Hodges well at depth of 10,620 feet in sec. 23, T. 21 N., R. 10 W., Webster Parish, Louisiana.

Family PERISPHINCTIDAE Steinmann, 1890
Genus PERISPHINCTES Waagen, 1869
Subgenus DICHOTOMOSPHINCTES Buckman, 1926
Perisphinctes (Dichotomosphinctes?) sp.
Plate 1, Figure 23

One ammonite fragment represents part of an outer whorl of a species that has very sharp, fairly high, widely spaced ribs of which most are simple. It shows some resemblance to a specimen from Oxfordian beds in Cuba that was originally described as *Perisphinctes plicatiloides* O'Connell (1920, p. 670-680, pl. 36, figs. 1a-b). It shows more resemblance to other specimens from Cuba that were referred to the same species by Jaworski (1940, p. 118, pl. 4, fig. 4, pl. 6, figs. 1a, b) but described under the subgeneric name *Dichotomosphinctes*. *P. (D.) plicatiloides* O'Connell, according to Jawroski (1940, p. 118), includes specimens from Mexico that were illustrated by Burckhardt (1912, pl. 3, figs. 3-6), but described under other names. It also includes specimens from Cuba, according to Arkell (1956, p. 573), that were originally described under other specific names by Sanchez Roig (1951, p. 75, pl. 19, figs. 1, 2, p. 90, pl. 22, figs. 3, 4).

Figured specimen: USNM 340029.

Occurrence: Smackover Formation, 657 feet below top, and 95 feet above base. Union Producing and Carruthers *et al.* No. 1 Kilgore well at depth of 11,234 feet, sec. 20, T. 21 N., R. 5 W., Claiborne Parish, Louisiana.

Subgenus DISCOSPHINCTES Dacque, 1914
Perisphinctes (Discosphinctes) **cf.**
P. (D.) carribeanus **Jaworski**
Plate 1, Figures 14-17

cf. *Perisphinctes virgulatus* Quenstedt in Burckhardt, 1912, Inst. Geol. Mexico Bol. 29, p. 35, pl. 7 figs. 4, 14.

cf. *Ataxioceras virgulatus* (Quenstedt) in O'Connell, 1920, Am. Mus. Nat. Hist. Bull., vol. 42, art. 16, pl. 38, figs. 4, 5.

cf. *Ataxioceras virgulatus* Sanchez Roig, 1920, Cuba, Secretario de Agr., Bol. Especial, p. 23, pl. 8, figs. 1-3.

Discosphinctes cf. *D. virgulatus* (Quenstedt) in Imlay, 1945, Jour. Pal., v. 19, p. 274, pl. 41, figs. 9-11.

cf. *Perisphinctes (Planites) virgulatus* Quenstedt var. *carribeanus* n. var. Jaworski, 1940, Neues Jahrb., Beilage, Bd. 83, Abt. B, no. 1, p. 109, pl. 3, figs. 1-2, pl. 4, fig. 5, pl. 7, fig. 6.

Nine small ammonites from the subsurface Jurassic of Texas and Louisiana are characterized by having fine, densely-spaced, nearly straight ribs that incline adorally appreciably on the flanks and become slightly stronger ventrally. Many of the ribs remain simple, but some bifurcate at, or a little above, the middle of the flanks. Some simple ribs begin freely at, or above, the middle of the flanks. In addition, narrow, deep constrictions are common and are usually bordered adapically by a forked rib.

These ammonites are identical with many small Mexican ammonites that were illustrated by Burckhardt (1912, pl. 7, figs. 4-14) under the name *Perisphinctes virgulatus* Quenstedt and were later assigned the subspecific name ***carribeanus*** by Jaworski (1940, p. 109). They were still later assigned to the subgenus ***Discosphinctes*** by Arkell (1956, p. 573).

Figured specimens: USMN 340030, 340031.

Occurrences: (1) Smackover Formation, near or below middle and in Union Producing Company No. 1 Carruthers *et al.* Kilgore well at depth of 11,210 feet, Claiborne Parish, Louisiana; (2) Smackover Formation, near or at top in the M. H. Marr, No. 1 Bennett Est. well at depth of 11,123 feet, Union Parish, Louisiana; (3) Bossier Formation, 208 feet above base, in the Exxon (Humble), No. 1 East Wright Mtn. Gas Unit at depth of 12,432 feet, Gregg County, Texas; (4) Smackover Formation in the Phillips No. 1 Kendrick well at depth of 8,741-56 feet, Bossier Parish, Louisiana; and (5) Smackover Formation in the Carter No. 1 Hope Fee well at depth of 6,570-6,571 feet, Morehouse Parish, Louisiana.

Subgenus ORTHOSPHINCTES Schindewolf, 1925
Perisphinctes (Orthosphinctes) cf. *P. (O.) rutteni* Jaworski
Plate 1, Figure 19

One ammonite fragment bears ribbing similar to that on *Perisphinctes (Orthosphinctes) rutteni* Jaworski (1940, Neues Jahrb., Beilage - B. 83, pt. B, no. 1, p. 105, figs. 2a, b, pl. 7, figs. 1, 2) from Oxfordian beds in Cuba. Its primary ribs are sharp moderately spaced, incline forward on the flanks, and generally bifurcate high on the flanks. Most forked ribs are separated by single sharp ribs that arise high on the flanks at, or a little above, the forked ribs.

Figured specimen: USNM 340033.

Occurrence: Smackover Formation, near middle, and 418 feet below top of formation. Ohio Oil Comapny No. H-1 Waller well at depth of 10,748 feet, sec. 3, T. 23 N., R. 8 W., Claiborne Parish, Louisiana.

Genus IDOCERAS Burckhardt, 1906
Idoceras cajense Burckhardt
Plate 2, Figures 4-8

Idoceras cajense Burckhardt, 1906, Inst. Geol. Mexico Bol. 23, p. 64, pl. 15, figs. 1-3.

This species is similar in shape, coiling, and ribbing to *I. zacatecanum* Burckhardt (1906, p. 42, pl. 9 figs. 1-4). It differs by its ribs inclining adorally more strongly on its flanks and bifurcating a little above instead of at the middle of its flanks, by the presence of more secondary ribs per primary rib, and by the ribs on its body chamber weakening considerably near the middle of the flanks.

Hypotypes: USNM 340034 to 340037.

Occurrences: (1) Top of Gilmer Limestone, Texas Oil and Gas Company No. 1-A Holmes well at depths of 13,789-13,802 feet, Freestone County, Texas. The species is possibly represented in the same well at depths of 13,756-13,783 feet; (2) Bossier Formation, lower part, 1,440 feet below top of formation, Carter Oil Company No. 3 Nora Davis well at depth of 12,770 feet, sec. 34, T. 18 N., R. 8 W., Webster Parish, Louisiana.

Idoceras densicostatum Imlay
Plate 2, Figure 10

Idoceras densicostatum Imlay, 1939, Geol. Soc. America Bull., vol. 50, p. 40, pl. 8, figs. 3-5.

This species is represented by one crushed specimen which closely resembles the holotype in its fairly involute coiling, very fine ribbing, and ribs bifurcating mainly on the upper third of the flanks. Compared with *I. zacatecanum* Burckhardt (1906, p. 42, pl. 9, figs. 1-4), it has finer ribs that bifurcate higher on the flanks and is more involute.

Hypotype: USNM 340038.

Occurrence: Gilmer Limestone, Texas Oil and Gas Company No. 1-A Holmes well at depth of 13,799 feet, Freestone County, Texas.

Idoceras cf. *I. durangense* Burckhardt
Plate 2, Figures 9, 11, 12

cf. *Idoceras durangense* Burckhardt, 1912, Inst. Geol. Mexico Bol. 29, p. 107, pl. 26, figs. 1-6, pl. 27, figs. 1, 2.

Idoceras cf. *I. durangense* Burckhardt. Imlay, 1945, Jour. Pal., v. 19, no. 3, p. 27, pl. 41, fig. 23.

Seven laterally crushed fragmentary ammonites show the features characteristic of *Idoceras durangense* Burckhardt (1912, p. 107, pl. 26, figs. 1-6, pl. 27, figs. 1, 2), of the similar species *I. tuttlei* Burckhardt (1912, p. 109, pl. 27, figs. 3-6), and of *I. neohispanicum* Burckhardt (1912, p. 111, pl. 29, figs. 1-4, 6, 8). The innermost exposed whorls, as preserved on four of these six specimens, bear fairly sharp, straight, forwardly inclined, moderately spaced ribs of which at least half bifurcate at, or above, the middle of the flanks. The next outer whorl bears slightly weaker primary ribs. The outermost whorl, preserved on all six specimens, bears slightly flexuous, widely spaced primary ribs that are much weaker near the middle of the flanks but divide higher up into two to four stronger and narrower secondary ribs that curve slightly backward and then are strongly curved forward.

Figured specimens: USNM 340038-340041.

Occurrences: (1) Bossier Formation, basal beds, Phillips Petroleum Company No. 1 Kendrick well at depths of 8,048-8,063 feet, sec. 22, T. 19 N., R. 11 W., Bossier Parish, Louisiana; (2) Gilmer Limestone, Texas Oil and Gas Company No. 1-A Holmes well at depths of 13,755, 13,763, 13,784.5, 13,785, and 13,788 feet, Freestone County, Texas; (3) Bossier Formation, basal beds in the Shell Oil Company No. 1 McNair well at depth of 22,363 feet, sec. 13, T. 4 N., R. 4 W., Hinds County, Mississippi.

Idoceras santarosanum Burckhardt
Plate 2, Figure 13

Idoceras santarosanum Burckhardt, 1906, Inst. Geol. Mexico Bol. 23, p. 58, pl. 14, figs. 5-7.

This species is represented by one specimen that includes parts of three whorls and is comparable in coiling and ribbing with the whorls near the adapical end of the body chamber of the holotype. The outer whorl in particular bears strong, fairly wide primary ribs that incline gently forward, that become a little weaker and wider near the middle of the flanks, and that bifurcate a little above the middle of the flanks. The secondary ribs are flexuous, fairly sharp, incline gently forward on the margin of the venter, and outnumber the primary ribs about 2 1/2 to 1. Some ribs arise singly between the forked ribs on the upper third of the whorl and some are connected faintly with the forked ribs.

Hypotype: USNM 340042 (a plaster replica of a specimen that was returned to Roy Hazzard of the Gulf Oil Company at his request).

Occurrences: Bossier Formation, lower part. (1) Union Producing Company No. B-1 Caddo Levee test well at depth of 10,150-10,170 feet, Caddo Parish, Louisiana; (2) The species is possibly represented by one fragmentary specimen from the lower part of the Bossier Formation in the Texaco Oil Company, No. 1 White well at depths of 11,547-11,647 feet, Limestone County, Texas. None the less, that specimen (No. 472 in the W. S. Adkins collection at the University of Texas) differs from *I. santarosanum* by being more involute and by having much fainter ribbing on its flanks. Overall it more closely resembles *I. sanlazarense* Imlay (1939, p. 39, pl. 11, figs. 3-6) from eastern Mexico.

Idoceras zacatecanum Burckhardt
Plate 2, Figures 1-3

Idoceras zacatecanum Burckhardt, 1906, Inst. Geol. Mexico Bol. 23, p.42, pl. 7, figs. 1-4.

Ten laterally crushed specimens in ribbing and coiling closely resemble the holotype of *Idoceras zacatecanum* Burckhardt (1906, p. 42, pl. 9, figs. 1-4). The whorls embrace about half of the preceding whorl. The ribs are fine, sharp, rather closely spaced, inclined gently forward on the flanks, become slightly stronger adorally, and form gentle chevrons on the venter. The outer whorl on the largest specimens bears several weak constrictions.

Hypotype: USNM 340043 to 340045.

Occurrences: Bossier Formation, basal part. Texas Oil and Gas Company No. 1-A Holmes well at depths of 13,760-13,763 feet and in the underlying Gilmer Limestone at depths of 13,787.5-13,800 feet, Freestone County, Texas.

Family ATAXIOCERATIDAE Buckman, 1921
Genus PROCRASPEDITES Spath, 1925
Procraspedites praecursor (Burckhardt)
Plate 1, Figure 22

Craspedites praecursor Burckhardt, 1906, Inst. Geol. Mexico Bol. 23, P. 98, figs. 1-3.

Procraspedites praecursor (Burckhardt). Spath, 1930, Monog. Geol. Dept. Hunterian Museum Univ. Glasgow, v. 4, p. 50.

One whorl fragment is identical in shape and ribbing with the adapical part of the outermost septate whorl of the genotype of *Procraspedites praecursor* (Burckhardt). It is discoidal and compressed, and has a small umbilicus. Several broad, very low, widely spaced ribs begin at the umbilical edge and trend radially to the middle of the flanks where each rib passes into several narrow but fairly strong ribs. These in turn pass into ever narrower, paired ribs on the upper third of the flanks. In addition, one or more equally narrow ribs arise between the paired ribs on the upper part of the flanks. Thus, three low, broad ribs pass ventrally into eleven pairs of fairly strong ribs which are succeeded near the venter by 32 very narrow ribs.

Hypotype: USNM 340046.

Occurrence: Bossier Formation, lower fourth. Texaco Oil Company No. 1 White well at depth of 11,547-11,647 feet, Limestone County, Texas.

Genus ATAXIOCERAS Fontannes, 1879
Ataxioceras sp.
Plate 1, Figure 18

Ataxioceras sp., Imlay, 1945, Jour. Paleontology, vol. 19, no. 3, p. 273, pl. 41, figs. 16, 17.

Several specimens are assigned to *Ataxioceras* because of the combination of sharp, prominent, widely spaced, prorsiradiate primary ribs; of bi-, tri-, and multiplicate rib branching; of equidistant secondary ribs that are strongest on the venter, and of pronounced constrictions (Imlay, 1945; p. 273-274).

Figured specimen: USNM 103667 and 103668.

Occurrence: Bossier? Formation, Phillips Petroleum Company No. 1 Kendrick well at a depth of 8,279-97 and 8,377-92 feet in sec. 22, T. 19 N., R. 11 W., Bossier Parish, Louisiana. The beds at depths between 8,230 to 8,404 feet probably represent a basinal facies of the Buckner Formation. The Bossier Formation is definitely represented at depths between 6,523 and 8,230 feet.

Genus VIRGATOSPHINCTES Uhlig, 1910
Virgatosphinctes cf. *V. aguilari* Burckhardt
Plate 2, Figure 14

One laterally compressed, fairly evolute ammonite bears very fine, closely spaced ribs that trend nearly radially on the flanks and become slightly stronger ventrally. Most of the ribs bifurcate near, or a little above the middle of the flanks, but a few remain simple.

This specimen in ribbing and coiling appears to be identical with the holotype of *Virgatosphinctes aguilari* (Burckhardt) (1906, p. 110, pl.27, figs. 6-9). It is much finer ribbed than *Aulacosphinctoides lauri* (Aguilera) as illustrated by Verma and Westermann (1973, p. 179-180, pl. 27, figs. 2-5).

This specimen of *Virgatosphinctes* is probably of early rather than late Tithonian age based on its occurrence in the same well and the same 100 feet of beds as *Procraspe-*

dites praecursor (Burckhardt), which is of Kimmeridgian age and is not known to occur above the middle part of the late Kimmeridgian (see figure 4). Presumably the specimen of *Virgatosphinctes* was collected from a slightly higher stratigraphic level than the specimen of *Procraspedites*.

Figured specimen: USNM 340047.

Occurrence: Bossier Formation. Texaco Oil Company No. 1 White well at a depth of 11,547-11,647 feet, Limestone County, Texas.

Family ASPIDOCERATIDAE Zittel, 1895
Genus EUASPIDOCERAS Spath, 1931
Euaspidoceras cf. *E. o'connelli* Sanchez Roig
Plate 1, figure 21

cf. *Aspidoceras o'connelli* Sanchez Roig, 1920, Cuba, Secretaria de Agricultura, Comercio y Trabajo, Habana, Bol., Especial, p. 30, pl. 13, figs. 1, 1A.

cf. *Euaspidoceras o'connelli* Sanchez Roig, 1951, Anales de la Academia de Ciencias Medicas, Fisicas y Naturales de la Habana, vol. 89, fasc. II, p. 70, pl. 8, fig. 3, pl. 11, pl. 12, pl. 13, figs. 1,2.

One laterally compressed internal mold of an inner whorl bears low broad widely spaced ribs that incline slightly forward on the flanks, bear weak tubercles at the umbilical margin, and terminate on the ventral margin in prominent, round tubercles. Both ribs and tubercles appear to be a little weaker and more closely spaced than on some comparable sized specimens of *A. o'connelli* from Cuba (Sanchez Roig, 1951, pl. 12, fig.1, pl. 13, fig. 1).

Figured specimen: USNM 340048.

Occurrence: Smackover Formation, near base of upper third. Ohio Oil Company No. H-1 Waller well at depth of 10,777 feet, Claiborne Parish, Louisiana.

Genus NEBRODITES Burckhardt, 1910
Nebrodites cf. *N. flexuosus* Burckhardt
Plate 2, Figures 16, 17

Two small specimens resemble *Nebrodites flexuosus* Burckhardt (1912, p. 92, pl. 21, figs. 4, 5, 10, 11) in their evolute coiling and rib pattern. The ribs are fairly strong, are nearly as wide as the interspaces, trend radially or slightly backward on the lower and middle parts of the flanks, and curve forward appreciably on the margin of the venter. On the smaller whorls all ribs appear to be simple. On the larger two whorls some of the ribs bifurcate at, or a little above, the middle of the flanks.

Figured specimens: USNM 340049 and 340059

Occurrence. Bossier Formation, lower sixth. Shell Oil Company No. 1 Temple well at depths of 14,361 and 14,391 feet, Sabine County, Texas.

Family OLCOSTEPHANIDAE Haug, 1910
Genus PRONICERAS Burckhardt, 1919
Proniceras? sp.
Plate 1, Figure 20

One crushed specimen has ribbing similar to that on the outer whorl of *Proniceras subpronum* Burckhardt (1919, p. 48; 1921, pl. 16, figs. 10, 21, 29) and *Proniceras jimulcense* Imlay (1939, p. 55, pl. 18, fig. 3). That specimen on the lower part of the flanks exhibits fairly strong radially elongate tubercles from which pass bundles of three much weaker ribs. Between those bundles arise one or two simple ribs. All ribs trend radially on the lower half of the flanks and then curve slightly forward on the upper half.

This specimen is associated with a small ammonite fragement that bears ribbing and constrictions similar to that on the *Aulacosphinctes torresianus* Burckhardt (1919, p. 52; 1921, pl. 17, figs. 4-6).

Figured specimen: USNM 340051.

Occurrence: Bossier Formation, middle part, Shell Oil Company No. 1 Boise Southern well at depth of 11,638-11,650 feet (Core 3) in sec. 7, T. 7 N., R. 13 W., Sabine Parish, Louisiana.

Family BERRIASELLIDAE Spath, 1922
Genus SUBSTEUEROCERAS Spath, 1923
Substeuroceras cf. *S. alticostatum* Imlay
Plate 2, Figure 15

One specimen resembles *S. alticostatum* Imlay (1939, p. 51, pl. 15, figs. 1-7) by having similarly coarse, widely spaced ribs of which most fork near the middle of the flanks. Its ribbing resembles that of *S. santarosanum* (Burckhardt)(1906, p. 129, pl. 35, figs. 1-4) except for the presence of fewer simple ribs. *Substeuroceras subquadratum* Imlay (1939, p. 49, pl. 15, figs. 8, 12-15) has finer ribbing on its inner whorls.

Figured specimen: USNM 340052.

Occurrence: Bossier Formation, lower part of upper third, 239 feet below top and 548 feet above base. Humble Oil and Gas Company No. 1 Benevides well at depth of 17,317 feet, Webb County, Texas.

Genus DURANGITES Burckhardt, 1912
Durangites? cf. *D. vulgaris* Burckhardt
Plate 1, Figure 24

Two ammonite fragments were recovered from one core and bear fine ribs and tubercles similar to those in some small specimens of *Durangites vulgaris* Burckhardt (1912, Inst. geol. Mexico Bol. 29, p. 150, pl. 37, figs. 3-15, 28-36; Imlay, 1942, Geol. Soc. America Bull., vol. 53, p. 1453, pl. 3, fig. 1). The smallest ammonite fragment represents only the outer part of a whorl, and bears both forked and single sharp ribs that pass into tubercles on the margin of the venter. The largest ammonite fragment represents the entire side of a whorl, bears somewhat stronger forked and single ribs that trend radially or slightly adorally on the lower part of the flanks, curve slightly adapically (backward) on the upper part of the flanks, and then pass into fairly prominent tubercles on the margins of the venter.

The specimens differ from any described species of *Micracanthoceras* by having much finer ribs on their flanks and much stronger tubercles on their venter.

Figured specimen: USNM 340053.

Occurrence: Bossier Formation, lower part of upper third, Humble Oil and Gas No. 1 Benevides well at depth of 17,803 feet, Webb County, Texas.

ACKNOWLEDGMENTS

This report concerning Jurassic ammonites from the subsurface of the Gulf Region of the United States is based mostly on fossils collected from well cores by many oil company geologists since 1945 (Imlay, 1945, p. 255). The resulting studies have made possible fairly accurate age determinations of the Jurassic formations from the lower fifth of the Smackover Formation to near to the top of the Bossier Formation.

REFERENCES

Arkell, W. J.,1956, Jurassic geology of the world: London, Oliver & Boyd, 806 p., 102 figs., 46 pls.

____________, and others, 1957, Mesozoic Ammonoidea, in Moore, R. C., ed., Treatise on invertebrate paleontology pt. L, Mollusca 4 - Cephalopoda Ammonoidea: New York and Lawrence, Kans., Geological Society of America and University of Kansas Press, 490 p.

Badon, C. L., 1975, Stratigraphy and petrology of Jurassic Norphlet Formation, Clarke County, Mississippi: American Association of Petroleum Geologists Bulletin, v. 59, p. 377-392, 9 figs.

Burckhardt, Carlos, 1906, La faune jurassique de Mazapil avec un appendice sur les fossiles du cretacique inferieur: Mexico Instituto Geologico Boletin 23, 216 p., 43 pls.

________, 1912, Faunes jurassiques et cretaciques de San Pedro del Gallo: Mexico Instituto Geologico Boletin 29, 264 p., 46 pls.

________, 1919, 1921, Faunes Jurasicas de Symon (Zacatecas): Mexico Instituto Geologico Boletin 33, 135 p., 32 pls.

Cantu, A. C., 1968, Sobre una Asociacion *Proniceras-Durangites-"Hildoglochiceras"* del Noreste de Mexico: Instituto Mexicano del Petroleo, Seccion Geologia, Monografia no. 2, p. 19-26, pls. 1-5.

____, 1982, The Jurassic-Cretaceous boundary in the subsurface of eastern Mexico: Journal of Petroleum Geology, vol. 4, pt. 3, p. 311-318, pl. 1, text-figures 1-7.

Dickinson, K. A., 1968, Upper Jurassic stratigraphy of some adjacent parts of Texas, Louisiana, and Arkansas: U. S. Geological Survey Professional Paper 594-E., p. E1-E25, 1 pl., 13 text figs.

Forgotson, J. M., Jr., 1954, Regional stratigraphic analysis of the Cotton Valley Group of Upper Gulf Coastal Plain: American Association of Petroleum Geologists Bulletin, v. 38, no. 12, p. 2476-2499.

Forgotson, J. M., and J. M. Forgotson, Jr., 1976, Definition of the Gilmer Limestone, Upper Jurassic formation, northeastern Texas: American Association of Petroleum Geologists Bull., v. 60, p. 1119-1123, figs. 1-3.

Hazzard, R. T., 1939, Notes on the Comanche and pre-Comanche? Mesozoic formations of Arkansas-Louisiana-Texas area, and a suggested correlation with northern Mexico: Shreveport Geological Society Guidebook 14th Annual Field Trip, p. 155-165.

Imlay, R. W., 1939, Jurassic ammonites from Mexico: Geological Society of America Bulletin, v. 50, p. 1-78, 18 pls., 7 figs.

______, 1940, Lower Cretaceous and Jurassic formations of southern Arkansas and their oil and gas possibilities: Arkansas Geological Survey Information Circular 12, 64 p., 26 pls.

_____, 1942, Late Jurassic fossils from Cuba and their economic significance: Geological Society of America Bulletin, v. 53, p. 1417-1478, 12 pls. 4 figs. of Gulf Region: American Association of Petroleum Geologists Bulletin, v. 27, no. 11, p. 1407-1533, 14 figs.

_____, 1945, Jurassic fossils from the Southern States, No. 2: Journal of Paleontology, v. 19, no. 3, p. 253-276, pls. 39-41, 1 fig.

Jaworski, E., 1940, Oxford-Ammoniten von Cuba: Neues Jahrb., Beilage-Band 83, pt. B, no. 1, p. 87-137, pls. 3-7.

Mann, C. J., and W. A. Thomas, 1964, Cotton Valley Group (Jurassic) nomenclature, Louisiana and Arkansas: Gulf Coast Association of Geological Societies, Transactions, v. 14, p. 143-152.

Moore, C. H., and Y. Druckman, 1981, Burial diagenesis and porosity evolution, Upper Jurassic Smackover, Arkansas and Louisiana: American Association of Petroleum Geologists Bulletin, v. 65, p. 597-628, 13 figs.

O'Connell, Marjorie, 1920, The Jurassic ammonite fauna of Cuba: American Museum of Natural History Bulletin, v. 42, art. 16, p. 643-692, pls. 34-38.

Oxley, M. L., E. D. Minihan, and J. M. Ridgeway, 1968a, A study of the Jurassic sediments in portions of Mississippi and Alabama: Gulf Coast Association of Geological Societies, Transactions, v. 17, p. 24-48, 15 text figs., Mississippi Geological, Economic and Topographical Survey Bulletin 109, p. 24-39, 15 text figs.

______, 1968b, Jurassic stratigraphy of central and southern Mississippi: Mississippi Geological , Economic and Topographical Survey Bulletin 109, p. 9-37, 1 fig.

Sanchez Roig, Mario, 1920, La fauna Jurasica de Vinales: Cuba, Secretaria de Agr., Bol. Especial, 61 p., 23 pls.

____________, 1951, La fauna Jurasica de Vinales: Acad. Cienc. Mexicas, Fisicas y Naturales Habana Anales, v. 89, p. 2, p. 46-94, 23 pls.

Spath, L. F., 1930, The Jurassic ammonite faunas of the neighborhood of Mombasa: Mon. Geol. Dept. Hunterian Mus., Univ. Glasgow, v. 4, p. 13-70, pls. 1-8.

Swain, F. K., 1944, Stratigraphy of the Cotton valley beds of northern Gulf Coastal Plain: American Association of Petroleum Geologists Bulletin, v. 28, no. 5, p. 577-614.

Verma, H. K., and G. E. G. Westermann, 1973, The Tithonian (Jurassic) ammonite fauna and stratigraphy of Sierra Catorce, San Luis Potosi, Mexico: American Paleontology Bulletin, v. 63, no. 277, p. 107-320, pls. 22-56, 32 text figs.

Weeks, W. B., 1938, South Arkansas stratigraphy with emphasis on the older coastal plain beds: American Association of Petroleum Geologists Bulletin, v. 22, no. 8, p.953-983, 3 text figs.

STRATIGRAPHIC STUDIES IN THE JURASSIC OF NORTHEASTERN MEXICO: EVIDENCE FOR THE ORIGIN OF THE SABINAS BASIN

JOSE F. LONGORIA

Geosciences Program
The University of Texas at Dallas
P.O. Box 688
Richardson, Texas 75080

ABSTRACT

The Jurassic sequence of northeastern Mexico includes Oxfordian through Tithonian-aged marine sediments that were deposited in eight environments: (1) sabkha, (2) nearshore, (3) lagoonal, (4) shoal, (5) inner-shelf, (6) restricted-shelf, (7) open-shelf, and (8) pelagic. Lithologic correlations relate the sections studied with the type sections of coeval rocks described from the Sierra de Parras. Based on their lithic characteristics they are assigned to the Minas Viejas, Zuloaga, La Gloria, La Caja, La Casita, and San Angel Formations.

Three biohorizons are established: (1) the first stratigraphic occurrence of the ammonite *Idoceras* in the early Kimmeridgian, (2) the first appearance of calpionellids in the middle Tithonian, and (3) the first occurrence of *Calpionella elliptica* Cadish at the Jurassic/Cretaceous boundary. Two lowerings of sea level are identified in the sections studied, the first in early Kimmeridgian and the second in the early Tithonian. These two drops of sea level are separated by a sea-level rise in the middle Tithonian. Eustatic changes of sea level are associated with tectonism in the area.

Active transform faulting in the Gulf of Mexico region resulted in the juxtaposition of Oxfordian-Kimmeridgian shallow- and deep-water facies. The translations along transcurrent faults result in a SE-NW displacement of about 300 kilometers, which caused the formation of the Sabinas Basin of northeastern Mexico. Late Jurassic sea-floor spreading in the central Gulf of Mexico is regarded as the driving mechanism.

The Oxfordian-Kimmeridgian and late Tithonian paleogeographies of northeastern Mexico are reconstructed, and parallel lineal facies belts are envisioned which account for the sporadic distribution of the Jurassic shallow-water facies in northeastern Mexico.

INTRODUCTION

The Jurassic deposits of northern Mexico consist of about 1,500 meters of marine sediments, representing the best known record of Jurassic history in the entire circum-Gulf area. Most of these rocks are well exposed in the cores of overturned anticlines (Fig. 1). They were first reported by Bose (1923) who wrote: "On the following pages I shall try to show that these mountain ranges and canyons, which sometimes are of real alpine beauty and grandeur, interest us not only because of their complicated structure, but also in that their stratigraphy, are of importance in the paleogeography of North America, and especially for northern Mexico and Texas."

Burckhardt (1930a) presented his excellent synthesis on the Mexican Mesozoic. In addition to the great number of Jurassic outcrops he discovered, he established the chrono and lithostratigraphic basis for the Jurassic of northern Mexico. Furthermore, he postulated the existence of an angular uncon-

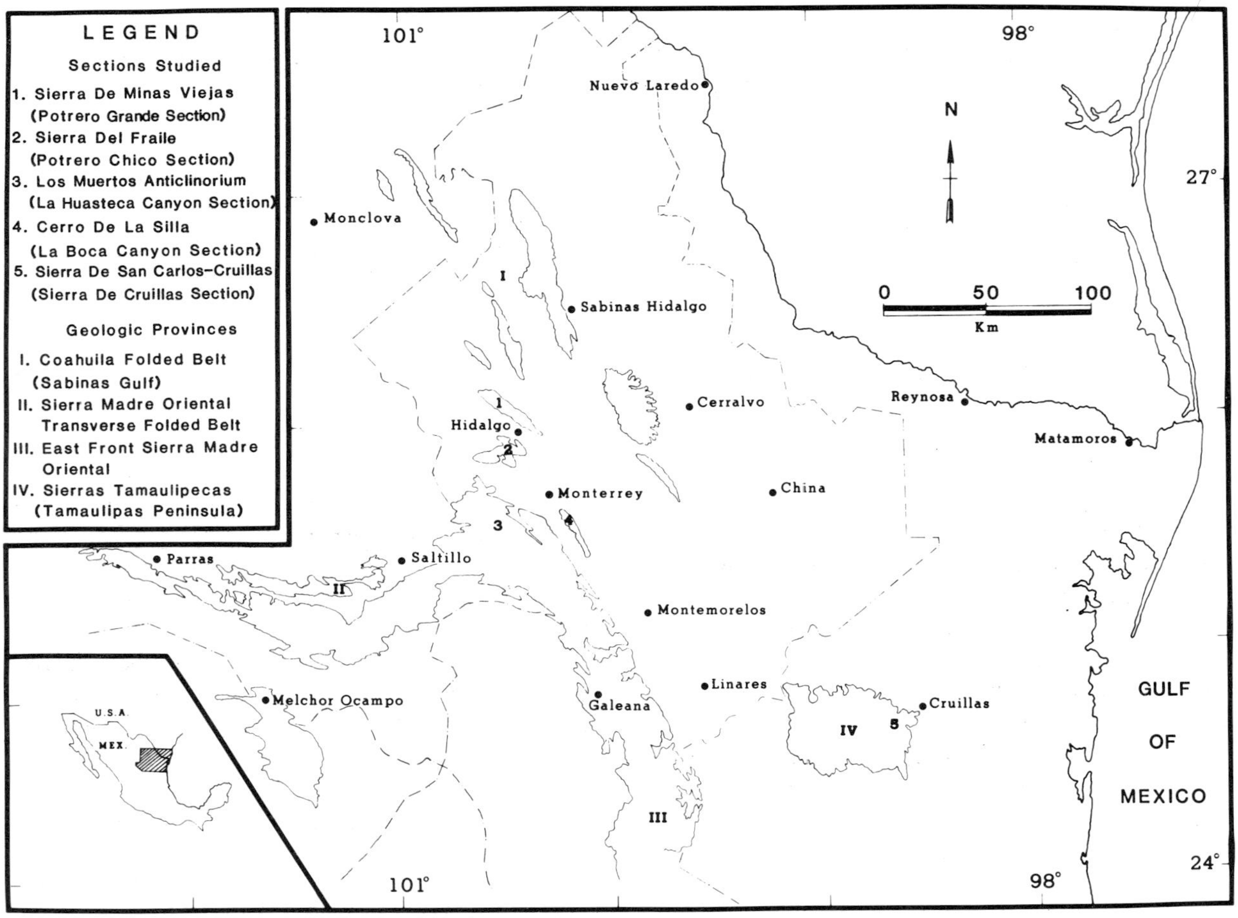

Figure 1. Index map of northeastern Mexico showing the geologic provinces and location of the stratigraphic sections studied.

formity between Portlandian (= Tithonian) and Early Cretaceous sediments of San Pedro del Gallo (Durango state).

Imlay (1936, 1937, 1938, 1943) established the basis for a formal Jurassic lithostratigraphy for northern Mexico in which he coined most of the lithostratigraphic terms in use today.

Barker (1941) presented a general view on the Late Jurassic deposits of the Monterrey-Saltillo area, describing some of the most important geologic structures of this region.

Bonet (1956) established the basis for a formal Jurassic biostratigraphy using pelagic microfossils and presented the first systematic description of some Jurassic rocks in thin section.

During the past three decades, very little has been added to the pioneering geological work of these authors. Furthermore, some fundamental aspects of their discoveries have remained neglected. For instance, there is no further reference to the Late Jurassic or Early Cretaceous tectonism in northern Mexico postulated by Burckhardt (1930a, p. 73-74).

Surprisingly enough, in 1983, sixty years after Bose's paper was published very little is yet understood on the Jurassic stratigraphy of northeastern Mexico and the significance of these rocks in the paleogeographic setting of North America. In this paper, I shall try to show that stratigraphic studies of the Jurassic deposits of northeastern Mexico are vital to the reconstruction of the paleogeography of North America;

furthermore, I shall show that Jurassic stratigraphy of northeastern Mexico has direct implications on the geodynamics of the Gulf of Mexico.

GEOLOGICAL SETTING

The area under consideration is located between the Mexican cities of Saltillo, Coahuila; Monterrey, Nuevo Leon; and Cruillas, Tamaulipas (Fig. 1). Geologically, the area includes part of the Sierras Tamaulipecas, Coahuila Ridge and Basin (=Coahuila Folded Belt), and Sierra Madre Oriental physiographic provinces of northern Mexico as defined by Humphrey (1956). The paleogeographic setting of these provinces, however, is unclear and remains purely speculative. It has been postulated, for example, that the Coahuila Ridge and Basin province corresponds to the area of the Mesozoic Sabinas Basin; whereas the Sierra Madre Oriental is believed to coincide with the Mesozoic Mexican Geosyncline. Neither of these two paleogeographic elements have actually been proven to have existed during Mesozoic time (Weide, 1968).

The oldest known rocks in the region are the Huizachal red beds of the Galeana area (about 30 km south of Monterrey). The oldest Jurassic rocks known within the studied area are the Minas Viejas evaporites whose stratigraphic relationships indicate that they are of Oxfordian age. In the Ciudad Victoria area (Peregrina Canyon), the evaporite sequence uncomformably overlies the Huizachal red beds.

Jurassic rocks of northeastern Mexico are intensely folded. The timing of the folding, however, is controversial. Earlier workers such as Bose (1923), Burckhardt (1930b), and Imlay (1943) had postulated a pre-Early Cretaceous tectonic episode, more likely related to the Nevadian or Palisade orogenies of North America. No reference to this tectonic episode in northern Mexico has appeared in later publications. According to Cserna (1956), the main deformation of the Mesozoic sequence occurred during the Laramide Orogeny. On the other hand, Tardy (1980) postulated that Mesozoic rocks of this region underwent two major orogenic events, the first of which took place in early Eocene time, and the second in Miocene time. According to Tardy (1980), the later episode is responsible for the NW-SE oriented folds in the entire region.

The disruption of Paleozoic and Mesozoic terranes in northern Mexico has led some authors to postulate the existence of megashears such as the Torreon-Monterrey Fault (Cserna, 1970), the Mojave-Sonora Megashear (Silver and Anderson, 1974; Anderson and Schmidt, 1983), and the Caltam Megashear (Lineament) proposed by Tardy (1980). The lack of stratigraphic studies, however, impedes the recognition of the influence that these paleotectonic elements had in the paleogeographic evolution of southern North America in Late Jurassic time.

LITHOSTRATIGRAPHY

The Jurassic rocks of northeastern Mexico are assigned to the following lithostratigraphic units: (1) Minas Viejas Gypsum, (2) Zuloaga Limestone, (3) La Gloria Formation, (4) La Caja Formation, (5) La Casita Formation, and (6) San Angel Limestone (Fig. 2.)

The oldest marine Jurassic rocks are assigned to the Oxfordian Stage (Minas Viejas Gypsum and Zuloaga Limestone). The presence of *Nerinea*-bearing limestones in northern Mexico was earlier noted by Bose (1923) and Burckhardt (1930a). Later, these rocks were referred to as the Zuloaga Limestone by Imlay (1938), and the Novillo Formation by Heim (1940).

The apparently isolated occurrence of Jurassic rocks in the western part of the state of Tamaulipas led Heim (1940) to introduce the terms "Olvido" and "Novillo" to describe the Oxfordian gypsiferous and carbonate rocks, respectively, exposed in the Huizachal-Peregrina Anticlinorium. The term "Zuloaga Group" was introduced by Humphrey (1956) to include the Zuloaga and Minas Viejas formations, and he also coined the term "La Casita Group" to include Jurassic strata

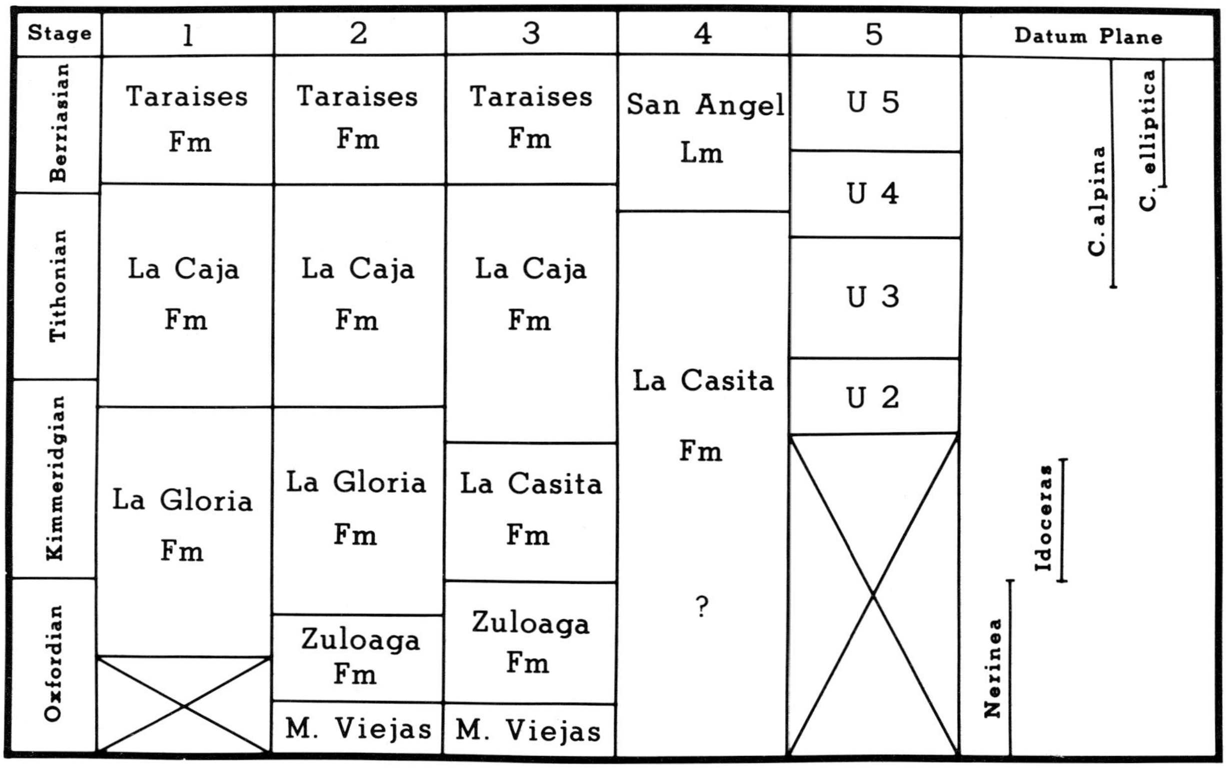

Figure 2. Bio- and chronostratigraphic position of the sections studied: (1) La Huasteca Section, (2) Potrero Chico Section, (3) PotreroGrande Section, (4) La Boca Canyon Section, (5) Sierra de Cruillas Section.

that overlie the Zuloaga Group. Both "groups" were included in the Sabinas Series. Later, Murray (1961) introduced the use of the provincial time-stratigraphic (chronostratigraphic) units, the Zuloaga and la Casita stages. Consequently, the lithic characteristics of these originally lithostratigraphic units are no longer the diagnostic feature used to recognize the formations.

The Sabinas Series as originally defined by Humphrey (1956) and later adopted by Murray (1961) represents a complete sedimentary cycle, reflecting the tectostratigraphic conditions that prevailed in the region. It corresponds to the Late Jurassic Series (Oxfordian through Tithonian). Humphrey (1956) noted that the Jurassic-Cretaceous boundary is at the top of the La Casita Formation. He also pointed out that in several ranges in the Sierra Madre Oriental the boundary is marked by an erosional uncomformity, in contrast to the eastern sequences, such as the Cerro de la Silla, where the Late Jurassic strata represent a continuous sedimentary cycle without apparent interruption.

Minas Viejas Gypsum
(Wall, Murray and Diaz, 1961)

The oldest Jurassic marine sediments in northeastern Mexico are referred to as the Minas Viejas Gypsum. This unit was first described by Wall *et al.*, (1961) as an informal unit to include the gypsiferous beds underlying the La Gloria and Zuloaga formations in the Minas Viejas Anticline (Potrero Grande), about 30 km north of Monterrey, Nuevo Leon.

The Minas Viejas Gypsum is a lateral equivalent of both the Zuloaga and La Gloria formations. The gypsum is believed to have been deposited in a semi-restricted environment. The stratigraphic relationships of the Minas Viejas with other evaporite units in northern Mexico, such as the Olvido Gypsum and the halite and anhydrite bodies found in Tamaulipas state and in the subsurface of eastern Mexico, remain unclear. The Olvido Formation described by Heim (1940) from the Victoria area is a synonym of this unit.

La Gloria Formation (Imlay, 1936)

The La Gloria Formation is predominantly a clastic unit, consisting of thick- to medium-bedded coarse conglomerate and coarse arkosic sandstone. Only a few limestone, mainly packstone, layers are often observed in the upper part. The La Gloria is in part the time equivalent of the Zuloaga Limestone, and represents a nearshore deposit. It is likely that in the Huasteca Canyon the La Gloria Formation overlies the Paleozoic basement. Its upper contact with the La Casita Formation is gradational. Oxfordian ammonites were reported by Burckhardt (1930a) and Imlay (1938) from the La Gloria near San Pedro del Gallo, Durango state.

Zuloaga Limestone (Imlay, 1938)

The depositional environments of the Zuloaga Limestone were established by Oivanki (1974). This unit is characterized by its thick- to massive-bedded dark-brown limestones, and by its lack of shales. The abundance of ooid packstones-grainstones and mollusk wackestones in the Zuloaga permit one to postulate that this unit was deposited as a calcareous off-shore bank or shoal.

In the majority of the breached anticlines of the Monterrey area, the Zuloaga Limestone forms the core of the structures and constitutes isolated bluffs. At La Huasteca Canyon (SW of Monterrey), the Zuloaga is overlain by the La Gloria Formation. This contact, which is difficult to observe in the field, is probably represented by a minor uncomformity. Basinward, the Zuloaga Limestone interfingers with the La Casita Formation. Field relationships show a gradual migration (onlapping) of the Zuloaga Limestone in a ESE-WNW direction. Thus, the Zuloaga represents one of the best units for reconstructing the Jurassic paleogeography of northeastern Mexico because it indicates linear depositional belts which parallel the paleo shoreline.

La Casita Formation (Imlay, 1936)

The La Casita Formation can easily be identified in the field because it consists of irregular alternating layers of highly bituminous and carbonaceous limestone, sandstone, shale, and conglomerate. Several authors have misidentified this unit and have referred to it strata that are outside the lithologic limits of variation that can be allowed for this unit. The La Casita as exposed along the La Boca Canyon (Cerro de la Silla)(Fig. 7) is considered to be one of the more typical exposures of this unit outside of the type area.

The predominantly shaly facies of the La Casita Formation is referred to as the La Caja Formation. The highly terrigenous character of the La Casita led Imlay (1938) to postulate that the La Casita represents the nearshore equivalent of the La Caja Formation. This interpretation needs, however, to be documented with more field studies. It is likely, nonetheless, that both units reflect changes in sea level. Abundant Kimmeridgian ammonites occur in the middle part of the La Casita and planktonic microfossils are common in the uppermost part of this unit.

La Caja Formation (Imlay, 1938)

La Caja Formation is predominantly terrigenous in character, but consists of homogeneous, alternating thin- to medium-bedded calcareous shales and shaly limestones, which contain abundant phosphatic nodules. The La Caja exposures are readily recognized from a distance because of their dusty weathering and tan to light brown color. The La Caja Formation contains abundant Tithonian ammonites, which date the formation as time equivalent of the upper part of the La Casita Formation. The La Caja Formation can be easily distinguished from the La Casita by the lack of black shales and bituminous limestone in the former.

San Angel Limestone
(Longoria and Davila, 1979)

The San Angel Limestone is characterized by its homogeneous sequence of black, medium- to thick-bedded limestone, mainly mudstone to wackestone. Locally, this unit contains black chert nodules and lenses. The San Angel Limestone was proposed by Longoria and Davila (1979) to represent the deeper-water, Late Jurassic strata in northern Mexico. The Jurasic/Cretaceous boundary occurs within the San Angel at the type locality. The San Angel was previously included in the Cupido or Tamaulipas formations, but it differs from those units by containing black bituminous carbonates. The San Angel Limestone has a wide geographic distribution in northern Mexico and is regarded to have been deposited in bathyal (outer shelf) water depths. Planktonic microfossils (calpionellids, radiolarians, and nannoconids), and ammonites are the main faunal components of San Angel Limestone.

STRATIGRAPHIC SECTIONS STUDIED

Stratigraphic studies in northern Mexico, like those in many other regions of the world, encounter serious nomenclatural problems. A first look at the geological literature from northeastern Mexico reveals the constant misuse of lithocorrelation, whereby the lithic characterization of some of the Jurassic formations has been modified from author to author. Furthermore, contrary to the fundamental procedures of lithostratigraphy (North American Commission on Stratigraphic Nomenclature; 1983, Article 22), the formations are no longer recognized based upon their lithic characteristics but rather on their assumed chronostratigraphic position. As an example, the strata referred to as the La Casita Formation by Vokes (1963) can not be assigned to this formation because it does not display the lithic features that serve to define it (Imlay, 1938). Consequently, serious limitations arise in using lithocorrelations in paleogeographic reconstructions of northern Mexico.

During the present study, field work was carried-out at selected localities and several sections were measured. The Jurassic sections so studied were subdivided into lithic units on the basis of their physical attributes (lithic characteristics) observed in the field. These units were later correlated with the type sections (stratotypes) of the Jurassic formations in northern Mexico. The descriptions of the measured stratigraphic sections will be given in a separate paper. A brief field characterization of the sections studied and their lithocorrelation with the Jurassic sequence established by Imlay (1936, 1937, 1938) are presented below (Figs. 3-8).

La Huasteca Canyon Section (Los Muertos Anticlinorium, Sierra Madre Oriental)

Locality

La Huasteca Canyon is located in the Los Muertos Anticlinorium of the northern front of the Sierra Madre Oriental. The canyon is located on the northern flank of the Santa Catarina Anticline (Fig. 1). The Santa Catarina is a narrow, elongate, east-west trending breached anticline about 25 km long and 5 km wide. The Santa Catarina Anticline was mapped by Vokes (1963), who gave a detailed description of several sections within the La Huasteca Canyon region.

The Jurassic section studied is exposed on the central portion of the structure. The breached nature of this structure exposes the basal part of the section, and the northern flank of the anticline was measured.

Description

The Jurassic sequences of the La Huasteca Canyon are about 740 meters thick, and are characterized by the predominance of siliciclastic sediments. The following lithic units were recognized, from bottom to top in this section (Fig. 3):

Unit 1: Thick-bedded brown to cream colored micritic limestone (mudstone, wacke-

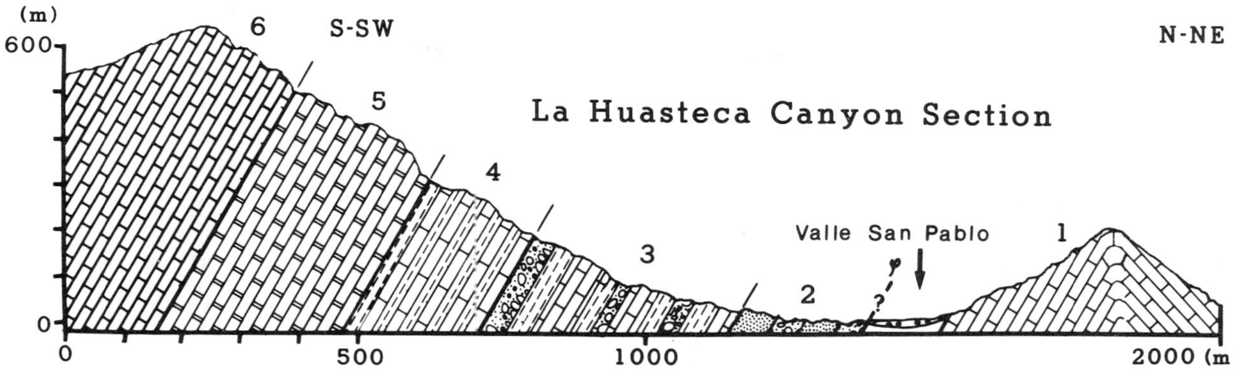

Figure 3. Field stratigraphic profile of the La Huasteca Canyon Section, showing the lithic units established: (1) Zuloaga Limestone, (2) and (3) La Gloria Formation (4) La Caja Formation, (5) Taraises Formation, (6) Cupido Formation. The lithic characterization of each unit is given in the text.

stone, and packstone). The middle part of this unit contains specimens of *Nerinea*.

Unit 2: Thick-bedded, tan to yellowish, coarse arkosic conglomerate and arkosic sandstone.

Unit 3: Irregular, alternating, medium-bedded (10 to 30 cm thick) limestone, shale, and conglomerate, weathering tan to yellowish. Limestone beds contain abundant mollusk fragments (coquinas).

Unit 4: Evenly bedded, alternating dark-brown shale and limestone.

Unit 5: Medium to thick-bedded dark-gray limestone (mudstone and wackestone), containing specimens of the ammonite genus *Olcostephanus*.

Lithocorrelation

Unit 1 corresponds to the Zuloaga Limestone. It forms isolated bluffs in the center of the breached anticline. Units 2 to 4 have been assigned to the La Casita Formation by Humphrey (1956), Humphrey and Diaz (1956), and Vokes (1963). However, my comparative studies of this part of the section with the type exposures of the La Casita Formation from the western part of the Sierra de Parras demonstrate that the La Huasteca section can not be assigned to the La Casita Formation because it lacks the bituminous character typical of latter unit. Vokes (1963) subdivided the La Casita into two formations: the Muralla Shale, and San Paulo (sic.!) Sandstone. It is likely that the Muralla Shale as described by Vokes (op. cit.) corresponds to my Unit 4. Thus, it is referred to the La Caja Formation. The San Paulo (sic.!) Sandstone described by Vokes (1963) corresponds to my Units 2 and 3 and are assigned to the La Gloria Formation mainly because of their clastic character. Unit 4 is referred to the La Caja Formation because of their clastic character. Unit 4 is referred to the La Caja Formation because this predominantly shale unit resembles the latter formation exposed in the Sierra de Parras area (Sierra de la Caja, Mazapil, Zacatecas state), and locally contains abundant ammonites assigned to the genus *Idoceras*. Unit 5 corresponds to the Taraises Formation.

Potrero Chico Section
(Sierra del Fraile)

Locality

The Sierra del Fraile is located 30 km to the northwest of Monterrey, Nuevo Leon, and lies between the towns of Hidalgo to the north and Villa de Garcia to the south (Fig. 1). The Sierra del Fraile consists of two doubly plunging breached anticlines separated by narrow synclines which plunge toward the east and west. The whole range has a maximum length of about 30 km. The northern anticline is known as Sierra San Miguel and contains in its middle part the deep Potrero Chico canyon. The southern anticline includes the Fraile Peak and the Garcia Caves.

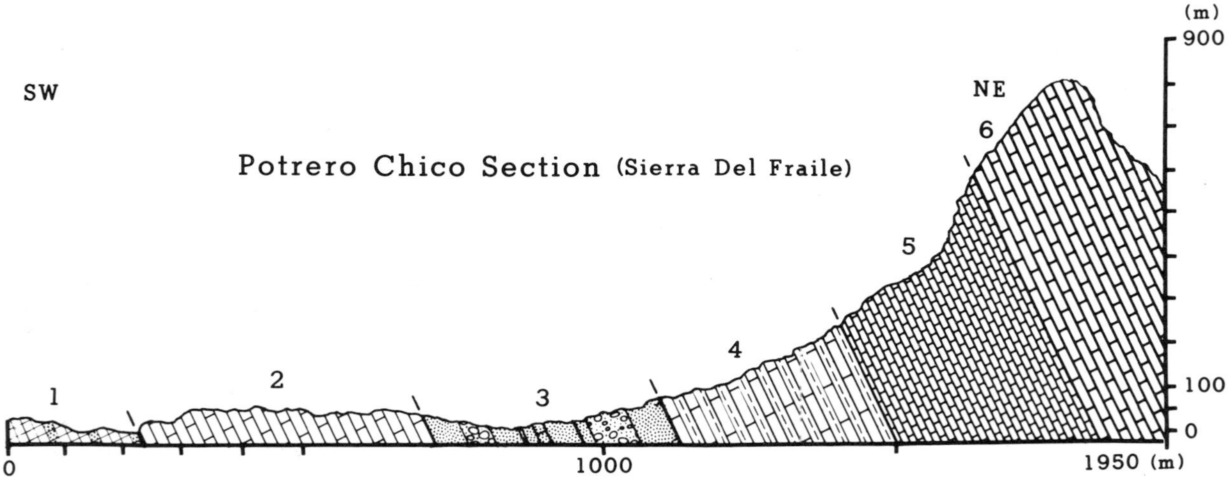

Figure 4. Field stratigraphic profile of the Potrero Chico Section: (1) Minas Viejas Gypsum, (2) Zuloaga Limestone, (3) La Gloria Formation, (4) La Caja Formation, (5) Taraises Formation, (6) Cupido Formation. The lithic characterization of each unit is given in the text.

Description

The section was studied on both the southern (Grutas de Garcia) and northern (Sierra de San Miguel) sides of this structure. At Potrero Chico (Fig. 4), about 700 meters of Jurassic sediments are exposed. The sequence consists of the following lithic units, from bottom to top:

Unit 1: White to brownish, massive gypsum, and gypsiferous sandstone.

Unit 2: Thick- to massive-bedded, brown, fossiliferous (mainly mollusks), and ooid limestones (wackestone and packstone).

Unit 3: Medium- to thick-bedded, tan to yellowish, coarse arkosic sandstone and conglomerate, with a few limestone beds.

Unit 4: Regularly alternating shale and clayey limestone (mudstone and wackestone) beds, dark-gray to brown in color.

Unit 5: Evenly bedded, medium-bedded, dark-gray limestone (mudstone and wackestone).

Lithocorrelation

The lithic units established from the Jurassic sequence of the Potrero Chico section are correlated with coeval rocks of the Parras area as follows: Unit 1 represents the Minas Viejas Gypsum; Unit 2 is remarkably similar to the Zuloaga Limestone in the Melchor Ocampo region; Unit 3 represents the La Gloria Formation; Unit 4 corresponds lithologically to the La Caja Formation; Unit 5 is lithocorrelated with the Taraises Formation. The upper part of Unit 4 represents a lithic transition between 4 and 5, which could easily be considered as a separate formation.

Potrero Grande Section
(Sierra de Minas Viejas)

Locality

The Sierra de Minas Viejas is a long, SE-NW trending, isolated, breached anticline, about 35 km long and 8 km wide, located north of the towns of Hidalgo and Minas, Nuevo Leon state (Fig. 1). Excellent exposures of Jurassic rocks can be studied in the central part of this structure which is known as Potrero Grande. Active mining in the gypsum quarries has uncovered the oldest Jurassic section known in the region.

The Jurassic section was studied at several localities within the Minas Viejas Anticline with the Potrero Grande section being the more complete exposure. The section was measured along the Arroyo Palos Blancos which cuts the structure in a north-south transect.

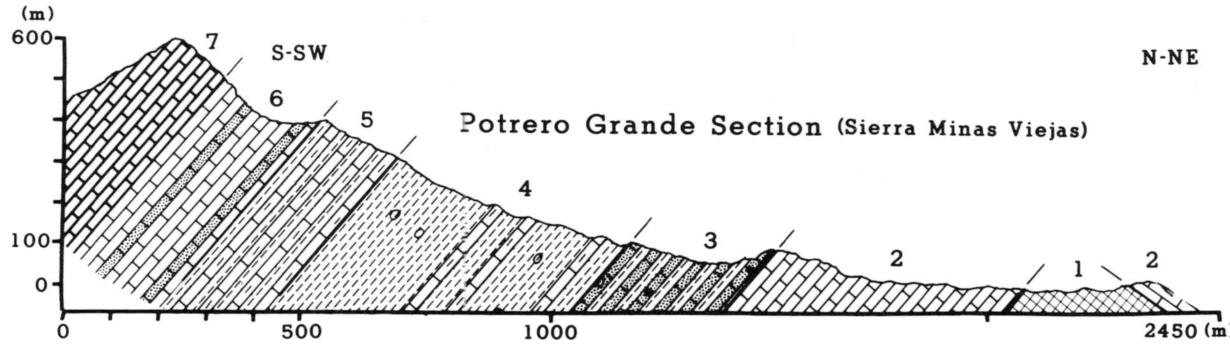

Figure 5. Field stratigraphic profile of the Potrero Grande Section: (1) Minas Viejas Gypsum, (2) Zuloaga Limestone, (3) La Casita Formation, (4) La Caja Formation, (5) and (6) Taraises Formation, (7) Cupido Formation. Lithic characterizations of each unit are given in the text.

Description

In the Potrero Grande section, the following lithologic units are recognized, from bottom to top (Fig. 5):

Unit 1: Massive, somewhat compact gypsum.

Unit 2: Thick-bedded, brown limestone (mudstone, wackestone to packstone), with stylolites.

Unit 3: Homogeneous, alternating medium- to thick-bedded, black bituminous, sandy limestone and carbonaceous shale. The limestone beds contain calcareous nodules about 20 cm in length.

Unit 4: Brown to yellowish-tan, evenly bedded calcareous shale, containing phosphatic concretions and thin- to medium-bedded clayey limestone.

Unit 5: Homogeneous, rhythmic alternation of medium-bedded, dark-gray calcareous shale and clayey limestone (mainly wackestone).

Unit 6: Evenly-bedded, medium-bedded, gray limestone (mudstone and wackestone), containing a few irregular layers of sandy limestone.

Lithocorrelation

Unit 1 corresponds to the Minas Viejas

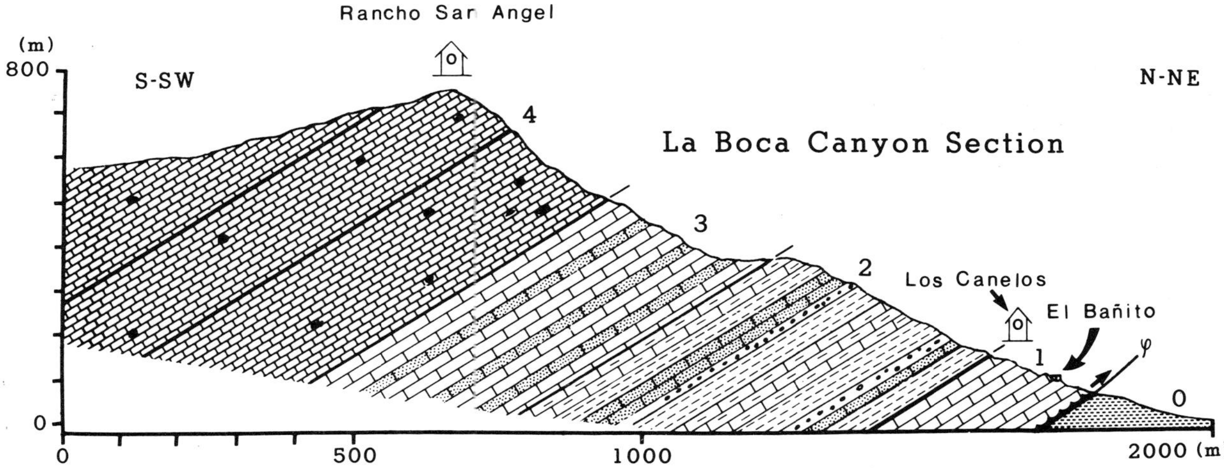

Figure 6. Field stratigraphic profile of the La Boca Canyon Section: (0) El Cercado Formation, (1), (2), and (3) La Casita Formation, (4) San Angel Limestone. The lithic characterization of each unit is given in the text.

Gypsum, and Unit 2 is referred to the Zuloaga Limestone. Unit 3 lithologically corresponds to the La Casita Formation, and Unit 4 corresponds lithologically to the La Caja. Units 5 and 6 are referred to as the Taraises; however, it is likely that Unit 5 could be assigned to a new formation.

La Boca Canyon Section (Cerro de la Silla)

Locality

The Cerro de la Silla, which is the northeasternmost anticline of the Sierra Madre Oriental, is an elongate, SE-NW trending anticlinal structure located S-SE of Monterrey (Fig. 1). The structure is about 30 km long and about 7.5 km wide. Its northern part is symmetric whereas its southern part is asymmetric to overturned. The southern half of the Cerro de la Silla Anticline is cut by La Boca Canyon, along which the most complete Jurassic section is exposed.

The La Boca Canyon section is located in the southern half of the Cerro de la Silla Anticline, approximately 30 km southeast of the city of Monterrey. The Jurassic section is exposed in the core of this structure, and the base of the Jurassic section is a faulted contact with Late Cretaceous sediments. The section studied is exposed along the Rio La Boca, and the base of the section starts at the El Banito hot spring, east of the San Angel Ranch.

Description

The Jurassic sequence at La Boca Canyon is about 800 meters thick, and is characterized by black coloration, resulting from abundant bituminous and carbonaceous matter. The section consists of the following units (Fig. 6):

Unit 1: Thick- to medium-bedded black, sandy, micritic limestone, with local patches of conglomerate and coarse sandstone.

Unit 2: Irregular alternation of black, sandy limestone, shale, marls, and fine conglomerates. Limestone beds contain sandy nodules.

Unit 3: Homogeneous interval of thick- to medium-bedded, black, bituminous limestone and sandy limestone. Locally, limestone beds contain light-gray, calcareous nodules about 20 cm in maximum diameter. The upper part of this interval is rich in terrigenous material.

Unit 4: Evenly-bedded, medium- to thin-bedded, black limestone with black chert nodules to gray limestone (mudstone) interbedded with layers of black chert. The upper part of this interval displays abundant stylolites.

Lithocorrelation

The first three units are assigned to the La Casita Formation, whereas Unit 4 is referred to the San Angel Limestone. The Jurassic sequence of the La Boca Canyon is

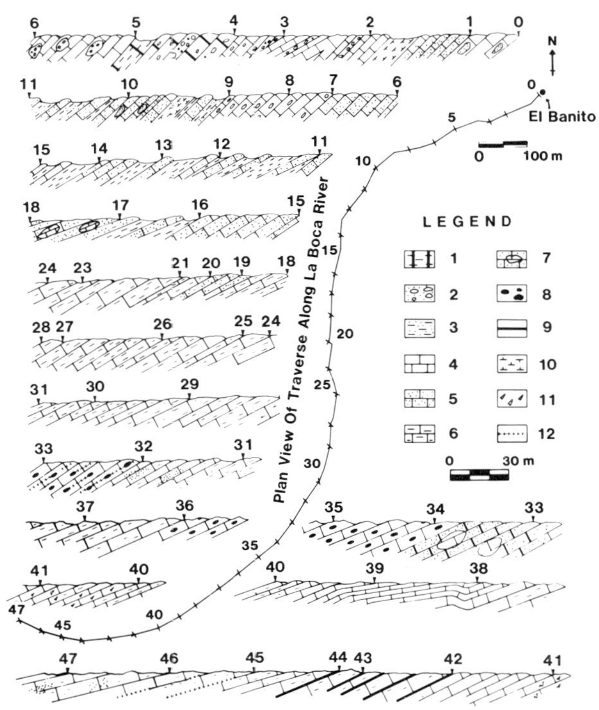

Figure 7. Hypostratotype of the La Casita Formation exposed along the La Boca Canyon. Plan view of traverse and field profiles: (1) Carbonaceous Mudstone, (2) Conglomerate, (3) Sandy Shale, (4) Micritic Limestone (mudstone and wackestone), (5) Sandy Limestone, (6) Clayey Limestone, (7) Calcareous Nodules, (8) Black Chert Nodules, (9) Black Chert Layer, (10) Marls, (11) Belemnites, (12) Stylolites.

remarkably similar to the type section of the La Casita Formation from the Sierra de Parras and may represent one of the most complete sections outside the type area. The hypostratotype of the La Casita Formation exposed at La Boca Canyon is shown in Figure 7.

The section of the San Angel Limestone (Unit 4), exposed along the La Boca Canyon, was selected by Longoria and Davila (1979) as the holostratotype of this unit. At this locality, the middle part of the La Casita Formation contains numerous ammonites including the genus *Idoceras*. The upper part of the La Casita contains abundant to common Tithonian calpionellids. The Jurassic/Cretaceous boundary occurs within the first 80 meters of the San Angel Limestone. Thus, this section contains a complete Late Jurassic (Oxfordian to Tithonian) succession. The presence of conglomeratic layers at some intervals within the La Casita and San Angel formations is regarded as an indication of tectonic activity in the region contemporaneous with the deposition of these units. As pointed out later, it is likely that the influx of terrigenous material may also be related to eustatic changes of sea level.

Sierra de Cruillas Section

Locality

The Sierra de Cruillas is the easternmost isolated range in northern Mexico, and occurs in the Sierra de San Carlos-Cruillas area (Fig. 1). The northeastern part of this range, referred to as Sierra La Virgen, contains a unique Jurassic section among those in northeastern Mexico.

The Jurassic section studied was measured in the central part of the Sierra La Virgen, about 5 km southeast of the town of Cruillas, and was described earlier by Imlay (1980). This section is exposed on the top of the ridges forming the Sierra La Virgen.

Description

The Sierra de Cruillas section consists of about 535 meters of carbonate, siliciclastic, and volcanic rocks (Fig. 8). The following lithic units are recognized, from bottom to top:

Unit 1: Whitish to yellowish tuff, which weathers red and forms the lower reliefs.

Unit 2: Tuffaceous, light-gray, evenly-bedded limestone, containing poorly preserved early Tithonian ammonites.

Unit 3: Regular alternation of thin- to medium-bedded, black limestone (mudstone and wackestone), calcareous shale and a few tuffaceous layers.

Unit 4: Evenly-bedded, medium-bedded, light-brown (cream) limestone (mudstone and wackstone).

Unit 5: Evenly-bedded, thick-bedded, light-brown (cream) limestone (wackestone), with brown chert nodules and lenses.

Lithocorrelation

It is difficult to refer the Sierra de Cruillas section to any of the known formal lithostratigraphic schemes from northeastern Mexico. It differs considerably from coeval

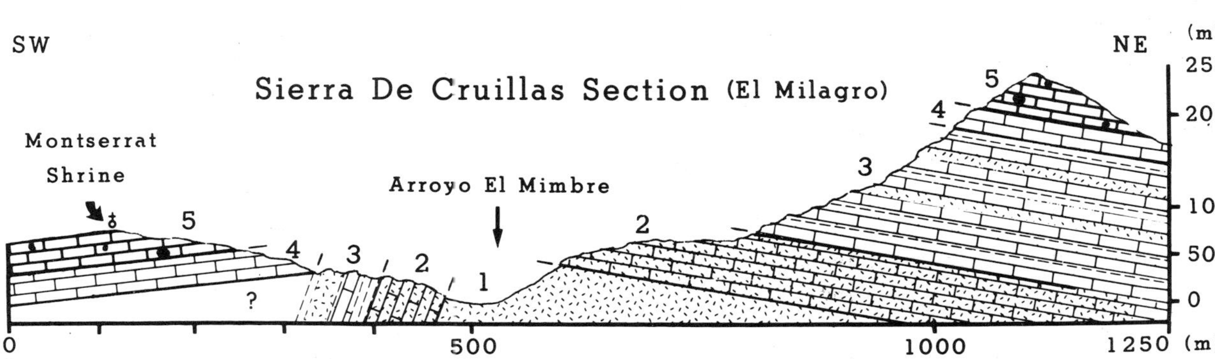

Figure 8. Field stratigraphic profile of the Cruillas Section. The lithic characterization of each unit is given in the text.

sequences such as these described from the Parras area. The only lithologically similar sections are located about 300 km to the south in the Huauchinango-Villa Juarez area in Puebla state, where an almost identical succession occurs. Imlay (1980) reported abundant Tithonian ammonites from a section measured by Humphrey and Diaz (1956) at the Sierra del Milagro (= La Virgen).

FOSSILS AND BIOSTRATIGRAPHIC IMPLICATIONS

The Jurassic of northeast Mexico contains abundant ammonites, which have been reported by Burckhardt (1930) and Imlay (1936, 1937, 1938).

Burckhardt (1930) established a biostratigraphic scheme for the Late Jurassic of northeast Mexico which has been the basis for chronostratigraphic subdivision of the Jurassic succession of this region. Recently, Imlay (1980) summarized the available ammonite date for northern Mexico and re-defined some of the previously established ammonite biozones.

On the other hand, very little is known about the microfossil content of the Jurassic rocks of northeast Mexico. A preliminary study demonstrates that planktonic microfossils are common to abundant in the Jurassic sections studied. The following biohorizons are established:

(1) The *Nerinea* biohorizon. This interval is characterized by the presence of species of the genus *Nerinea*. The precise stratigraphic range of this gastropod is not well-defined, but Burckhardt (1930) regarded the "*Nerinea*-bearing Limestone" of Coahuila state as Oxfordian, because of the associated ammonites *Perisphinctes* and *Ochetosceras*.

Microfossils in this interval include some miliolids of unknown affinity, and *Favreina* fecal pellets.

The *Nerinea* biohorizon is present in the La Huasteca, Potrero Chico, and Potrero Grande sections.

(2) *Idoceras* first stratigraphic occurrence. The genus *Idoceras* includes a number of species characterized by having dense, fine ribbing that projects on to the shoulders and is interrupted on the venter. It is restricted to the early and middle Kimmeridgian.

Very few microfossils have been found in this biohorizon, but samples from the La Boca Canyon contain benthic foraminifers assignable to the genera *Lenticulina* and *Epistomina*. Poorly preserved radiolarians also occur in this interval.

(3) The callpionellids-first occurrence. The base of this interval is defined by the first stratigraphic occurrence of members of the genus *Calpionella*. This biohorizon contains common to abundant ammonites, and it is likely that it is the age equivalent of the *Kossmatia* interval described by Burckhardt (1930) and Imlay (1980).

(4) The *Calpionella alpina/Calpionella elliptica*-interval. This interval is defined by the first stratigraphic occurrence of the nominal species. It contains abundant planktonic foraminifers (caucasellids) and common radiolarians. The top of this zone defines the Jurassic-Cretaceous boundary in northeast Mexico (Longoria, 1983a).

This inverval was observed in the Cruillas, La Boca, La Huasteca, and Potrero Grande sections.

ENVIRONMENTS OF DEPOSITION

The carbonate and siliciclastic facies of the Jurassic of northern Mexico (Fig. 9) were deposited in eight major environments of deposition:

(1) Sabkha (Supratidal)
(2) Nearshore
(3) Lagoonal (Back-reef)
(4) Shoal-water
(5) Inner-shelf
(6) Restricted-shelf
(7) Open-shelf
(8) Pelagic

Sabkha Facies. The sabkha facies are characterized by evaporites in which both gypsum and anhydrite were deposited in isolated basins. In the sections studied, this

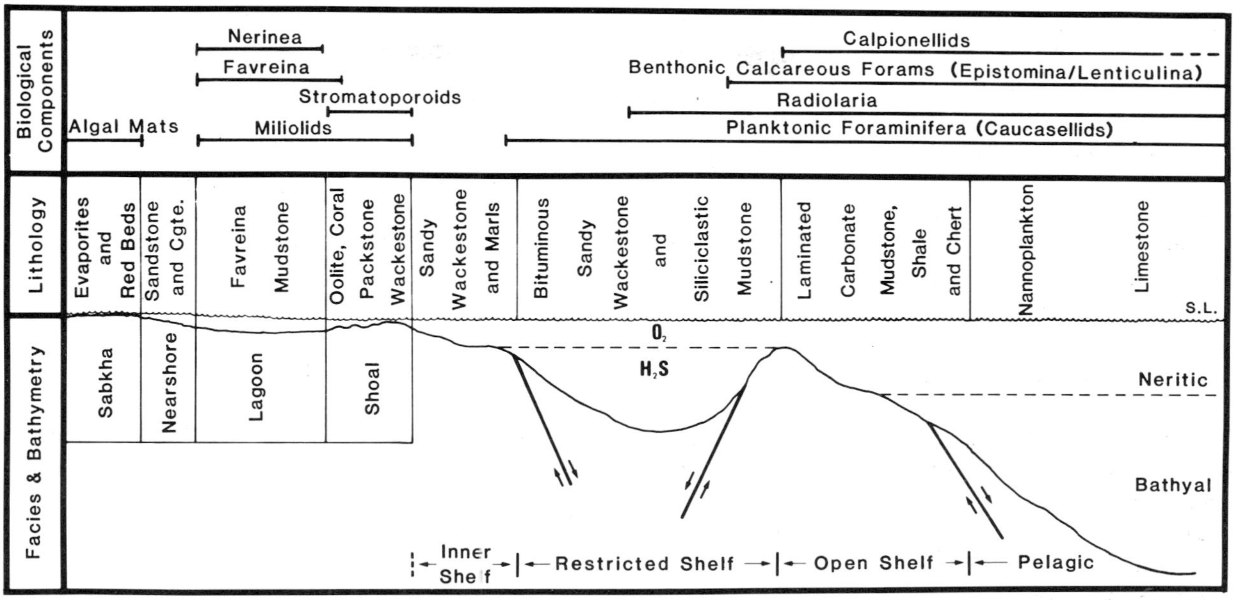

Figure 9. Environmental model postulated for the deposition of Jurassic facies in northeastern Mexico.

facies is represented by the Minas Viejas Gypsum.

Nearshore Facies. The Nearshore Facies are mainly composed of coarse conglomerate and arkosic sandstones. A few mollusk-bearing limestone (coquina beds) interbeds are associated with the clastic sediments and thin sections show that these limestones consist of mollusk packstones. In the studied area, this facies is represented by the La Gloria Formation.

Lagoonal Facies. The presence of *Nerinea* fragments, *Favreina* pellets and miliolids in mud-supported rocks are indicative of low-energy (lagoonal) conditions. The rocks from this facies suggest a well-oxygenated environment. The Lagoonal Facies are represented by the lower part of the Zuloaga Limestone.

Shoal-water Facies. The Shoal-water facies are characterized by grain-supported rocks, mainly ooid packstones and grainstones, and can also be recognized by the abundance of stromatoporids and coral fragments that they contain. Most of the rock from this facies are light in color. The Shoal-water Facies are represented by the Zuloaga Limestone.

Inner-Shelf Facies. The Inner-Shelf facies are characterized by the predominance of terrigenous fine-grained rocks which are irregular alternations of shales, marls, and clayey limestones. The limestones are mainly light-colored wackestones which lack significant quantities of bituminous material as a result of deposition under oxygenated conditions; planktonic foraminifers are common. These limestones are characterized by the abundance of shallow-water debris, such as coral and stromatoporoids fragments and were deposited in middle-neritic water depths. In the sections studied, the Inner-Shelf Facies is represented by the La Caja Formation.

Restricted-Shelf Facies. The Restricted-Shelf facies are characterized by the predominance of dark-colored bituminous, sandy wackestones and siliciclastic mudstones. Layers of carbonaceous shales vary from 5 to 15 meters in thickness and pyrite is abundant. The wackestones and mudstones are organic-rich (bituminous) and contain common to abundant planktonic foraminifers, and radiolaria. It is clear that these facies were deposited under restricted (euxinic) conditions in outer-neritic water depths. However, the surface waters maintained an "open" circulation in order to account for the presence of planktonic faunas. The Restricted-Shelf Facies are represented by the La Casita and San Angel formations.

Open-Shelf Facies. The Open-Shelf facies are characterized by the predominance of laminated lime mudstones which contain minor amounts of shale. Rocks of this facies may be light colored and contain diverse amounts of chert. Faunal components include planktonic foraminifers (caucasellids), radiolaria, calcareous benthic foraminifers and calpionellids. The relative abundance of these faunal components may be regulated by local oceanographic conditions. In the sections studied, this facies is represented by the San Angel Limestone.

Pelagic Facies. The Pelagic facies are haracterized by carbonate and volcaniclastic rocks, and a biological content that is basically the same as in the previous facies. Carbonate rocks of the Pelagic Facies are more micritic than the Open-shelf Facies, and commonly contain tuffaceous material. Thus, skeletal wackestones and nannoplankton mudstones are characteristic. The abundance of pelagic microfaunas separates the Pelagic Facies from the Open Shelf facies. In general terms, radiolaria and calpionellids predominate over the planktonic foraminifers, and benthic, calcareous foraminifers are rare or absent. It is quite likely that the Pelagic Facies of northern Mexico were deposited in bathyal water depths. The rocks of the Cruillas section represent the Pelagic Facies in the studied area.

TECTONIC AND PALEOGEOGRAPHIC IMPLICATIONS

Tectonic Setting of Northeastern Mexico

It is generally accepted that the Mesozoic (mainly Late Jurassic and Cretaceous) sedimentary rocks of northern Mexico underwent strong tectonic deformation during the Laramide Orogeny (Late Eocene). Following Humphrey's original hypothesis (1956), it is also widely accepted that Late Jurassic paleogeographic elements controlled the resulting structures. Thus, it is believed that:

(1) The Sierra Madre Oriental (Transverse Folded Belt) corresponds to the sites of major downwarping of the Mesozoic Mexican geosyncline, which resulted in a high-relief, east-west oriented mountain belt of intensely folded and faulted anticlines and synclines that contain Jurassic and Cretaceous marine sediments. The Santa Catarina and Cerro de la Silla anticlines represent typical examples of this morphotectonic unit.

(2) The Coahuila Folded Belt (also known as the Coahuila Ridge and Basin Province) is the site of the central Mesozoic Sabinas Basin. This tectonic event resulted in numerous isolated, SE-NW trending, elongated anticlines, separated by broad synclinal valleys formed of Jurassic and Cretaceous marine rocks. The Potrero Chico (Sierra del Fraile) and Potrero Grande (Sierra de Minas Viejas) anticlines are included in this belt.

(3) The Sierras Tamaulipecas, of which the Sierra de San Carlos-Cruillas is a typical example, is the area occupied by the Tamaulipas Peninsula, and is a series of isolated SE-NW trending anticlines composed primarily of Cretaceous sedimentary rocks.

From the structural viewpoint, the main differences between these three morphostructural units of northeastern Mexico are the style and intensity of the deformation:

In the Sierra Madre Oriental (Transverse Folded Belt), the anticlines are commonly isoclinally folded and are often doubly recumbent (fan-folded).

In the Coahuila Folded Belt, the anticlines are described as broad, gentle, domelike or elongate, normally faulted anticlinal structures, complicated by igneous intrusions.

In the Sierras Tamaulipecas (Tamaulipas Peninsula), the anticlines are closely related to those of the Sabinas Basin to the northwest. They are gently folded and appar-

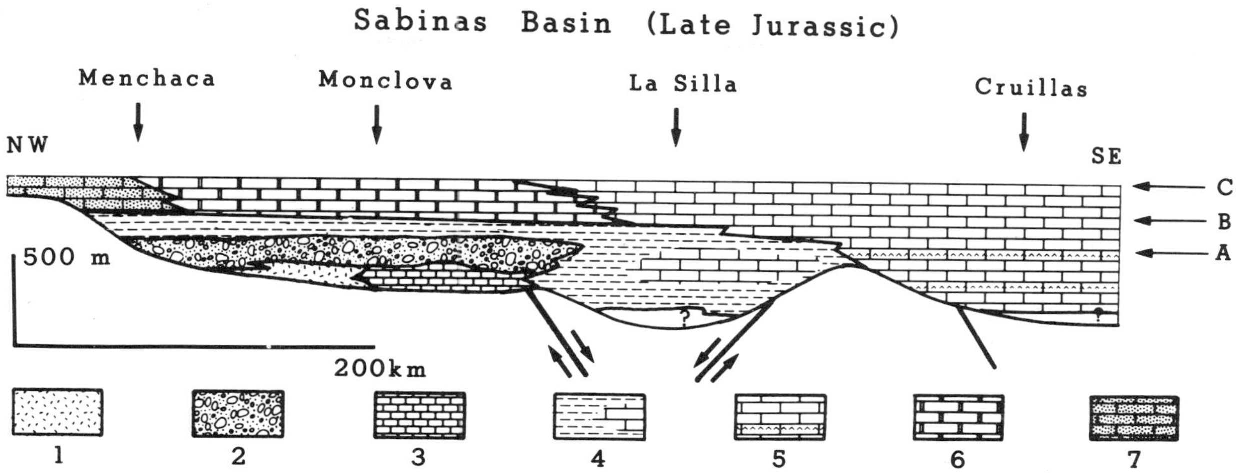

Figure 10. SE-NW (down-dip) cross section of the Sabinas Basin, from Cruillas to Monclova. (see Figure 1 for geographic location). (1) Minas Viejas Gypsum, (2) La Gloria Formation, (3) Zuloaga Limestone, (4) La Caja and Casita formations, (5) San Angel Limestone, (6) Taraises Formation, (7) Menchaca Formation. (A) First occurrence of *Idoceras,* (B) first occurrence of *Calpionella alpina* Cadish, (C) first appearence of *Calpionella elliptica* Cadish.

ently normal-faulted structures (e.g. Sierra de Picachos, Sierra de Sabinas Lampazos, Sierra de Peyotes, Serrania del Burro, Sierra de San Carlos-Cruillas).

The idea of Jurassic tectonism in the Gulf of Mexico region has been controversial. Earlier workers (Burckhardt, 1930; Imlay, 1943; Humphrey, 1956) believed that a tectonic episode occurred in this area during the Late Jurassic.

Burckhardt (1930) demonstrated the uncomformable contact of the Jurassic sequences of the San Pedro del Gallo area (Durango state). Imlay (1943) commented on the influence of tectonic activity in the Portlandian (= Tithonian) sediments of northeast Mexico.

Interesting enough, on the other side of the Gulf of Mexico, in the Pinar del Rio province of Cuba, Khudoley (1967a, b) postulated that this province of Cuba underwent tectonic activity in the Tithonian, related to the Nevadian Orogeny of western North America. Khudoley's tectonic interpretations, however, were strongly rejected by Hatten (1967) who stressed the lack of an uncomformity between the Late Jurassic and Early Cretaceous sediments of the Pinar del Rio area The controversy, thus, remains unresolved.

Late Jurassic Paleogeographic Reconstruction of Northeastern Mexico

Stratigraphic Observations

From the foregoing stratigraphic analysis, it is clear that the Jurassic (Oxfordian-Kimmeridgian) sections studied maintain lateral relationships in a SE-NW direction (down-dip cross section, Fig. 10): the Cruillas section (Figs. 1, 8) corresponding to the deep-water (Pelagic) facies; the La Boca Section (Figs. 1, 6) to the Restricted-Shelf Facies; and the Monclova area to the shallow-water (Shoal Water) facies (Figs. 1, 5). On the other hand, it also becomes evident that the Kimmeridgian sections studied are abruptly truncated in an east-west direction (strike cross section, Fig. 12), whereby the Nearshore Facies (La Gloria Formation) of the La Huasteca are juxtaposed against the Restricted-Shelf Facies (La Casita Formation) of the La Boca Canyon. The juxtaposition of Oxfordian-Kimmeridgian marine sequences is considered to be of fundamental importance in reconstructing the Late Jurassic paleogeography of this region because it has direct tectonic implications in understanding the origin of the Sabinas Basin.

Proposed Paleogeographic Evolution of Northeastern Mexico During Oxfordian, Kimmeridgian, and Tithonian Times

Figure 11 illustrates the postulated paleogeography of northeast Mexico in Oxfordian-Kimmeridgian time. The presence of carbonate banks and shoal-water facies in the Zuloaga Limestone indicates that these sediments were deposited in shallow-marine environments more likely parallel to the shoreline. At the same time, stratigraphic field relationships of the Zuloaga-La Gloria-La Caja sequence demonstrate that in part they are time equivalents (late Oxfordian-early Kimmeridgian) and grade one into the other laterally. Then, parallel facies belts are also postulated for the deposition of these units (Fig. 11). The presence of the La Casita Formation is clearly controlled by the bottom

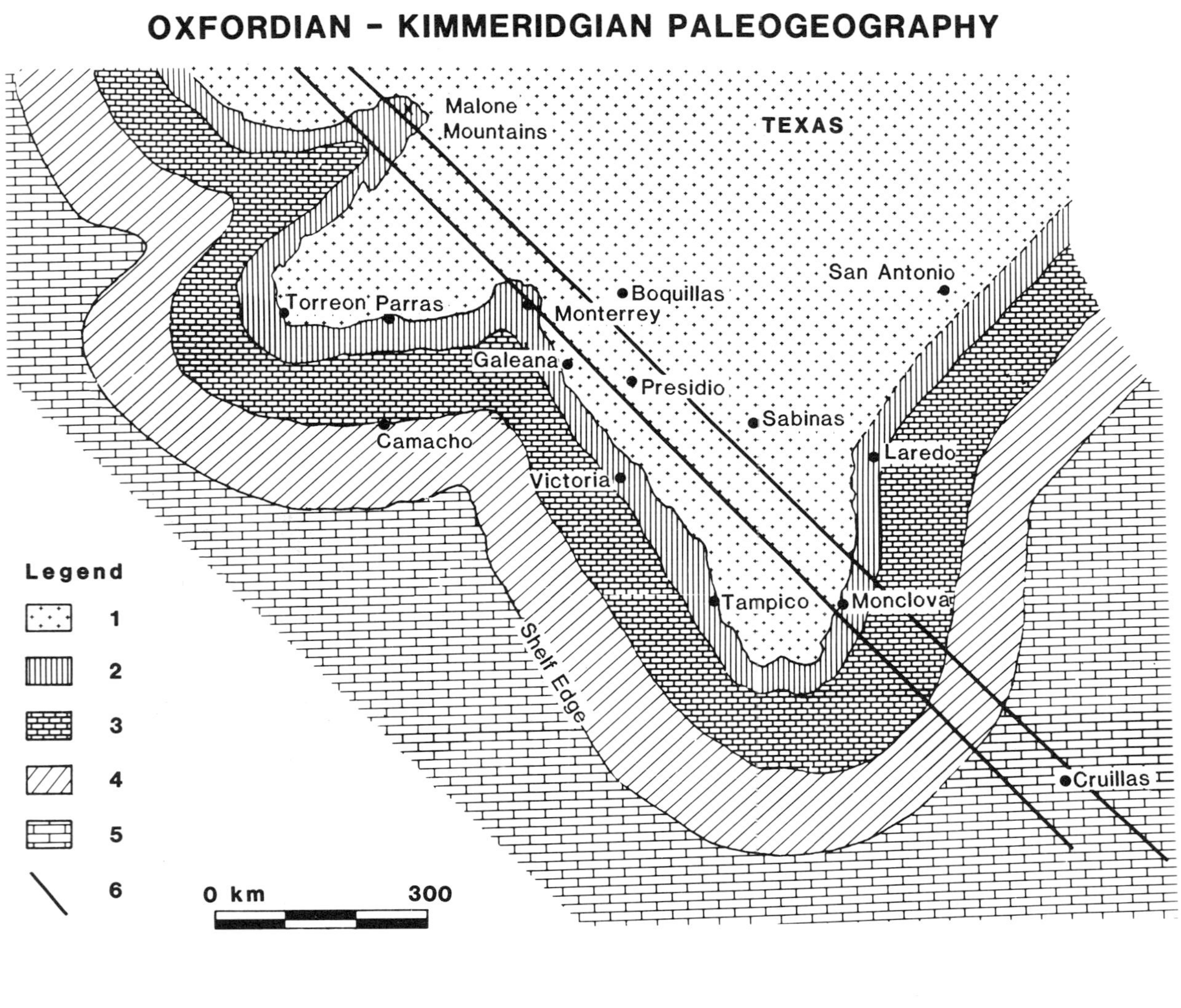

Figure 11. Proposed paleogeographic reconstruction of northeastern Mexico during Oxfordian-Kimmeridgian times. (1) Continental area, mainly folded Paleozoic rocks and continental Triassic deposits, (2) Sabkha and Nearshore Facies, (3) Lagoonal and Shoal Facies, (4) Inner-, Restricted- and Open-Shelf Facies, (5) Pelagic Facies, (6) Trace of postulated late Tithonian transcurrent faults. See Figure 9 for bathymetric and ecological profile.

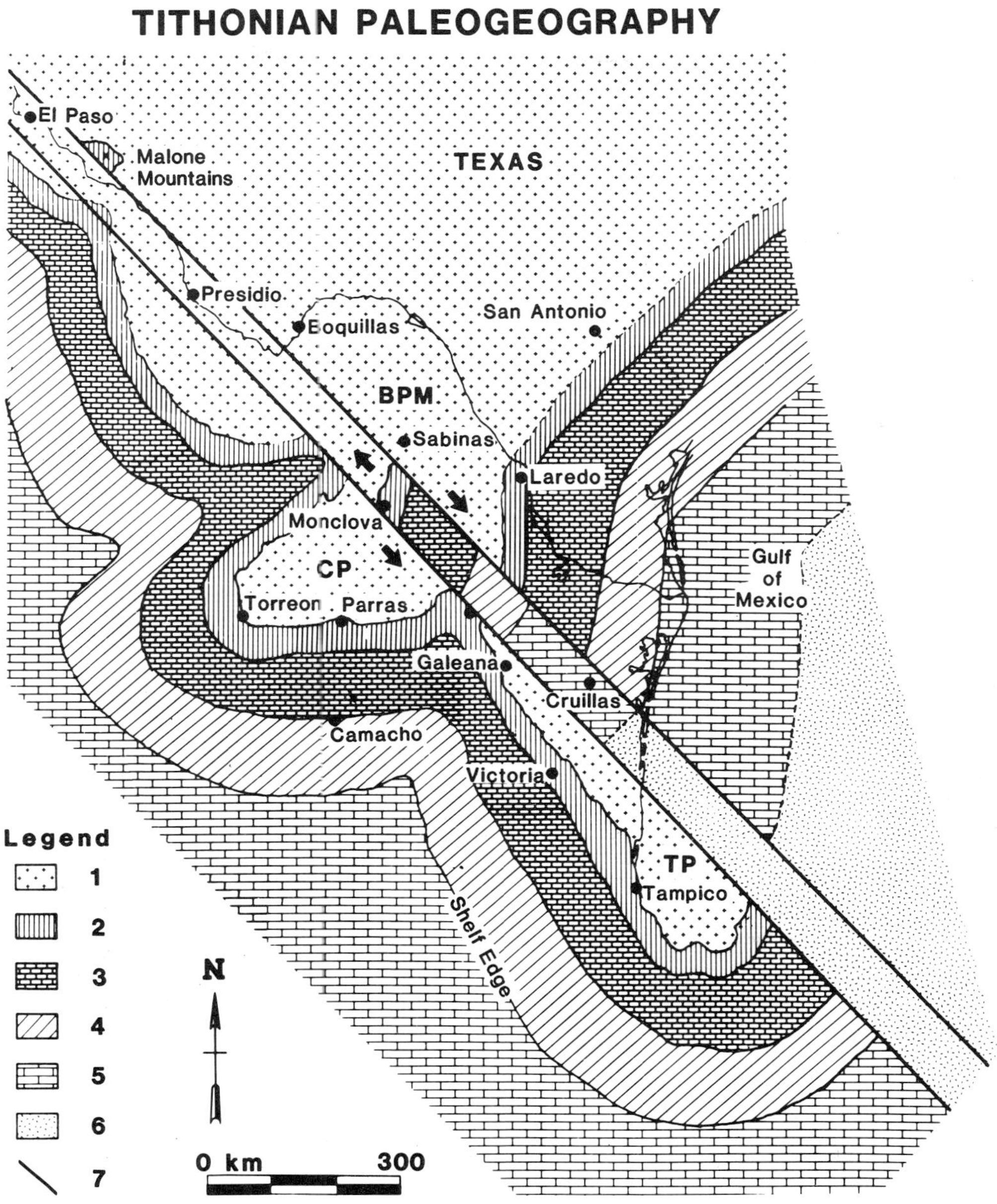

Figure 12. Proposed late Tithonian paleogeographic reconstruction of northeastern Mexico. Active sea-floor spreading in the Gulf of Mexico is accompanied by transform faulting (see Figure 14) reaching the continental area. Accumulated translation along the transform faults accounts for a northwesterly displacement of a 100 km-wide slab, reaching the continent at a high angle, and giving rise to the Sabinas Basin. The accumulated horizontal translation is estimated to be about 350 kilometers, resulting in the juxtaposition of pre-late Tithonian facies. See Figure 11 legend for 1-5; (6) Oceanic Crust, (7) Trace of transcurrent faults, relative movement of slabs is indicated by the arrows. Positive paleoelements are: (BPM) Burro-Picachos Masif, (CP) Coahuila Peninsula, (TP) Tamaulipas Peninsula. The negative paleogeographic element between the first two landmasses corresponds to the area occupied by the Sabinas Basin.

topography which existed at the site of deposition. Its bituminous character is evidence of the restricted environment of deposition of this unit. The deposition of the Late Jurassic sediments of northeast Mexico was largely controlled by: (1) eustatic changes of sea level, (2) bottom topography, and (3) tectonic activity.

In early Oxfordian time, marine waters reached the continental area over an irregular topography. Thus, semi-closed lagoonal conditions were formed at the margin of the continent, and normal-marine conditions prevailed offshore, allowing the shoals to develop (Fig. 11). Restricted-shelf conditions were also formed in deeper waters and caused the development of stagnant water masses. The deposition of euxinic facies in late Oxfordian-early Kimmeridgian time was accompanied by a fall in sea level resulting in the accumulation of dark-colored bituminous limestone and shale (La Casita Formation), as found in the Cerro de la Silla section.

By middle Kimmeridgian time, a marine transgression took place which caused a migration of deeper water (open-shelf) facies toward the continent (Fig. 10). At the same time, open circulation of the water masses was reestablished and carbonates accumulated in the areas of restricted circulation, such as the site of La Boca Canyon.

In early Tithonian time, the accumulation of a thick terrigenous sequence (La Casita Formation) indicates a major drop in sea level. At the end of the Tithonian, the entire region attained tectonic stability and a gradual transgression is evident from the lateral relationships between the Jurassic section of Cruillas and Minas Viejas.

The bituminous and carbonaceous facies of the Jurassic of northern Mexico (La Casita Formation and San Angel Limestone) are the result of sedimentation in stagnent bodies of water during early Kimmeridgian time. These rocks are indicative of restricted circulation of bottom waters and reducing conditions. The restricted circulation of the surface waters are evidence of the accumulation of sediments on a restricted shelf bounded by faults ("Barred Basin", Fig. 9). The lack of benthic fossils in the bituminous, sandy wackestone of this environment reflects the anoxic conditions here during late Kimmeridgian time.

THE ORIGIN OF THE SABINAS BASIN

Sabinas Basin versus Sabinas Gulf

Alvarez (1949) introduced the term "Sabinas Basin" to refer to, "that negative feature which has existed in the area between the Coahuila and Tamaulipas peninsulas since the Early Jurassic." In 1956, Humphrey used the term "Sabinas Gulf" to refer to the early Late Jurassic paleogeographic element which was, according to him, the site of deposition of the area now known as the Coahuila Ridge and Basin Province of northeastern Mexico.

The precise extent of this paleogeographic element is still unknown; however, it extends northwest from the vicinity of Monterrey for approximately 350 km and averages 100 km in width. According to Murray (1961), the Jurassic and Early Cretaceous depocenters seem to have been in the central and western part of the basin, the Late Cretaceous depocenters, which constitute the "Sabinas Coal Basin", lie to the east.

Although the origin of the Sabinas Basin is unknown, I believe that the postulated juxtaposition of Late Jurassic marine facies in the vicinity of Monterrey has a direct implication in interpreting its origin and subsequent paleogeographic evolution.

Proposed Tectonic Models for the Origin of the Sabinas Basin

I propose two alternate tectonic models to account for the observed juxtaposition of Jurassic rocks in northeast Mexico and to explain the origin of the Sabinas Gulf (Figs. 13, 14):

(1) Block-faulting ("Aulacogen") Model. Aulacogens form where continental basins intersect a continental margin at high angles, and they correspond to the failed arm of an RRR-triple junction (Hoffman, Burkey and Dew-

ey, 1974). The geographic configuration of the Sabinas Basin suggests that this paleotectonic element may have corresponded to an aulacogen in Late Jurassic time.

The existence of the Sabinas Basin (=Sabinas Gulf) in northern Mexico has been traced back to Late Jurassic time, as the late Oxfordian Minas Viejas evaporities are the oldest marine rocks found in this area. The outcrop distribution of Late Jurassic rocks within the basin, however, seems somewhat chaotic and without a definite paleogeographic pattern. Some authors (Gonzalez, 1979; Tardy, 1980) have explained the sporadic distribution of the shoal facies in northern Mexico by postulating the existence of a paleo island during Late Jurassic and Early Cretaceous time. The positive paleo elements would allow the formation of shallow-water facies during times of transgression as marine waters encroach on the land. Thus, the positive areas would correspond to faulted blocks (horsts). According to this interpretation, the Sabinas Gulf was formed as a rift-valley, and a failed rift (Aulacogen) was generated for a short span of time (Fig. 13).

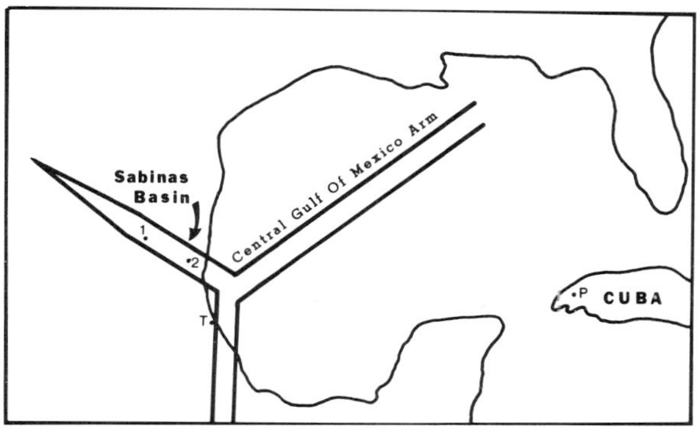

Figure 13: The Sabinas Basin as a failed arm of a hypothetical Late Jurassic RRR-triple junction in the Gulf of Mexico. This figure depicts the position of the Cerro de la Silla (1) and Sierra de Cruillas (2) sections. Transcurrent faults bounding the Sabinas Basin may have resulted as the final stage of the rifted margin. The central Gulf of Mexico arm of the RRR-triple junctions is the origin of the oceanic crust of the Gulf Basin. (T) Tampico, (P) Pinar el Rio.

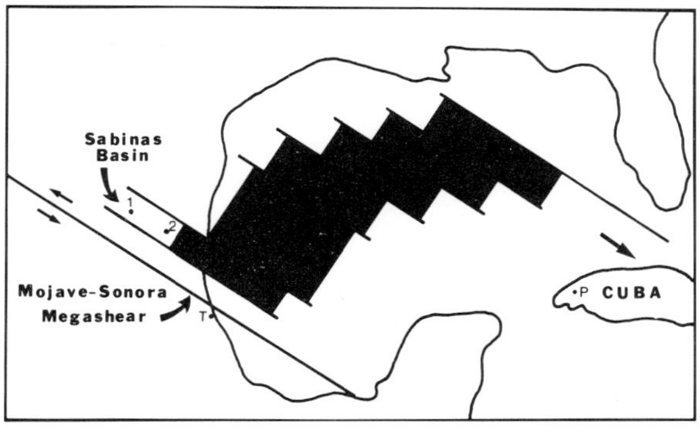

Figure 14. Origin of the Sabinas Gulf by transform faulting. This figure depicts the inferred plate boundaries and direction of Late Jurassic (Tithonian) sea-floor spreading in the central Gulf of Mexico basin (modified after Buffler, *et al.*, 1980). The accumulated northwesterly translation of a transform fault during Late Jurassic time accounts for the origin of the Sabinas Basin of northeastern Mexico and the resulting juxtaposition of pre-Tithonian facies (see Figures 11 and 12). The extension of the Mojave-Sonora megashear, *sensu* Longoria (1983b) into northeastern Mexico is regarded as the southern plate boundary. The location of the La Boca (1), and Cruillas (2) sections is shown. (T) Tampico, (P) Pinar del Rio.

Although it is clear that the rifting in the Gulf of Mexico began in the Late Triassic-Early Jurassic with the separation of South America from North America, the resulting aulacogen stratigraphy has not been observed anywhere in northern Mexico; that is, the base of the Mesozoic sequence of the Sabinas Basin lacks evidence of magmatic activity and continental clastic deposition that typically characterize an aulacogen sequence (Walper, 1977). On the other hand, the rift-valley (aulacogen) model requires that the faults of block-faulted areas trend in a SE-NW direction, because this would be the trend of the suspected aulacogen. Thus, high-angle faults in the Sabinas Basin would trend SE-NW, and NNW-SSE lateral facies relationships would then be expected. Although the aulacogen model encounters serious limitations, it provides a way of linking the Sabinas Basin to the early rifting phase of the Gulf region.

(2) Transcurrent (Strike-slip) Model. An alternate model to account for the juxtaposition of Jurassic facies in northern Mexico is to relate the Sabinas Gulf's origin to the geodynamics of the Gulf of Mexico (Fig. 14).

A pair of SE-NW trending, parallel faults, which disrupt the Paleozoic basement for a distance of about 300 km, is envisioned. This northwesterly displacement of the Jurassic sediments would account for the juxtaposition of Jurassic shallow-water facies of the Sierra Madre Oriental (Santa Catarina Anticline) and the deep-water facies of the Cerro de la Silla and Cruillas sections. At the same time, it would leave space for the presence of the continental Triassic rocks in the Galeana area, south of Monterrey (Fig. 14). Furthermore, northwestwardly displacement would relate the disruption of the Marathon folded belt to the geodynamics of the Gulf of Mexico.

Transform faulting took place in Late Jurassic time, although it is likely that it began in early Kimmeridgian time; and the major displacement took place in the Tithonian as can be interpreted from the stratigraphic relationships of the rocks involved (see above).

This tectonic episode is directly linked to active sea-floor spreading in the Gulf of Mexico, according to Buffler, *et al.* (1980), and is the more likely mechanism to explain the formation of the oceanic crust in the Gulf of Mexico.

Relation of the Proposed Tectonic Event To Previous Tectonic Models

Several tectonic models involving transcurrent faulting have been proposed in northern Mexico, north of 22 degrees north latitude. Cserna (1970, 1976) first introduced the idea of Triassic segmentation in northern Mexico. According to Cserna, three major east-west transcurrent (left-lateral) faults segmented continental Mexico in the Triassic. Cserna's model was based on an earlier hypothesis by Woollard *et al.* (1969) who interpreted their gravimetric data to indicate that continental Mexico was subdivided into three large segments. The northernmost of these Triassic segments lies in northern Mexico, approximately south of Monterrey and was named the Torreon-Monterrey Fault by Cserna (1970), who considered that the movement along this fault was in the Early Triassic. Furthermore, he related this lineament to the Texas Lineament described by Muehlberger (1965).

The idea of an early Mesozoic megashear in northwestern Mexico, displacing Precambrian terranes in southern California and northern Sonora, was first postulated by Silver and Anderson (1974). Recently, Tardy (1980) suggested the existence of the Caltam Lineament, which, according to him, is the continuation of the Mojave-Sonora Megashear into northeastern Mexico. Tardy (op. cit.) stated that the transcurrent movement along the Caltam Lineament took place in the Late Jurassic and could have continued during the Neocomian. The same easterly extension of the Mojave-Sonora Megashear (left-lateral movement) has been projected into northeastern Mexico by Anderson and Schmidt (1983). Anderson and Schmidt believe that the easternmost extension of the megashear approximately coincides with the Campeche escarpment, and they also noticed that the extended Mojave-Sonora megashear cuts across the Torreon-Monterrey fault at a low angle.

Longoria (1983b) postulated that the extension into northeastern Mexico of the Mojave-Sonora Megashear trends north of Ciudad Victoria. The main evidence for this is the lithocorrelation between Early Jurassic marine rocks of eastern Mexico and coeval sediments of northeastern Sonora. According to him (op. cit.), the major displacement along the Mojave-Sonora Megashear took place in the Middle Jurassic, corresponding to the opening of the Gulf of Mexico in Callovian time.

Alfonso (1978) postulated the existence of transcurrent left-lateral faults in northern Mexico in an attempt to unravel the disruption of the Marathon-Ouachita trend. He postulated three SE-NW trending transcurrent faults north of Ciudad Victoria.

Charleston (1981) explained the tectonic deformation of the Coahuila Folded Belt by postulating shearing along the La Babia and San Marcos left-lateral transcurrent faults that bound this province. According to him (op. cit.), this mechanism produced the required compression to account for the Coahuila Folded Belt.

The integration of the proposed tectonic event into previous models is difficult, if not impossible at this time. All the previously proposed models have some features in common: (1). they imply left-lateral displacements, (2). their general trend is SE-NW, with great variation in trend angle, (3). the timing of the event is in many cases speculative or rather broad, and (4). physical stratigraphic evidence has not been given which conclusively demonstrates their existence. At a regional scale, the transcurrent model herein proposed to account for the origin of the Sabinas Basin agrees well with the spreading phase postulated by Buffler *et al.* (1980) for the Gulf of Mexico.

On the other hand, I suggest that the La Babia and San Marcos transcurrent faults as described by Charleston (1981) are the reactivation of the ancient (Late Jurassic) transform faults herein proposed, corresponding in turn to reactivated Paleozoic faults, such as those described by Cebull *et al.* (1976).

ACKNOWLEDGMENTS

The present investigation was funded in part by the National Science Foundation (Grant EAR-8121550). I would like to express my gratitude to the Programs in Geosciences of the University of Texas at Dallas for all the facilities during the preparation of this report. Special thanks are due to Mr. Manuel Valenzuela for his assistance in the field and in the drafting of the figures.

REFERENCES

Alfonso, Z.J., 1978, Geología regional del sistema sedimentario cupido: Asociación Mexicana de Geólogos Petroleros, Boletin, v. 30, p. 1-55.

Alvarez, M. Jr., 1949, Tectónica de la Republica Mexicana: Sociedad Geologica Mexicana, Boletín v. 33, p. 1319-1335.

Anderson, T.H. and V.A. Schmidt, 1983, The evolution of Middle America and the Gulf of Mexico-Caribbean Sea region during Mesozoic time: Geological Society of America Bulletin, v. 94, p. 941-966.

Barker, C.L., 1941, Upper Jurassic deposits and structures of the Monterrey-Saltillo area: South Texas Geological Society, Guidebook, 13th. Ann. Mtg. Monterrey, Mexico., 4 p.

Bonet, F., 1965, Zonificación Microfaunística de las Calizas Cretácicas del Este de México: Asociación Mexicana de Geológos Petroleros, Boletín 8 (2-3), p. 389-488.

Bose, E., 1923, Vestiges of an ancient continent in northeast Mexico: American Journal of Science v. 5, n. 6.

_______, and O.A. Cavins, 1928, The Cretaceous and Tertiary of southern Texas and northern Mexico: University of Texas, Bulletin 2748, p. 7-142.

Buffler, R.T., J.S. Watkins, F.J. Shaub, and J.L. Worzel, 1980, Structure and early geologic history of the deep central Gulf of Mexico basin. *In,* The Origin of the Gulf of Mexico and the early opening of the Central North Atlantic Ocean (Pilger, Jr. R.H., ed): Proceedings of a symposium at Louisiana State University, Baton Rouge (March, 1980), p. 3-16.

Burckhardt, C., 1930a, Etude synthétique sur le Mésozoïque mexicaine. Première Partie: Mémoires de la Société Paléontologique Suisse vol. IL. p. 1-123.

__________, 1930b, Etude synthétique sur le Mésozoïque mexicaine. Second partie: Mémories de la Société Paléontologique Suisse vol. L, p. 125-280.

Cebull, S.E., D.H. Shurbet, G.R. Keller, and L. R. Rusell, 1976, Possible role of transform faults in the development of apparent offsets in the Ouachita-southern Appalachian tectonic belt: Journal of Geology, v. 84, p. 107-114.

Charleston, S., 1981, A summary of the structural geology and tectonics of the state of Coahuila, Mexico: *In* Katz, S.R. and Smith, C.I. (eds.), Field Trip Guidebook, West Texas Geological Society, Pub. no. 81-74, p. 28-36.

Cserna, Z. De, 1956, Tectónica de la Sierra

Madre Oriental, entre Torreon y Monterrey: XX Congreso Geológico Internacional (México), Monografía, 87 p.

__________, 1970, Tectonics of northern México (Mesozoic sedimentation, and magmatic activity and deformation in northern Mexico). *In* Seewald, K. and Sundeen, D., eds., The Geologic framework of the Chihuahua tectonic belt: West Texas Geological Society Publication 71-59, p. 99-118.

__________, 1976, Mexico-Geotectonics and mineral deposits: New Mexico Geological Society, Special Publication no. 6, p. 18-25.

Gonzalez, R., 1976, Bosquejo geológico de la Zona Noreste: Asociación Mexicana de Geólogos Petroleros, Boletín XXVII, n. 1-2, p. 1-49.

Hatten, C.W., 1967, Principal features of Cuban Geology: Discussion: The American Association of Petroleum Geologists, Bulletin v. 51, no. 5, p. 780-803, 6 figs.

Heim, A., 1940, The front ranges of the Sierra Madre Oriental, Mexico, from Ciudad Victoria to Tamazunchale: Eclogae Geologicae Helvetiae, v. 33.

Hoffman, P.B., K.C.A. Burkey, and J.F. Dewey, 1974, Aulacogens and their genetic relation to geosynclines, with a Proterozoic example from Great Slave Lake, Canada: *In* Dott, R.H. Jr., and R. H. Shaver, (eds.), Society of Economic Paleontologists and Mineralogists, Special Pub. no. 18, p. 38-55.

Humphrey, W.E., 1956, Tectonic framework of northeast Mexico: Gulf Coast Association Geological Societies, Trans. v. 6, p. 25-30.

__________, and T. Diaz-Gonzales, 1956, Correlation of the Mesozoic strata of northeast Mexico: Gulf Coast Association of Geological Societies, Trans. v. 6, p. 32-34.

Imlay, R.W., 1936, Evolution of the Coahuila Peninsula, Mexico. Part IV. Geology of the western part of the Sierra de Parras: Geological Society of America, Bulletin, v. 47, p. 1091-1152. 10 pls. 3 figs.

__________, 1937, Geology of the middle part of the Sierra de Parras, Coahuila, Mexico: Geological Society of America, Bulletin v. 48, p. 587-630, 14 pls. 4 figs.

__________, 1938, Studies of the Mexican Geosyncline: Geological Society of America, Bulletin 49, p. 1651-1694.

__________, 1943, Jurassic formations of Gulf Region: American Association of Petroleum Geologists, Bulletin 27, no. 11 p. 1407-1544,

__________, 1980, Jurassic Paleobiogeography of the Conterminous United States in its continental setting: U.S. Geological Survey Professional Paper No. 1062, 134 p.

Khudoley, K. M., 1967a, Principal features of Cuban Geology: The American Association of Petroleum Geologists, Bulletin v. 51, no. 5, p. 668-677. 3 figs.

__________, 1967b, Reply: The American Association of Petroleum Geologists, Bulletin, v. 51, no. 5, p. 789.

Longoria, J.F., 1983a, Cretaceous biochronology from the Gulf of Mexico Region based on planktonic microfossils: Micropaleontology (in press).

__________, 1983b, Mesozoic tectostratigraphic domains in east-central Mexico: Canadian Journal of Earth Sciences (in press).

__________, and V.M. Davila, 1979, Estratigrafia y microfacies del Cerro de la Silla, SE de Monterrey: Departamento de Geologia de la Universidad de Sonora, Boletin 2, no. 1, p. 65-95.

Muehlberger, W.R., 1965, Late Paleozoic movement along the Texas lineament, Trans. New York Academy of Science, Ser. 2, v. 27, p. 385-392.

Murray, G.E., 1961, Geology of the Atlantic and Gulf Coastal Province of North America. Harper & Brothers, Publishers, New York, p. 692.

North American Commission on Stratigraphic Nomenclature, 1983, North American Stratigraphic Code: The American Association of Petroleum Geologist Bulletin, v. 67, no. 5, p. 841-875, 11 figs., 2 tables.

Oivanki, S.M., 1974, Paleodepositional environments in the Upper Jurassic Zuloaga Formation (Smackover), northeastern Mexico: Gulf Coast Association of Geological Societies, Transactions v. XXIV, p. 258-278.

Silver, L.T. and T.H. Anderson, 1974, Possible left-lateral early to middle Mesozoic disruption of the Southwestern North American craton margin: Geological Society of America, Abstracts with Programs v. 6, p. 955.

Tardy, M., 1980, Contribution a l'etude geologique de la Sierra Madre Oriental de Mexique: Department de Geotectonique de l'Universite Pierre et Marie Curie, Travaous, 459 p.

Vokes, H.E., 1963, Geology of the Canon de la Huasteca area in the Sierra Madre Oriental, Nuevo Leon, Mexico. Tulane Studies in Geology vol. 1, no. 4, pl 125-148.

Wall, J.R., G.E. Murray and T. Diaz, 1961, Geology of the Monterrey area, Nuevo Lion, Mexico: Gulf Coast Association of Geological Societies, Transactions, Vol. XI, p. 57-71.

Walper, J. L., 1977, Paleozoic tectonics of the southern margin of North America: Gulf

Coast Association of Geological Societies, Transactions vol. XXVII, p. 230-241.

Weide, A.E., 1968, The myth of the Mesozoic Mexican Geosyncline: Geological Society of America, Abstract with Programs, Ann. Mtg. p. 314.

Woollard, G.P., J. Monges-Caldera, and L. Machesky, 1969, Gravity relations in Mexico: Pan-American Symposium on the Upper Mantle (Mexico, 1968), v. 1, p. 43-48.

DIAGENESIS AND RESERVOIR POTENTIAL OF THE UPPER JURASSIC SMACKOVER FORMATION OF SOUTH TEXAS

R.G. LOUCKS

D.A. BUDD

ARCO Oil and Gas Company
Exploration and Production Research
P.O. Box 2819
Dallas, Texas 75221

ABSTRACT

The hydrocarbon potential of the Smackover Formation in South Texas is virtually untested. Much of the section penetrated is impermeable; however, reservoirs as thick as 33 feet (10 m), with porosity ranging from 4 to 26 percent and permeabilities ranging from 0.1 to 6.5 millidarcys, have been cored at depths below 18,000 feet (5,500m). Nearly complete dolomitization has resulted in the development of intercrystalline porosity in innershelf wackestones and shoal-complex grainstones. In addition, some grainstones have subsurface-derived oomoldic porosity.

In the grainstone facies, four general stages of diagenesis affected porosity: Stage 1 (marine-phreatic environment)-precipitation of an isopachous carbonate cement and extensive grain micritization; Stage 2 (shallow-meteoric environment)-precipitation of very coarse-crystalline syntaxial calcite and fine-crystalline equant calcite, dissolution of aragonitic skeletal grains, and incipient solution-compaction; Stage 3 (regional fluid-mixing environment)-intrapore precipitation of and grain/matrix replacement by fine- to medium-crystalline rhombic dolomite; and Stage 4 (subsurface environment associated with basinal fluid expulsion)- dissolution of ooids and dolomite, microstylolitization by solution compaction resulting from decarboxylation of kerogen, and precipitation of coarse-crystalline calcite and baroque dolomite. The magnitude of each general diagenetic stage varies regionally.

Porosity development in dolomitized wackestones is sporadic but within grainstones porosity is in part facies related. Hydraulically equivalent oolitic grainstone and oolitic quartzarenite facies are commonly dolomitized and have associated porosity development. In addition, dolomitization of oolitic facies is more extensive in the updip half of the shoal-complex trend. These facies/diagenetic relationships can be used as an exploration tool in the Smackover Formation in South Texas.

Reprinted with permission from Transactions of the Gulf Coast Association of Geological Societies, v. 31, p. 339-346, 1981.

INTRODUCTION

The Smackover Formation is economically significant throughout the Gulf Coast region. Reservoir properties within Smackover carbonates were affected by depositional and diagenetic processes. Hydrocarbons are primarily produced from shoal-water grainstones and less commonly from dolomitized lagoonal mudstones. In South Texas, where the Smackover interval is virtually untested, it is necessary to evaluate both depositional and diagenetic patterns in assessing reservoir potential.

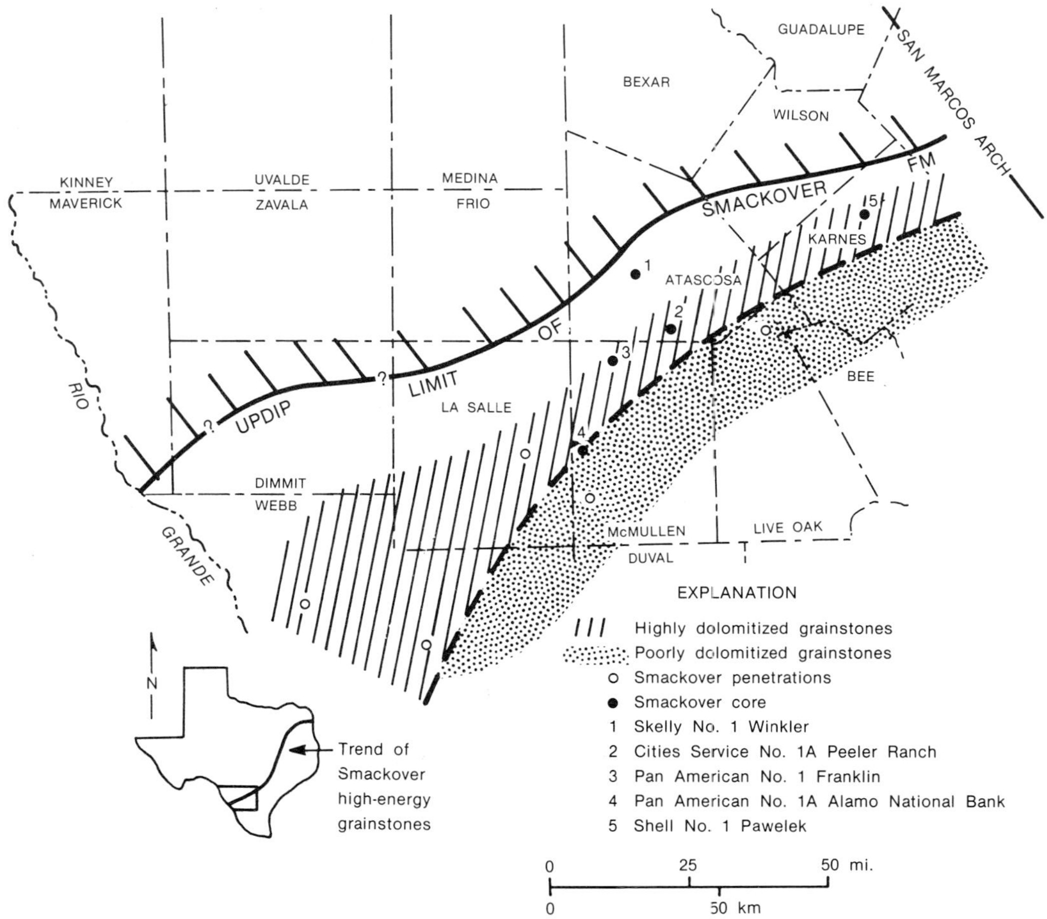

Figure 1. Index map of South Texas study area showing Smackover penetrations. Projected trend of highly dolomitized Smackover grainstones is also depicted. The continuation of the trend into Mexico is uncertain but it probably turns south. Modified from Budd and Loucks (1981).

Budd and Loucks (1981) have evaluated the depositional facies in the Smackover and lower part of the Buckner Formations in South Texas. Therefore, it is the objective of this paper to correlate the general diagenetic history of the Smackover Formation with porosity and permeability development and patterns that may aid in hydrocarbon exploration of this deep (15,000 to 24,000 feet; 4,570 to 7,010 m) carbonate trend. Figure 1 shows location of well data used in this study.

GENERALIZED DEPOSITIONAL MODEL

The Smackover Formation of South Texas can be divided into informal lower, middle and upper parts (Budd and Loucks, 1981), each of which originated in a different depositional setting (Fig. 2). This is similar to the tripartite subdivision of the Smackover Formation in Northeast Texas and Louisiana by Dickinson (1968). The lateral tract of basinal, open-shelf, and shoal depositional systems is interpreted to have been deposited on a gently sloping carbonate ramp. Seaward advance of these depositional systems produced a single, large progradational sequence and the tripartite subdivision of the formation.

Sediments in the lower part of the Smackover Formation were deposited in a basinal system and consist of laminae of dark brown to black, organic-rich carbonate mudstone alternating with laminae of pyrite and siliciclastic silt or sand. The laminae grade upward in section from very thin parallel laminae to slightly thicker irregular laminae. The middle part of the formation consists of sediments deposited on a low-energy, open-shelf system. Two lithofacies are present: a wispy-laminated, crustacean-pellet carbonate mudstone containing a sparse pelagic fauna and a highly burrowed skeletal-pellet wackestone containing a benthic fauna. The upper part of the formation consists of several grain-supported lithofacies deposited within a shoal system. Burrowed pellet and oncolite packstone formed the moderate-energy outer shoal, and a sequence of cross-bedded mixed-allochem, oolite and oolite-intraclast grainstones and oolitic sandstones composed the higher-energy inner shoal.

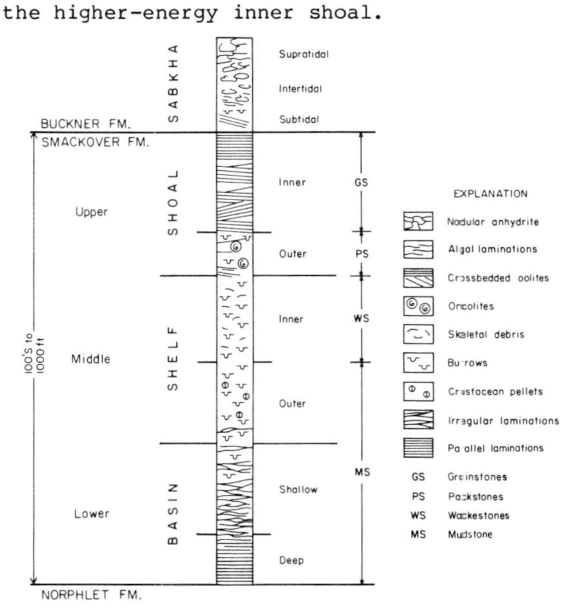

Figure 2. Idealized vertical section of the Smackover Formation in South Texas showing depositional systems and characteristic features of the different lithofacies. Modified from Budd and Loucks (1981).

POROSITY AND PERMEABILITY DISTRIBUTION

A minimum of four percent porosity and 0.1 millidarcy permeability in the Smackover carbonates is considered to be reservoir grade (Becher and Moore, 1976). Reservoirs with these minimum parameters in the South Texas Smackover cores occur only within the lithofacies deposited in the shoal system and the inner-shelf system (Fig. 2). Low permeabilities, however, characterize the Smackover Formation, even where porosities exceed four percent (Fig. 3).

Dolomitization is necessary for development of reservoirs in the inner-shelf system. A minimum of 80 percent dolomite appears necessary for consistently good reservoir-grade porosity. However, rocks which are 100 percent dolomite do not necessarily have good porosity. Potential reservoir zones of the

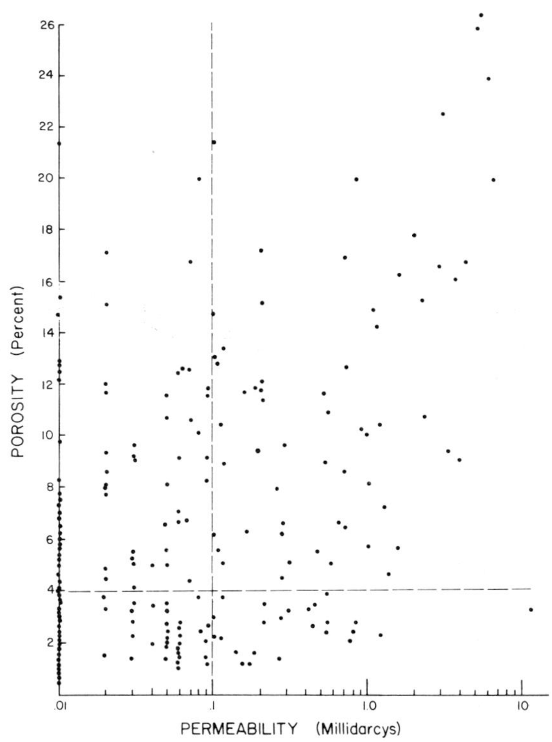

Figure 3. Porosity versus permeability plot for the Smackover Formation in South Texas. Facies from each depositional system are included in core analyses (n=202). Permeability greater than 0.1 millidarcy is considered reservoir grade.

dolomitized inner-shelf wackestone range in thickness from 3 to 27 feet (1.0 to 8.25 m), but are internally quite variable.

Reservoirs are best developed in shoal-water grainstones which have been almost completely dolomitized and have had ooids subsequently leached. The thickest reservoir encountered has 33 feet (10 m) of section and porosity as high as 26 percent and permeability as great as 6.5 millidarcys. Within the shoal system, oolite grainstone, oolite-intraclast grainstone and oolitic sandstone are the only porous lithofacies, which suggests that reservoir distribution may be controlled in part by depositional environment. Not all oolitic grainstones and sandstones, however, are of reservoir grade. Therefore, diagenetic processes are probably the ultimate control over reservoir development.

DIAGENESIS OF SMACKOVER RESERVOIR ROCKS

Inner-Shelf Dolostones

Dolomitized reservoir rocks in the inner-shelf lithofacies consist of iron-poor, fine- to medium-crystalline euhedral dolomite turbid appearance. These abundant inclusions of matrix suggest that the dolomite is replacement in origin. Depositional textures are commonly destroyed by the dolomitization.

Porosity and permeability are intercrystalline (Fig. 4). In some samples very clear, anhedral crystals of megaquartz and anhydrite partially fill pore spaces between dolomite rhombs (Fig. 5). Nonporous dolostones consist of interlocking rhombs with remaining intercrystalline space filled with lime micrite.

Inner-Shoal Oolitic Grainstone

Original porosity in well-sorted oolitic Smackover sands was probably as high as 35 percent (Coogan, 1970), but later diagenesis modified the original pore system and reduced porosity. Smackover grainstones in the deep subsurface are interpreted to have undergone four general stages of diagenesis: marine-phreatic, shallow-meteoric, regional fluid-mixing and basinal-fluid expulsion (Fig. 6). In general, the aerial distribution and extent of development of diagenetic features are similar to the "Southern" and "Transition" diagenetic zones delineated by Moore and Druckman (1981) for Smackover grainstones in southern Arkansas and northern Louisiana.

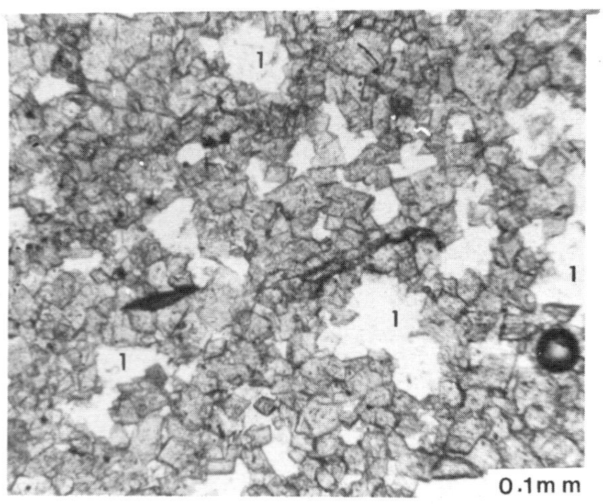

Figure 4. Fine-crystalline dolomite with 4.7 percent intercrystalline porosity (1) and 1.4 millidarcys permeability. Note turbid appearance of rhombs caused by inclusions. Skelly No. 1 Winkler, 15,219 feet (4,639m).

Figure 5. Fine-crystalline dolomite with porosity occluded by anhydrite (1) and megaquartz (2). Crossed Nicols. Skelly No. 1 Winkler; 15,203 feet (4,633 m).

Figure 6. General diagenetic stages and events in the South Texas Smackover grainstones. Solution-compaction event is modified by inferred origin and features produced. Time and depth increase towards the right.

Marine-phreatic stage of diagenesis affected sediments in the marine environment from the sediment/water interface to a few feet below the surface (Fig. 6). Mechanical and biological breakage and rounding of shells produced sand-size allochems and completely eliminated intraparticle porosity. Allochems were extensively micritized by algal and fungal microborings (Figs. 7, 8, 9). Complete micritization of grains produced abundant microporosity, thus enhancing its susceptability to replacement and dissolution. Rare zones of possible submarine "hard-ground" cementation are present in the Cities Service No. 1A Peeler Ranch core. Grains are cemented by bladed, isopachous calcite crusts, 0.03 mm thick, that intersect in polygonal junctions. These crusts contain the first cement to be precipitated and were not affected by later solution compaction. The crusts are thought to have originally been magnesium-calcite, a common modern marine-phreatic cement. General lack of marine-phreatic cements indicates that interparticle porosity was still very high at the end of the marine-phreatic stage of diagenesis.

Shallow-meteoric stage of diagenesis affected sediments from the surface to several hundreds of feet in depth (Fig. 6). Meteoric water may have entered the sediments through localized island-related ground-water systems or by a regional ground-water system. Meteoric waters were probably undersaturated with respect to calcite and had a low Mg:Ca ratio, as indicated by the dissolution of skeletal aragonite allochems (Fig. 8) and solution compaction (microstylolitization) of micritized oolites before meteoric-stage cementation (Figs. 8, 9). Dissolution of aragonite allochems has left hollow grains preserved only by micrite envelopes, which were later broken by mechanical compaction (Fig. 8). Mechanical compaction was the result of overburden pressure created by continued sedimentation and by solution compaction of supporting grains. Solution compaction started in the shallowmeteoric stage of diagenesis and continued through the deep burial stage, culminating with megastylolitization (Fig. 6). Evidence of the continued history of solution compaction is given by the paragenetic relationships between cements from different diagenetic stages and stylolites (Fig. 6).

In the No. 1 Franklin and No. 1A Alamo National Bank cores calcite pseudomorphs after an evaporite mineral are present (Figs.

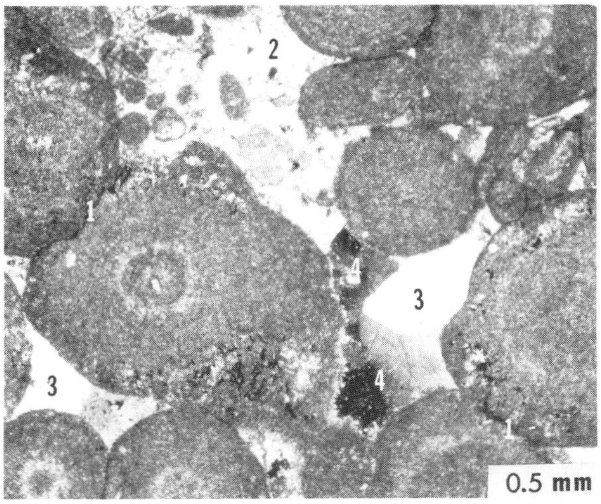

Figure 7. Micritized oolite grainstone. Solution compaction produced organic-rich microstylolite seams (1). Cements consist of early fine-crystalline (2) and late, coarse-crystalline (3) calcite and baroque dolomite (4). Crossed Nicols. Pan Am No. 1A Alamo National Bank; 21,188 feet (6,458 m).

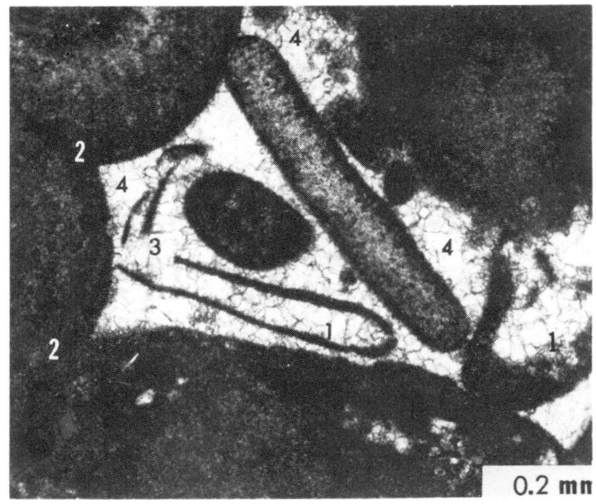

Figure 8. Micritized oolite grainstone. Sequence of diagenesis: a) dissolution of aragonite allochems (1), b) solution (2) and mechanical(3) compaction, and c) precipitation of fine- to medium-crystalline calcite (4). Pan Am. No. 1A Alamo National Bank; 21,179 feet (6,455 m)

10, 11). Rectangular fan-shaped to lenticular evaporite minerals filled pore spaces and replaced compacted grains. Remnants of microstylolites were preserved (Fig. 10). The evaporite crystals must have formed after early compaction because they cross microstylolites and replace parts of juxtaposed grains, but occurred before the shallow-meteoric, fine-crystalline equant and coarse-crystalline syntaxial calcite cements that filled evaporite-dissolution molds (Figs. 10, 11).

Fine- to medium-crystalline equant calcite precedes all cements, with the exception of the bladed isopachous calcite crust, and is not incorporated into microstylolite residue seams (Figs. 7, 8, 9). In most samples this cement is unevenly distributed, filling small intergranular pores, but only occurring at grain contacts in larger pores (Figs. 7, 9). Such a distribution, especially grain-contact cementation, is characteristic of vadose-meteoric cements (Land, 1970; Halley and Harris, 1979). In a few samples, predominantly from the No. 1A Peeler Ranch core, this equant cement is more uniformly distributed around all grains which is characteristic of phreatic-meteoric cements (Land, 1970; Steinen, 1974; Halley and Harris, 1979).

Very coarse-crystalline syntaxial calcite cement, which precipitated in optical continuity around echinoderm fragments, commonly occurs in poikilotopic patches (Fig. 11). Where the echinoderm "seed" is not obvious, such as out of the plan of the thin section, it may be difficult to distinguish coarse-crystalline syntaxial cement from a later-formed, coarse-crystalline equant calcite cement. The syntaxial cement occurred after early solution compaction and has inhibited later compaction (Fig. 11). This type of cement most commonly precipitates in the shallow-subsurface meteoric environment (Land, 1970; Longman, 1980).

Porosity was greatly reduced during the shallow-meteoric stage of diagenesis because of solution compaction. Early calcite cements are not abundant and had little affect on reducing porosity.

Regional fluid-mixing stage of diagenesis affected sediments from tens of feet to hundreds of feet in depth where meteoric fluids mixed with marine-phreatic and/or basinal brines (Fig. 6). Dolomitizaton, which produced intercrystalline porosity, was the dominant diagenetic process in this stage. Iron-poor, fine- to medium-crystalline euhedral rhombic dolomite precipitated in pore spaces and replaced allochems, especially micritized ooids (Figs. 12, 13).

Ooids in the highly dolomitized zones were not affected by solution compaction which indicates that dolomitization started early or that some matrix may have inhibited fluid flow and prevented dissolution. In areas of little dolomitization individual rhombs precipitated directly onto grains and coarse-crystalline equant calcite later filled remaining pores (Fig. 12).

A regional mixing model (Hanshaw *et al.*, 1971; Land, 1973) is suggested for dolomitization because the aerial distribution of the dolomite shows a definite regional pattern. Dolomite is concentrated updip in the area of the No. 1A Peeler Ranch and No. 1 Franklin cores, and it occurs only rarely downdip in the area of the No. 1A Alamo National Bank core (Fig. 1).

At the end of the regional fluid-mixing stage of diagenesis intercrystalline porosity in the updip, highly dolomitized lithofacies ranged from low to moderate. In the downdip, poorly dolomitized lithofacies only small amounts of interparticle porosity was still preserved.

Basinal fluid-expulsion stage of diagenesis affected the Smackover strata after several thousand feet of burial. This stage of diagenesis can be divided into two substages: a dissolution-dominated substage and a later cementation-dominated substage. Both substages are controlled by fluids generated within the deeper parts of the basin and expelled through permeable pathways to shallower, lower pressured strata. Subsurface-generated fluids may have included compaction and diagenetic waters from shales (Burst,

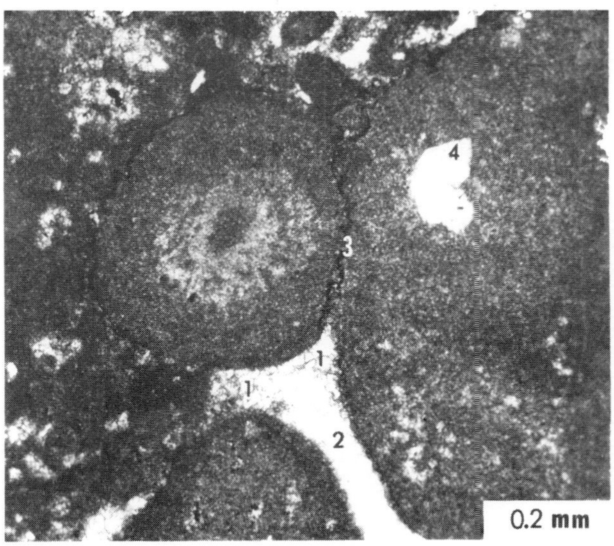

Figure 9. Micritized oolite grainstone. Fine-crystalline calcite concentrated at grain contact (1) with later coarse-cyrstalline calcite (2) filling pore. Note calcite cement is not included in the microstylolite (3) indicating solution compaction occurred first. Euhedral quartz overgrowth replacing ooid (4). Pan Am. No. 1A Alamo National Bank; 21,179 feet (6,455 m).

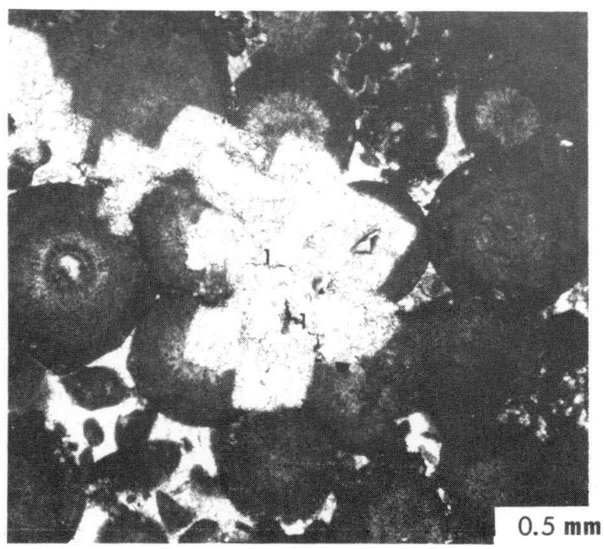

Figure 10. Micritized oolite grainstone. Sequence of diagenesis: a) solution compaction, b) replacement by evaporite mineral with subsequent dissolution and c) precipitation of fine- to medium-crystalline calcite in mold. Note preservation of microstylolite (1) in former evaporite. Pan Am. No. 1A Alamo National Bank; 21,179 feet (6,455 m).

1969) and brines squeezed out of deeply buried evaporite strata (Moore and Druckman, 1981; Land and Prezbindowski, 1981). These waters also contained products of organic maturation (decarboxylation and hydrocarbon generation; Tissot and Welte, 1978) and H_2S generated by thermal interaction of hydrocarbons and anhydrite (Orr, 1977).

During the dissolution substage micritized lime ooids from the highly dolomitized lithofacies were dissolved (Fig. 13), and the dolomite framework underwent minor solution compaction as indicated by sutured and embayed contacts between rhombs. However, the relatively stable dolomite framework inhibited the solution compaction of grains.

In the downdip dolomite-poor lithofacies solution compaction was reactivated. Also, rhombic dolomite cement was corroded (Fig. 12), and some rhombs were incorporated into rejuvenated microstylolites (Fig. 12).

Water charged with CO_2, related to thermal decarboxylation of kerogen-rich rocks in the lower Smackover basinal facies, may have promoted the subsurface dissolution stage of diagenesis. Solution compaction could have been reactivated by CO_2 generation in seams of organic-rich residue along previously formed microstylolites.

The dissolution substage was an important period of subsurface reservoir development. Relatively tight rocks developed into quality reservoirs by the dissolution of micritized ooids. This process occurred before hydrocarbon generation and migration (Fig. 6) as evidenced by bitumen within the oomoldic porosity (Fig. 13).

Euhedral quartz overgrowths (Figs. 9, 12, 13) on detrital quartz grains are paragenetically related to the early part of the dissolution substage. Quartz overgrowths, formed on quartz grains in ooid nuclei, replaced micritized ooid rinds and incorporated

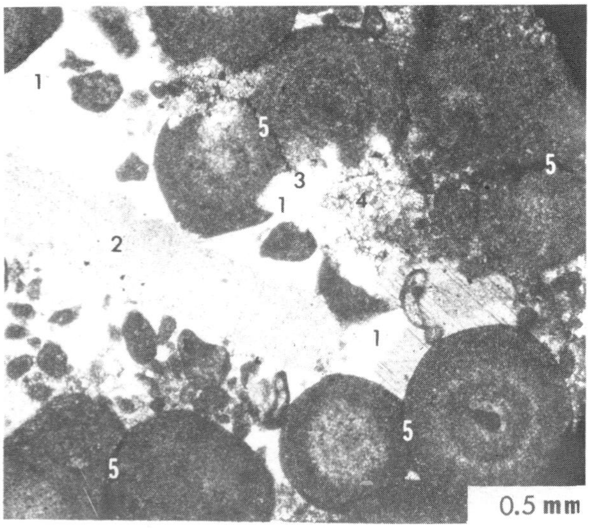

Figure 11. Poikilotopic, very coarse-crystalline syntaxial calcite (1) formed on an echinoderm fragment (2) and filled in the mold of a former evaporite crystal (3). Note fine-crystalline calcite (4) in evaporite mold and microstylolites between micritized ooids (5). Pan Am. No. 1A Alamo National Bank; 21,188 feet (6,458 m).

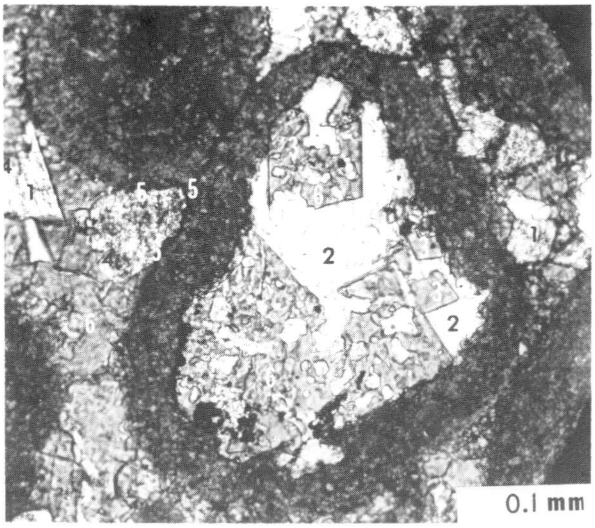

Figure 12. Micritized oolite grainstone. Sequence of diagenesis: a) precipitation of medium-crystalline rhombic dolomite (1) in former interparticle and moldic pore spaces, b) precipitation of quartz (2) which incorporated dolomite rhombs, c) dissolution (3), corrosion (4) and incorporation of dolomite into late solution compaction seams (5) and d) remaining pore space filled with Fe-poor, coarse-crystalline equant calcite (6). Pan Am No. 1A Alamo National Bank; 21, 275 feet (6,485 m).

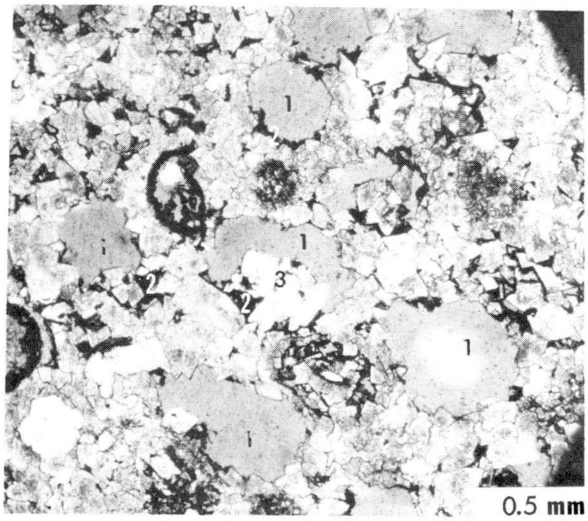

Figure 13. Oomoldic porosity (1) in dolomitized oolite grainstone. Note geopetal residue of bitumen (2) and quartz overgrowths containing dolomite rhombs (3). Quartz overgrowths and dolomite rhombs are replacement products of the micritized ooids. Pan Am. No. 1 Franklin; 18,745 feet (5,713 m).

dolomite rhombs that previously had replaced ooids (Figs. 12, 13). However, euhedral quartz overgrowths which dropped to the bottom of ooid molds indicate dissolution occurred after quartz diagenesis (Fig. 13).

Abundant coarse- to very coarse-crystalline equant calcite (Figs. 7, 9, 12) and minor amounts of coarse-crystalline baroque dolomite (Fig. 7) were precipitated during the basinal fluid-expulsion cementation substage. These cements are most abundant in the downdip No. 1A Alamo National Bank core. They were the last cements to precipitate and they occluded the remaining pore space.

Coarse-crystalline equant calcite overlies fine-crystalline equant calcite (Fig. 9) and rhombic dolomite cements (Fig. 12) and occurs in molds of dissolved ooids, which suggests a late origin for the coarse-crystalline equant calcite. Coarse (greater than 0.25 mm) calcite crystals, excluding syntaxial calcite crystals, are rare in Holocene and Pleistocene meteoric-phreatic environments. This also supports a late-stage origin for the coarse-crystalline equant calcite. The coarse-crystal size, straight-crystal edges as seen in thin section, and abrupt size change with earlier cements are characteristic of calcite cements that formed late in the deeper subsurface (Loucks, 1977; Badiozamani *et al.*, 1977). Moore and Druckman (1981) also suggest that coarse-crystalline calcite in the Smackover grainstone of their "Southern" diagenetic zone was precipitated in the deeper subsurface. Their conclusions are based on fluid-inclusion temperature and stable-isotope data. The late coarse-crystalline equant calcite cement in the Smackover Formation may have formed from products produced by solution compaction.

The coarse-crystalline baroque dolomite formed in close association with the coarse-crystalline equant calcite (Fig. 7). Baroque dolomite has been shown in other areas to be a later subsurface cement in both carbonates and sandstones (Loucks, 1977; Loucks *et al.*, 1977; Brock, 1980; Radke and Mathis, 1981). Radke and Mathis (1981) have proposed that baroque (saddle) dolomite forms between temperatures of 60° to 150° C. Loucks (1977) has suggested that the baroque dolomite in the Lower Cretaceous section in South Texas formed from mixing of subsurface brines and diagenetic waters from clays. Further study may show this mechanism to be applicable to the Smackover Formation.

Porosity in the dolomite-poor lithofacies was totally occluded by the end of the cementation substage. Secondary dissolution porosity, formed at the expense of late equant calcite cement, as noted by Moore and Druckman (1981) in their "Southern" diagenetic zone, was not observed in South Texas. This style of secondary porosity, however, may yet be encountered in future wells.

The final diagenetic process was the formation of megastylolites, probably from the enlargement of earlier formed microstylolites through continued solution compaction. The megastylolites cut across all grain and cement types and incorporated insoluble material into a thick residue seam. Overburden pressure, and possibly the generation of thermally derived H_2S, promoted this process.

CONCLUSIONS

Reservoir development within the Smackover Formation in the deep subsurface of South Texas was greatly affected by diagenetic processes. Smackover strata went through four general diagenetic states (Fig. 6). The first stage, marine phreatic, had little effect on porosity except for the elimination of intraparticle porosity. The next stage of diagenesis, in the shallow-meteoric environment, was basically a porosity-reduction stage through chemical-solution compaction and minor cementation. Intercrystalline porosity developed during dolomitization in the regional fluid-mixing stage. Reservoirs containing oomoldic porosity developed in dolomitized grainstones during the dissolution substage of the basinal fluid-expulsion stage. Porosity was further reduced by rejuvenated chemical-solution compaction in poorly dolomitized lithofacies during this same substage. In the cementation substage porosity was occluded by the precipitation of coarse-cyrstalline calcite and baroque dolomite.

Hydrocarbon exploration in the Smackover in South Texas should be concentrated in the dolomitized oolitic grainstone and oolitic quartzarentie facies belt (Fig. 1). In these facies secondary oomoldic porosity associated with intercrystalline dolomite porosity is commonly well developed. Reservoirs can also be expected in dolomitized muddy facies, such as in the inner-shelf wackestones. A production test in the oolitic grainstone facies of the No. 1 Franklin well (Fig. 1) produced at a flow rate of 331 MCF of H_2S gas plus 168 barrels of salt water per day. An equivalent of 3,600 barrels of salt water per day was tested in a Smackover dolomite section in the Humble No. 1 Benavidies well in Webb County. This test gives some indication of flow rates that might be expected from Jurassic dolomitized reservoirs in South Texas.

ACKNOWLEDGMENTS

The authors wish to express their appreciation to C.R. Handford and D.G. Bebout who critically reviewed the manuscript. Cities Service Company, Amoco Production Company, and Shell Oil Company provided geophysical logs and loaned cores to the Bureau of Economic Geology. E. L. Stoudt of Getty Oil Company kindly provided thin sections from the Skelly No. 1 Winkler core. We thank Deborah Loucks for originally typing and editing this paper.

Publication authorized by the Director, Bureau of Economic Geology, The University of Texas at Austin, Austin, Texas.

REFERENCES

Badiozamini, K., F.T. Mackenzie and D.C. Thorstenson, 1977, Experimental carbonate cementation: salinity, temperature, and vadose-phreatic effects: Jour. Sed. Petrology, v. 47, p. 529-542.

Becher, J.W. and C.H. Moore, 1976, The Walker Creek Field: a Smackover diagenetic trap: Gulf Coast Assoc. Geol. Socs. Trans., v. 26, p. 34-56.

Brock, F.C.,Jr., 1980, Walker Creek revisited: a reinterpretation of the diagenesis of the Smackover Formation of Walker Creek Field, Arkansas; Louisiana State Univ. M.S. Thesis, 78 p.

Budd, D.A. and R.G. Loucks, 1981, The Smackover and lower Buckner Formations, Jurassic, South Texas: deposition on an ancient carbonate ramp: Univ. Texas, Austin, Bur. Econ. Geology Rept. of Inv. 112, 38 p.

Burst, J.F., 1969, Diagenesis of Gulf Coast clayey sediments and its possible relation to petroleum migration: Am. Assoc. Petroleum Geologists Bull., v. 53, p. 73-93.

Coogan, A.H., 1970, Measurements of compaction in oolitic grainstone, Jour. Sed. Petrology, v. 40, p. 921-929.

Dickinson, K.A., 1968, Upper Jurassic stratigraphy of some adjacent parts of Texas, Louisiana and Arkansas: U.S. Geol. Survey Prof. Paper, 594E, 25 p.

Halley, R.B., and P.M. Harris, 1979, Fresh-water cementation of a 1,000 year old oolite: Jour. Sed. Petrology, v. 49, p. 969-988.

Hanshaw, B.B., W. Back, and R.G. Deike, 1971, A geochemical hypothesis for dolomitization by ground water: Econ. Geology, v. 66, p. 710-724.

Land, L.S., 1970, Phreatic versus vadose meteoric diagenesis of limestone: evidence from a fossil water table: Sedimentology, v. 14, p. 175-185.

__________, 1983, Holocene meteoric dolomization of Pleistocene limestones, North Jamaica: Sedimentology, v. 20, p. 411-424.

__________, and D.R. Prezbindowski, 1981, The origin and evolution of saline formation water, Lower Cretaceous carbonates, south Central Texas: Jour. of Hydrology.

Longman, M.W., 1980, Carbonate diagenetic textures from nearsurface diagenetic environments: Am. Assoc. Petroleum Geologists Bull., v. 64, p. 461-487.

Loucks, R.G., 1977, Porosity development and distribution in shoal-water carbonate complexes-subsurface Pearsall Formation (Lower Cretaceous), South Texas, *in* D.G. Bebout, and R.G. Loucks, (eds.), Cretaceous carbonates of Texas and Mexico: application to subsurface exploration: Univ. Texas, Austin, Bur. Econ. Geology, Rep. of Inv. 89, p. 97-126.

__________, D.G. Bebout, and W.E. Galloway, 1977, Relationship of porosity formation and preservation to sandstone consolidation history-Gulf Coast lower Tertiary Frio Formaton: Gulf Coast Assoc. Geol. Socs. Trans., v. 27, p. 109-120.

Moore, C.H. and Y. Druckman, 1981, Burial diagenesis and porosity evolution, Upper Jurassic Smackover, Arkansas and Louisiana: Am. Assoc. Petroleum Geologists Bull. v. 65, p. 593-628.

Orr, W.L., 1977, Geologic and geochemical controls on the distribution of hydrogen sulfide in natural gas, in R. Campos and J. Goni, (eds.), Proceedings of the 7th International Meeting on Organic Geochemistry, Published by Revista Expanola de Micropaleontologia, p. 572-597.

Radke, B.M. and R.L. Mathis, 1980, On the formation and occurrence of saddle dolomite: Jour. Sed. Petrology, v. 50, p. 1149-1168.

Steinen, R.P., 1974, Phreatic and vadose diagenetic modification of Pleistocene limestone: petrographic observations from subsurface of Barbados, West Indies: Am. Assoc. Petroleum Geologists Bull., v. 58, p. 1008-1024.

Tissot, B.P. and D.H. Welte, 1978, Petroleum formation and occurrence: Springer-Verlag, 527 p.

JOHNS FIELD, UPPER SMACKOVER, MISSISSIPPI

HERSCHELL J. McCUNN
G. DAN WARD
CAMERON COX

Tomlinson Interests, Inc.
925 Houston National Gas Building
Houston, Texas 77002

The Johns Field is located in Rankin County, Mississippi approximately 40 miles southeast of the city of Jackson. It produces sour gas from the Upper Smackover Formation of Jurassic age. The structure is a deep-seated, early-growth salt anticline which is faulted on the south side (Fig. 1). The structure was first detected by improved seismic methods (CDP) which led to the discovery (Figs. 1 and 2).

The pay zones in the field are geopressured and require special drilling and completion techniques. Permeabilities range from .01 to 10 millidarcies; however, porosities are low due to the presence of solid hydrocarbon and calcareous cement in the sandstones.

The following information is submitted on the Johns Field:

DISCOVERY DATA

Exploration Method: Seismic
Well: No. 1 Caley T. Jones 33-10
Total Depth: 19,863 feet
Perforations and Slotted Liner:
18,795 - 810 (4 JSPF)
18,803 - 19,260
1/2" hole - 6 per foot
Initial Potential: 7,850 MCFPD
19/64" CK
FTP 3700#

STRUCTURE. Salt-related, west-northwest to east-southeast trending anticline associated with a large down-to-the-south fault (Fig. 4).

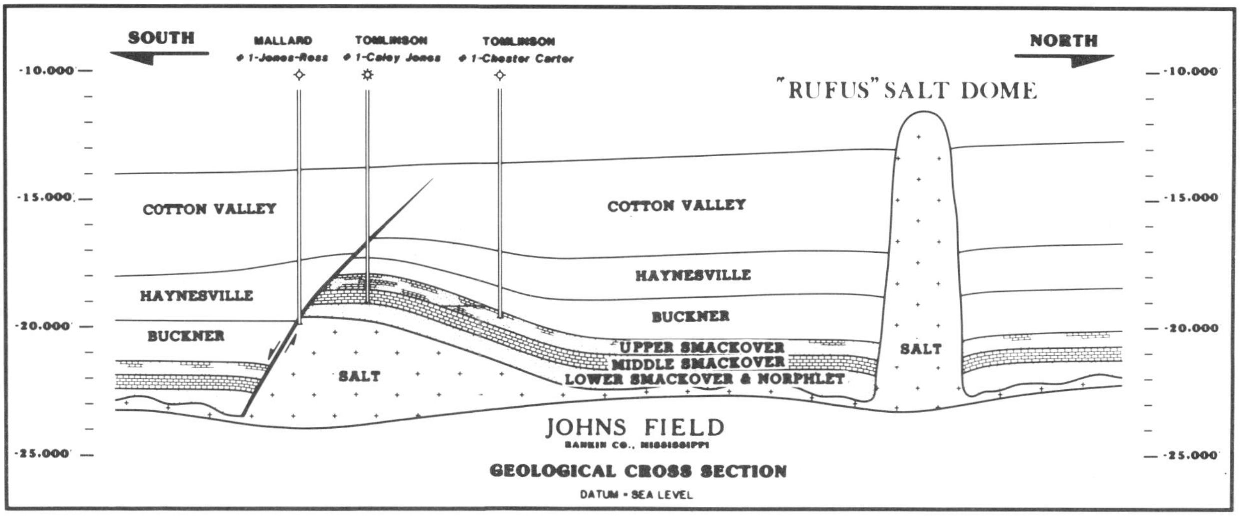

Figure 1. Geological cross section of Johns Field, Rankin County, Mississippi.

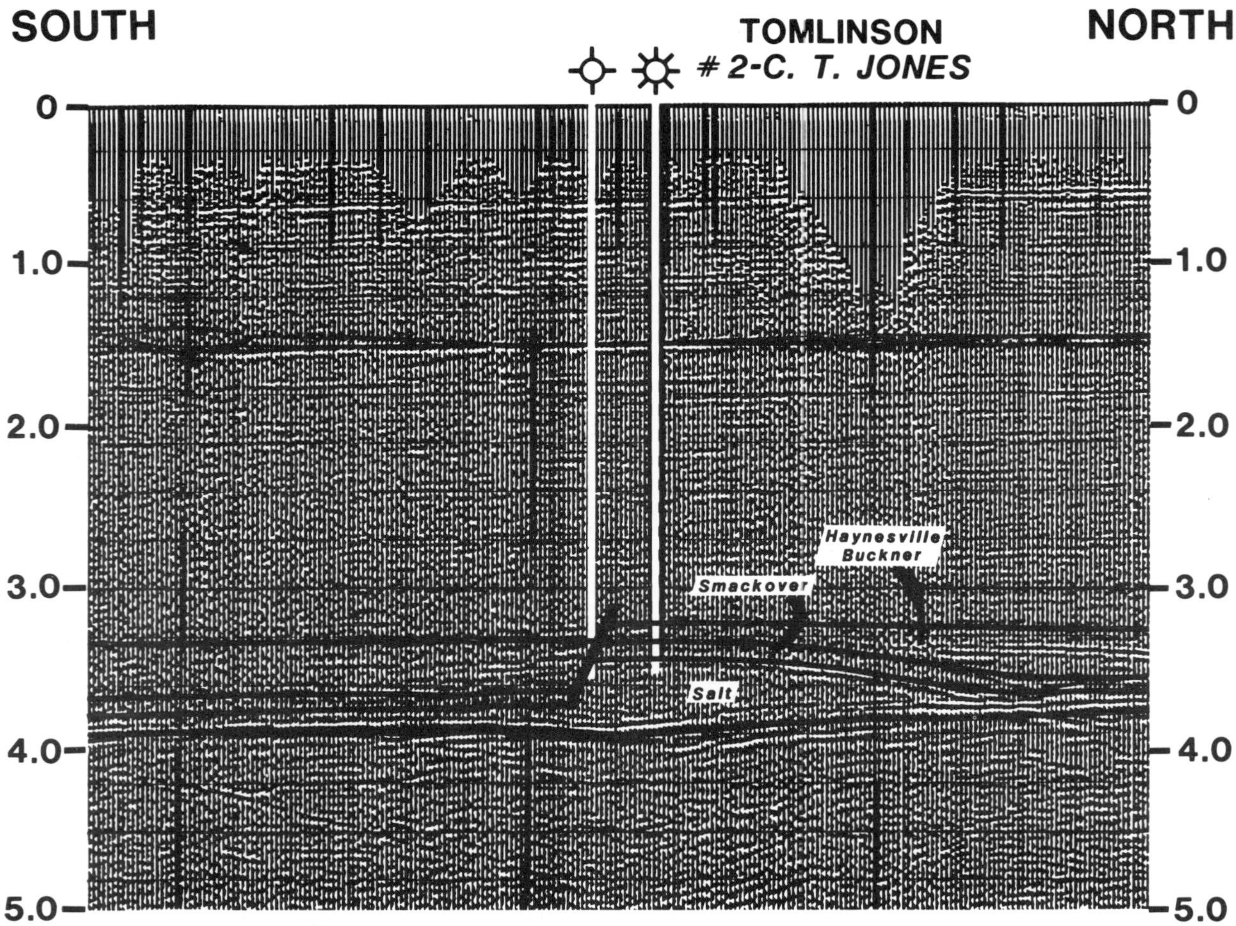

Figure 2. North-South seismic section Johns Field.

STRATIGRAPHY OF RESERVOIR ROCK. Fine- to medium-grained calcareous and dolomitic sandstones with abundant bitumen in the matrix.

OLDEST STRATIGRAPHIC HORIZON PENETRATED

Middle Smackover No. 1 Jones 33-10

RESERVOIR DATA

Depth: -18,170 to -19,150 feet

Spacing: 640 acres (with option for 2 wells per 640)

Current Number of Wells: 3 plus 2 drilling (6/7/82)

Average Pay: 128 feet

Average Porosity: 6.6%

Average Permeability: .01 - 10 md (limited core data indicated that the reservoir is highly fractured. These fractures have been extensively filled with calcite. A small hydraulic fracture treatment of the reservoir in the No. 1 Vera Cox greatly enhanced flow from the well from 3,000 MCFPD to an excess of 17,000 MCFPD.)

Average Water Saturation: 17%

Drive Mechanism: Gas expansion

Production Acreage: 2,496

Gas - Water Contact: (-19,150 feet)

BHT: 350° F

IBHSIP: 17,000 psig

GOR: Dry

Production History: Waiting on plant facilities

Gas Analyses: On-site chromatographic anal-

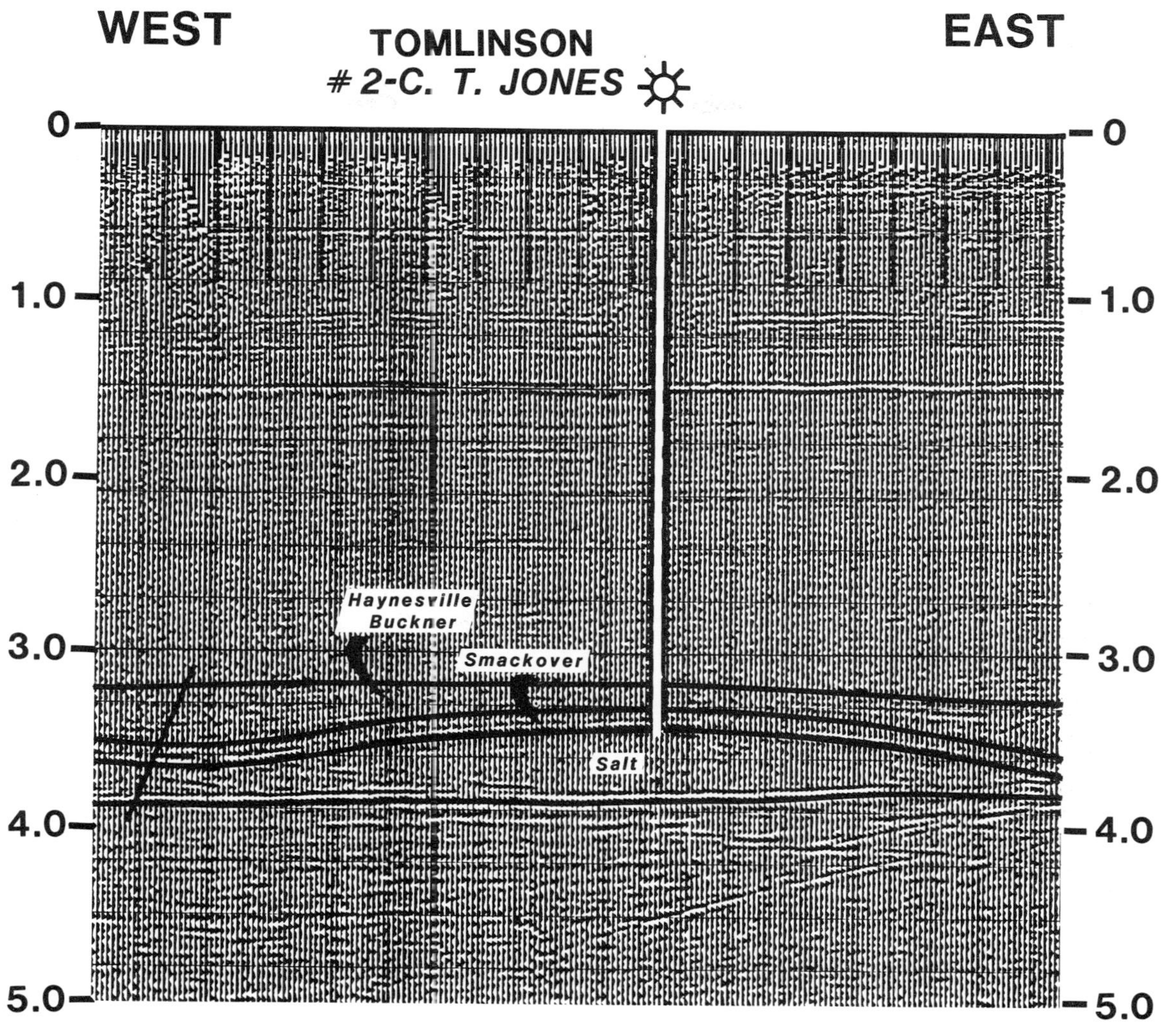

Figure 3. East-West seismic section Johns Field.

yses of gas showed:

Components	COMPOSITION No. 1 C.T.Jones	No. 1 V.Cox	Field Average
H_2S	22.88	28.62	25.74
CO_2	4.65	4.04	4.34
N_2	0.85	0.58	0.70
CH_4	71.57	66.81	69.17
C_2H_6	0.03	0.03	0.03
C_3H_8	0.02	0.02	0.02
	100.00	100.00	100.00
SP.Gr (60°F-Air=1)	0.7471	0.7738	0.7604
BTU/SCF @14.696psia and 60°F-Wet Basis	871	839	855

DRILLING AND COMPLETION PROGRAM. To date three wells have been drilled in the Johns Field and two wells are being drilled. The wells in the Field produce from the upper Smackover Formation. Based on the two wells which have been tested, the produced gas contains 24% to 29% hydrogen sulfide (H_2S). Due to the high concentrations of H_2S, all mechanical equipment exposed to the produced gas has to meet or exceed NACE specifica-

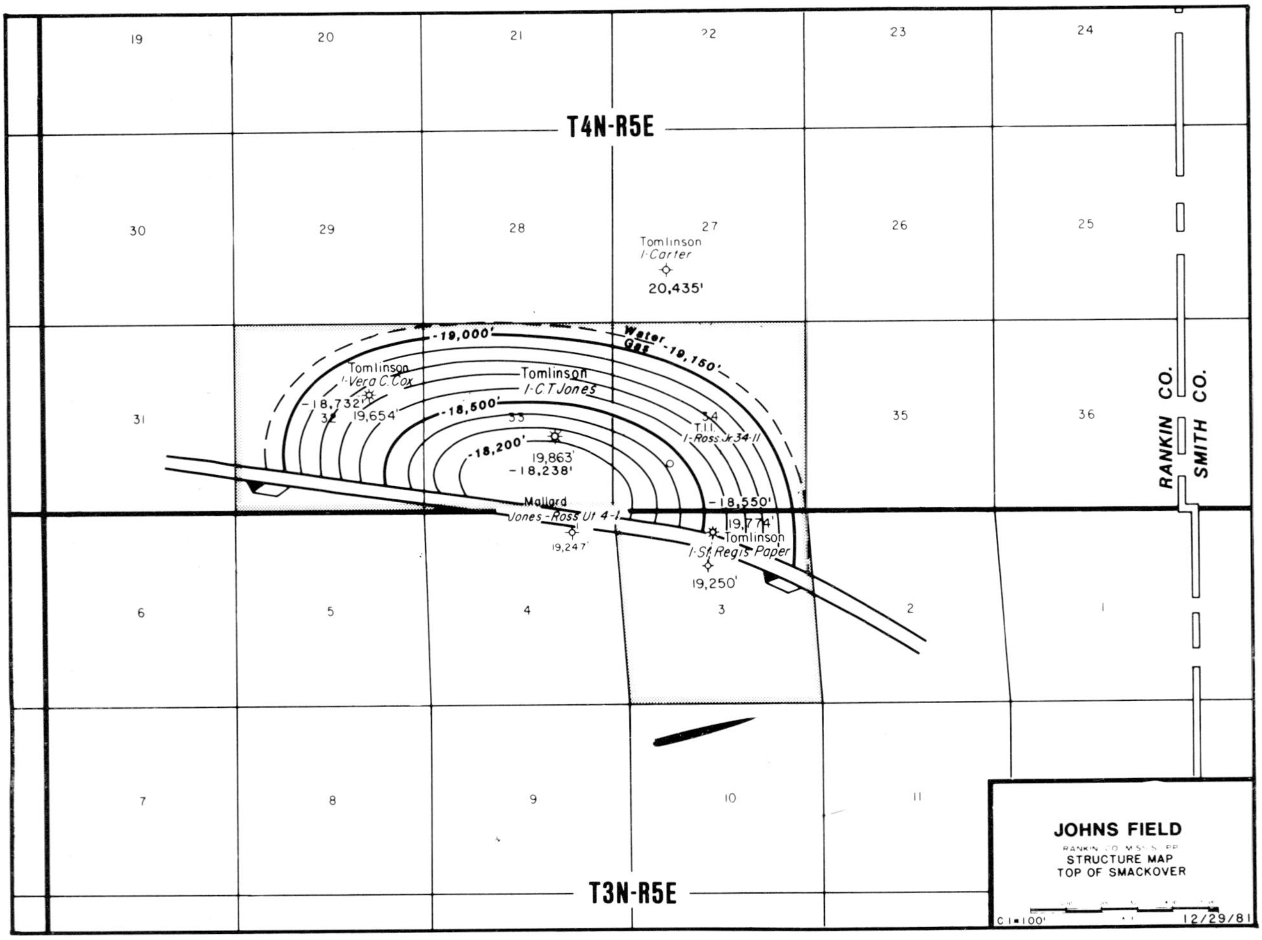

Figure 4. Structure map on top of Smackover Formation Johns Field.

tions. In order to meet these specifications, tubular goods and well heads made of exotic metals are used. These pieces of equipment have to be ordered 12 to 24 months prior to delivery, and should any of this equipment mechanically fail, contingency programs must be implemented quickly. Taking these factors into account, the drilling of a high-pressure, sour gas test requires considerable lead-time planning. The following is a brief discussion of the drilling program used in the John Field (Figs. 5, 6, and 7).

Prior to moving in the drilling rig, a 20-inch conductor pipe is set at 60 feet. A 17-1/2 inch hole is then drilled with 8.9 ppg mud to approximately 5,600 feet in 6 days, where 13-3/8 inch casing is set. Next a 12-1/4 inch hole is drilled to about 14,400

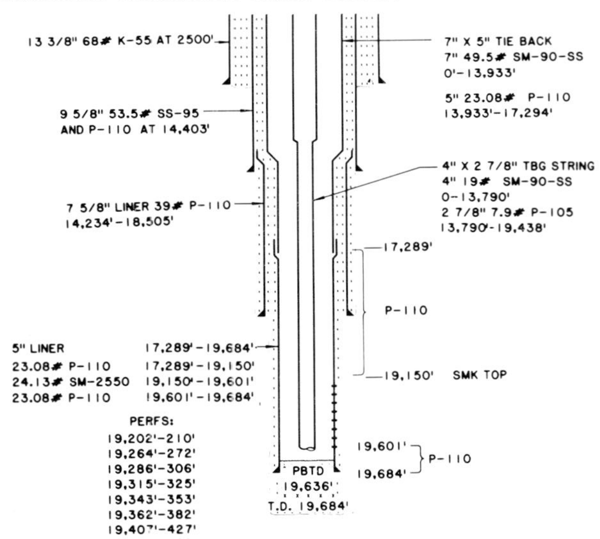

Figure 5. Well completion diagram for No. 1 Vera Cox (32-7) No. 1 well Johns Field.

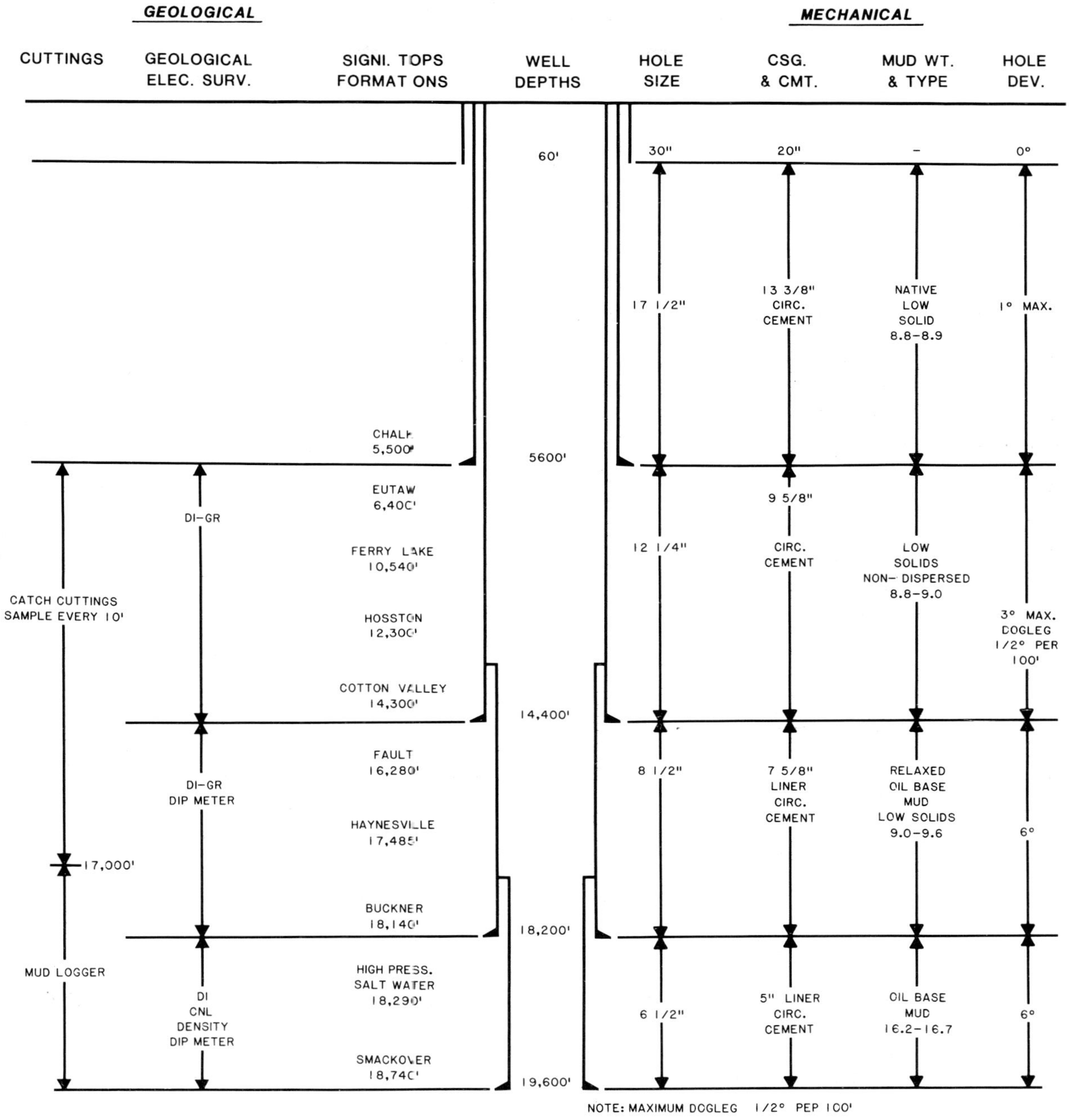

Figure 6. Mud and logging program for Johns Field.

feet using 9 ppg mud where 9-5/8 inch casing is set. The 9-5/8 inch casing is set at this depth in order to accomodate the 7-inch by 5-inch tie-back string. To reach the 9-5/8 inch casing point requires an additional 40 days or 46 days total drilling time. Next an 8-1/2 inch hole is drilled to approximately 18,200 feet using a relaxed oil base mud (9 - 9.6 ppg), where a 7-5/8 inch liner is set. A relaxed oil base mud has a high water to oil ratio. To reach the 7-5/8 inch liner depth involves an additional 36 days to 38 days of total drilling time. The well is then drilled to total depth with 17.0 - 17.4

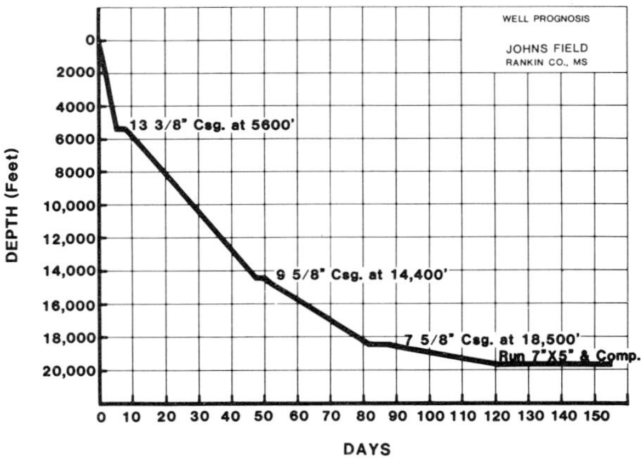

Figure 7. Well prognosis for Johns Fields.

ppg oil base mud and a 6-1/2 inch bit. Between about 18,200 feet and the top of the Smackover Formation, a Buckner high-pressure water flow sometimes is encountered. At total depth a 5-inch liner is set. Drilling of this interval is slow and takes an additional 30 days. Approximately 120 days of total drilling time are required to reach total depth. Safety equipment and contingency programs must be ready prior to penetrating the deepest interval because the Smackover will be opened and is the source of H_2S. The 5-inch liner is the first string of pipe which has to be resistant to the effects of H_2S. A tapered 7-inch by 5-inch tie-back string is run in the well, which ties into the top of the 5-inch liner. The tie-back string acts as the production casing and also is made of corrosion resistant metals. The well is then perforated and the tubing string is run. The tubing is a tapered 4-inch by 2-7/8 inch string which also must be made of corrosion resistant metals. The well is completed with a live annulus, which allows inhibitor oil to be circulated down the annulus and up the tubing during production. The end of the tubing is hung below bottom perforations to ensure complete inhibitor oil circulation during production.

PRODUCTION PLANT. The Johns Plant utilizes the Shell Sulfinol Process, Shell "SCOT Tail Gas Clean-up" Process and the Amoco/Ortloff Sulfur Recovery Process. The plant has an inlet gas design capacity of 50 MMSCFD nominal. The plant at design capacity delivers 33.7 MMSCFD of pipeline residue sales gas and 421 long tons per day of "Texas Bright" sulfur (Fig. 8).

Technology is now available for the detection, safe completion and production of future, deep-seated Smackover sour gas prospects.

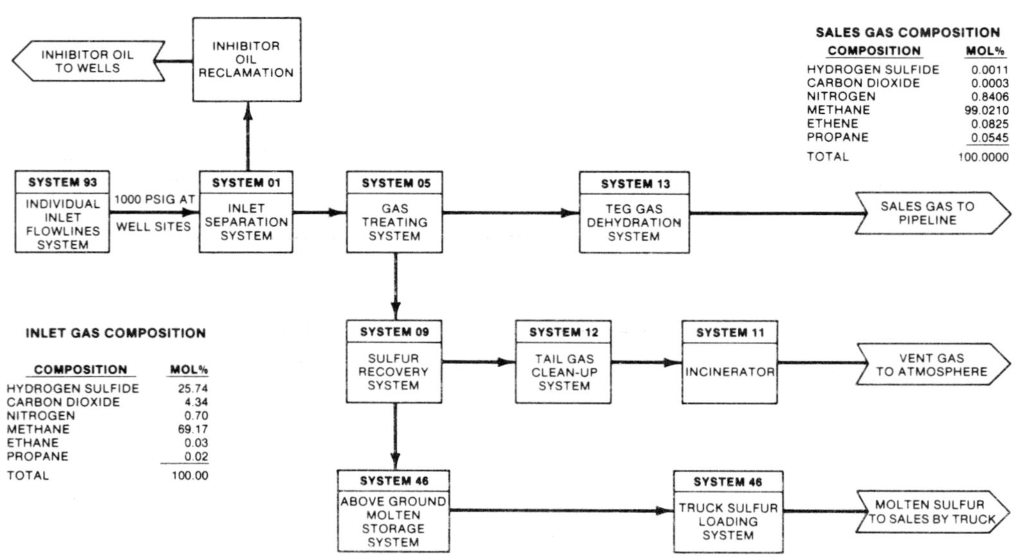

Figure 8. Processing flow diagram for Johns Field, processing and sulfur recovery facility.

COTTON VALLEY (UPPER JURASSIC) AND HOSSTON (LOWER CRETACEOUS) DEPOSITIONAL SYSTEMS AND THEIR INFLUENCE ON SALT TECTONICS IN THE EAST TEXAS BASIN

MARY K. McGOWEN

ARCO Oil and Gas Company
P.O. Box 2819
Dallas, Texas 75221

DAVID W. HARRIS

Marathon Oil Company
P.O. Box 2659
Casper, Wyoming 82602

ABSTRACT

Correct interpretation of the effect of basin infilling on salt mobilization is critical to understanding salt dome growth and stability. The size of salt structures in the East Texas Basin is determined by the original thickness of the underlying Louann Salt (Middle Jurassic): that is, salt structures distinctly increase in size toward the interior of the basin. Initial movement of salt apparently occurred in the marginal areas of the basin during Smackover (Late Jurassic) deposition. This movement seems to have resulted from downward creep that was induced by loading of carbonate units and was enhanced by basinward tilting.

During a major shift from carbonate to clastic sedimentation in the Late Jurassic, salt movement became more extensive. This salt migration was caused by uneven sediment loading of fluvial-deltaic systems in the Cotton Valley Group (Upper Jurassic) and the Hosston Formation (Lower Cretaceous). Terrigenous source areas to the west and north persisted throughout Cotton Valley and Hosston time. Clastics were delivered to the East Texas Basin by many small streams, rather than by one major stream, because a mature drainage system had not yet formed.

The Cotton Valley Group, which is thought to be a fan-delta system, can be subdivided into three types of facies: prodelta deposits, delta-front deposits, and braided fluvial deposits. Fan deltas, supplied by braided streams, prograded from the north, northwest, and west. Dip-oriented sandstone trends dominate in the northwestern part of the basin and change basinward to northeast to southwest strike-oriented trends.

During Hosston time, sedimentation in the northwestern part of the basin was dominantly fluvial. The depositional characteristics of sediments in this area are typical of braided streams. In the study area, parallel net-sandstone and sediment thicks are clearly defined in the distal part of the Cotton Valley but are not as well defined in the Hosston. This suggests that most deltaic sedimentation during Hosston time occurred basinward of the study area. A major transgression at the end of Hosston time resulted in deposition of the Pettet Limestone.

Apparently, the location of salt domes and salt anticlines was controlled by the position of the Smackover-Gilmer carbonate platform. This platform impeded local subsidence to the extent that fan-delta sediments of the Cotton Valley Group spread laterally across the shelf rather than stacked vertically. Sediment depocenters formed preferentially basinward of the platform, resulting in migration of the underlying salt into ridges that fronted the prograding sediment wedge. As the salt was depleted under these depocenters, subsidence slowed and thereby allowed the fan deltas to override the salt ridges. This resulted in a basinward progradation of deltaic depocenters and produced younger depocenters toward the interior of the basin. Further salt migration and differentiation of salt ridges produced the present complex array of salt domes and anticlines of the East Texas Basin. Seismic and gravity data clearly demonstrate the existence of these salt ridges and intervening sediment thicks.

INTRODUCTION

The Cotton Valley Group (Upper Jurassic) and Hosston Formation (Lower Cretaceous) were studied as part of the East Texas Waste Isolation project being conducted by the Bureau of Economic Geology for the U. S. Department of Energy. The purpose of the project is to assess the suitability of salt domes in the East Texas Basin as potential repositories for nuclear waste; this suitability is contingent on the tectonic stability of the domes. The objective of the present analysis was to investigate the effect of early basin infilling on salt mobilization in the East Texas Basin. Understanding the mechanisms responsible for early salt movement is essential to predicting domal growth evolution and ultimate stability.

An area in the northwestern part of the East Texas Basin consisting of seven counties--Hunt, Hopkins, Wood, Rains, Kaufman, Van Zandt, and Henderson--was selected for the study of the relationship between salt movement and the influx of Upper Jurassic terrigenous clastic sediment. The study area was chosen for two reasons: first, because preliminary studies indicated the presence of a fan-delta system prograding from the northwest and north into the basin (McGowen and Harris, 1981) and second, because deep-well-control, seismic, and gravity data were available.

Salt movement began at different times in different parts of the basin. The earliest movement occurred around the margins of the basin during Smackover deposition (Jackson and Harris, 1981). At that time, increased subsidence toward the center of the basin caused basinward tilting that, induced by downward creep, mobilized salt.

More extensive salt movement occurred after the influx of Cotton Valley clastic sediment during the Late Jurassic (Fig. 1). Before that time, deposition in the East Texas Basin was dominated by carbonates, evaporites, and marine mudstones and claystones. Salt movement apparently was controlled by differential loading of Upper Jurassic and Lower Cretaceous fluvial-deltaic systems, as well as by the position of the subjacent Smackover-Gilmer carbonate shelf complex (Jackson and Harris, 1981; McGowen and Harris, 1981).

Data Base

Electric logs from 232 wells (Fig. 2), supplemented by Bouguer residual gravity maps and two dip-oriented, six-fold conventional CDP seismic profiles, served as a data base for this study. When possible, well data were integrated with seismic data by using velocity conversion tables. Five seismic reflectors within the Mesozoic were used, including the base of the Louann Salt, the top of the Louann Salt, the top of the Gilmer Limestone (Cotton Valley Limestone)(Forgotson and Forgotson, 1976), and the top of the Pettet Limestone (Table 1). The fifth reflector, which we believe is the top of the Massive Anhydrite, was used in the northern part of the basin, where the Pettet Formation changes lithologically from a limestone facies to a sandy facies and thereby loses its character as a distinct seismic reflector. The Louann Salt is characterized by prominent boundary reflections (Jackson and Harris, 1981). Its inferred thickness, based on

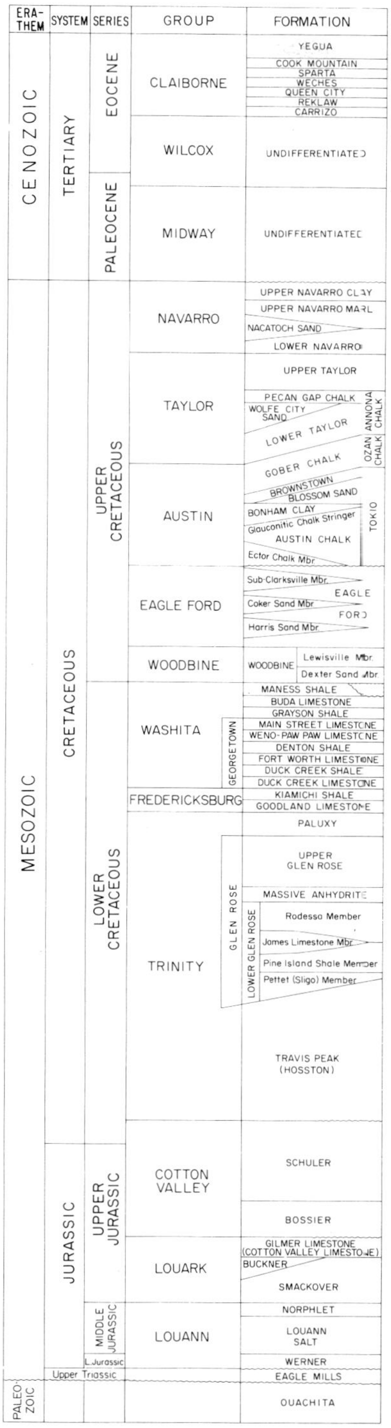

Figure 1. Stratigraphic succession and nomenclature in the East Texas Basin (Wood, 1981).

seismic interpretation, was correlated with

Table 1. Seismic Reflectors and Seismic Units In the Northwestern Part Of the East Texas Basin

SEISMIC REFLECTOR	SEISMIC UNIT
Upper Navarro Marl	
Top of the Pecan Gap Chalk	
Top of the Austin Chalk	
Top of the Buda Limestone	
*Top of the Massive Anhydrite?	
*Top of the Pettet Limestone	D
*Top of the Gilmer Limestone	C
*Top of the Louann Salt	B
*Base of the Louann Salt	A

*Seismic reflectors used in this study.

gravity data. Zones of thicker salt generally coincide with gravity lows, whereas areas of thinner salt correspond to gravity highs (Jackson and Harris, 1981) (Fig. 3).

Isopach, net-sandstone, and sandstone-percent maps of the Cotton Valley Group and the Hosston Formation were prepared. The boundary between the two was based on scout card information and regional correlations within the East Texas Basin. Using the Pettet Limestone as a datum, nine stratigraphic cross sections were constructed within the study area; selected sections are included in this report (Fig. 2).

Limitations of this data base include the following: First, although well spacing within individual oil and gas fields is good, overall spacing is poor, precluding detailed mapping of the Cotton Valley Group and Hosston Formation on a regional scale. Second, because conventional-core data were not available to verify environmental interpretations, facies designations were based entirely on electric log response and on sand-body geometry determined from net-sandstone maps, sandstone-percent maps, and textural and compositional features observed in well cuttings. And third, because Jurassic formations in northeast Texas are restricted to the subsurface, facies relationships of surface exposures could not be examined.

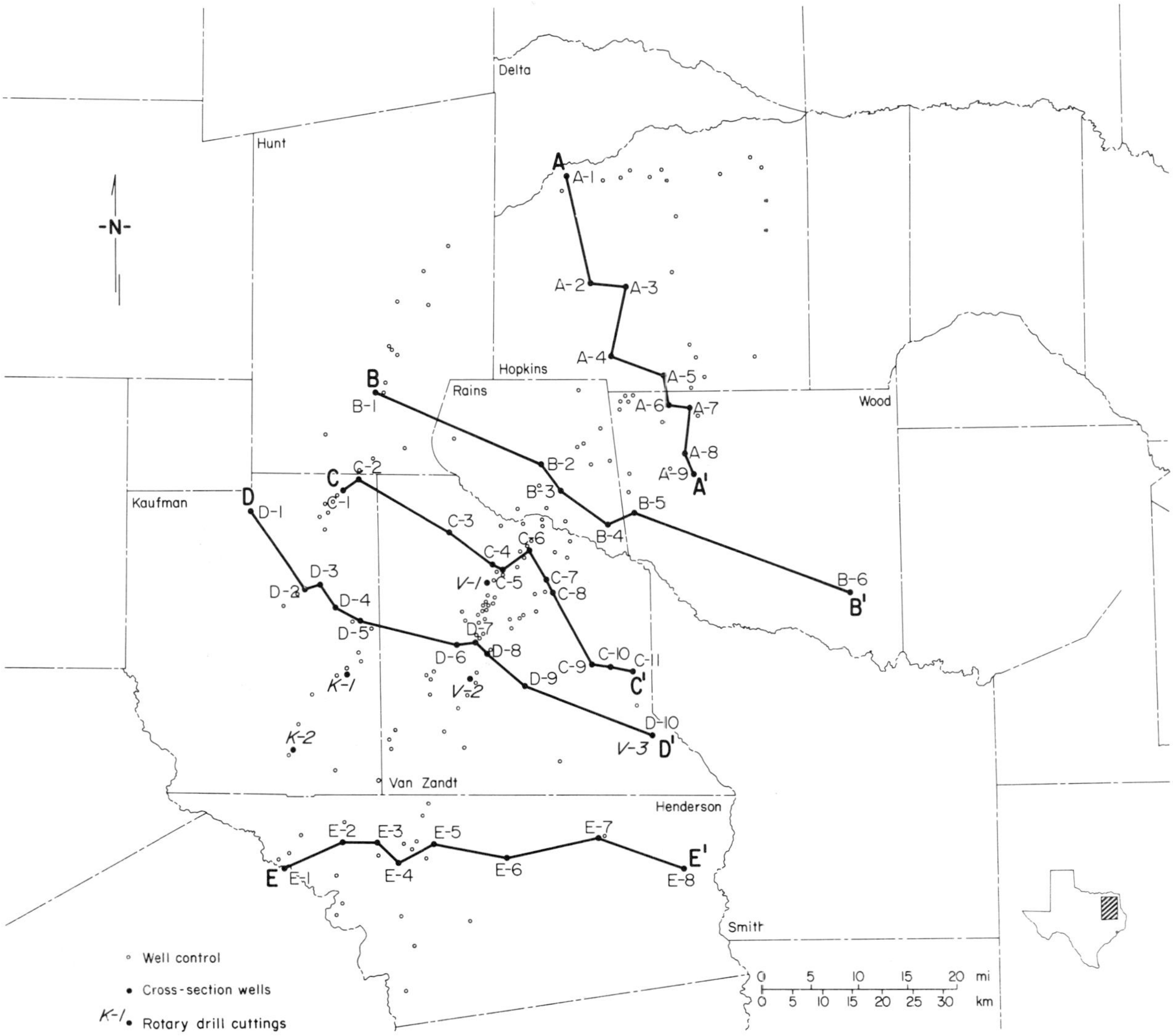

Figure 2. Index map showing well control, well cuttings, and location of stratigraphic cross sections.

Previous Work

Early work on the Cotton Valley Group (Upper Jurassic) and the Hosston Formation (Lower Cretaceous) in the East Texas Basin emphasized regional stratigraphic and environmental synthesis; the limited number of wells penetrating the Jurassic section precluded more detailed studies. Regional studies that provide excellent background material include Imlay (1943), Swain (1949), Forgotson (1954), Bushaw (1968), Dickinson (1968), Nichols *et al.*, (1968), and Newkirk (1971). Todd and Mitchum (1977) were the first to present seismic data on the Jurassic section in East Texas and identified several distinct seismic sequences within the section by integrating seismic data with lithologic, environmental facies, biostratigraphic, radiometric, and well-log information.

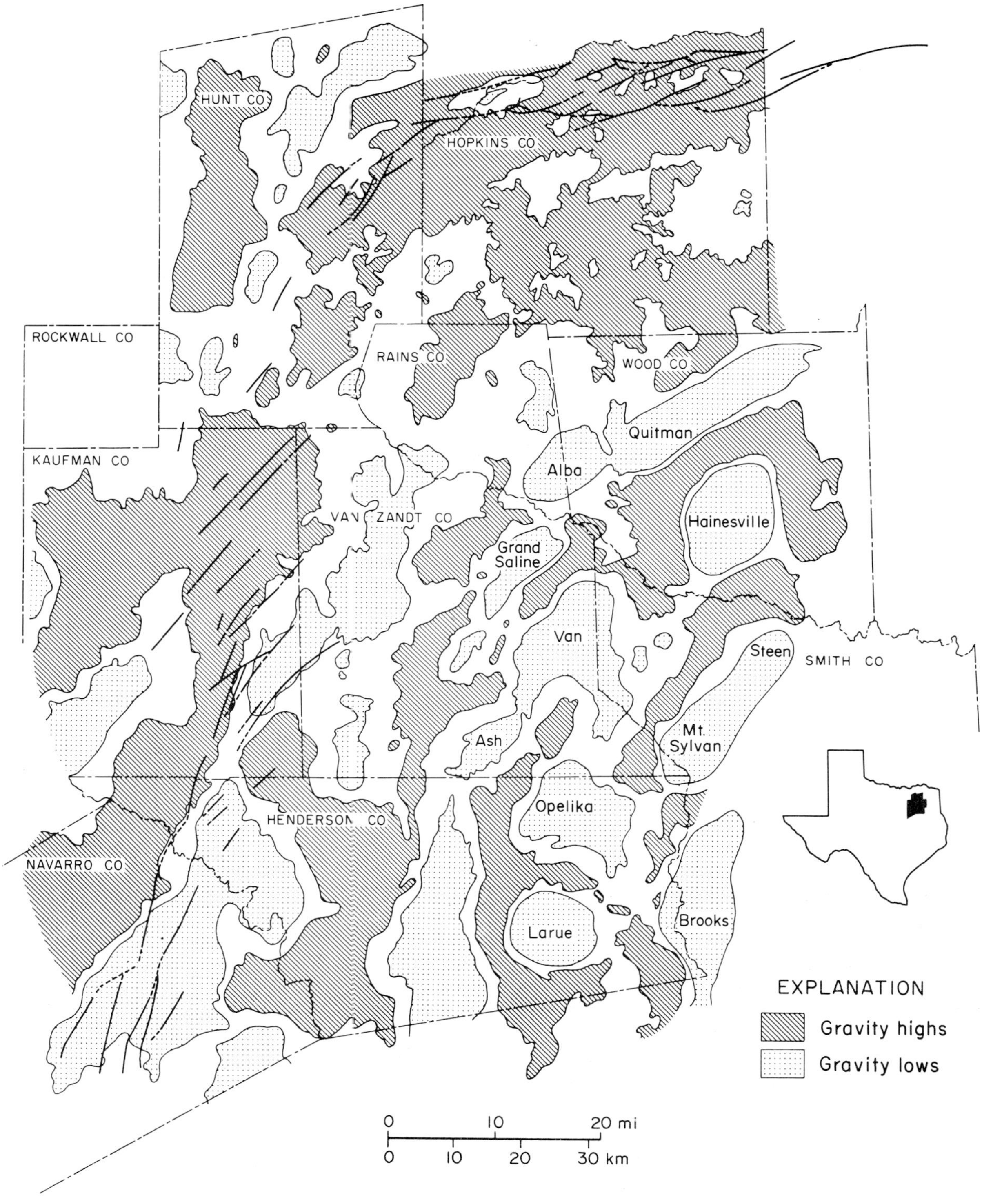

Figure 3. Generalized residual gravity map (modified from map done by Exploration Techniques, Inc.).

TECTONIC FRAMEWORK

The East Texas Basin is recognized as a subbasin, or reentrant, of the larger Gulf Coast Basin (Wood and Walper, 1974; Walper, 1980). Most researchers agree that the East Texas Basin was formed from one of either the megashear zones, the rift grabens, or the aulacogens that formed along the margins of the Gulf of Mexico, probably coincident with the breakup of Pangea and the separation of North and South America during the Triassic (Kehle, 1971; Burke and Dewey, 1973; Moore and Del Castillo, 1974; Wood and Walper, 1974; Beall, 1975; Salvador and Green, 1980). Kehle (1971) suggested that the interior salt basins of Mississippi and northern Louisiana also are foundered grabens that are marginal to the ancestral Gulf Coast Basin. Major tectonic elements in and around the East Texas Basin are shown in Figure 4.

Tectonic History

Kehle (1971) and Wood and Walper (1974) maintained that the development of the interior salt basins resulted from the opening of the Gulf of Mexico. They described the history of these interior salt basins as follows: The interior salt basins (Mississippi, North Louisiana, East Texas, and Salinas Basins) represent the most marginal grabens that are associated with continental rifting. These grabens were initially filled with alluvial-fan deposits of the Eagle Mills Formation. With prolonged spreading, however, the interior grabens continued to founder. The southern margin of the East Texas Basin may have been elevated, thereby restricting circulation of sea water between the basin and the Gulf of Mexico. Evaporites of the Werner Anhydrite and Louann Salt were precipitated, possibly by the brine-mixing process (Fig. 7 of Raup, 1970).

Continued subsidence resulted in open marine conditions; this is evidenced in the widespread occurrence of carbonates in the Smackover and Gilmer Formations. One of the characteristics of rifting is that during the early stages the bounding crustal blocks tilt away from the incipient rift, allowing large quantities of terrigenous clastics to enter the basin only when the dip of the rift margin reverses. The major influx of terrigenous clastics into the East Texas Basin during the Late Jurassic (Cotton Valley) and Early Cretaceous (Hosston) may reflect this dip reversal.

Salt Tectonics

Two general observations about salt structures in the East Texas Basin can be made: First, the size and type of salt-related structures seem to be directly controlled by the thickness of the underlying salt. This relationship was also observed in the interior Mississippi salt basin by Hughes (1968). Second, salt apparently migrated at different times in different parts of the East Texas Basin through several mechanisms: faulting (common in the marginal areas of the basin) (Parker and McDowell, 1955; Rosenkrans and Marr, 1967; Hughes, 1968); gravity gliding of post-Louann strata, which was caused by basinward tilting (Kehle, 1971; Jackson and Harris, 1981); and mass imbalance caused by uneven sediment loading (Rogers, 1967; Turk, Kehle, and Associates, 1978; McGowen and Harris, 1981).

Dome Growth Mechanisms. Kehle(1971) maintained that uneven sediment loading is the dominant mechanism responsible for salt dome initiation. He observed that the density inversion caused by uneven sediment loading is accentuated in the resulting salt flow because the viscosity of salt is highly sensitive to shear stress. Kehle concluded that the unequal pressure gradient set up within a salt mass because of uneven sediment loading is dependent on the slope of the overlying sediment, which in the present study is a deltaic lobe.

Loocke (1978) applied this principle to the East Texas Basin, suggesting that the growth of the Hainesville salt dome (south-central Wood County) was initiated by mass imbalance caused by the progradation of Cotton Valley deltas into that part of the ba-

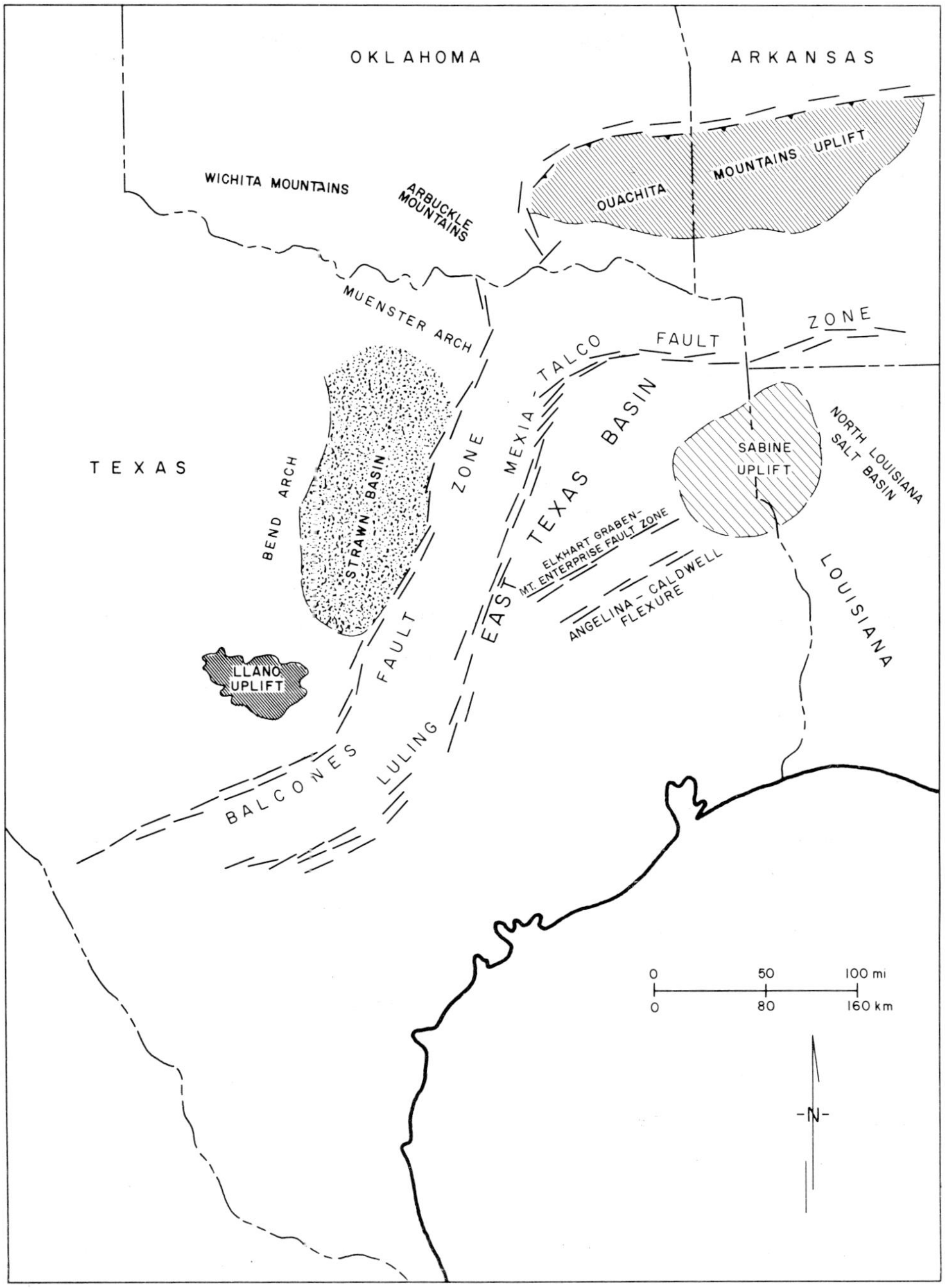

Figure 4. Tectonic elements in and around the East Texas Basin.

sin. By the end of Hosston (Early Cretaceous) sedimentation, the structure had grown into a topographically high salt pillow, or anticline (Loocke, 1978). The evolution of initial salt anticlines into salt domes depends on continued sediment loading (Kehle, 1971) as well as on the thickness of the underlying salt, which controls the supply of salt for continued growth.

Timing of Initial Salt Movement. Two seismic profiles available in the study area document the timing of salt mobilization (Figs. 5 and 6). Profile S-1 extends northwest to

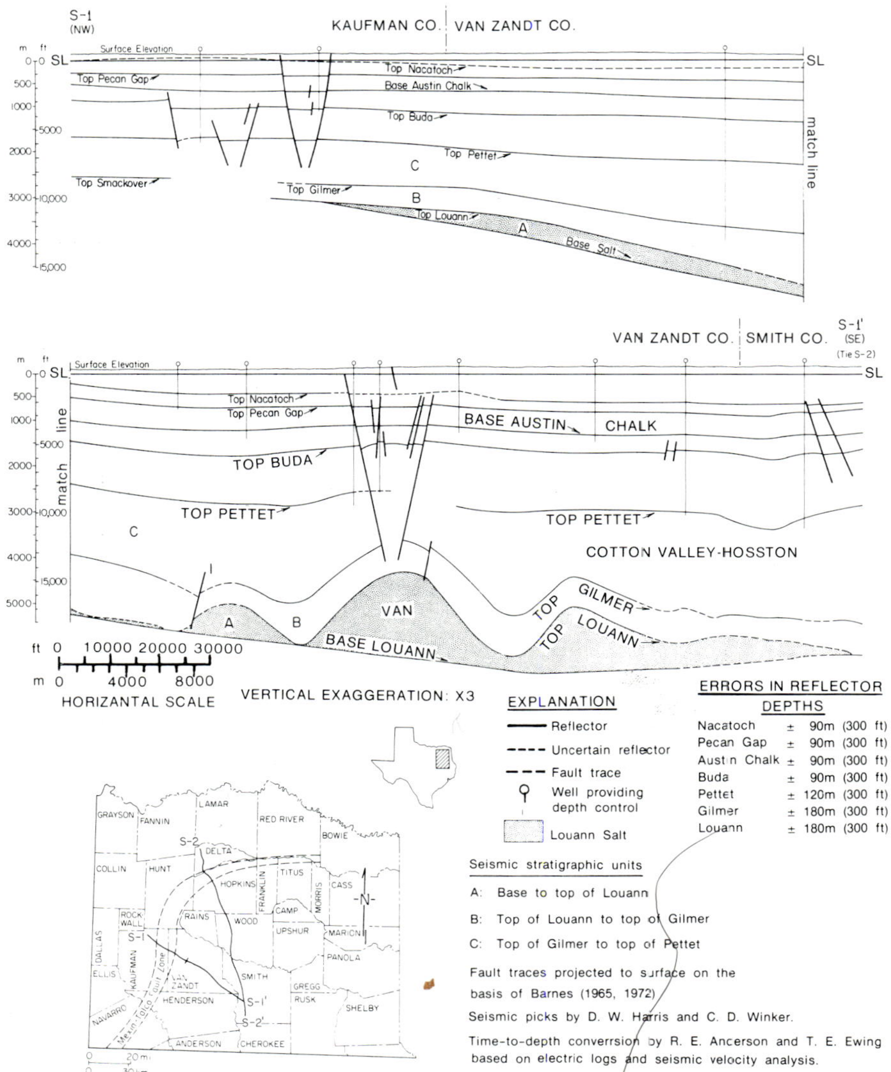

Figure 5. Time-to-depth conversion section S-1. Seismic units A, B, and C refer to Table 1. Southeastern end of line shows salt-withdrawal feature associated with Mount Sylvan salt dome.

southeast through Kaufman, Van Zandt, northeastern Henderson, and western Smith Counties; profile S-2 trends north to south through Delta, Hopkins, Wood, and Smith Counties.

Seismic data suggest that salt movement was both pre-Gilmer and coeval with Cotton Valley-Hosston deposition (Jackson and Harris, 1981; McGowen and Harris, 1981). Pre-Gilmer salt migration occurred in an area north and west of a line through central Wood, eastern Rains, central Van Zandt, and central Henderson Counties (Figs. 5 and 6). Time-thickness variations are apparent between reflections at the top of the Louann Salt and the top of the Gilmer Limestone (seismic unit B, Table 1). No thickness variations caused by salt movement were observed in the seismic interval between the top of the Gilmer Limestone and the top of the Pettet Limestone (seismic Unit C, Fig. 5) or between the top of the Gilmer Limestone and the top of the Massive Anhydrite (seismic unit D, Fig. 6). Thus, it can be concluded that salt

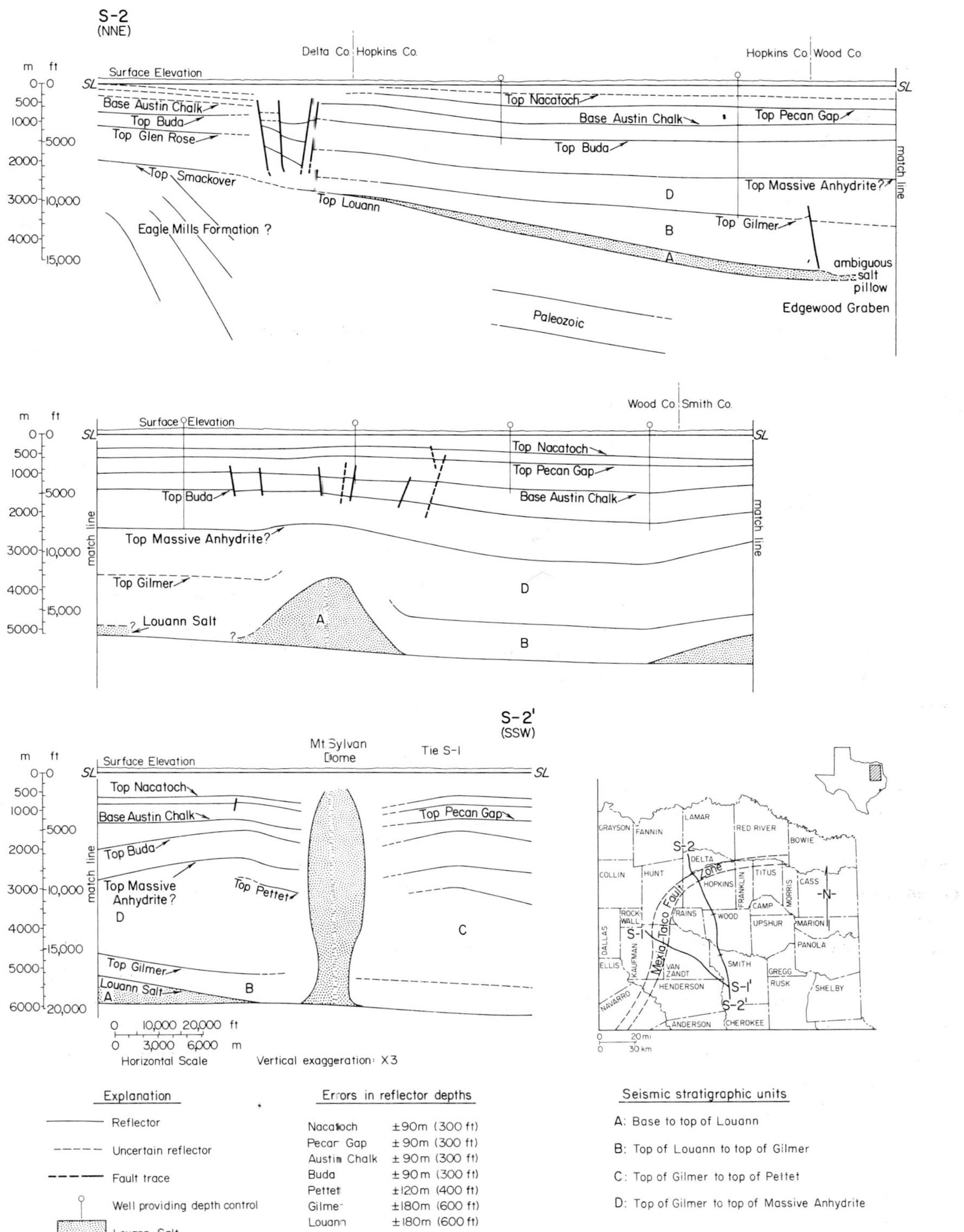

Figure 6. Time-to-depth conversion section S-2. Seismic units A, B, C, and D refer to Table 1. Fault traces projected to surface on the basis of Barnes (1965, 1966). Seismic picks by D.W. Harris and C.D. Winker. Time-to-depth conversion by R.E. Anderson and T.E. Ewing based on electric logs and seismic velocity analysis.

moved during Smackover deposition, which preceded Cotton Valley-Hosston deposition.

During Cotton Valley-Hosston deposition, salt moved south and east of a line through central Wood, eastern Rains, central Van Zandt, and central Henderson Counties. The seismic interval between the top of the Gilmer Limestone and the top of the Pettet Limestone (seismic unit C, Fig. 5) apparently thins over salt structures and thickens within synclines caused by salt withdrawal. In contrast, the seismic interval between the reflections at the top of the Louann Salt and top of the Gilmer Limestone (seismic unit B, Fig. 5) appears to be planar. This relationship is not as discernible on profile S-1 (Fig. 5) as on profile S-2 (Fig. 6) because seismic resolution is poor on the deep stratigraphic horizons in southern Wood County.

Structural Styles

Structural styles within the East Texas Basin can be grouped into four general categories: a peripheral graben system; a deformation-free zone; low- to intermediate-amplitude salt structures, which show pre-Cotton Valley salt movement; and salt anticlines and domes, which show movement coincident with Cotton Valley-Hosston deposition. These features have been discussed by Kehle (1971) and Jackson and Harris (1981), and similar structures have been observed in the Mississippi (Hughes, 1968) and the North Louisiana (Kehle, 1971) salt basins.

Peripheral Graben System. The Mexia-Talco fault zone bounds the study area on the north and west. The position of the fault system marks the updip depositional limit of the Louann Salt (Kehle, 1971; Agagu *et al.*, 1980; Jackson and Harris, 1981). The fault zone is a series of en echelon normal faults and grabens that formed early in the history of the basin. However, there is no evidence on seismic lines observed during this study that the peripheral faults extend into the basement. Rather, evidence suggests that the grabens are based in the Louann Salt (Jackson, 1982). Turk, Kehle, and Associates (1978) maintained that major displacement along the fault zone was largely caused by downdip creep of the Louann Salt. Jackson and Harris (1981) suggested that fault displacement was also caused, in part, by basinward creep of clastic strata above the Louann decollement zone. Movement along the fault affected strata of Mesozoic through early Tertiary age.

Basinward of the Mexia-Talco fault zone, a second graben system (the Edgewood Graben) developed parallel to the Mexia-Talco trend (Rosenkrans and Marr, 1967) (Fig. 7). These faults were active during the Jurassic and became inactive during the Early Cretaceous before deposition of the Pettet Limestone (Fig. 6). These faults and related structures are discussed in more detail in the section titled "Low- to Intermediate-Amplitude Salt Structures."

Deformation-Free Zone. On both seismic profiles S-1 and S-2, a deformation-free zone between the peripheral graben system and the first occurrence of low-amplitude salt swells (Figs. 5 and 6) underlies Delta and Hopkins Counties to the north and Kaufman and western Van Zandt Counties to the west. The Louann Salt is recognized by prominent boundary reflections and lack of internal reflections (Jackson and Harris, 1981). Within the deformation-free zone, boundary reflections are planar and diverge basinward, indicating a thickening of the salt wedge (approximately 1,020 to 1,920 feet, or 340 to 640 m) (Jackson *et al.*, 1982). Existence of the deformation-free zone suggests that a critical thickness of salt (approximately 1,500 feet, or 500 m) must be present to initiate flow (Jackson *et al.*, 1982). Kehle (1971) maintained that the deformation-free zone is discontinuous along the full length of the basin margin.

Low- to Intermediate-Amplitude Salt Structures. Low-amplitude salt structures occur basinward of the deformation-free zone in southern Hopkins, southern Rains, and western Van Zandt Counties. Amplitudes, indicated by relief on the top of the salt anticlines, in-

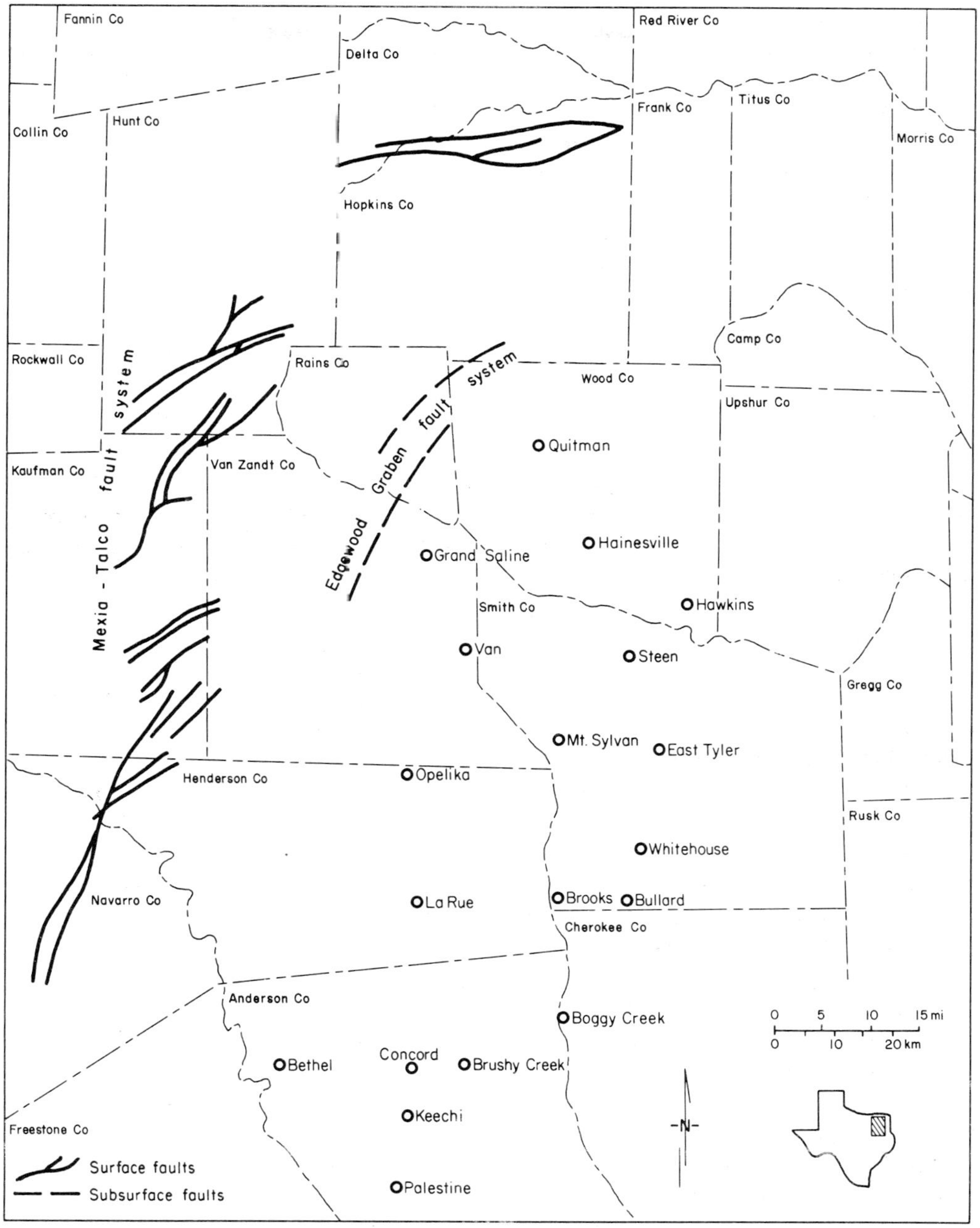

Figure 7. Salt domes and major fault systems, northwestern part of the East Texas Basin.

crease basinward as the salt wedge thickens. Basinward, folds are commonly tighter and salt has pierced some anticlines (Hughes, 1968; Kehle, 1971). Seismic profiles indicate that these low- to intermediate-amplitude structures formed early in the development of the basin. Salt movement began during deposition of the Smackover Formation; this is evidenced by time-thickness variations between reflections on the top of the Louann Salt and the top of the Gilmer Limestone (Figs. 5 and 6). Salt migration was probably initiated by downward creep that was induced by sedimentary loading of carbonate deposits

(Rogers, 1967) and was enhanced by basinward tilting. Residual gravity maps suggest that these initial salt structures are aligned roughly parallel to the Mexia-Talco fault zone (Figs. 3 and 4). Hughes (1968) deduced that salt movement occurred as early as deposition of the Norphlet in the Mississippi salt basin. He also observed that low-amplitude salt structures in marginal areas of the basin are commonly arranged in ridges that parallel the peripheral faults.

Most growth of low-amplitude salt structures apparently occurred before deposition of Cotton Valley clastic sediment. This is evidenced on seismic profiles, which show little variation in time-thickness between the reflections at the top of the Gilmer Limestone and the top of the overlying Pettet Limestone. The low-amplitude salt features exhibit little to no structural expression above the Gilmer Limestone. The thin underlying salt wedge appears to have largely controlled the size of the structures that formed in this part of the basin.

Near the Edgewood Graben (Fig. 7), salt structures of intermediate size are truncated by the fault system. Rosenkrans and Marr (1967) maintained that the salt structures were generated by faulting, and Jackson and Harris (1981) concluded that salt withdrawal by downdip creep, coincident with sediment loading within the graben, may have caused the faults. Fault movement along the system displaced strata of Late Jurassic and Early Cretaceous age but ended in the Early Cretaceous. Seismic profile S-2(Fig. 6) indicates that faulting was contemporaneous with deposition of Cotton Valley clastic sediment.

Seismic profile S-2 (Fig. 6) also shows evidence of displaced reflections along the Edgewood graben during deposition of the Cotton Valley-Hosston strata. The following relationships were observed: First, a salt structure underlies the Edgewood fault, but the exact relationship of the structure to the fault is not clear on the seismic profile. Second, reflection of the Gilmer Limestone is offset down to the basin without appreciable time-thickness variations within seismic unit B (Table 1), indicating that most of the movement occurred after deposition of the Gilmer Limestone. And third, seismic unit D thickens on the downthrown side of the fault, not on the upthrown side, indicating that movement along the fault was contemporaneous with deposition of the Cotton Valley Group. The fault terminates within seismic unit D (Hosston Formation).

Evidence of faults that were active northwest of Van Dome during Cotton Valley-Hosston deposition is shown on seismic profiles S-1 (top structure, Fig. 5). First, updip of the fault, seismic unit A (Table 1) is distorted, and the configuration of the Louann Salt cannot be discerned. Second, seismic unit B thickens slightly basinward of the fault, suggesting that salt movement occurred during Smackover deposition. Third, the Gilmer Limestone reflection is clearly offset at the fault. Fourth, seismic unit C thickens appreciably on the downthrown side of the fault, indicating movement contemporaneous with deposition of Cotton Valley clastic sediment. And fifth, the fault terminates either in the upper part of the Cotton Valley or in the Hosston below the Pettet reflection, which is planar.

Salt Anticlines and Domes. Within the study area, salt anticlines and domes occur south and east of a line through southern Hopkins, eastern Rains, central Van Zandt, and central Henderson Counties (Fig. 8). The location of salt anticlines and domes in the basin appears to have been controlled largely by the location of the Smackover-Gilmer carbonate platform, particularly where Gilmer carbonate shelf-edge strata overlie Smackover deposits. Rogers (1967) maintained that a carbonate shelf-edge facies began to form during Smackover deposition and continued in some areas during Gilmer deposition, reaching a maximum combined thickness of 3,500 feet (1,166 m) (Fig. 9). A carbonate shelf of this thickness could provide a stable platform, upon which fan-delta sediments of the Cotton Valley Formation would tend to spread laterally rather than to stack vertically. The isopach map of the Cotton Valley Group indicates a close correlation between the position of the

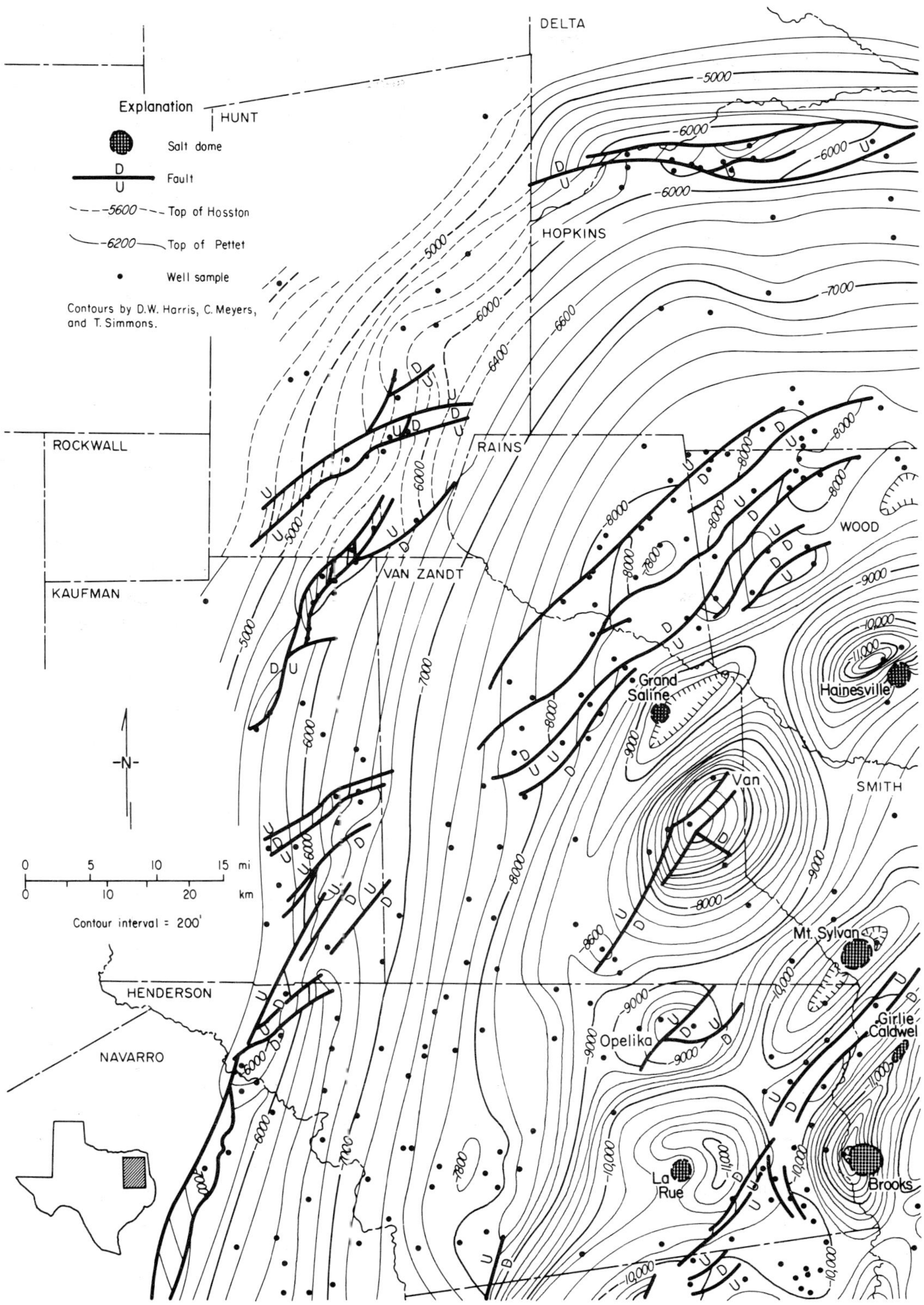

Figure 8. Structure map of the top of the Pettet Formation and the top of the Hosston Formation (Hunt County only).

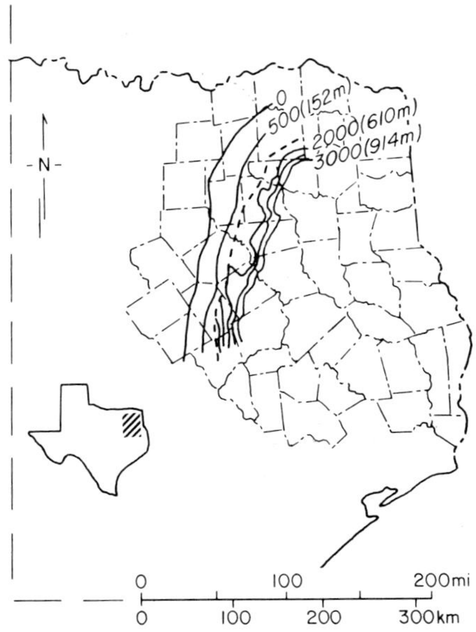

Figure 9. Isopach map of Smackover-Gilmer carbonate shelf facies (Rogers, 1967). Isopach of Smackover northwest of dashed line, combined isopach of Smackover and Gilmer southeast of dashed line.

Smackover carbonate shelf edge, as mapped by Eaton (1961) and Rogers (1967), and the occurrence of Cotton Valley deltaic depocenters (Fig. 10). Superposition of deltaic sediments occurred in synclines that were located between salt ridges and were immediately basinward of the older carbonate wedge. Salt mobilization was initiated by basinward migration of the salt into ridges that fronted the progradational fan-delta system. The salt ridges probably were bathymetric highs that acted as effective sediment dams to perpetuate the synclines as depocenters until either the underlying salt was depleted or the rate of sedimentation was greater than the rate of subsidence. This allowed the delta to overrun the salt ridge.

During continued progradation, the fan-delta complex probably overran the initial syncline and salt ridge and established a new depocenter basinward of and parallel to the original salt ridges. Consequently, sediments in depocenters should be progressively younger toward the center of the basin. Initial salt ridges that developed during the Late Jurassic and Early Cretaceous were later modified by continued salt flow, which resulted from uneven sediment loading by younger sedimentary units. Johnson (1980) showed that a preexisting salt ridge on the outer continental shelf and upper slope of the Gulf Basin evolved into individual salt domes when buried by an influx of fluvial-deltaic sediments during the Pliocene.

A residual gravity map of the study area indicates a parallel arrangement of lows and highs (Fig. 3). A comparison of the Cotton Valley isopach map (Fig. 10) with the residual gravity map shows that sediment thicks generally coincide with gravity highs and salt thins, whereas sediment thins coincide with gravity lows and areas of thicker salt. The approximate parallel alignment of residual gravity lows again suggests that the original salt structures may have been a series of parallel salt ridges that evolved through time into salt anticlines and domes.

Seismic lines across the salt anticlines and domes in the study area indicate that in most cases, salt moved after deposition of the Gilmer Limestone, whereas low-amplitude structures located in the marginal areas indicate pre-Gilmer movement. Deeper seismic reflections near Quitman Dome (Wood County) are distorted, and thus obscure the relationship between the salt structure and the overlying sedimentary units. However, a seismic survey profile across Mount Sylvan Dome (Smith County) shows virtually no evidence of pre-Cotton Valley salt movement (Fig. 6). West Tyler, to the southeast of Mount Sylvan Dome, and Boynton Field, to the northwest, are turtle structures created by early salt withdrawal that started coeval with Cotton Valley deposition (Jackson *et al.*, 1982). Turtle structures are anticlinal structures without salt cores that were created by sediment thicks in primary withdrawal basins formed by early salt withdrawal (Trusheim, 1960; Kehle, 1971; Wood, 1981). Turtle structures probably were depocenters during deltaic sedimentation of the Late Jurassic.

Seismic profile S-1 passes to the south of Grand Saline and Van Domes but crosses two parallel, lateral salt ridges that extend southeastward from Van Dome (Figs. 3 and 5)

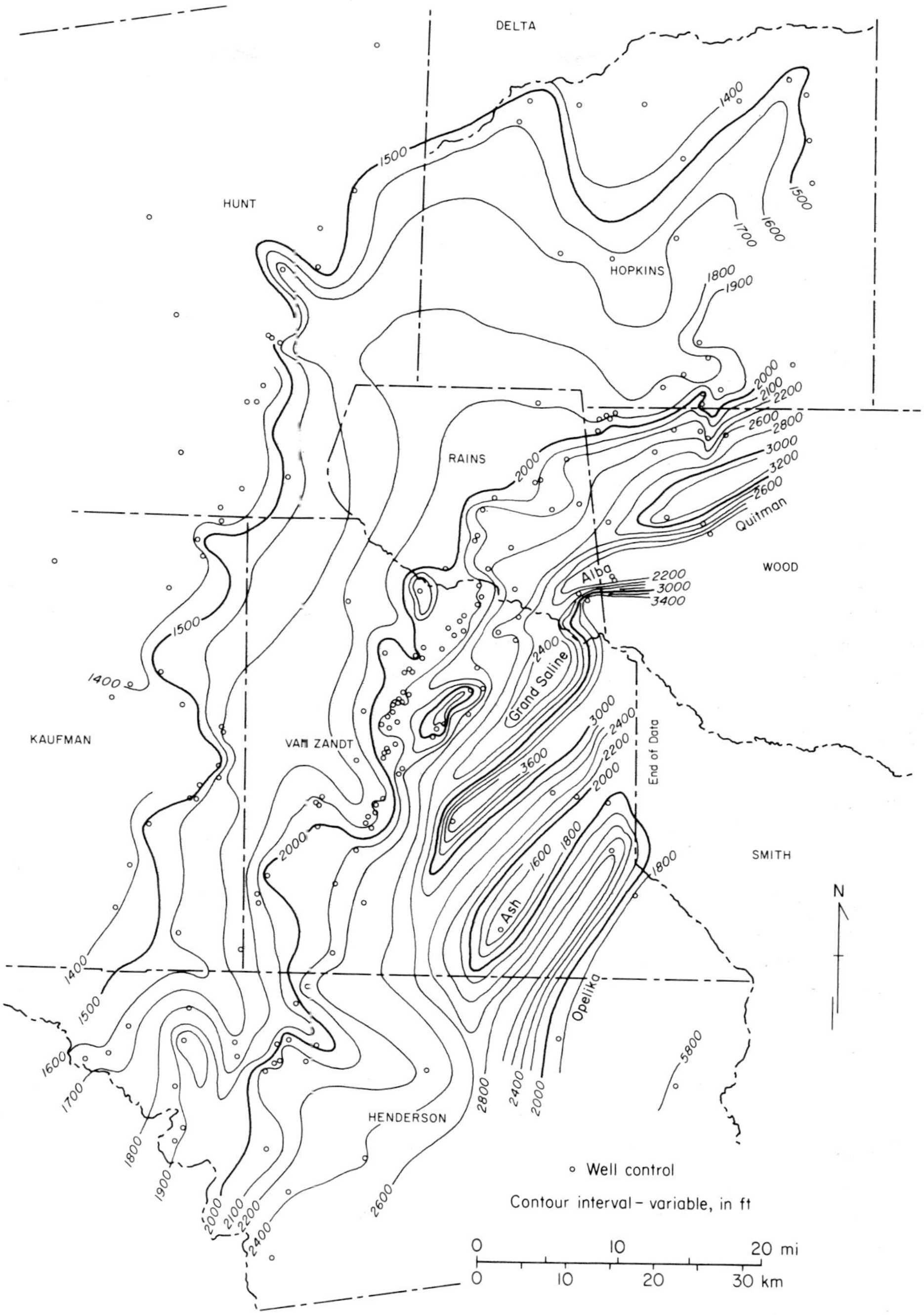

Figure 10. Isopach map of the Cotton Valley Group.

(Jackson and Harris, 1981). The seismic survey across the western ridge indicates that some pre-Gilmer salt movement occurred (seismic unit B, Fig. 5). Major movement, however, was coincident with deposition of Cotton Valley deltaic sediments (seismic unit C, Fig. 5). Basinward of the western ridge, no pre-Gilmer salt movement apparently occurred.

SUMMARY

Seismic profiles indicate that salt movement occurred at different times in different parts of the East Texas Basin and demonstrate that more than one mechanism caused movement. Within marginal parts of the basin, low-amplitude salt structures formed during Smackover (pre-Gilmer) time. Salt migration was caused by loading of the Smackover carbonate deposits in conjunction with basinward tilting. Time-thickness variations caused by low-amplitude salt structures are not indicated on Cotton Valley-Hosston isopach maps. The sizes of salt structures increase basinward, coincident with thickening of the underlying salt wedge.

Basinward of a line running through central Henderson, Van Zandt, and Wood Counties, salt movement occurred later than it did in the marginal parts of the basin. Salt migrated during deposition of Cotton Valley-Hosston strata as a result of mass imbalance, which was caused by uneven sediment loading. Time-thickness variations (divergence) are evident between seismic reflections at the top of the Gilmer Limestone and the top of the Pettet Limestone. In contrast, reflections at the top of the Louann Salt and the top of the Gilmer Limestone are parallel, suggesting that salt did not move in this part of the basin until Cotton Valley time.

SEDIMENTOLOGIC FRAMEWORK

Methodology

Early Jurassic sedimentation was dominated by the deposition of carbonate, evaporite, and mudstone facies. The first major influx of terrigenous clastic sediment into the East Texas Basin occurred during the Late Jurassic (Cotton Valley) and continued into the Early Cretaceous (Hosston).

Basin infilling during the Late Jurassic and Early Cretaceous produced a variety of complexly distributed facies in the East Texas Basin and the adjacent North Louisiana Basin. As a result, a complex nomenlature has evolved to describe these deposits. Because this study deals primarily with sedimentation during the Late Jurassic and Early Cretaceous, a rock-stratigraphic classification restricted to those times has been used. No attempt has been made to classify the Cotton Valley Group in the northwestern part of the basin using the stratigraphic nomenclature that has been applied to the eastern part of the basin and to Louisiana.

In this study, the boundary between the Cotton Valley Group and the Hosston Formation is arbitrary, based on regional correlations and information of geologic picks from scout cards. In the northwestern part of the East Texas Basin, the top of the upper Cotton Valley Group is difficult to pick on electric logs and seismic profiles because Hosston sandstones overlie upper Cotton Valley sandstones. Nichols *et al.* (1968) maintained that deposition was continuous through Late Jurassic and Early Cretaceous time except around the western and northern margins of the basin, where an erosional contact was recognized. In these marginal areas, a basal Hosston conglomerate, ranging in thickness from 50 to 100 feet (16 to 33 m), separates the two formations (Nichols *et al.,* 1968). It has not been determined whether this conglomerate bed occurs regionally or whether it consists of unrelated local facies, such as proximal alluvial-fan deposits or braided-stream channel-fill deposits. On the basis of seismic interpretations, Todd and Mitchem (1977) recognized a major unconformity between the Cotton Valley Group (Upper Jurassic) and the Hosston Formation (Lower Cretaceous) in the East Texas Basin. However, we did not recognize a major unconformity on seismic sections in the northwestern part of the basin. The depositional sequence between either the Gilmer Limestone (Cotton Valley Limestone) or the base of the Buckner Anhydrite and the top of the Hosston Formation is thought to represent one regressive sequence with minor interruptions. The regression was terminated by a major transgression, which is evidenced by the transition of the uppermost Hosston Formation into the Pettet Limestone.

The volume, texture, and composition of sediment, the nature of the sediment-dispersal system, and the geometry of sands of the Cotton Valley Group and Hosston Formation were studied using sandstone-percent and net-sandstone maps. These were supplemented by regional stratigraphic cross sections, electric logs, isopach maps, and well cuttings. A sandstone-percent map is an effective way to reveal sand-body geometry because it minimizes the effect of thickness variations within a mappable unit (Krumbien and Sloss, 1959; Kaiser *et al.*, 1978). Stratigraphic markers are virtually absent in the Cotton Valley Group and Hosston Formation within the northwestern part of the basin; therefore, sandstone maps represent the composite sandstone thickness of superposed facies. The cumulative values do not reflect the geometry of individual sand bodies, so interpretation of individual depositional systems from these maps should be done cautiously. Brown (1969) observed that some depositional systems appear to remain in approximately the same geographic position through time; when this occurs, similar systems would tend to stack vertically and thus be revealed in a gross lithologic map.

Source Area

The same source area supplied terrigenous clastics to the East Texas Basin during both the Late Jurassic (Cotton Valley) and the Early Cretaceous (Hosston). Deposition of Cotton Valley terrigenous clastics in the East Texas Basin suggests that a reactivation of this source area along the northern and western margins of the basin began as early as the Late Jurassic. The Central Mineral Region and the Ouachita, Arbuckle, and Wichita Mountains were all highlands during the Late Jurassic and Early Cretaceous (Imlay, 1943) (Fig. 4).

Patterns of the net-sandstone and sandstone-percent maps (Figs. 19 and 20) suggest that during the Late Jurassic, terrigenous clastics were delivered to the East Texas Basin by many small streams and rivers. This differs from the major river and tributary system in Louisiana and Mississippi, where an ancestral Mississippi River was the principal fluvial system entering the North Louisiana Basin (Thomas and Mann, 1966).

Tongues of quartz conglomerate along the northwestern and northern margins of the basin in the Cotton Valley Group and throughout the Hosston Formation were probably derived from the Ouachita, Arbuckle, and Wichita highlands. Textural maturity and the dominance of very fine grained to fine-grained sandstone suggest that older sedimentary rock surrounding the basin during the Late Jurassic and Early Cretaceous was also an important source of terrigenous clastic sediment for the East Texas Basin. Most of the sandstone is quartzarenite and subarkose.

Depositional Systems

Fan-Delta Processes and Environments. The Cotton Valley Group and the Hosston Formation are thought to be a large fan-delta depositional system. A fan delta is an alluvial fan that progrades into a body of water from an adjacent highland (Holmes, 1965; McGowen, 1970). Fan-delta deposits exhibit a higher ratio of coarse-grained to fine-grained sediment than either lobate or elongate deltas (Erxleben, 1975). Characteristically, fan deltas have relatively small drainage areas and flashy runoff, and they are supplied by bed-load streams braided to the toe of the delta (McGowen, 1970). Aggradation and progradation occur only during periods of high discharge (McGowen, 1970).

Rates of progradation are controlled by sediment supply, by discharge rates of the fluvial system supplying the fan delta with sediment, by intensity of marine processes (whether the fan delta is debouching into a low-energy or high-energy marine environment), and by depth of water. Heavy seasonal rains or floods are essential to providing discharge rates capable of transporting large quantities of sand-size sediment to the toe of the fan delta. Sediment is stored in channels during periods of low rainfall, whereas sediment is entrained during floods, when rapid progradation occurs (Casey, 1980).

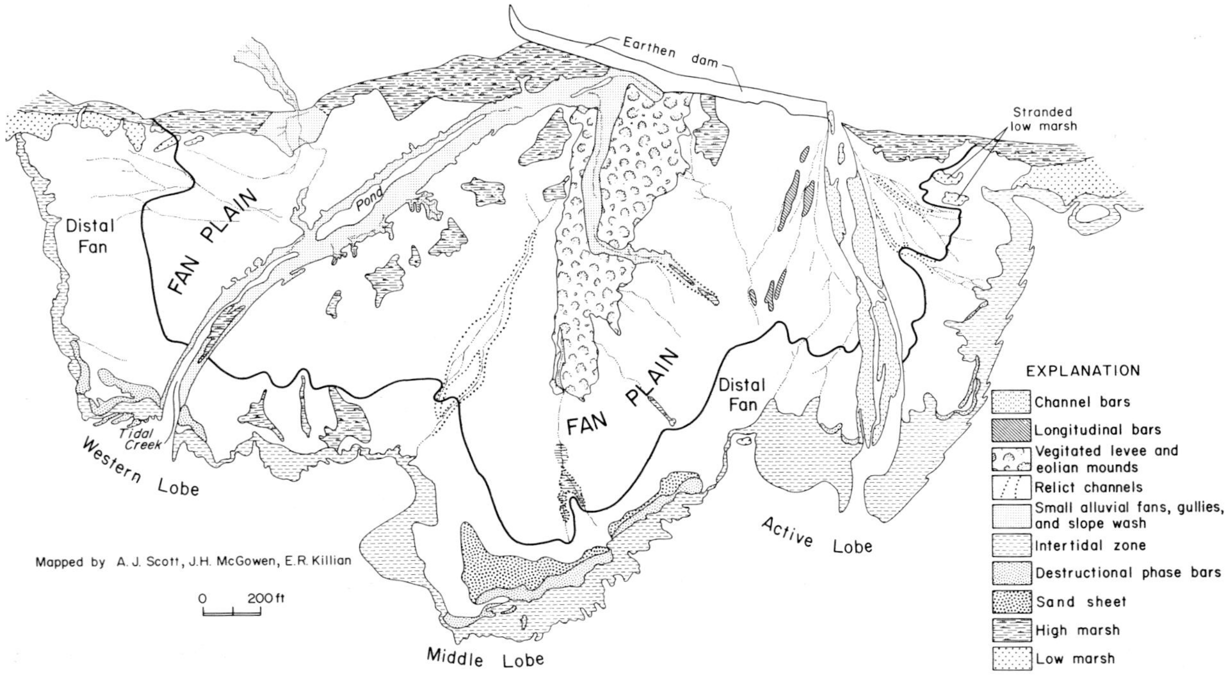

Figure 11. Gum Hollow fan-delta model (McGowen, 1970).

McGowen (1970) attributed major growth of the Gum Hollow fan delta to three principal depositional events: The heavy spring rains of 1966 and the heavy rains associated with Hurricane Beulah in 1967 and with Tropical Storm Candy in 1968. During periods of low fluvial discharge, distal parts of the active fan deltas are modified by marine processes. Inactive lobes are subject to destructional processes year round. The greatest rates of progradation occur when fan deltas debouch into shallow-water low-energy environments. Progration is accomplished by lateral shifts in sites of deposition, resulting in a complex overlapping of fan-delta lobes (Casey, 1980).

McGowen (1970) subdivided fan deltas into two subenvironments--fan plains and distal fans--based on their dominant sedimentary process (Fig. 11). The fan plain includes the subaerial part of the fan delta and is dominated by braided-stream fluvial processes. A braided stream is defined by Leopold *et al.* (1964) as a channel divided into several channels that successively meet and redivide. A braided-stream system is composed of bed-load streams with rapid discharge fluctuations. Finer sediment is transported through the system without accumulating; constructional features include both longitudinal and transverse bars. With the absence of muddy levees and top strata, the channel banks are easily eroded; thus, bars tend to laterally coalesce, forming continuous and extensive sand sheets (Walker and Cant, 1979)(Fig. 12).

The distal fan includes the transitional zone between the subaerial fan plain and the subaqueous part of the fan delta (McGowen, 1970). Fluvial and marine interaction produces a complex environment. Marine processes dominate this part of the fan delta except during periods of high discharge, when fluvial influence is evident. During periods of low discharge, delta-front sediments are reworked into bars, spits, and shoals (Lucchi *et al.*, 1981)(Fig. 13). Inactive fan delta lobes are continuously modified. Characteristic environments include marshes, destructional bars, intertidal zones, and eolian mounds (McGowen, 1970). Benthic fauna may migrate into the delta front and the aban-

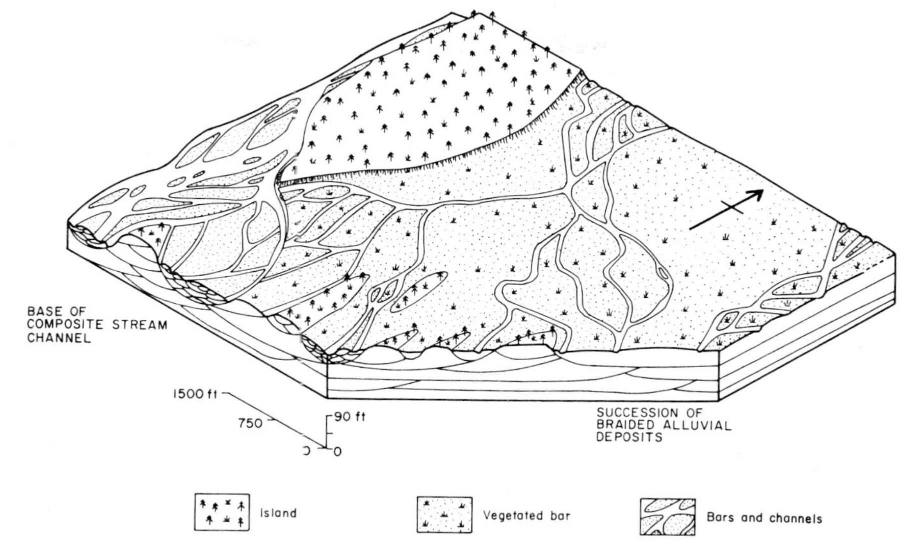

Figure 12. Braided-stream model (Rust, 1978).

doned channel areas (Casey, 1980).

Wescott and Ethridge (1980) presented an excellent synopsis of modern and ancient deposits that are thought to be fan delta systems. Examples of modern fan deltas that debouch into marine environments include the Gum Hollow fan delta (McGowen, 1970), the Yallahs fan delta (Wescott and Ethridge, 1980), and the fan deltas associated with glacial outwash along the coast of Iceland (Ward *et al.*, 1976; Boothroyd and Nummendal, 1978) and the coast of Alaska (Boothroyd and Ashley, 1975; Boothroyd, 1976; Galloway, 1976; Boothroyd and Nummedal, 1978). Temperatures vary considerably between these geographic areas; the intensity of marine processes affecting the distal margins of the fan deltas also varies. However, many of the processes controlling sedimentation are similar. For example, the fan deltas are fed by braided-stream fluvial systems that exhibit rapidly fluctuating discharge rates, which are controlled by seasonal events; most of the fans prograde into relatively shallow water (the width of the shelf varies with each geographic location).

Cotton Valley-Hosston Fan-Delta System. Deposits of the Cotton Valley Group and Hosston Formation are thought to be a system of coalescing fan deltas that prograded from the west, northwest, and north. Facies interpretations are based on sandstone-percent values, sandstone distribution, electric log re-

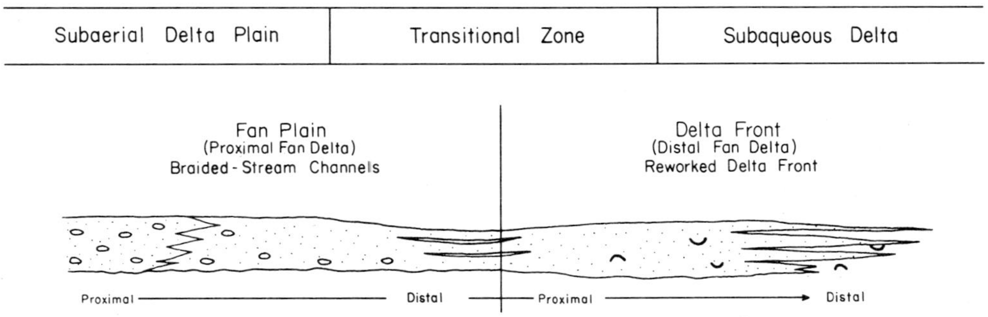

Figure 13. Diagrammatic cross section of fan-delta subenvironments (after Lucchi *et al.*, 1981).

sponse patterns, descriptions of well cuttings (appendix A), and lithologic descriptions from other studies. Lack of core data precludes the study both of sedimentary structures and of detailed relationships among vertical sequences of facies within the Cotton Valley Group and Hosston Formation.

Cotton Valley Group. *Facies*.--Facies of the Cotton Valley Group can be divided into three general categories: prodelta deposits, delta-front deposits, and braided stream deposits, all components of a fan-delta system (Figs. 2, 14 through 18). Prodelta deposits compose a thin subordinate facies in the study area; however, the facies thickens basinward. It consists of black or green calcareous, fossiliferous mudstone interbedded with light-gray to light-brown crystalline limestone and shelly, sandy limestone. Minor amounts of very fine-grained sandstone and siltstone occur within the facies.

Delta-front deposits consist of alternating beds of sandstone and mudstone and a few thin beds of sandy limestone. Sandstone, the dominant lithology, is white to light gray and very fine-grained to fine-grained, containing glauconite, shells, and finely disseminated carbonized plant fragments. Interbedded mudstone is dark gray to black to green. Thick beds of conglomeratic sandstone occur in the proximal areas.

Braided-stream fluvial deposits consist of white to light red, very fine-grained to fine-grained sandstone, conglomeratic sandstone, and chert-gravel quartz and conglomerate. Thin beds of red and green mudstone occur as a subordinate lithology. Conglomerate is more common updip, extending basinward as tongues (Newkirk, 1971). This facies exhibits blocky electric log patterns with sharp (erosional) bases and tops, characteristic of braided-stream deposits (Erxleben, 1975; Galloway *et al.*, 1979). Well-developed upward-coarsening or upward-fining sections are rare to absent.

Sandstone Distribution.--Sandstone distribution in the Cotton Valley Group is indicated on the sandstone-percent map (Fig. 19). Higher sandstone-percent values are restricted to the basin margin, and values decrease basinward. Many dip-oriented sandstone-rich belts extend across Kaufman, Hunt, Hopkins, and northwestern Van Zandt Counties, suggesting the presence of many smaller streams in this part of the study area. Basinward, the dip-oriented sandstone geometry appears to change into a northeast to southwest strike-oriented trend. The change in orientation of sandstone trends marks the fluvial-marine interface. Seaward of this area, dominant marine processes had reworked sandstones into strike-oriented facies. This change in orientation seems to coincide with the position of an older Smackover-Gilmer carbonate shelf described by Eaton (1961) and Rogers (1967) (Fig 9).

The Cotton Valley Group gradually thickens basinward toward a line that runs through central Henderson, central Van Zandt, and north-central Wood Counties (Fig. 10). Basinward of this line, the isopach contours outline parallel-aligned, strike-oriented thicks and thins. When compared with the residual gravity map (Fig. 3), sediment thicks correspond to salt-poor (gravity highs), and sediment thins overlie salt structures (gravity lows). Parallel sandstone thicks indicate successive seaward depocenters of prograding fan deltas.

The area where the Cotton Valley Group gradually thickens basinward is coincident with subjacent Smackover-Gilmer carbonate shelf facies. The combined thickness of the Smackover-Gilmer deposits is approximately 3,500 feet (1,067 m). The isopach map of the Cotton Valley suggests that the Smackover-Gilmer shelf was a stable platform over which the advancing Cotton Valley fan deltas prograded both laterally and basinward. Superposition of deltaic sediments occurred in synclines that were located between salt ridges immediately basinward of the Smackover-Gilmer carbonate shelf edge. The thickness of the Louann Salt, which directly determined the amount of salt available for the formation of salt structures, also controlled the rate and amount of subsidence that occurred coeval with sedimentation.

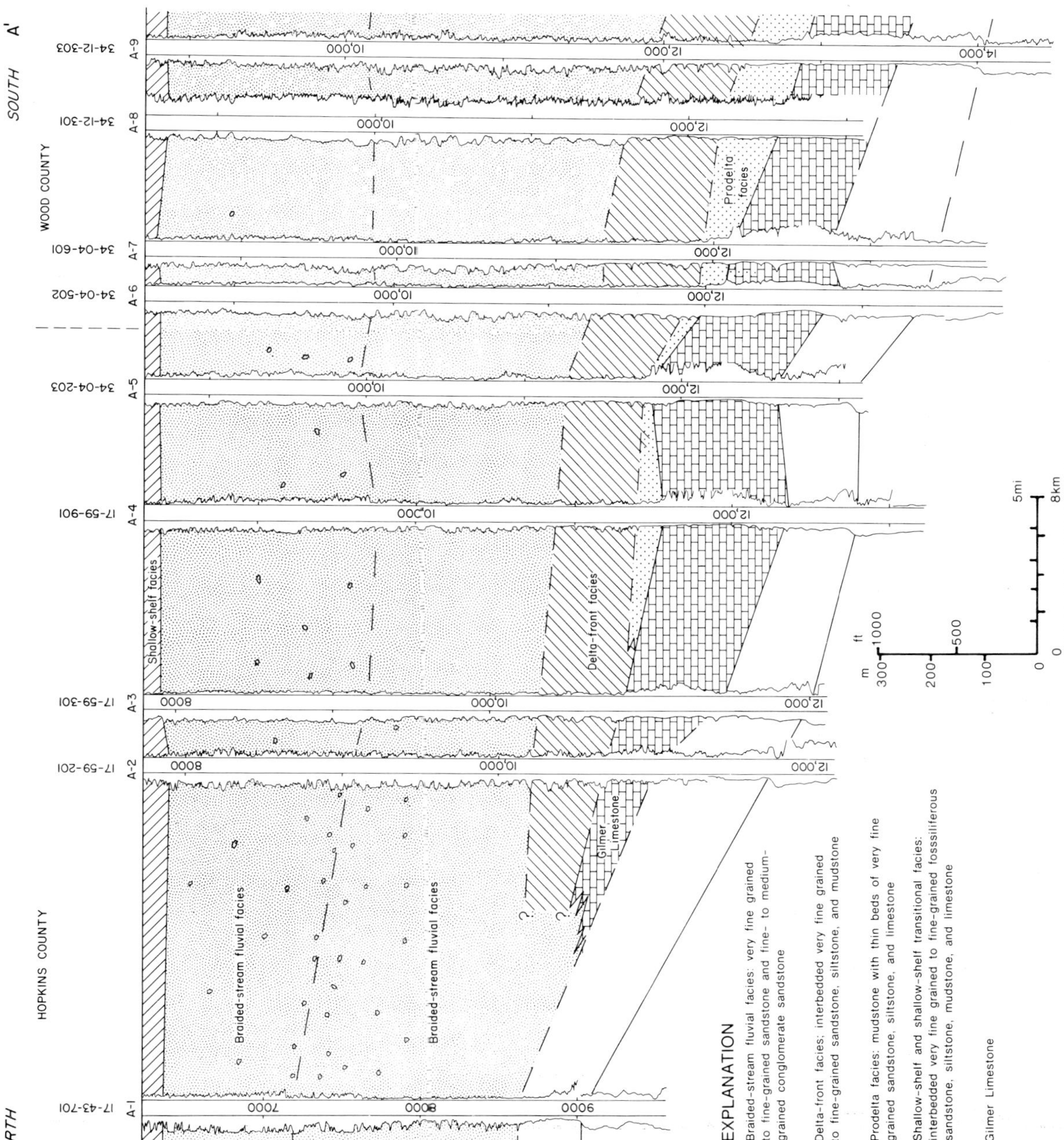

Figure 14. Stratigraphic cross section A-A' through Hopkins and Wood Counties.

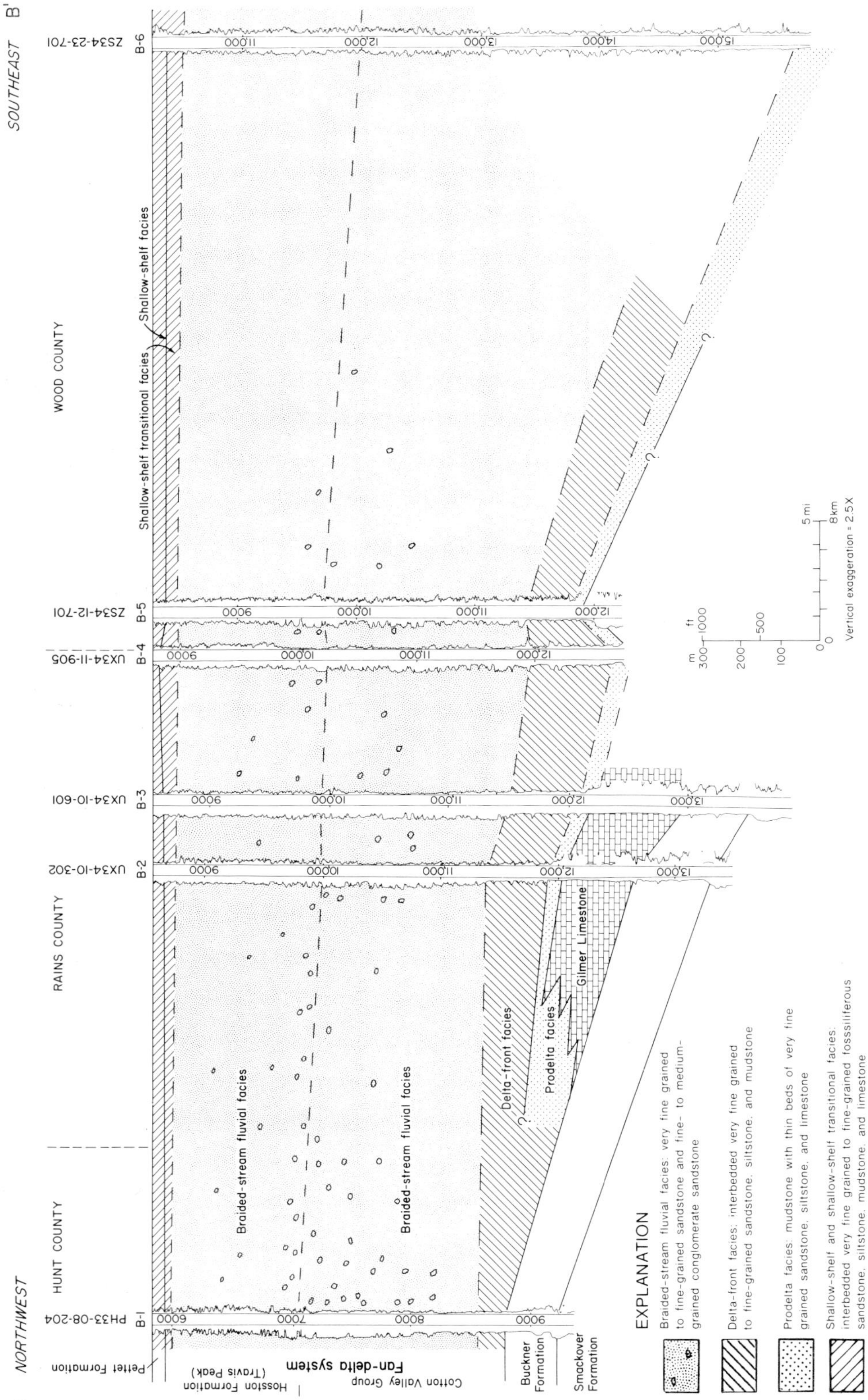

Figure 15. Stratigraphic cross section B-B' through Hunt, Rains, and Wood Counties.

The net-sandstone map indicates that sandstone-rich belts coincide with thick isopach trends of the Cotton Valley Group (Fig. 20). Net-sandstone values increase gradually across the Smackover-Gilmer shelf; beyond the shelf edge, net-sandstone highs correspond to sediment thicks indicated on the Cotton Valley isopach map (Fig. 10). Net-sandstone thicks coincide with salt-poor areas (Fig. 3).

Sediment accumulation around the northern and western margins of the basin appears to have been controlled, in part, by contemporaneous faulting in the Mexia-Talco fault zone. In Kaufman County, for example, massive sandstone units in the upper Cotton Valley Group are up to 600 feet (183 m) thick. These units are thought to be superposed braided-stream deposits that accumulated early in grabens of the Mexia-Talco fault zone. Subsidence was accelerated by accumulation of thick braided-stream deposits composed of quartz and pebble conglomerate and fine- to medium-grained sandstone.

Hosston Formation. *Facies.*--Two facies constitute the Hosston Formation: a braided-stream facies, which is dominant, and a transitional shallow marine facies, which is subordinate. The braided-stream deposits represent the subaerial fan-plain facies, a component facies of the larger fan-delta system (Figs. 11 through 13). Fluvial sedimentation ended with a major transgression reflected by a transitional shallow marine facies restricted to the upper 100 to 200 feet (33 to 67 m) of the Hosston Formation. The transitional deposits grade into the overlying Pettet Formation (Limestone).

The fluvial facies is composed of conglomeratic sandstone and white to light-red, fine- to medium-grained sandstone; light-gray to red muddy sandstone; white to pale-red siltstone; and thin beds of red and grayish-green mudstone. Carbonaceous and lignitic material occurs in some sandstone units. Quartz and pebble conglomerate beds occur throughout the section but are more common in the lower part of the formation (Bushaw, 1968; Nichols *et al.*, 1968) (Appendix A). Electric log patterns of individual sand beds are blocky and have sharp (erosional) lower and upper boundaries. Upward-fining and upward-coarsening log responses are rare. Superposition of braided-stream deposits, occurring contemporaneous with faulting, is not as apparent here as it is in the Cotton Valley Group.

The upper 100 to 200 feet (33 to 67 m) of the Hosston Formation consists of interbedded white to light-red, very fine-grained to fine-grained sandstone; gray mudstone; and gray to light-tan sandy, fossiliferous, oolitic limestone. Uppermost Hosston deposits reflect a decrease in the sediment supply to the East Texas Basin because of a shift from dominantly fluvial deposits to dominantly marine deposits.

Sandstone Distribution.--Sandstone distribution within the Hosston Formation is characterized by dip-oriented sandstone-rich belts in proximal areas (Fig. 21). There is, however, a change from many dip-oriented high-sandstone-percent trends in the Cotton Valley to a single dominant dip-oriented sandstone trend in the Hosston centered in Hunt County (Figs. 19 and 21). Narrower sandstone-percent trends in the Hosston occur around the northern margin of the basin (Fig. 22). The sandstone-percent map suggests that the Cotton Valley drainage system had evolved, by Hosston time, from an immature fluvial complex made up of many smaller streams into a more mature system, the principal drainage system being located in the northwestern part of the study area.

In general, net-sandstone and isopach trends in the Hosston Formation (Figs. 22 and 23) resemble those of the Cotton Valley Group. Thickness and net-sandstone values of the Hosston Formation gradually increase basinward across the subjacent Smackover-Gilmer stable carbonate platform. Basinward of the platform, the parallel high-net-sandstone trends and sediment thicks in the Cotton Valley Group coincide with those of the Hosston Formation. Thickness variations, however, are not as great within the Hosston, suggesting that most of the Hosston deltaic sedimen-

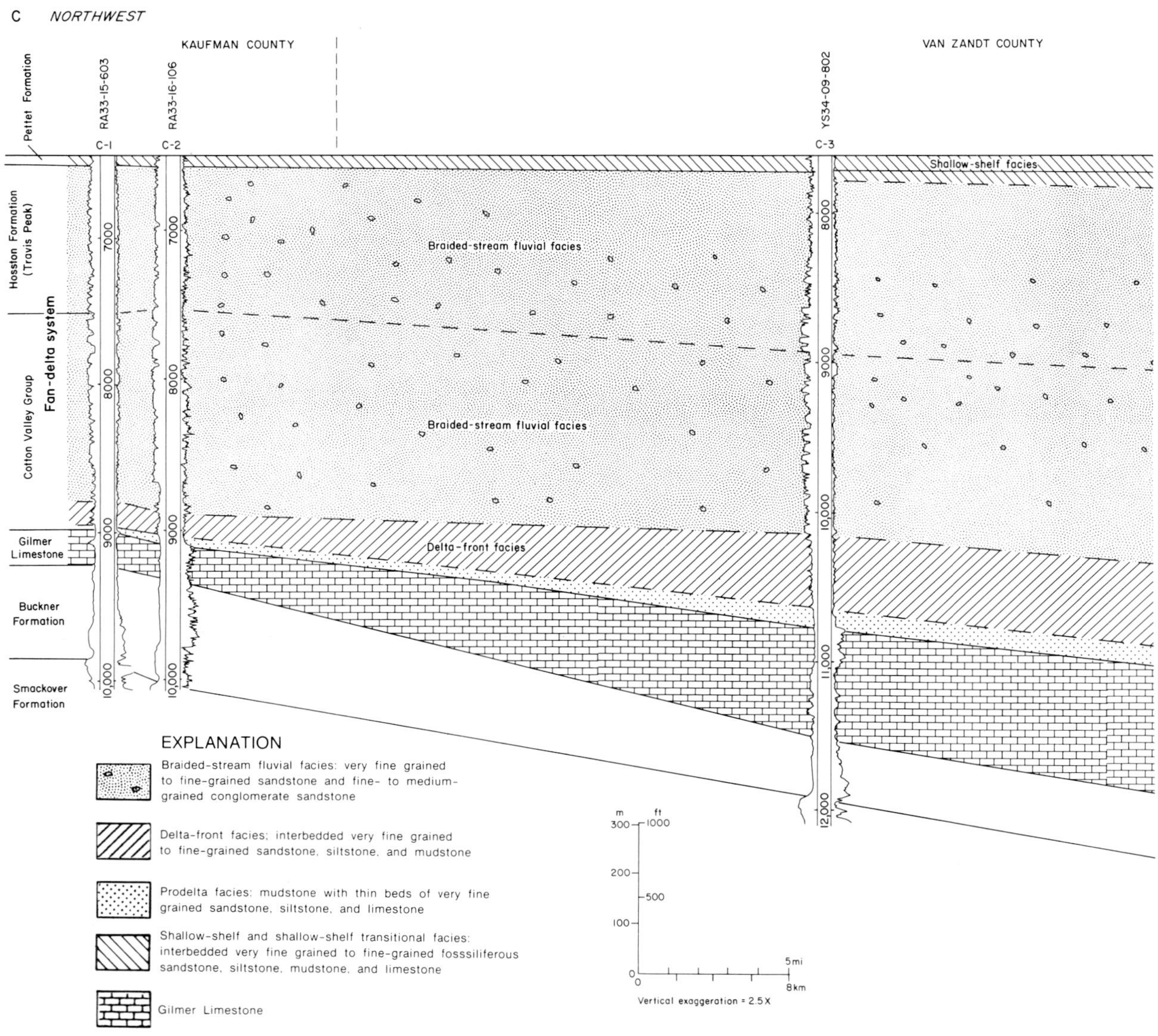

Figure 16. Stratigraphic cross section C-C'

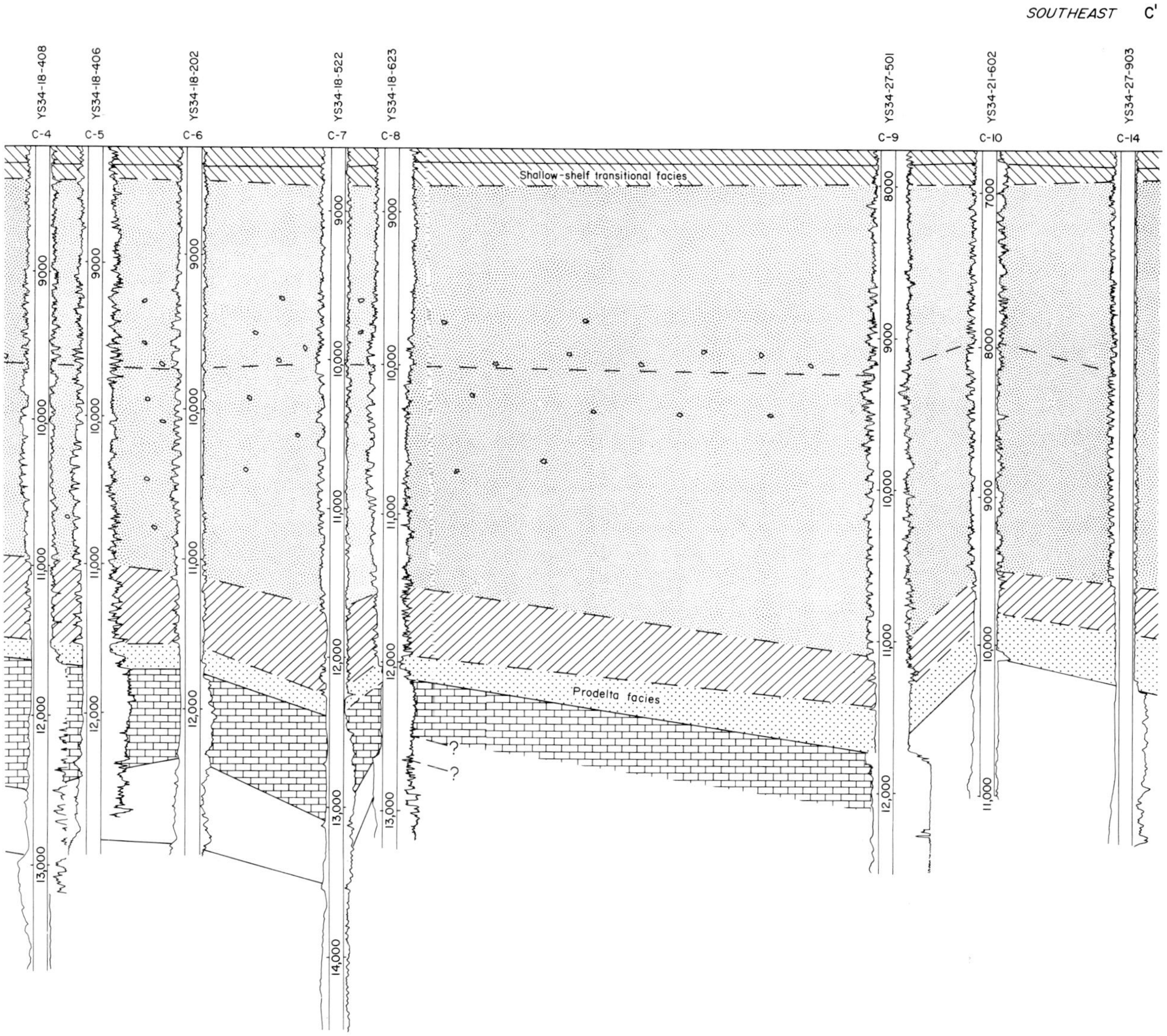

through Kaufman and Van Zandt Counties.

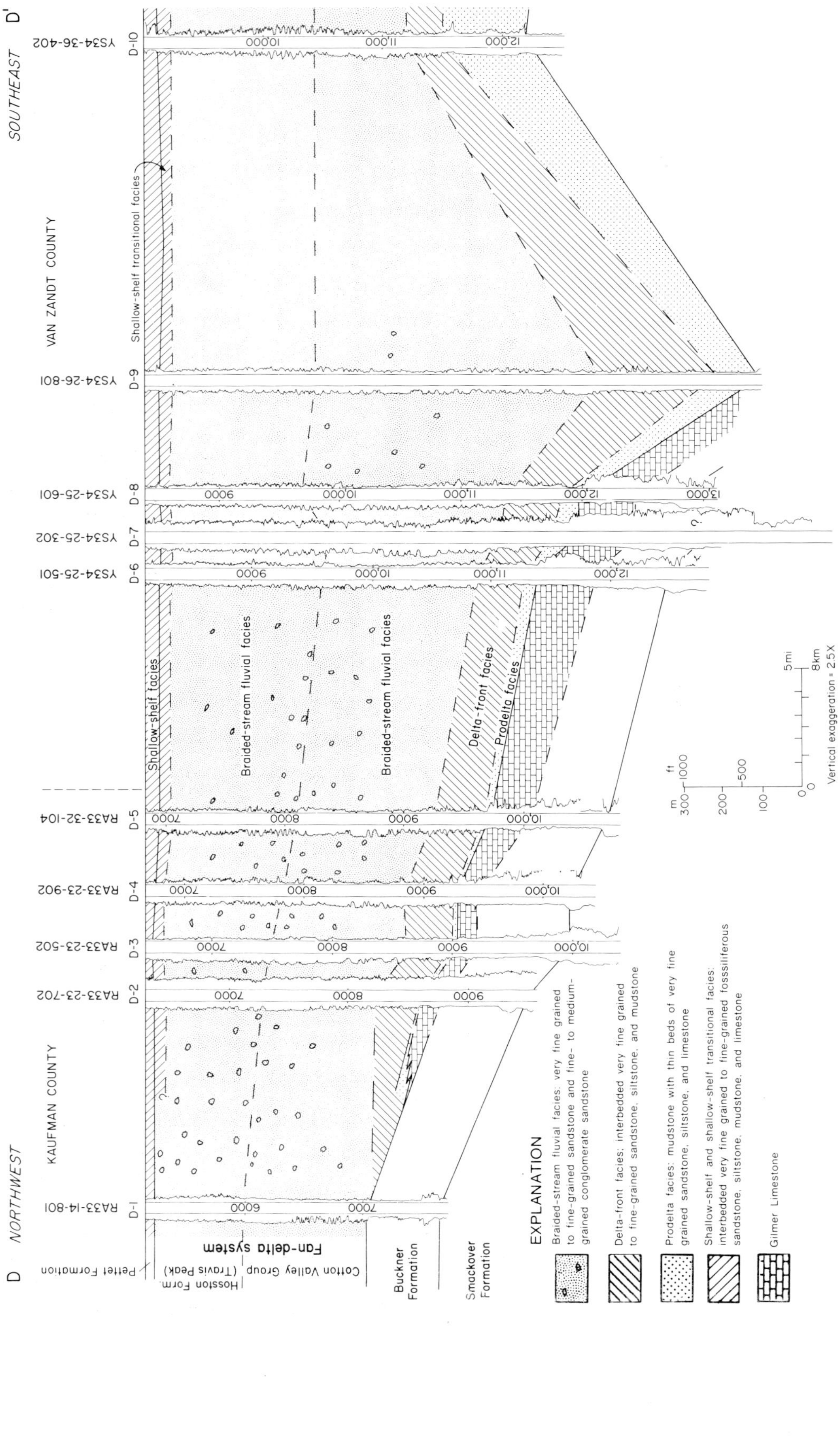

Figure 17. Stratigraphic cross section D-D' through Kaufman and Van Zandt Counties.

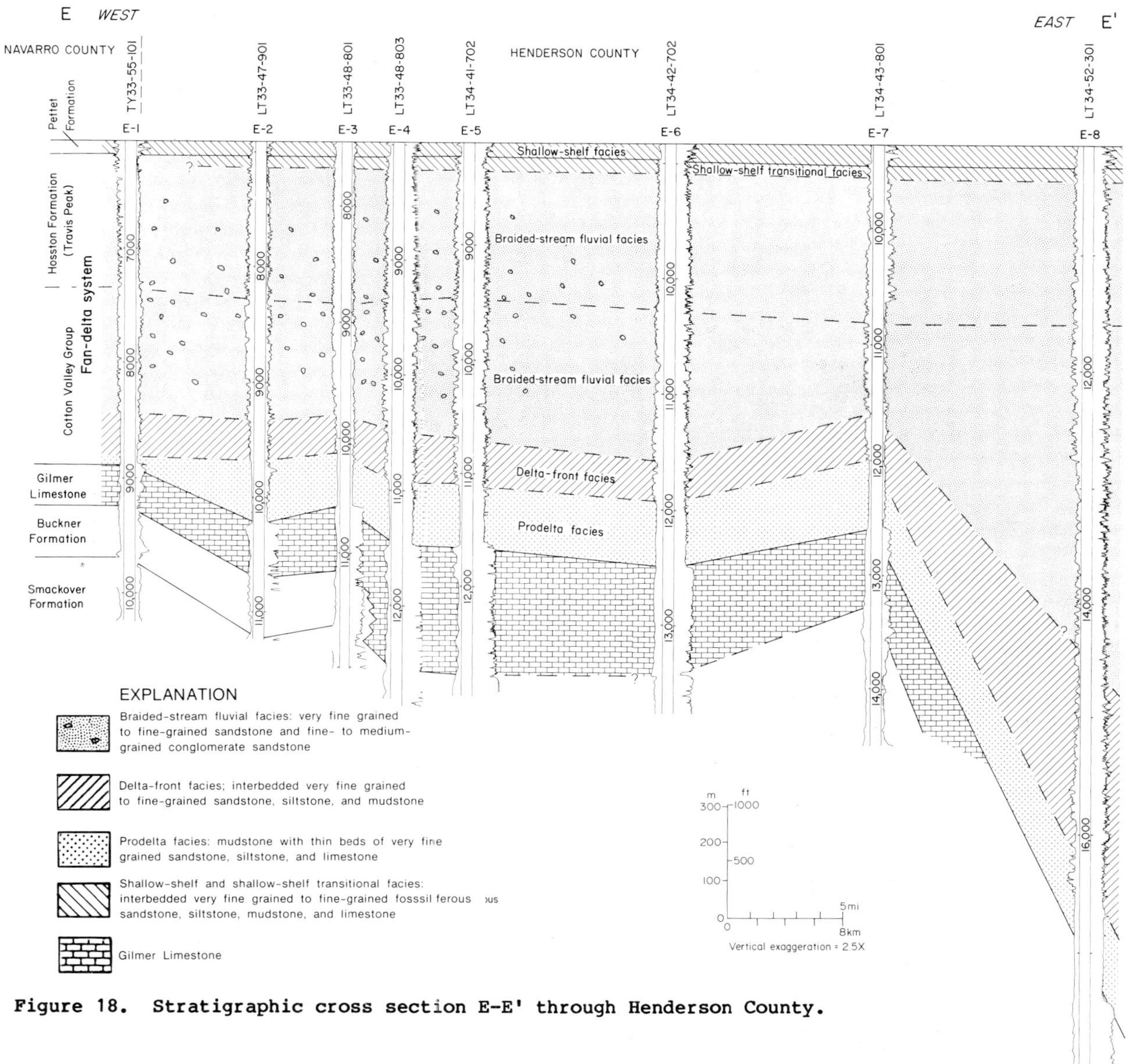

Figure 18. Stratigraphic cross section E-E' through Henderson County.

tation was basinward of the study area. A map showing depositional environments during middle Hosston time indicates a similar trend (Fig. 8 of Bushaw, 1968). Basinward, the Hosston Formation increases in thickness from 800 to 1,350 feet (244 to 412 m); the Cotton Valley Group exhibits a similar increase, from 1,400 to 5,800 feet (427 to 1,700 m).

DEPOSITIONAL AND STRUCTURAL MODEL

The model selected for use in this study was proposed by Kehle (1971) to explain the initiation of salt movement on an originally flat salt surface. Kehle suggested that mass imbalance, resulting from uneven sediment loading, has greater impact on initial salt migration than any other mechanism. He attributed uneven sediment loading to several sedimentary processes, such as reef development, formation of submarine fans at the base of a continental slope, and progradation of deltaic and strandline systems. According to this model, the underlying salt flows laterally away from the sedimentary anomaly, thereby forming an initial withdrawal basin. With continued accumulation of sediment, the withdrawal basin enlarges so that salt ridges

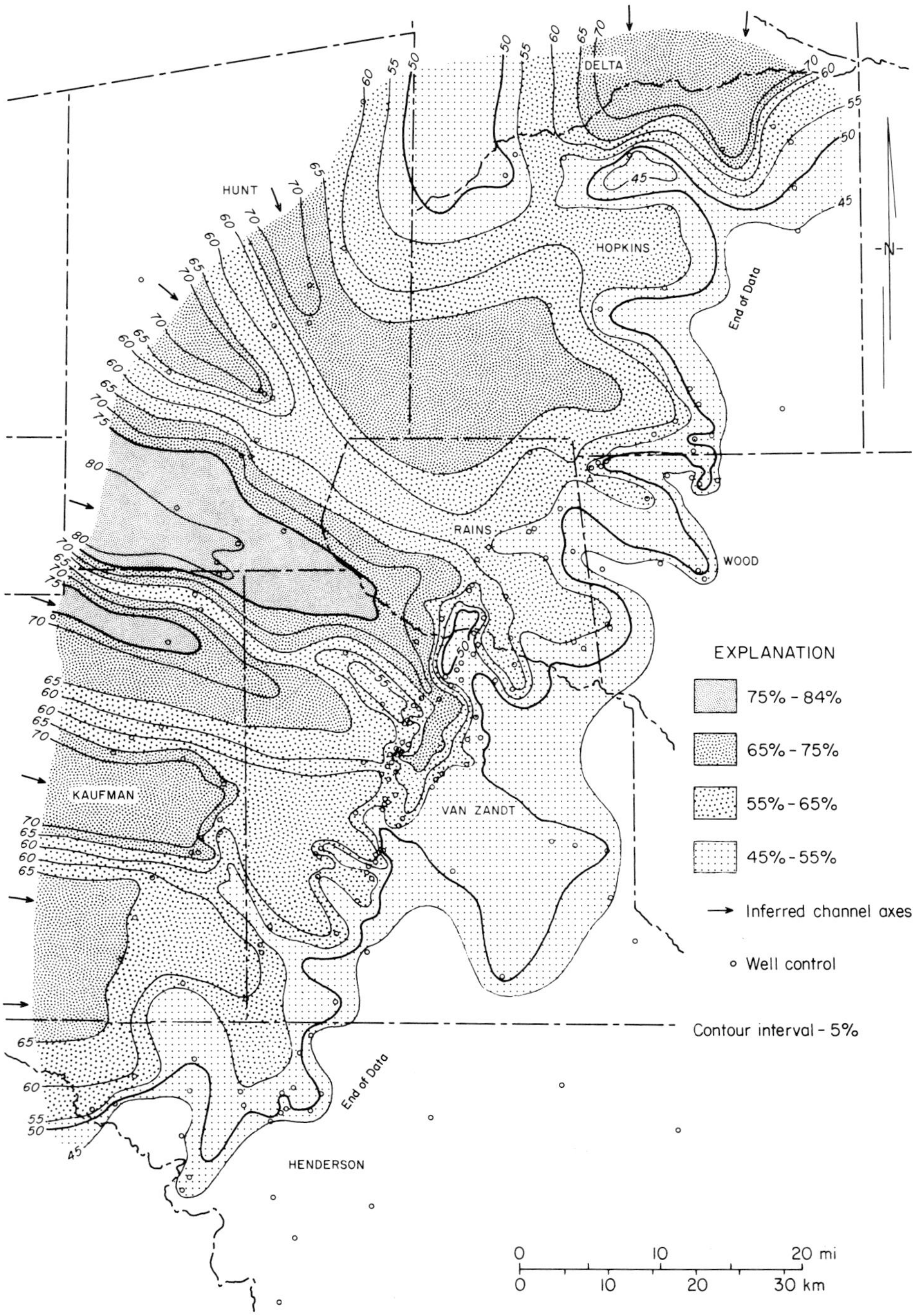

Figure 19. Sandstone-percent map of the Cotton Valley Group.

are formed basinward of the sediment thicks. Lehner (1969) and Martin (1973) documented the presence of analogous features on the abyssal plain and lower continental slope in the northern Gulf of Mexico.

Bishop (1978) proposed a similar model to explain the emplacement of piercement diapirs. His model emphasized the importance of uneven sediment loading but also suggested additional factors that might contribute to

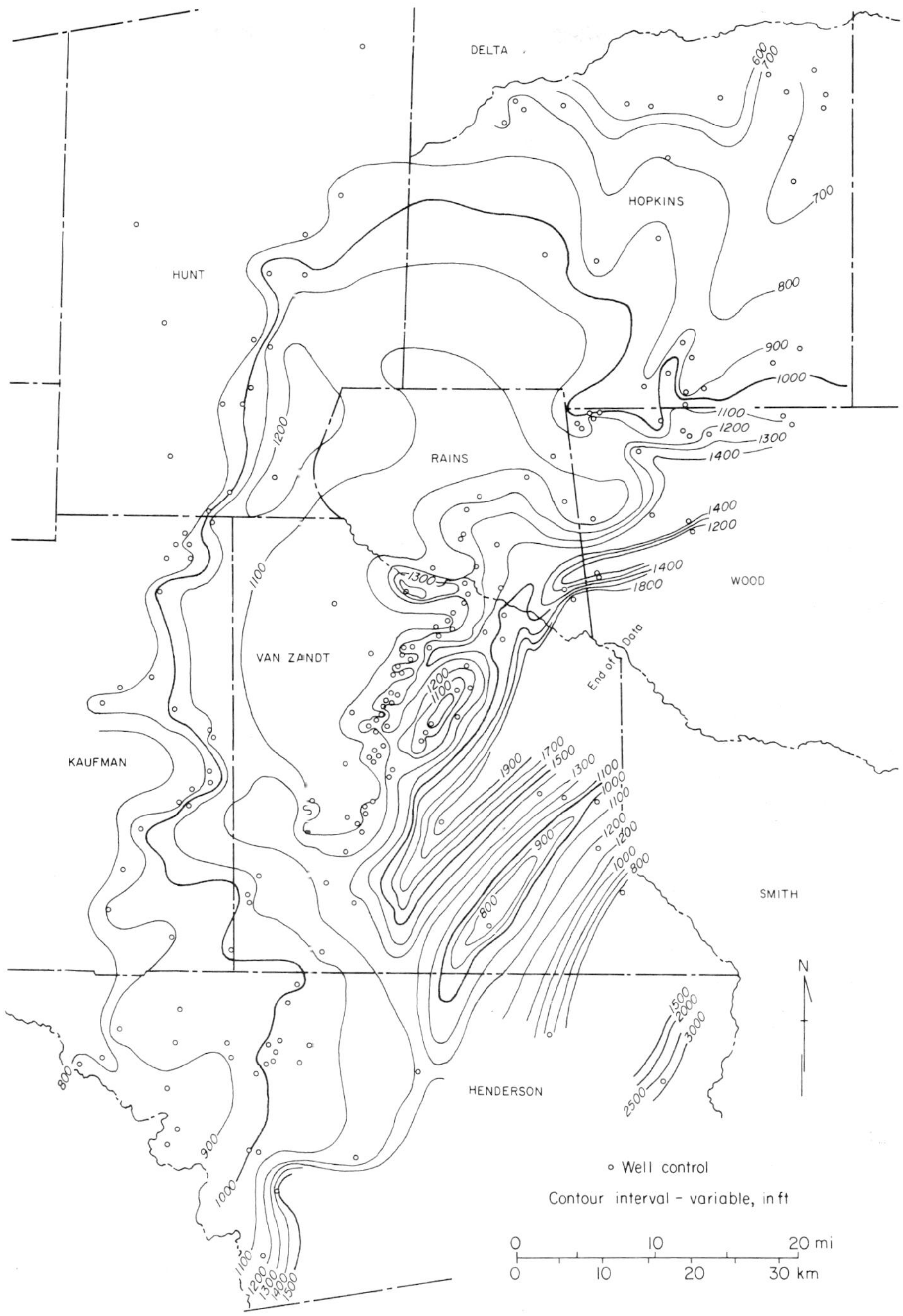

Figure 20. Net-sandstone map of the Cotton Valley Group.

the formation of incipient salt structures; these include regional dip, sedimentation rate, progradation rate, sediment density, thickness of overburden, and thickness of the underlying salt.

Fisher (1973) proposed a direct relationship between sedimentation and salt tectonics in the western Gulf Basin. He distinguished two main types of sedimentation styles and their related salt structures. In the first type, high-constructive lobate deltaic systems initiate mobilization of salt

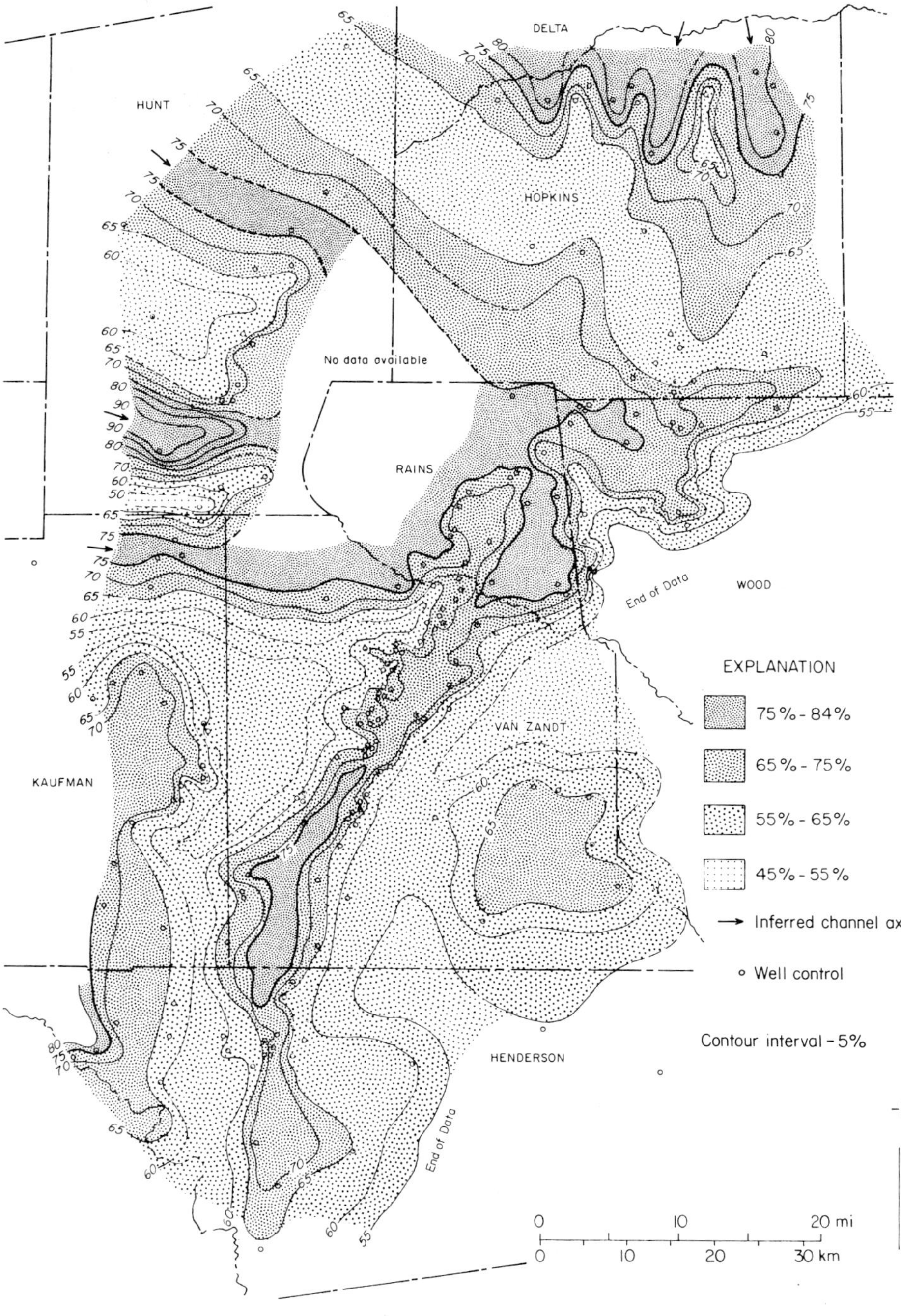

Figure 21. Sandstone-percent map of the Hosston Formation.

and the salt migrates laterally into interdeltaic areas, resulting in the formation of diapirs between the major deltaic lobes. In the second type, strike depositional systems generally induce broad salt ridges, rather than discrete salt structures, in the incipient stages. The second type is closer than the first to the salt structures that are thought to have formed during deposition of the Late Jurassic fan delta system.

In using these two models to explain how Late Jurassic depositional systems effected

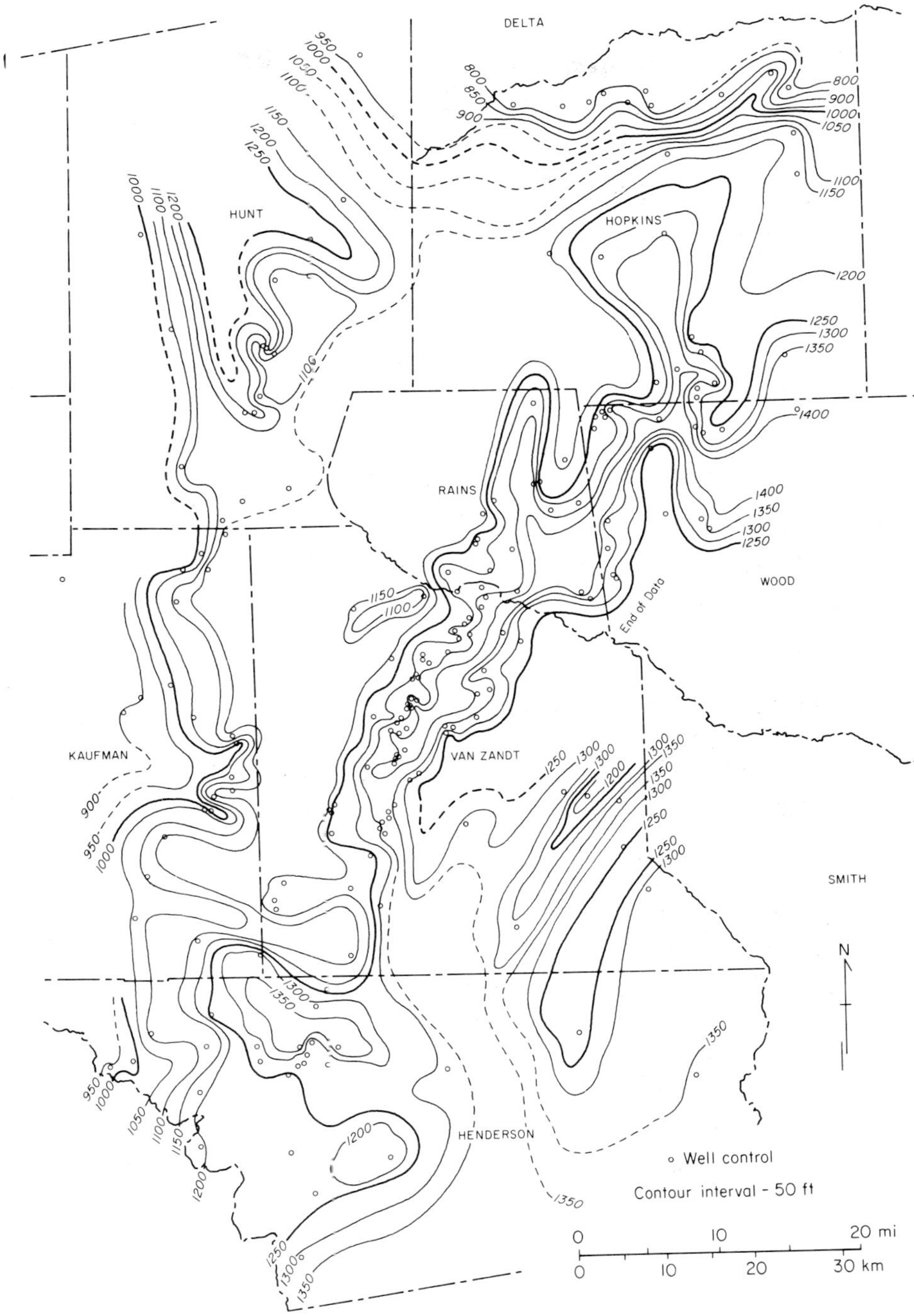

Figure 22. Isopach map of the Hosston Formation.

salt migration, the following sequence of events is suggested: Reversal of ċip along the rift margin allowed an influx of terrigenous clastics into the East Texas Basin during the Late Jurassic and Early Cretaceous. Before that time, main drainage along the western margins of the basin flowed into the Triassic basins to the west. The Ouachita Mountains to the north were low-lying and thus supplied only a small quantity of terrigenous clastic sediment, which was limited mostly to the northern periphery of the basin.

As the supply of terrigenous clastic

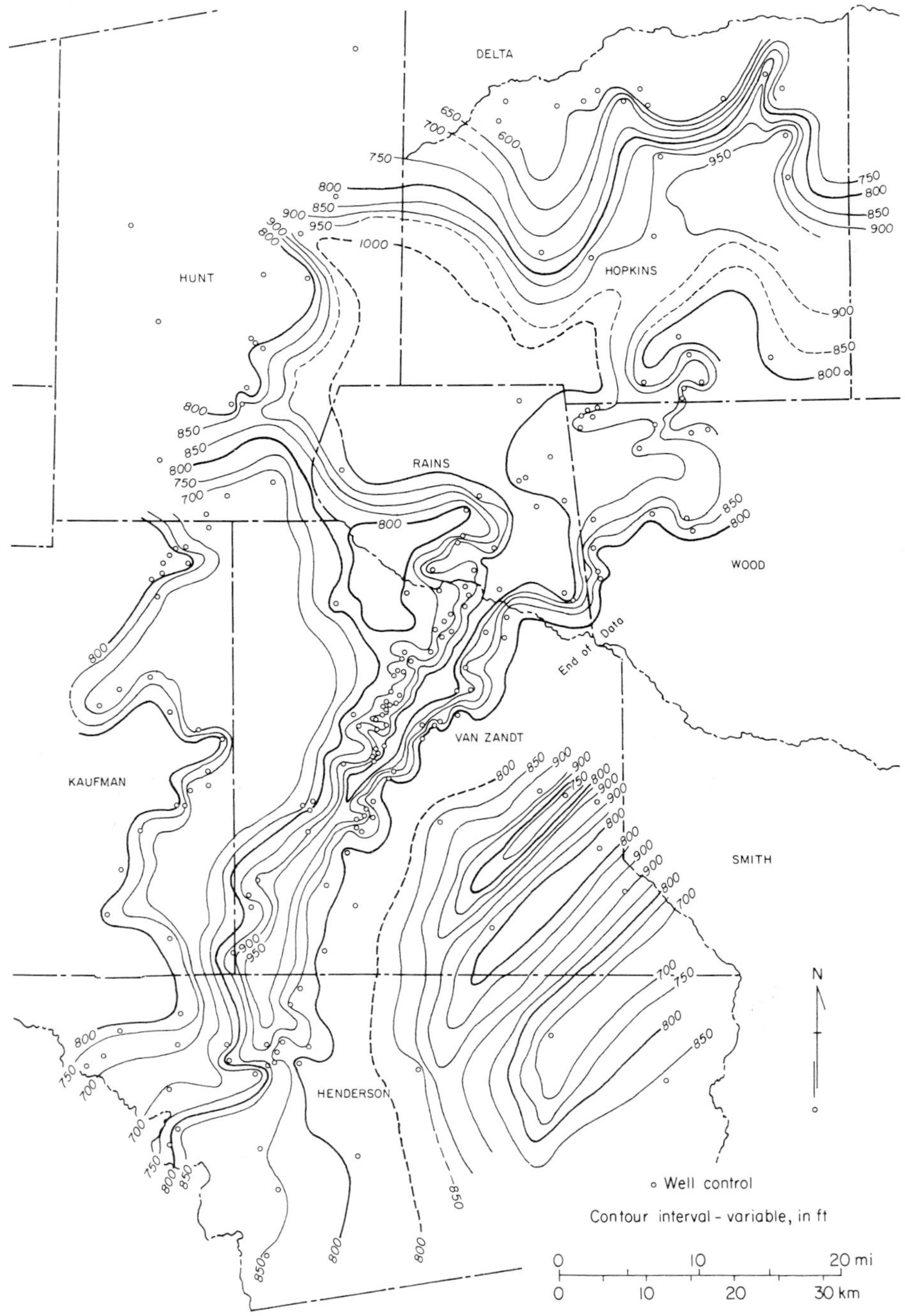

Figure 23. Net-sandstone map of the Hosston Formation.

sediment increased, Cotton Valley and Hosston fan-delta systems, which were advancing southward and eastward and were supplied by a braided-stream complex, began prograding across the Smackover-Gilmer carbonate shelf in the East Texas Basin (Fig. 24). The shelf was a stable platform, and although regional subsidence occurred, it impeded the formation of local depocenters. Most likely, the seas over the carbonate platform were shallow and the rate of progradation of the advancing deltaic complex was rapid. Low-amplitude

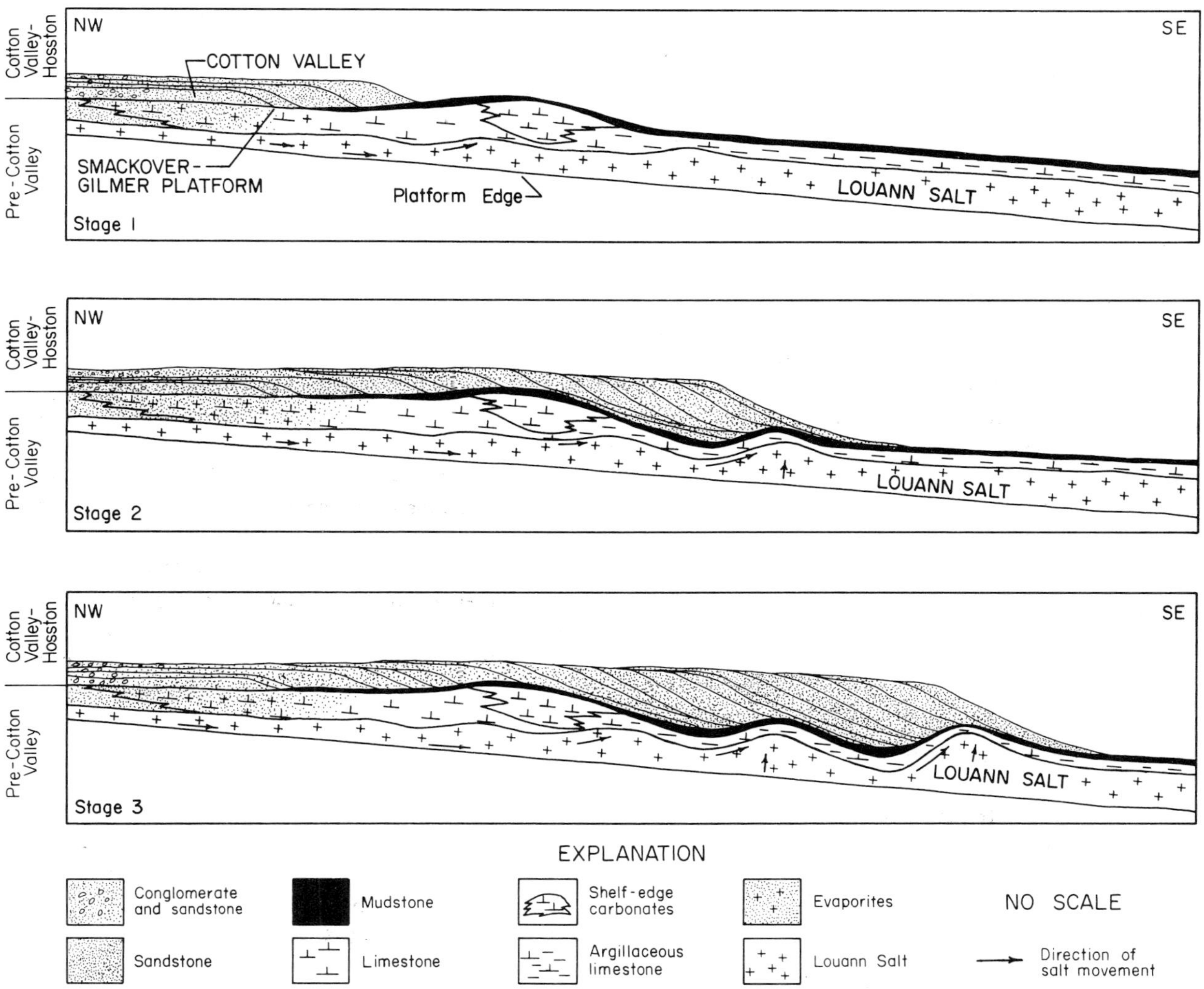

Figure 24. Structural and depositional model of salt migration in the northwestern part of the East Texas Basin during Late Jurassic and Early Cretaceous time.

salt structures occur under the Smackover-Gilmer shelf complex, apparently the result of downward creep that was caused by loading of carbonate deposits (Rogers, 1967; Jackson and Harris, 1981). These low-amplitude salt swells are not evident in the overlying Cotton Valley-Hosston sediments.

Basinward of the ancient Jurassic shelf edge, the rate of progradation slowed, while subsidence caused by salt migration and increased water depth allowed the fan delta complex to stack up, forming elongate depocenters. Salt migrated basinward, forming a salt ridge in front of the sediment wedge. This ridge acted as a dam and thereby enhanced the effectiveness of the syncline as a sediment trap. As the Louann Salt became depleted through basinward migration, subsidence slowed and allowed aggradation to surpass it. Consequently, each subsequent depocenter and associated salt ridge shifted basinward, resulting in a series of parallel salt ridges and sediment thicks. As the older salt ridges became buried, continued salt migration evolved the original salt ridges into separate salt structures, such as salt anticlines and domes.

Toward the end of Early Cretaceous time,

the volume of terrigenous clastic sediment entering the East Texas Basin decreased. The transitional nature of the uppermost Hosston and marine deposits in overlying units indicate that a marine transgression occurred.

CONCLUSIONS

Subsurface mapping of the Cotton Valley and Hosston depositional systems and subjacent salt structures leads to four major conclusions regarding the role of basin infilling in the initiation of salt movement within the northwestern part of the East Texas Basin. Absence of data in the central part of the basin, however, precludes delineation of the basinward extent of the proposed depositional systems.

First, a prograding fan delta system comprises the Cotton Valley-Hosston stratigraphic units. The strata between the Gilmer and Pettet Limestones were deposited during a period of regression that was ended by a major post-Hosston transgression. The transgression is evidenced by transitional shallow marine strata in the upper 100 to 200 feet (33 to 66 m) of the Hosston Formation and by upward gradation into the overlying Pettet Limestone.

Second, the initial movement of salt that occurred in the proximal parts of the East Texas Basin during Smackover deposition was the result of downward creep that was induced by loading of carbonates and was enhanced by basinward tilting (Jackson and Harris, 1981). The resulting salt movement produced small salt structures. Relief on the salt structures increased basinward, coincident with thickening of the salt wedge.

Third, salt mobilized by mass imbalance was induced by Cotton Valley and Hosston deltaic deposition. Uneven sediment loading was controlled by the position of the underlying Smackover-Gilmer carbonate shelf. The Smackover-Gilmer carbonate shelf comprised a stable platform that impeded local subsidence and vertical aggradation of deltaic deposits. Therefore, advancing Cotton Valley-Hosston fan deltas spread laterally and basinward, depositing a fairly uniform wedge of sediments that gradually thickened basinward. Mass imbalance became a viable mechanism to initiate salt movement only when the advancing fan deltas prograded basinward of the stable carbonate platform. Salt migrated basinward of the prograding sediment wedge, forming a proximal incipient withdrawal basin and distal salt ridge. Subsequent depocenters and salt ridges shifted basinward, forming parallel sediment thicks and salt ridges.

Fourth, parallel salt ridges that formed during deposition of the Cotton Valley and Hosston apparently were the initial stage in the development of salt anticlines and domes. Through continued loading, the salt ridges evolved into discrete salt structures. Domal growth appears to depend directly on continued sediment loading, which occurs during periods of major deltaic deposition where the underlying salt wedge is adequately thick.

ACKNOWLEDGMENTS

Funding for this study was provided by the U. S. Department of Energy under Contract No. DE-AC97-80ET46617. Discussions with M. P. A. Jackson, C. W. Kreitler, J. H. McGowen and S. J. Seni were very helpful. L. F. Brown, Jr., R. J. Finley, W. E. Galloway, M. P. A. Jackson, C. W. Kreitler, S. J. Seni, and Noel Tyler reviewed the manuscript and made many helpful suggestions. Cynthia Lopez and Keith Pollman served as research assistants. The report was edited by Jean Trimble. Word processing was by Dottie Johnson and Margaret Chastain under the direction of Lucille C. Harrell. Figures were drafted under the direction of Dan F. Scranton by John T. Ames, Thomas M. Byrd, Micheline R. Davis, Margaret L. Evans, Richard P. Flores, Byron P. Holbert, Jeffrey Horowitz, and Jamie McClelland, Camerawork for illustrations was by James A. Morgan.

REFERENCES

Agagu, O. K., McGowen, M. K., Wood, D. H., Basciano, J. M., and Harris, D. W., 1980, Regional tectonic framework of the East Texas Basin, *in* Kreitler, C. W., and others, Geology and geohydrology of the East Texas Basin--a report on the progress of nuclear waste isolation feasibility studies (1979): The University of Texas at Austin, Bureau of Economic Geology Geological Circular 80-12, p. 11-19.

Barnes, V. E., 1965, Tyler sheet: University of Texas, Austin, Bureau of Economic Geology, Geologic Atlas of Texas, scale 1:250,000.

__________ 1966, Texarkana sheet: University of Texas, Austin, Bureau of Economic Geology, Geologic Atlas of Texas, scale 1:250,000.

__________ 1972, Dallas sheet: The University of Texas at Austin, Bureau of Economic Geology, Geologic Atlas of Texas, scale 1:250,000.

Beall, Robert, 1975, Plate tectonics and the origin of the Gulf of Mexico: Gulf Coast Association of Geological Societies Transactions, v. 23, p. 109-114.

Bishop, R. S., 1978, Mechanism for emplacement of piercement diapirs: American Association of Geological Societies Bulletin, v. 62, no. 9, p. 1561-1583.

Boothroyd, J. C., 1976, A model for alluvial fan - fan delta sedimentation in cold-temperate environments, *in* Recent and ancient sedimentary environments in Alaska: Alaska Geological Society Symposium Proceedings, N1-N3.

________________, and Ashley, G. M., 1975, Processes, bar morphology, and sedimentary structures on braided outwash fans, northeastern Gulf of Alaska, *in* Jopling, A. V., and McDonald, B. C., ed., Glaciofluvial and glaciolacustrine sedimentation: Society of Economic Paleontologists and Mineralogists Special Publication 23, p. 193-222.

________________, and Nummedal, Dag, 1978, Proglacial braided outwash: a model for humid alluvial-fan deposits, *in* Miall, A. D., ed., Fluvial sedimentology: Canadian Society of Petroleum Geologists Memoir 5, p. 641-688.

Brown, L. F., Jr., 1969, Geometry and distribution of fluvial and deltaic sandstones (Pennsylvanian and Permian), North-Central Texas: Gulf Coast Association of Geological Societies Transactions, v. 19, p. 23-47. reprinted in 1969, The University of Texas at Austin, Bureau of Economic Geology Geological Circular 69-4, 15 p.

Burke, Kevin, and Dewey, J. F., 1973, Plume-generated triple junctions: key indicators in applying plate tectonics to old rocks: Journal of Geology, v. 81, no. 4, p. 406-433.

Bushaw, D. J., 1968, Environmental synthesis of the East Texas Lower Cretaceous: Gulf Coast Association of Geological Societies Transactions, v. 18, p. 416-438.

Casey, J. M., 1980, Depositional systems and basin evolution of the late Paleozoic Taos Trough, northern New Mexico: The University of Texas at Austin, Texas Petroleum Research Committee Report No. UT 80-1, 236 p.

Dickinson, K. A., 1968, Upper Jurassic stratigraphy of some adjacent parts of Texas, Louisiana, and Arkansas: United States Geological Survey Professional Paper 594-E, p. E1-E25.

Eaton, R. W., 1961, Development and future possibilities of the Jurassic in northeast Texas: South Texas Geological Society Publication, 15 p., reprinted in 1964, South Texas Geological Society Bulletin, v. 4, no. 13, p. 4-13.

Erxleben, A. W., 1975, Depositional systems in Canyon Group (Pennsylvania System), North-Central Texas: The University of Texas at Austin, Bureau of Economic Geology Report of Investigations No. 82, 75 p.

Fisher, W. L., 1973, Deltaic sedimentation, salt mobilization, and growth faulting in Gulf Coast Basin (abs.): American Association of Petroleum Geologists Bulletin, v. 54, no. 4, p. 779.

Forgotson, J. M., Jr., 1954, Regional stratigraphic analysis of Cotton Valley Group of upper Gulf Coastal Plain: American Association of Petroleum Geologists Bulletin, v. 38, no. 12, p. 2476-2499.

________________, and Forgotson, J. M., Jr., 1976, Definition of Gilmer Limestone, Upper Jurassic formation: American Association of Petroleum Geologists Bulletin, v. 60, no. 7, p. 1119-1123.

Galloway, W. E., 1976, Sediments and stratigraphic framework of the Copper River fan delta, Alaska: Journal of Sedimentary Petrology, v.46, no. 3, p. 726-737.

________________, Kreitler, W. W., and McGowen, J. H., 1979, Depositional and ground-water flow systems in the exploration for uranium: The University of Texas at Austin, Bureau of Economic Geology Research Colloquium, 1978, 267 p.

Hobday, D. K., Woodruff, C. M., Jr., and McBride, M. W., 1981, Paleotopographic and structural controls on nonmarine sedimentation of the Lower Cretaceous Antlers

Formation and correlatives, north Texas and southeastern Oklahoma, *in* Recent and ancient nonmarine depositional environments: models for exploration: Society of Economic Paleontologists and Mineralogists Special Publication 31, p. 71-87.

Holmes, Arthur, 1965, Principles of physical geology (2nd ed.): New York, Roland Press, 1288 p.

Hughes, D. J., 1968, Salt tectonics as related to several Smackover fields along the northeast rim of the Gulf of Mexico Basin: Gulf Coast Association of Geological Societies Transactions, v. 18, p. 320-330.

Imlay, R. W., 1943, Jurassic formations of the Gulf region: American Association of Petroleum Geologists Bulletin, v. 27, no. 11, p. 1407-1533.

Jackson, M. P. A., 1982, Fault tectonics of the East Texas Basin: The University of Texas at Austin, Bureau of Economic Geology Geological Circular 82-4, 31 p.

________________, and Harris, D. W., 1981, Seismic stratigraphy and salt mobilization along the northwestern margin of the East Texas Basin, *in* Kreitler, C. W., and others, Geology and geohydrology of the East Texas Basin: a report of the progress of nuclear waste isolation feasibility studies (1980): The University of Texas at Austin, Bureau of Economic Geology Geological Circular 81-7, p. 28-32.

________________, Seni, S. J., and McGowen, M. K., 1982, Initiation of salt flow in the East Texas Basin (abs.): American Association of Petroleum Geologists Bulletin, v. 66, no. 5, p. 584-585.

Johnson, L. C., 1980, Structure and stratigraphy of an evolving salt ridge and basin complex, Louisiana continental shelf: The Univ. Tex. at Austin, Mast. Thesis, 120 p.

Kaiser, W. R., Johnston, J. E., and Bach, W. N., 1978, Sand-body geometry and the occurrence of lignite in the Eocene of Texas: The University of Texas at Austin, Bureau of Economic Geology Geological Circular 78-4, 19 p.

Kehle, R. O., 1971, Origin of the Gulf of Mexico: The University of Texas at Austin, Geology Library, call no. q557K260, unpublished report, unpaginated.

Krumbein, W. C., and Sloss, L. L., 1959, Principles of stratigraphy and sedimentation: San Francisco, W. H. Freeman and Co., 497 p.

Lehner, Peter, 1969, Salt tectonics and Pleistocene stratigraphy on continental slope of northern Gulf of Mexico: American Association of Petroleum Geologists Bulletin, v. 53, no. 12, p. 2431-2482.

Leopold, L. B., Wolman, M. G., and Miller, J. P., 1964, Fluvial Processes in geomorphology: San Francisco, W. H. Freeman and Co., 522 p.

Loocke, J. E., 1978, Growth history of the Hainesville salt dome, Wood County, Texas: The University of Texas at Austin, Master's thesis, 95 p.

Lucchi, F. R., Colella, A., Ori, G. G., and Oglian, F., 1981, Pliocene fan deltas of the Intra-Apenninic Basin, Bologna, *in* Lucchi, F. R., ed., International Association of Sedimentologists Excursion Guidebook 4, 162 p.

Lyons, P. L., 1957, Geology and geophysics of the Gulf of Mexico: Gulf Coast Association of Geological Societies Transactions, v. 7, p. 1-9.

Martin, R. G., 1973, Salt structure and sediment thickness, Texas-Louisiana continental slope, northwestern Gulf of Mexico: U.S. Geological Survey Open-File Report, 21 p.

McGowen, J. H., 1970, Gun Hollow fan delta, Nueces Bay, Texas: The University of Texas at Austin, Bureau of Economic Geology Report of Investigations No. 69, 91 p.

______________, Granata, G. E., and Seni, S. J., 1979, Depositional framework of the Lower Dockum Group (Triassic), Texas Panhandle: The University of Texas at Austin, Bureau of Economic Geology Report of Investigations No. 97, 60 p.

______________, and Harris, D. W., 1981, Preliminary study of the Upper Jurassic (Cotton Valley) and Lower Cretaceous (Hosston/Travis Peak) Formations of the East Texas Basin, *in* Kreitler, C. W., and others, Geology and geohydrology of the East Texas Basin--a report on the progress of nuclear waste isolation feasibility studies (1980): The University of Texas at Austin, Bureau of Economic Geology Geological Circular 81-7, p. 43-47.

Miall, A. D., 1977, A review of the braided-river depositional environment, *in* Earth Science Reviews No. 13: Amsterdam, Elsevier Scientific Publishing Co., p. 1-62.

Moore, R. W., and Del Castillo, Luis, 1974, Tectonic evolution of the southern Gulf of Mexico: Geological Society of America Bulletin, v. 85, no. 4, p. 607-618.

Newkirk, T. F., 1971, Possible future petroleum provinces of Jurassic, western Gulf Basin, *in* Cram, I. H., ed., Future petroleum provinces of the United States--their geology and potential: American Association of Petroleum Geologists Memoir 15, p. 927-953.

Nichols, P. H., Peterson, G. E., and Wuestner, C. E., 1968, Summary of subsurface geology of northeast Texas: American Association of Petroleum Geologists Memoir 9, v. 1, p. 982-1004.

Parker, T. J., and McDowell, A. M., 1955, Model studies of salt dome tectonics: American Association of Petroleum Geologists Bulletin, v. 39, no. 12, p. 2384-2470.

Raup, O. B., 1970, Brine mixing--an additional mechanism for formation of basin evaporites: American Association of Petroleum Geologists Bulletin, v. 54, no. 12, p. 2246-2259.

Rogers, J. K., 1967, Comparison of some Gulf Coast Mesozoic carbonate shelves: Gulf Coast Association of Geological Societies Transactions, v. 17, p. 49-60.

Rosenkrans, R. R., and Marr. J. D., 1967, Modern seismic exploration of the Gulf Coast Smackover trend: Geophysics, v. 32, no. 2, p. 184-206.

Rust, B. R., 1978, Depositional models for - braided alluvium, *in* Miall, A. D., ed., es Fluvial sedimentology: Canadian Society of Petroleum Geologists Memoir 5, p. 605-626.

Salvador, Amos, and Green, A. R., 1980, Opening of the Caribbean Tethys (origin and development of the Caribbean and the Gulf of Mexico), *in* Colloque C5: Geologie de chaines alpines issues de la Tethys: Bureau de Recherches Geologiques et Minieres Memoire 115, p. 224-229.

Swain, F. M., 1949, Upper Jurassic of northeastern Texas: American Association of Petroleum Geologists Bulletin, v. 33, no. 7, p. 1206-1250.

Thomas, W. A., and Mann, C. J., 1966, Late Jurassic depositional environments, Louisiana and Arkansas: American Association of Petroleum Geologists Bulletin, v. 50, no. 1, p. 178-182.

Todd, R. G., and Mitchum, R. M., Jr., 1977, Seismic stratigraphy and global changes of sea level, part 8: identification of Upper Triassic, Jurassic, and Lower Cretaceous seismic sequences in Gulf of Mexico and offshore West Africa, *in* Payton, C. E., ed., Seismic stratigraphy--applications to hydrocarbon exploration: American Association of Petroleum Geologists Memoir 26, p. 145-163.

Trusheim, F., 1960, Mechanisms of salt migration in northern Germany: American Association of Petroleum Geologists Bulletin, v. 44, no. 9, p. 1519-1540.

Turk, Kehle, and Associates, 1978, Tectonic framework and history, Gulf of Mexico region: Report for Law Engineering Testing Co., Marietta, Georgia, 28 p.

Walker, R. G., and Cant, D. J., 1979, Sandy fluvial systems, *in* Walker, R. G., ed., Facies models: Geological Association of Canada, Geoscience Canada Reprint Series 1, p. 23-32.

Walper, J. L., 1980, Tectonic evolution of the Gulf of Mexico, *in* Pilger, R. H., Jr., ed., The origin of the Gulf of Mexico and the early opening of the central North Atlantic Ocean: Louisiana State University Symposium Proceedings, Baton Rouge, 1980, p. 87-98.

Ward, L. G., Stephen, M. F., and Nummedal, Dag, 1976, Hydraulics and morphology of glacial outwash distributaries, Skeidararisandur, Iceland: Journal of Sedimentary Petrology, v. 46, no. 4, p. 770-777.

Wescott, W. A., and Ethridge, F. G., 1980, Fan-delta sedimentology and tectonic setting--Yallahs fan delta, southeast Jamaica: American Association of Petroleum Geologists Bulletin, v. 64, no. 3, p. 374-400.

Wood, D. H., 1981, Structural effects of salt movement in the East Texas Basin, *in* Kreitler, C. W., and others, Geology and geohydrology of the East Texas Basin--a report on the progress of nuclear waste isolation feasibility studies: The University of Texas at Austin, Bureau of Economic Geology Geological Circular 81-7, p. 21-27.

Wood, M. L., and Walper, J. L., 1974, The evolution of the interior western basins and the Gulf of Mexico: Gulf Coast Association of Geological Societies Transactions, v. 24, p. 31-41.

APPENDIX A

Electric log tracings and descriptions of rotary well cuttings (refer to Fig. 2 for location).

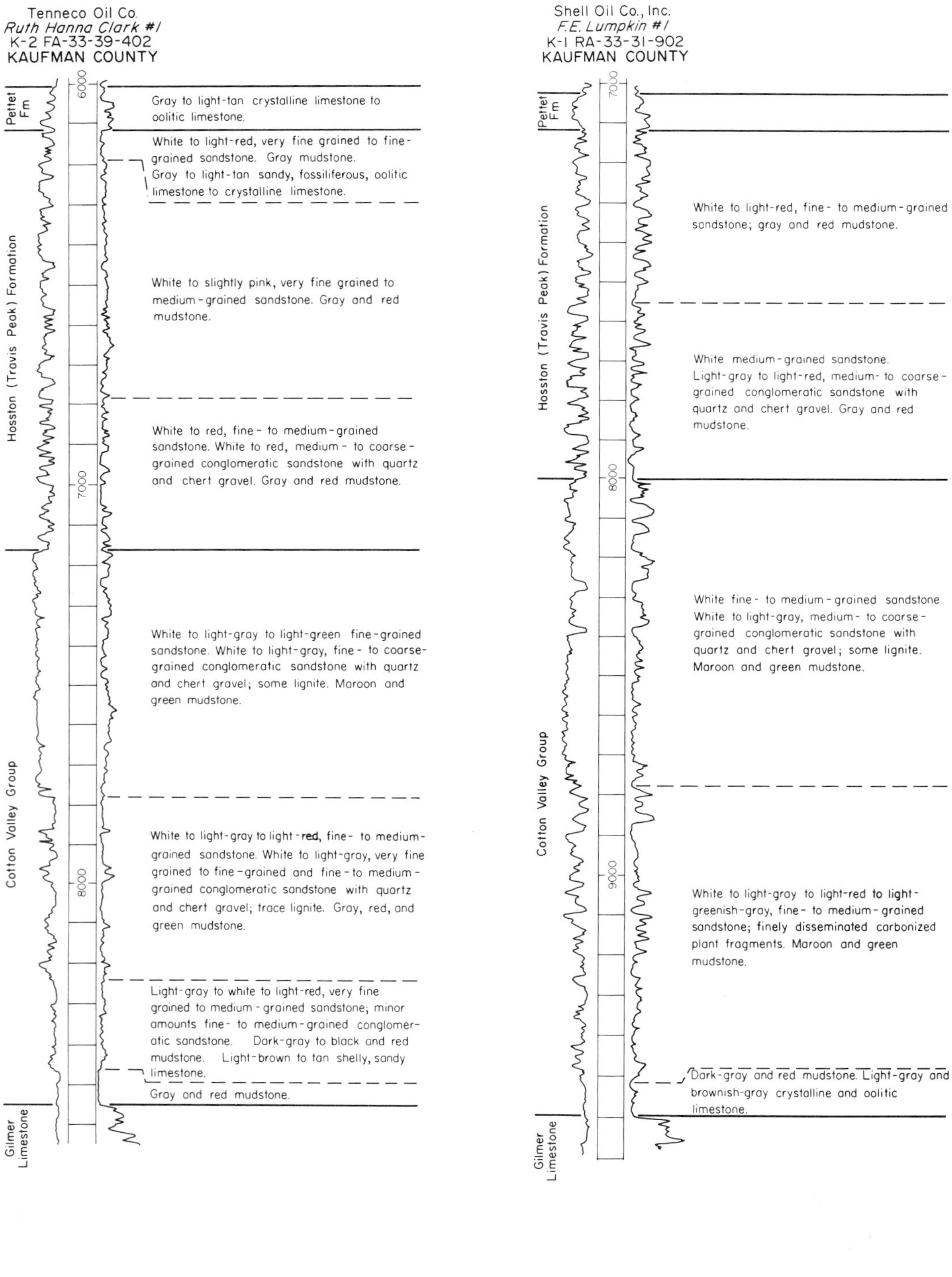

APPENDIX A - Continued

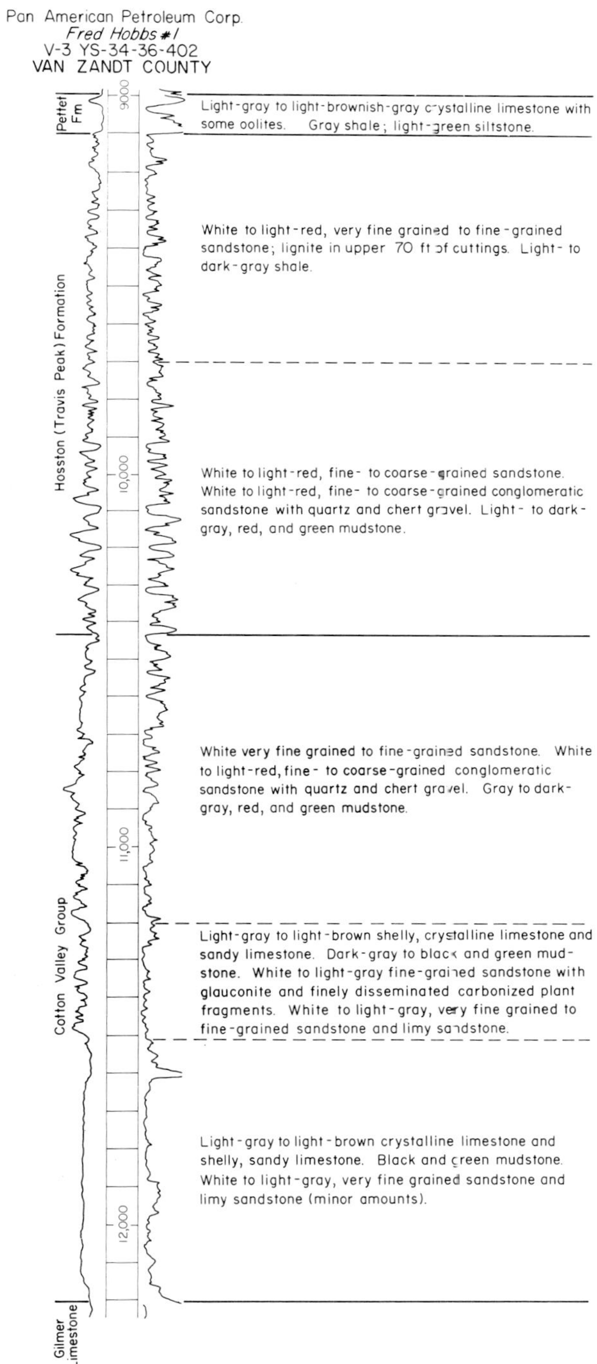

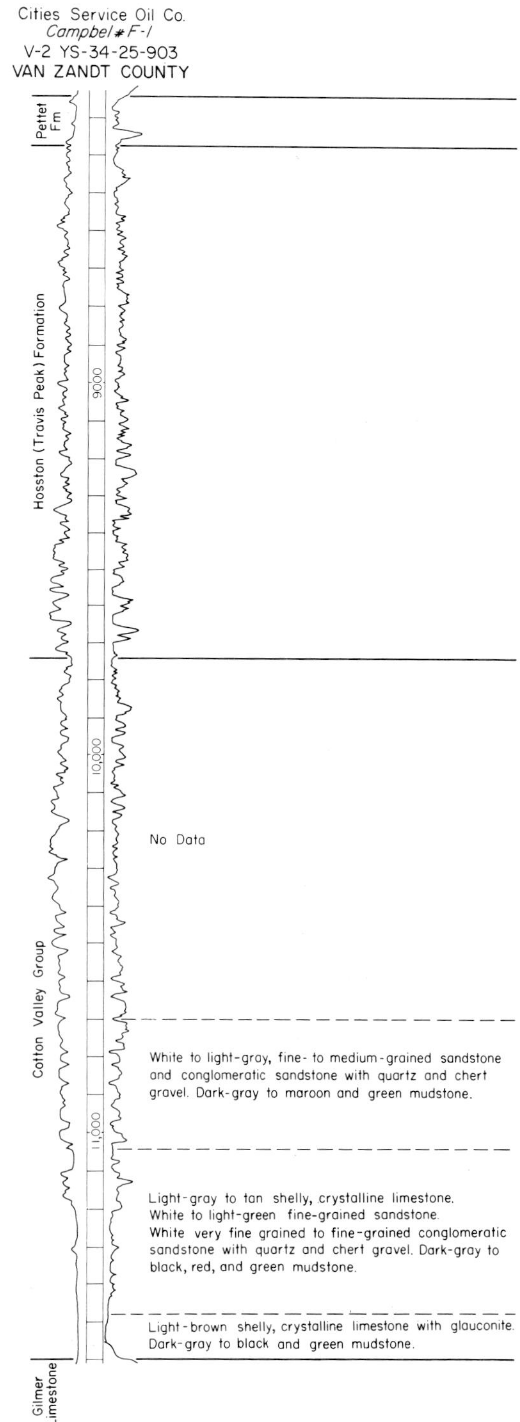

APPENDIX A - Continued

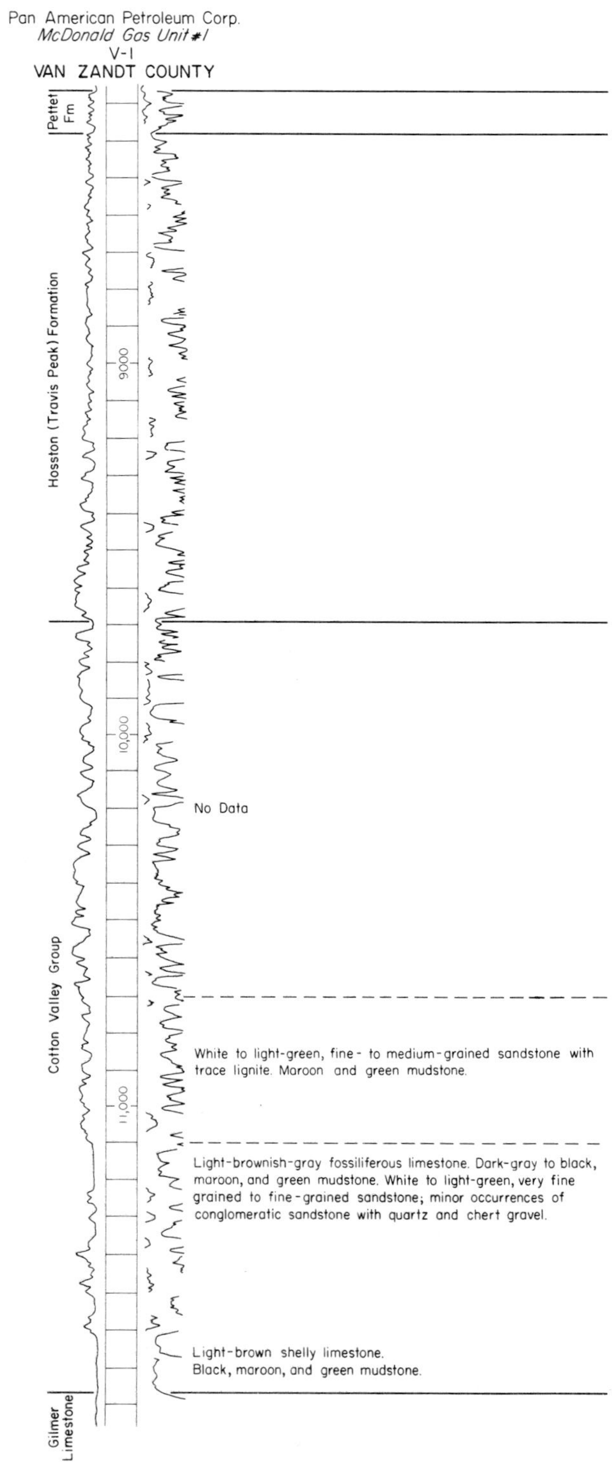

APPENDIX B

Cross-section wells

Well No.	Well Name	County
Line A-A'		
A-1 LZ 17-43-701	Humble Oil & Refining Co. No. 1 Dunham	Hopkins
A-2 LZ 17-59-201	Sunray Dx Oil Co. No. 1 Seaman	Hopkins
A-3 LZ 17-59-301	North Central Oil Co. No. 1 Moseley	Hopkins
A-4 LZ 17-59-901	Hinton No. 1 Walker	Hopkins
A-5 ZS 34-04-203	Forest Oil Corp. No. 1 Asher	Hopkins
A-6 ZS 34-04-502	Hughley Operating Co. and No. Am. Expl. Co.	Wood
A-7 ZS 34-04-601	Humble Oil & Refining Co. No. 1 Allen	Wood
A-8 ZS 34-12-301	Getty Oil No. 1 Blalock	Wood
A-9 ZS 34-12-303	Shell Oil Co. No. 1 Wright	Wood
Line B-B'		
B-1 PH 33-08-204	Ohio Oil Co. No. 1 Popper	Hunt
B-2 UX 34-10-302	Texaco, Inc. Irvine Gas Unit No. 1	Rains
B-3 UX 34-10-601	Delta Drilling No. 1 Hare	Rains
B-4 UX 34-11-905	Caraway & Smith No. 1 Gilley	Rains
B-5 ZS 34-12-701	Samedan Oil Corp. Buchanan No. 1	Wood
B-6 ZS 34-23-701	Humble Oil & Refining Co. NW Hawkin No. 1	Wood
Line C-C'		
C-1 RA 33-15-603	Schneider and Murray No. 1 Jones Estate	Kaufman
C-2 RA 33-16-106	Santa Fe Minerals, Inc. No. 1 Barrow Estate	Kaufman
C-3 YS 34-09-802	E. C. Johnston Co. No. 1 Martin Gas Unit	Van Zandt
C-4 YS 34-18-408	Pan American Petr. Corp. Brown Gas Unit B-1	Van Zandt
C-5 YS 34-18-406	Pan American Petr. Corp. No.1 Nichols Gas U.	Van Zandt
C-6 YS 34-18-202	Caraway and Smith Parker Gas Unit No. 2	Van Zandt
C-7 YS 34-18-522	R. J. Caraway No. 1 Stone (Fruitvale GU)	Van Zandt
C-8 YS 34-18-623	Continental Oil Co. No. 1 Elliot	Van Zandt
C-9 YS 34-27-501	Halbouty No. 1 Rowan	Van Zandt
C-10 YS 34-27-602	Midwest et al. No. 1 Clark	Van Zandt
C-11 YS 34-27-903	Pure Oil Co. D-8 Swain No. 11	Van Zandt
Line D-D'		
D-1 RA 33-14-801	Rockwall Exploration Co. No. 1 Wallace	Kaufman
D-2 RA 33-23-702	W. M. Hughes No. 1 Billings	Kaufman
D-3 RA 33-23-502	The T x 1 Oil Corp. No. 1 Liston	Kaufman
D-4 RA 33-23-902	Southland Royalty Co. No. 1 Frosch Unit	Kaufman
D-5 RA 33-32-104	Union Tex. Pet.-Lacal Pet. No. 1 Phillips	Kaufman
D-6 YS 34-25-501	R. J. Caraway No. 1 Yates	Van Zandt
D-7 YS 34-25-302	Superior Oil Co. No. 1 Porter	Van Zandt
D-8 YS 34-25-601	R. J. Caraway No. 1 Parker	Van Zandt
D-9 YS 34-26-801	R. J. Caraway No. 1 Gilmore	Van Zandt
D-10 YS 34-36-402	Pan American Petr. Corp. No. 1 Hobbs	Van Zandt
Line E-E'		
E-1 TY 33-55-101	Humble Oil & Refining Co.	Navarro
E-2 LT 33-47-0-1	Max Pray No. 1 Carter	Henderson
E-3 LT 33-48-801	Pan American Petr. Corp. No. 1 Sorrell	Henderson
E-4 LT 33-48-803	Rudman Resources, Inc. No. 1 Hammock	Henderson
E-5 LT 34-41-702	Lake Ronel Oil Co.-Bill Ross No. 1 Shaver	Henderson
E-6 LT 34-42-702	B. Smith, G. Lehnertz, W. Perryman No. 1 Lee	Henderson
E-7 LT 34-43-801	Lone Star Producing Co. 1-B Allyn	Henderson
E-8 LT 34-57-301	Texas Interstate Oil & Gas Co. No. 1 Cotton	Henderson

CARBONATE FACIES OF THE UPPER SMACKOVER FORMATION (JURASSIC), PAUP SPUR-MANDEVILLE FIELDS, MILLER COUNTY, ARKANSAS

MARYANN M. McGRAW

6558 East Cooper Street
Tucson, Arizona 85710

ABSTRACT

The Paup Spur-Mandeville Fields are located in Miller County, Arkansas, about six miles northeast of Texarkana. Of more than 20 wells drilled in this field to date, twelve have produced oil from the Smackover Formation and one has produced gas condensate.

Based on an analysis of cores from nine wells and porosity plugs from a tenth well from these fields the following carbonate facies have been defined:

(1) Laminated pelsparite with shaly partings. The sedimentary features of this unit are indicative of a shallow, subtidal environment, and the sedimentary structures in the lowest part of the section show a predominance of gravity-related features. Upsection, wave-generated currents were more prevalent, which indicates that deposition occurred in shallower water.

(2) Cross-bedded and burrowed pelsparite-pelmicrite. This unit is heavily burrowed near the center of the field and interfingers laterally with intertidal cross-beds and horizontal laminations. The burrowed zone is probably indicative of slightly deeper water.

(3) Laminated to burrowed silty pelsparite with shaly partings. This unit is characterized by cycles which exhibit a decrease in wave energy. Each cycle is cross-bedded or finely laminated and rippled at the base, and shows increasing bioturbation and decreasing silt and shale content upwards.

(4) Algal biolithite and intrasparite. The most outstanding features of this unit are the bluegreen algal structures. Stromatolitic structures (largest structure in the cores is 10 cm in height) are best developed in wells to the south; however, oncolites and algal-laminated sediments occur across the field.

(5) Cross-bedded oosparite. Sedimentary accumulations in this unit are a high-energy, oolitic sand belt. Bimodal cross-beds and horizontal laminations characterize this unit.

The Smackover sediments of the PS-M Fields were formed near the mouth of a shallow embayment on the ancestral Gulf of Mexico shelf. A depositional model for the area shows ooid shoals forming over pre-Smackover structural highs where maximum wave energy was expended over their southeast flank. Peripheral facies formed in protected areas and where wind-driven tidal variations had greater influence. Bluegreen algal structures modified environemnts to the west.

Early diagenetic features of the Upper Smackover indicate an active, freshwater phreatic environment. During early diagenesis partial to complete dissolution of allochems occurred, which created most of the field's porosity. Primary cements are finely crystalline calcite occurring as equant to bladed crusts and equant, interlocking mosaics which coarsen toward the pore centers.

Later diagenetic features include solution enhancement of pore space during hydrocarbon migration. Anhydrite and celestite are rare, but ubiquitous - the anhydrite oce curring as pore filling and replacement and the celestite occurring as pore and fracture fill. Locally, baroque dolomite occupies allochem molds.

INTRODUCTION

The Paup Spur-Mandeville Fields (PS-M Fields) occupy an area of approximately four square miles (10.4 sq km) in Miller County, Arkansas, about ten kilometers northeast of Texarkana. In this area, more than 20 wells have penetrated the Smackover Formation. Of these wells, twelve have produced oil and one has produced gas condensate from the upper Smackover. The remaining wells were dry. Total Smackover thickness in these fields is approximately 122 meters.

The regional distribution of the Smackover Formation is along the northern and western margins of the Gulf of Mexico basin (Fig. 1). The Smackover has been informally divided into two members, the upper and lower Smackover by Bishop (1968). The upper Smackover is basically sparite with the exception of the segment in western Mississippi, easternmost Arkansas, and Louisiana where it becomes more sandy and shaly (Bishop, 1968), and in some areas it is dolomitized. The lower Smackover is generally more argillaceous and carbonaceous and contains larger proportions of lime mud and pyrite. Regionally, oil and gas production has been from the upper Smackover. For this reason, most well material has come from the upper portion and Smackover study has concentrated on this part; study of the Smackover Formation in the PS-M Fields is no exception. This investigation of the Smackover will emphasize the upper member but the lower member will be considered where data exist.

METHODS OF STUDY

The Smackover Formation in the PS-M Fields is buried to a depth of about 1,830 meters. All data used for this study were gathered through information techniques associated with exploratory drilling for gas and oil. Study of the Smackover was initiated by the correlation of geophysical logs available from wells that penetrated the Smackover Formation in the PS-M Fields and surrounding areas (Fig. 2 and Table 1). Dual Induction Laterologs were cross-correlated to gain information on the thickness and structure of the unit and structural development in the area during Late Jurassic time. Cores are available from ten wells in these fields and porosity plugs are also available from one well in the Mandeville Field. The sampled intervals mainly cover the upper member and the upper part of the lower member. These cores and plugs were examined, and carbonate facies were identified based on lithology, sedimentary structures, and allochemical constituents. Thin-section observations served as the main technique for describing carbonate textures and diagenetic fabric as well as details of mineral composition.

STRATIGRAPHIC AND STRUCTURAL ELEMENTS OF THE PS-M FIELDS

The stratigraphic sequence in the vicinity of the PS-M Fields is shown in Figure 3. The PS-M Fields are situated near the northern updip limit of Smackover deposition (Fig. 1). In this area, sediments of the Smackover Formation occur north of the updip limit of the Louann Salt. Bishop (1967) reported the northern limit of Werner deposition as being south and east of the PS-M Fields as well (Fig. 4). However, six wells penetrated Werner sediments below the Smackover and Norphlet Formations in the PS-M Fields. In two other wells, (No. 21 and No. 41) the Werner Formation is absent, thus indicating a local topographic high during deposition of these sediments in the PS-M area. Two miles to the southeast of the PS-M Fields, Smackover sediments increase in thickness; and in the Lyons No. 1 Purifoy (No. 35) well the Smackover is underlain by thin sediments of the Norphlet Formation, 61 meters of Louann Salt, and more than 30 meters of Werner anhydrite.

Cross-section A - A' (Fig. 5) has been

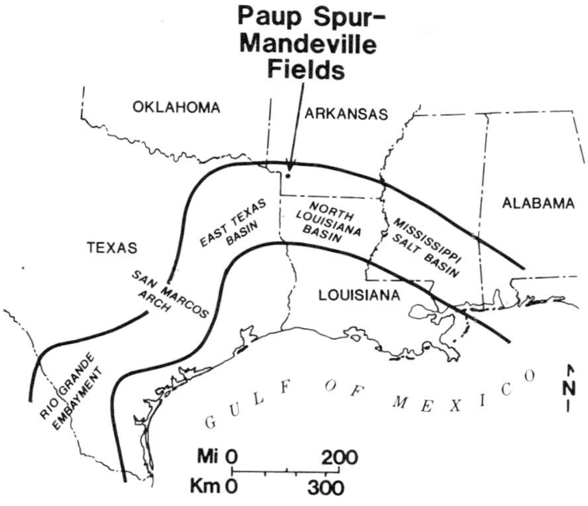

Figure 1. The regional distribution of the Smackover Formation along the northern and western margins of the Gulf of Mexico basin. The PS-M Fields are situated near the updip limit of Smackover deposition. Down dip, the Smackover Formation interfingers with basinal Bossier Formation sediments. (Modified from Rosenkrans and Marr, 1967.)

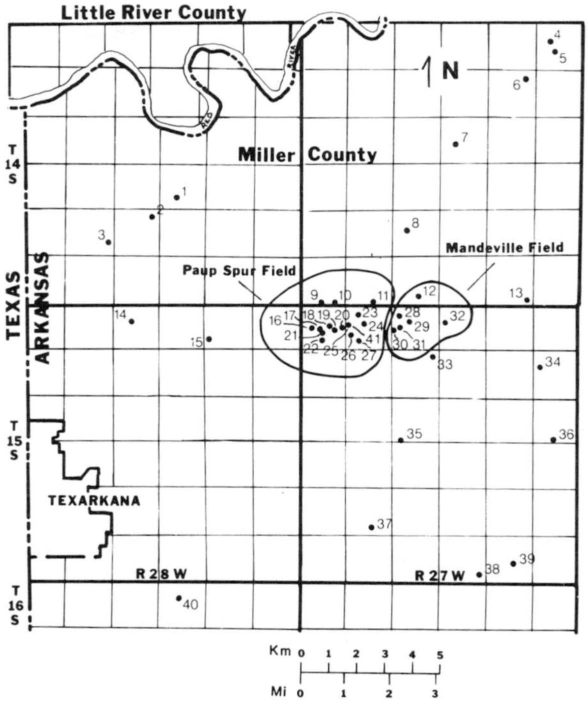

Figure 2. Location of the Paup Spur-Mandeville Fields and wells used in this study.

constructed from six wells that penetrate the entire Smackover Formation to show variations in thickness of the Smackover within the PS-M Fields. The top of the Smackover Formation is used as a datum and the thickness variations are assumed to reflect basement topography at the end of Smackover deposition. The cross section shows that the formation is thickest in the vicinity of well No. 17 to the east and in well No. 23, which is near the western edge of the Paup Spur Field. The Smackover Formation thins where Werner sediments are absent - an indication of the influence of basement structures on subsequent depositional trends.

Another cross section through the same wells (Fig. 6) has sea level as a datum. In this illustration, the configuration of the Werner and Eagle Mills Formations show that the area penetrated by wells No. 41, No. 21 and No. 17 in the Paup Spur Field is tilted toward the west. It also shows a downward movement of well No. 23 relative to well No. 41 to the west and to well No. 29 to the east, and defines the downthrown side of a fault which possibly occurs between wells No. 41 and No. 23. Also, well No. 16 is downthrown relative to this block. Smackover and Buckner sediments thicken in the downthrown wells indicating some movement was contemporaneous with deposition. However, vertical displacement is also apparent from the top of the Smackover and Buckner Formations showing that the topographic highs present during their deposition were also enhanced at some time after the deposition of the Buckner Formation.

A structure-contour map of the top of the Smackover Formation (Fig. 7a) shows two structural highs - one coinciding with the location of the Paup Spur Field, and the other with the Mandeville Field. A depression separates the two fields. An isopach map of the Buckner Formation has also been constructed for the PS-M Fields (Fig. 7b). In this area, thinning occurs along a linear east-west ridge roughly coinciding with present-day topographic highs in the field. This thinning serves as additional evidence that

Table 1. List of Wells In the PS-M Fields and the Surrounding Area That Were Used for This study

Numbers in the left column refer to well locations in Figure 2.

WELL NO.	WELL NAME AND LOCATION
1	Wadley No. 1 Shuffield; Sec. 22, T14S, R28W
2	Damson No. 1 Starks; Sec. 28, T14S, R28W
3	Anadarko No. 1 Jones; Sec. 29, T14S, R28W
4	Wadley No. 1 Smith; Sec. 1, T14S, R27W
5	Wadley No. 1 Stuart; Sec. 1, T14S, R27W
6	Wadley No. 2 Stuart; Sec. 11, T14S, R27W
7	Sneed No. 1 Ross: Sec. 15, T14S, R27W
8	Wadley No. 1 Shoptaw; Sec. 28, T14S, R27W
9	Mobley No. 1 Batchelor; Sec. 31, T14S, R27W
10	Weiser-Brown No. 1 Brannum; Sec. 31, T14S, R27W
11	Rustex No. 1 Blackshare; Sec. 32, T14S, R27W
12	Plains No. 1 Stricklin; Sec. 33, T14S, R27W
13	Horizon No. 1 Rhodes; Sec. 35, T14S, R27W
14	Wadley No. 1 Mann; Sec. 4, T15S, R28W
15	Hunt Energy No. 1 Brown; Sec. 2, T15S, R28W
16	Barret No. 1 Walls; Sec. 6, T15S, R27W
17	Barret No. 4 Kuhn; Sec. 6, T15S, R27W
18	Barret No. 2 Kuhn; Sec. 6, T15S, R27W
19	Barret No. 3 Kuhn; Sec. 6, T15S, R27W
20	Barret No. 1 Kuhn; Sec. 6, T15S, R27W
21	Barret No. 1 Gibson; Sec. 6, T15S, R27W
22	Grande No. 1 Yeager; Sec. 6 T15S, R27W
23	Johnson No. 1 Stricklin; Sec. 5, T15S, R27W
24	Scoggins No. 1 Gibson; Sec. 5, T15S, R27W
25	Barret No. 1 Stricklin; Sec. 5, T15S, 27W
26	Barret No. 1 Martindale; Sec. 5, T15S, R27W
27	Scoggins No. 1 Stricklin; Sec. 5, T15S, R27W
28	Scoggins No. 1 Collins; Sec. 4, T15S, R27W
29	Barnsdall No. 1 Grace; Sec. 4, T15S, R27W
30	Crystal No. 1 Collins; Sec. 4, T15S, R27W
31	Fortune No. 1 Collins; Sec. 4, T15S, R27W
32	Rotary No. 1 Turner; Sec. 3, T15S, R27W
33	Waldo No. 1 Grace; Sec. 9, T15S, R27W
34	Petrofina No. 1 Boyce; Sec. 12, T15S, R27W
35	Lyons No. 1 Purifoy; Sec. 16, T15S, R27W
36	Wadley No. 1 Choupique; Sec. 13, T15S, R27W
37	California No. 1 Vanderbilt; Sec. 29, T15S, R27W
38	Petrofina No. 1 Morrow; Sec. 34, T15S, R27W
39	Vanderlaan No. 1 Neustadt; Sec. 35, T15S, R27W
40	Carter No. 1 Orr; Sec. 3, T16S, R28W
41	Barret No. A-1 Kuhn; Sec. 6, T15S, R27W

these areas were topographically high during deposition of the Smackover Formation. Dickinson (1968) reported that the thickness and distribution of the Buckner Formation is related to local and regional structural features. These sediments are thickest to the north and to the far west, possibly indicating a source direction for the detrital fraction of Buckner sediments and/or the influence of a Smackover topographic high on water circulation behind it, thus accounting for thicker evaporitic accumulation. Continuing differential structural movement during the deposition of the Buckner Formation may also account for local, anomalous variations in sediment thickness.

An isopach map of the Smackover Formation (modified from Arper, 1953) (Fig. 8) shows that the Smackover sediments thicken in a northwest-southeast direction. This thickening of Smackover sediments possibly delineates a shallow linear embayment that was present on the shelf during the deposition of the Smackover Formation. The embayment was bounded on the north by the Ouachita structural rim. To the southwest, circulation was

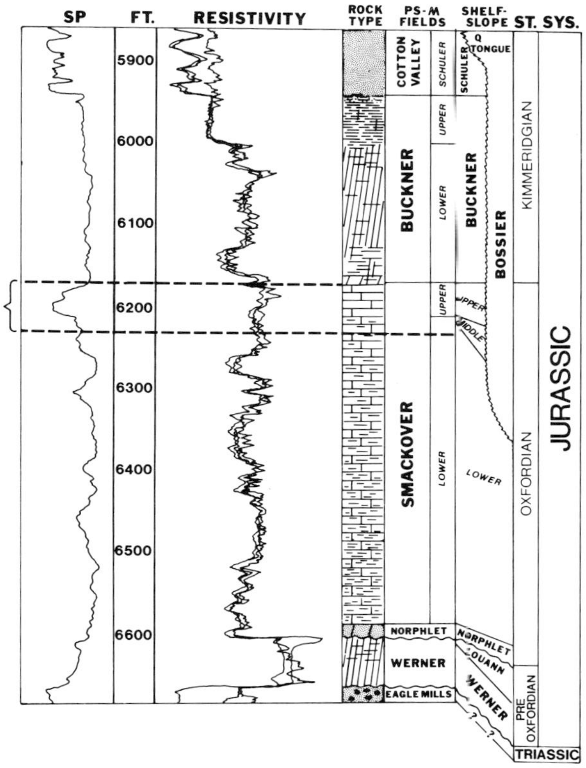

Figure 3. The rock types represented by a typical SP and resistivity log response [Barret No. 4 Kuhn (No. 17) well] and the corresponding stratigraphic sequence. The bracketed segment shows the interval discussed in this paper. To the right is the equivalent stratigraphic sequence found basinward of the PS-M Fields, and the European equivalent stages of the Late Jurassic system.

modified by a ridge (horst block) in the vicinity of the Texarkana-Eyeleu-Flowers Acres Fields (Lokke and Brent, 1966) and water circulation into the embayment was from the southeast. Wind direction and local topographic relief most likely modified water circulation on a finer scale in the Paup Spur-Mandeville area.

CARBONATE FACIES AND ASSOCIATED SEDIMENTARY STRUCTURES

Core examination identified the following five sedimentary facies. The carbonate facies names are based on the carbonate rock classification developed by Folk (1959). Horizontal and vertical facies relationships are illustrated in cross-sections B - B' and C - C' (Figs. 9 and 10).

Laminated Pelsparite with Shaly Partings

The features of this unit described below are indicative of deposition in a shallow-water subtidal, near-shore environment. It is composed of gray to dark-gray colored organic-rich carbonate and silty shale (Fig. 11a) and is actually the upper portion of the lower Smackover Formation. Its texture is

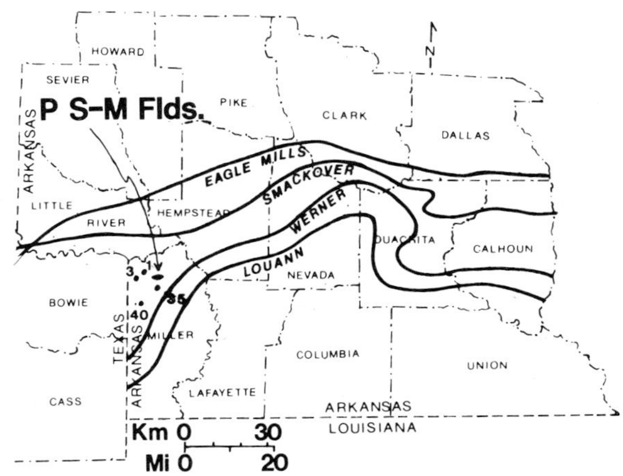

Figure 4. Northern limits of Late Jurassic and older sedimentary units of the northern Gulf of Mexico basin.

sorted to unsorted and locally is a poorly-washed pelsparite. Pellets are the predominant grain type, but fine detrital fossil fragments are common and crustacean pellets are common to abundant (Fig. 12a). Thin horizontal and wavy laminations are the dominant structures within this unit. Thin silty and shaly partings are numerous - some only one to three mm apart and one to three mm thick. Bioturbation is common in the top few feet of the unit. Lower in the unit, individual burrows or burrowed laminae occur. Micrite is patchy and normally occurs as burrow filling and fused pellets. Rounded to digitate oncolites are present. Stylolites are usually

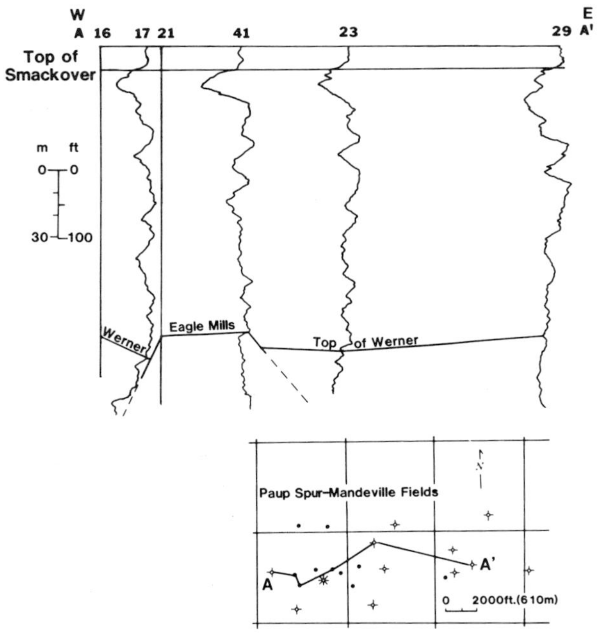

Figure 5. Cross section through wells in the PS-M Fields that penetrate the base of the Smackover Formation. The top of the Smackover is used as a datum. The names of these wells are given in Table 1. The SP response is used for correlation and is shown except for wells No. 16 and No. 21.

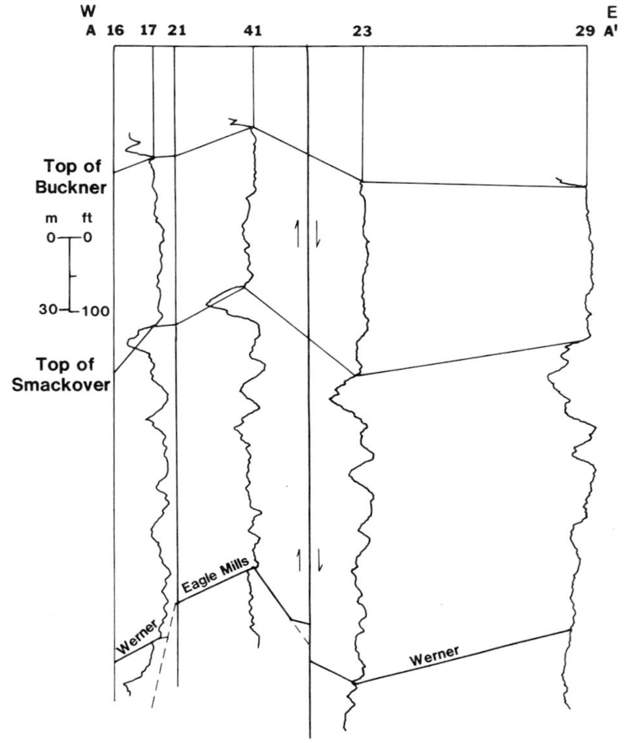

Figure 6. Cross section through the same wells as shown in Figure 6. Sea level is used as a datum.

common near the oncolite zone. Also, lower in the unit, there is an increase in shale content, and high-angle bedding and slump structures are more common, their convolutions highlighted by thin shale drapes. A reducing environment is indicated by ubiquitous specks of pyrite. Sediments at the base of the section are finer-grained and some structures may have been the result of downslope gravity currents. Up section, wave-generated currents show increasing influence, indicating decreasing water depth (Budd and Loucks, 1981).

Cross-bedded and Burrowed Pelsparite - Pelmicrite

The unit is extensively burrowed near the center of the field and interfingers laterally with intertidal cross beds and horizontal laminations (Figs. 11b and 11c). Lateral facies relationships are illustrated in cross-section B-B' (Fig. 9). Allochems (pellets, coated grains, and rounded intraclasts) show greater variety and admixture outside the burrowed zone (Fig. 12b). In the burrowed zone, pellets are the most abundant allochem but are generally represented by fine (<Ø.2 mm) moldic pore space (Fig. 12c). Irregular concentrations of allochems, silt, hydrocarbon residue, and micrite create a distinctive mottled appearance. The hydrocarbon residue and silty clay are responsible for the mottled medium-brown to cream coloration. The burrowed zone is probably indicative of slightly deeper water than the laterally adjacent cross-bedded portions. Ball (1967) noted that at Schooner Cays a slight increase in water depth stopped the daily reworking of tidal-bar sands by tidal currents.

Laminated to Burrowed Silty Pelsparite with Shaly Partings

This unit is characterized by cycles that are indicative of decreasing wave energy. Each cycle is cross-bedded or finely laminated and rippled at the base, and has more bioturbation and less silt and shale content towards the top (Fig. 11d). The cycles range from 15 cm to 1 m in thickness.

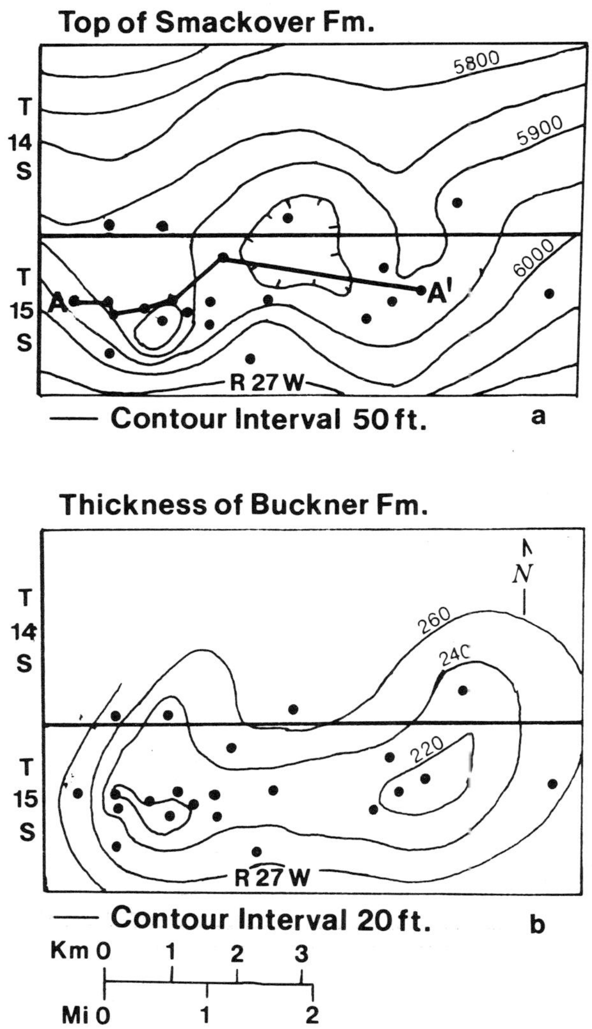

Figure 7. Structural contours of the top of the Smackover Formation (a) in the PS-M Fields compared with an isopach map of the Buckner Formation (b) for the same area. Cross-section A - A' is located on map a for comparison with Figures 5 and 6.

This unit is best developed in the Barret No. 1 Stricklin (No. 25) well (Fig. 9), which may have been the site of a localized depression where intermittent or seasonal currents reworked the sediments.

Algal Biolithite and Pelsparite

The most characteristic features of this unit are sedimentary structures attributed to blue-green algal associations. Stromatolitic structures, which form algal heads with irregular finger-like protrusions, are best developed in wells to the southwest - in the Grande No. 1 Yeager (No. 22) and in the Barret No. 3 Kuhn (No. 19) wells. The largest structure present in core samples is only 10 cm in length (Fig. 11e). Their small size and slender morphology may be a result of vertical accretion in a lower intertidal environment. The surrounding matrix is generally a finely-bedded pelsparite. In other areas (Johnson No. 1 Stricklin [No. 23] and Barret No. 1 Walls [No. 16] wells) crenulated laminae occur as mats (Fig. 11f). Oncolites

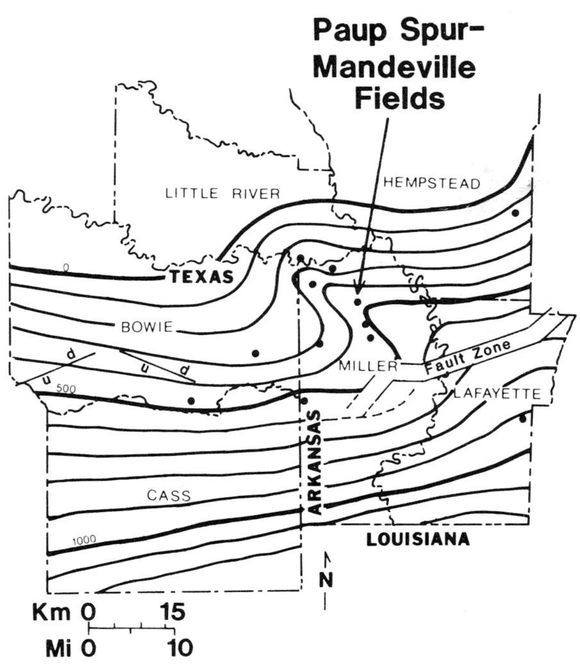

Figure 8. Isopach map of the Smackover Formation. (Modified from Arper, 1953.)

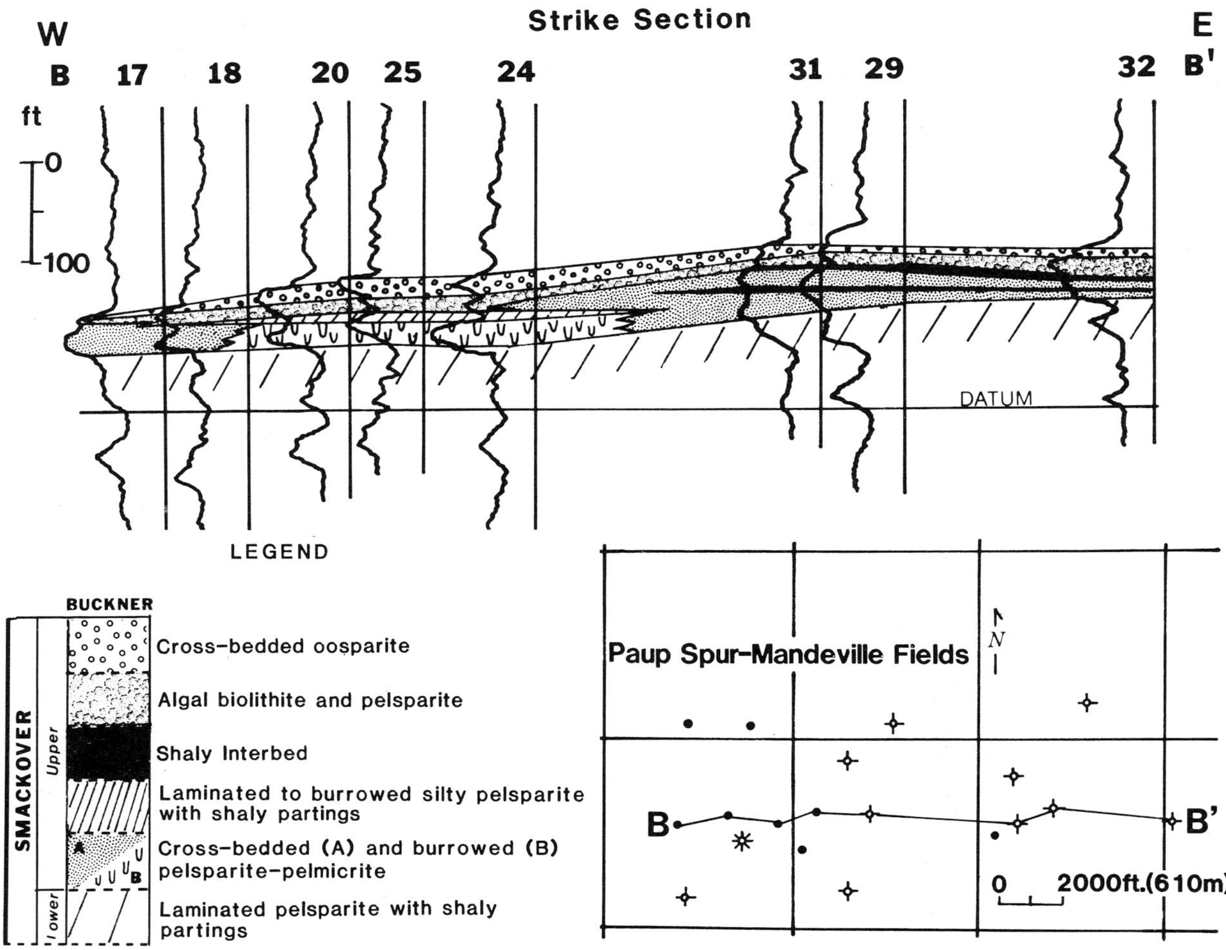

Figure 9. Strike section B - B' showing the lateral facies relationships and thinning of the upper Smackover to the east and west. SP log traces are used for correlation.

are present across the field in this facies and are another indication of a highly mobile substrate (Logan *et al.*, 1964) (Fig. 12d). Oncolites are usually associated with finely laminated, burrowed and structureless carbonate sediment (Fig. 11g). The accretion of oncolite laminae probably occurred in an environment of moderate and variable wave energy, where agitation was weak enough for algal laminae to colonize detrital carbonate fragments, but strong enough to roll the large fragments around during growth, producing the concentric algal form.

Allochems are dominantly peloids, fecal pellets, algal clasts (Fig. 13a), and intraclasts (Fig. 13b). A variety of fossils are present but none are abundant. Fossils include dasycladacean fragments (Fig. 13c) and other green algae, high-spired gastropods, ostracods, foraminifers, and pelecypod and echinoid fragments. The blue-green algal structures in the PS-M samples are indicative of shallow-water subtidal to intertidal environments.

Cross-bedded Oosparite

Sedimentary structures in this unit are representative of a high-energy, intertidal oolitic sand body such as that described by

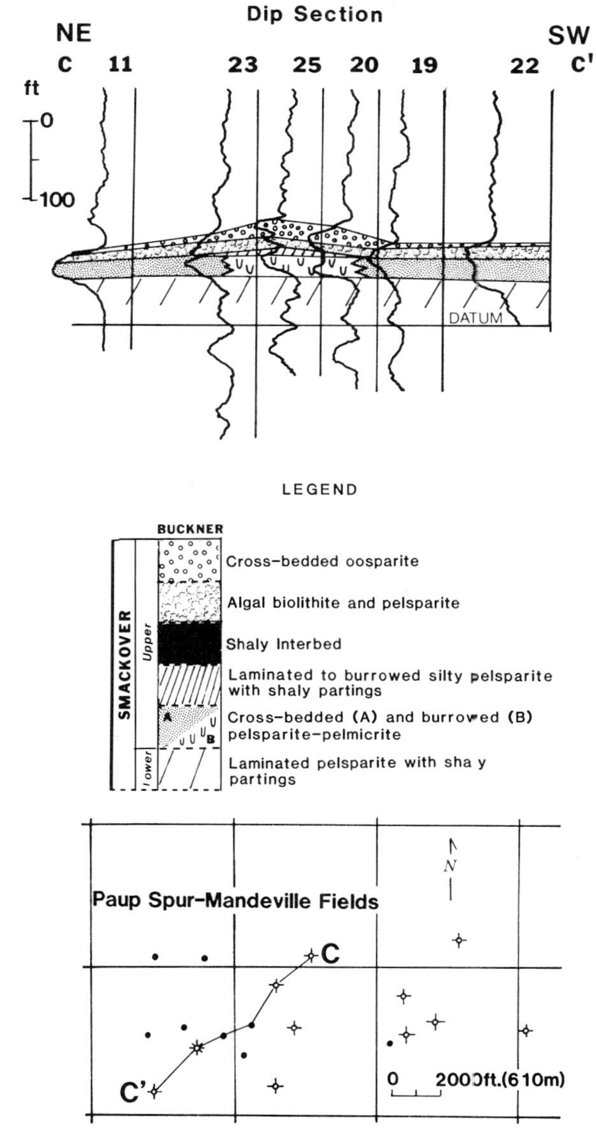

Figure 10. Dip section C - C' illustrating the upward shoaling sequence of facies.

Ball (1967) at Cat Cay. Bimodal cross-beds and horizontal laminations are characteristic (Fig. 11h) of this unit. Well-sorted (0.1-1 mm) ooids are the major allochemical constituents, but in places are mixed and interbedded wih rounded intraclasts and superficial ooids, pisoids, fossil fragments, and pellets (Fig. 13d). In the PS-M Fields, much of the originally oolitic grainstone is now represented by round to oval voids. Some voids contain remnant nuclei and preserved concentric laminae of finely crystalline calcite. The preserved allochem nuclei are generally carbonate pellets and cryptocrystalline grains (Fig. 14a).

Sedimentary structures are less distinct and the unit is thinner in the Barret No. 2 Kuhn (No. 18) well, indicating more protected conditions on the leeward side of the ooid shoal. The unit is absent in the Barret No. 1 Walls (No. 16) and Grande No. 1 Yeager (No. 22) wells. In these wells, the algal biolithite unit is dominant. In the Fortune No. 1 Collins (No. 31) well, the sediments are predominantly horizontally laminated, and well-sorted ooids grade into more mixed allochemical constituents and superficially coated grains (mobile grain flats) (Fig. 14b). Water currents in this area probably had a weaker bi-directional component but were still strong enough to mask the reworking of the sediments by burrowing organisms and to prevent the deposition of carbonate mud.

Analysis of Carbonate Facies

Smackover facies of the PS-M area are mainly representative of deposition in shallow-water subtidal to intertidal marine environments. The general vertical trend of sedimentary structures was the result of decreasing water depth with time. The upper portion of the lower Smackover represents uniform, low-energy subtidal conditions. Up section, these sediments grade into current-sorted coarser clastic debris that is characteristic of the littoral zone. Some of these sediments were reworked by burrowing organisms in deeper water or protected areas. Further up section, a variety of algal forms becomes dominant with increasing availability of sun light, and laterally-equivalent sedimentary facies become more variable and interdependent. The sediments that form the matrix surrounding the algal structures reflect moderate wave-energy conditions. This lowered wave energy could possibly be a result of restricted water circulation caused by the development of ooid shoals to the south in the lower shelf portion of the Smackover. Another possibility is that the

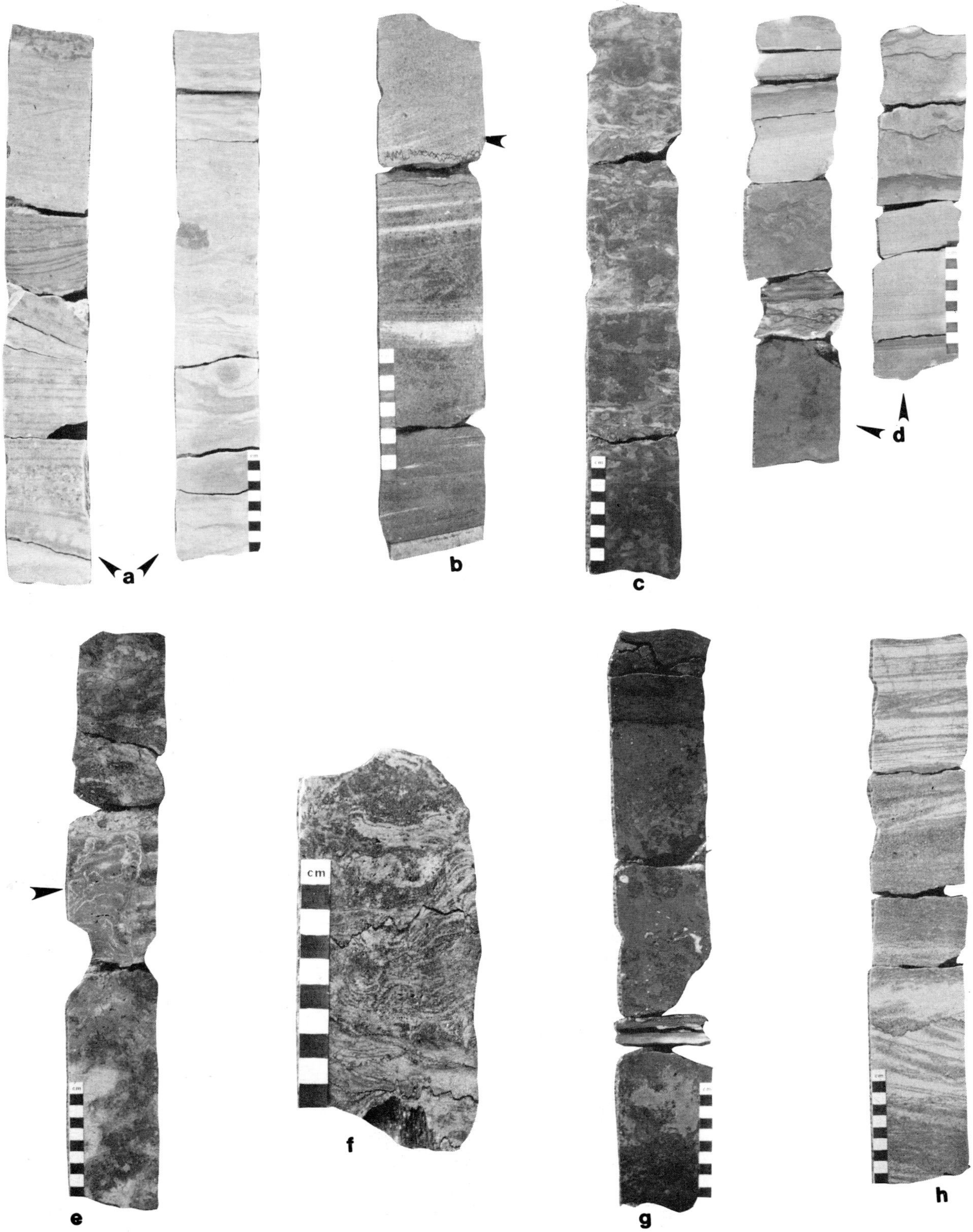
a
b
c
d
e
f
g
h
cm

algal structures stablize their own environment (Gebelein, 1969) so that sedimentary structures do not actually reflect current energy. High-energy conditions again prevail locally during the development of an ooid shoal over structural highs.

The paleotopographic setting for the deposition of the Smackover Formation in the PS-M area has already been interpreted from cross-correlations of geophysical logs (Section 3). During Smackover deposition, a shallow-water, trough-like embayment was present on the northwestern Smackover shelf. The PS-M area is situated to the southeast, near the mouth of this embayment (Fig. 8). The geometry of the Smackover carbonate sand body present in the PS-M area is the result of two main variables - bottom topography and direction of wave impingement on the topographic features present. The topographic highs are aligned in an east-west direction following the general Ouachita structural trend. However, lateral facies relationships show that the highest-energy facies (cross-bedded oosparite) is aligned in a southwest-northeast direction perpendicular to the long axis of the Smackover shelf embayment and is partly wrapped around topographic highs on the seaward edge (Fig. 15). This trend further supports the suggestion that at that time tidal currents were from the southeast. Wave energy was concentrated along the southeast sides of topographic highs and possibly was channeled between them.

DIAGENESIS

Figure 16 summarizes the diagenetic products observed by petrographic analysis. Evidence of four and possibly five major phases of diagenesis occurred during the postdepositional history of the upper Smackover in the PS-M Fields. Evidence of sea-floor diagenesis is largely micritization of allochems and local minor precipitation of evaporites. The presence of some evaporite minerals, without extensive sabkha development within the upper Smackover sediments of this study, suggests increased salinity of Jurassic marine waters or localized and intermittent hypersaline conditions in exposed areas.

The predominant diagenetic imprint was formed by solutions of meteoric water at near-surface conditions. Metastable aragonite allochems were dissolved by these solutions, and the dissolution was accompanied by the precipitation of finely crystalline (20-

Figure 11. Cores from the Smackover Formation of the Paup Spur-Mandeville Fields, Miller County, Arkansas.

a. Laminated pelsparite with shaly partings facies. Barret No. 1 Stricklin (No. 25) well, 6,239-6,241 feet (1,902.1-1,902.7 m).

b. Laminated portion of the cross-bedded and burrowed pelsparite-pelmicrite facies. Note faint cross-bedding at the top. Barret No. 2 Kuhn (No. 18) well, 6,172-6,173 feet (1,881.7-1,882.0 m).

c. Heavily burrowed portion of the cross-bedded and burrowed pelsparite-pelmicrite facies. Barret No. 1 Stricklin (No. 25) well, 6,219-6,220 feet (1,896.0-1,896.3 m).

d. Laminated to burrowed silty pelsparite with shaly partings facies. Barret No. 1 Stricklin (No. 25) well, 6,200-6,202 feet (1,890.2-1,890.9 m).

e. Algal biolithite and pelsparite facies. Note the large layered algal head in the center of the core. Grande No. 1 Yeager (No. 22) well, 6,314-6,316 feet (1,925.0-1,925.6 m).

f. Algal laminae with fenestral and small vuggy pore space. Johnson No. 1 Stricklin (No. 23) well, 6,226 feet (1,868 m).

g. Oncolite-bearing portion of the algal biolithite and pelsparite facies. Oncolites are interbedded with clay drapes at the top. Large oncolite in the center shows a digitate growth form. Barret No. 1 Stricklin (No. 25) well, 6,190-6,191 feet (1,887.2-1,887.5 m).

h. Cross-bedded oosparite facies. Barret No. 1 Stricklin (No. 25) well, 6,173-6,174 feet (1,882.0-1,882.3 m).

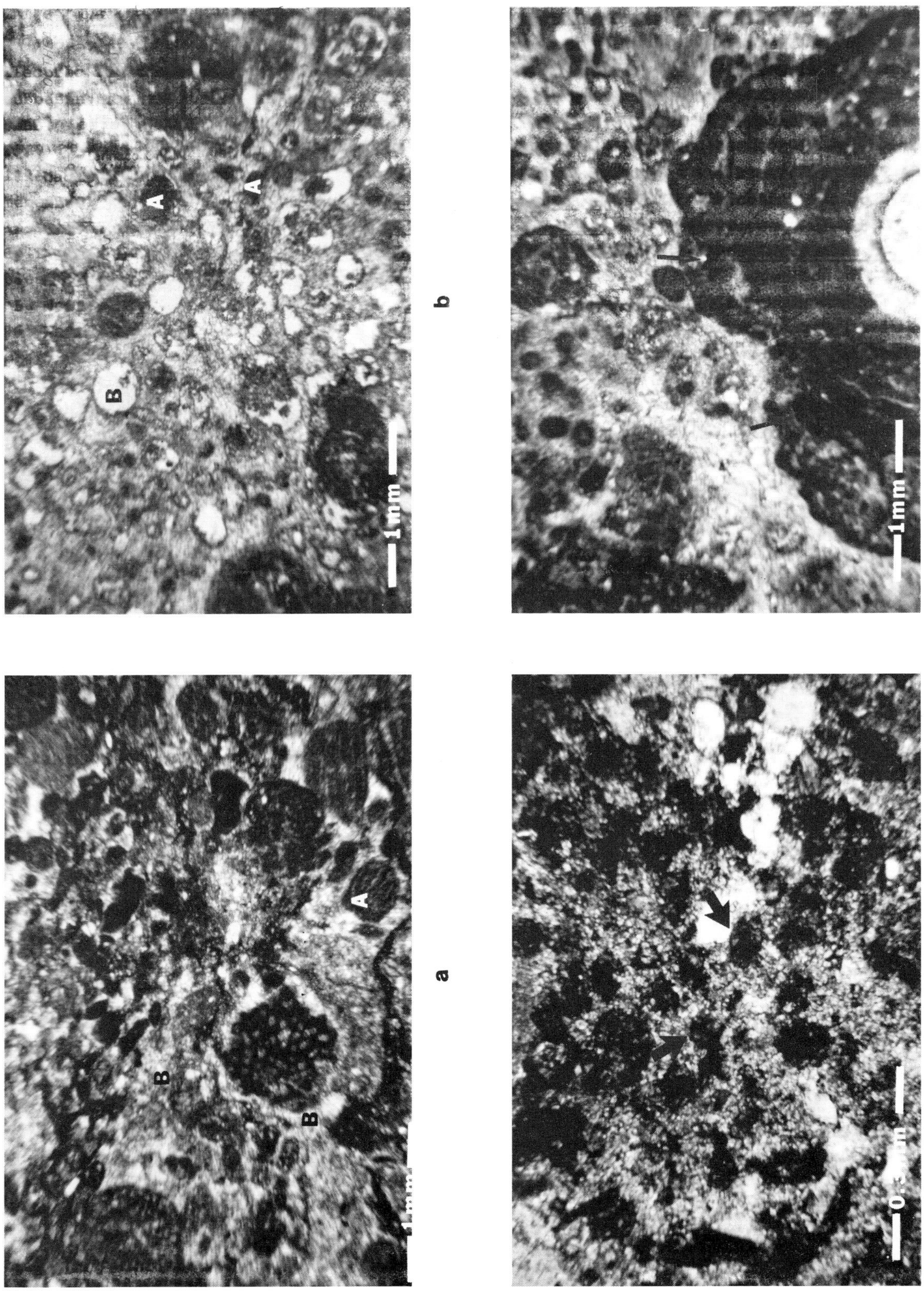

Figure 12

40 µm) calcite cement in nearby interparticle pore space (Figs. 14c and 14d). The bulk of the porosity found in the PS-M Fields is the product of this dissolution phase. Porosity loosely follows facies because each facies is generally represented by characteristic grain types and porosity preservation is determined by the solubility of these grains.

Petrographic recognition of meteoric-cement textures is essential to enable identification of the presence of a fresh-water phase - particularly because Smackover and underlying and overlying sediments have the characteristics of sediments deposited in marine to hypersaline environments, and present-day formation fluids of these sediments have high ionic concentrations. Moore and Druckman (1981) believe the development of this meteoric-water system is regional. More data from wells off structural highs would help clarify whether meteoric water tables developed locally in the PS-M Fields area, such as occur today in Bermuda and the Bahamas, or regionally, as in south Florida.

In one core (Barret No. 2 Kuhn [No. 18] well), thin-section samples showed an increase in dolomite crystals replacing allochems with increasing depth within the upper Smackover. Additional study of dolomite composition and distribution would help determine whether a mixing-zone of fresh and sea water was established in the PS-M area, or if the dolomite is a later diagenetic feature.

Upon further burial, several other minor phases of cementation and dissolution occurred in response to changing overburden pressure, temperature, pore-solution chemistry, and hydrocarbon migration. Post-burial diagenetic features include minor brittle collapse of grains and cement, and stylolitization. Fractures provide permeable pathways for fluid movement and are indicators of subtle tectonic movement after lithification of the sediments. The stylolites and microstylolite swarms are the largest visible means of porosity destruction. Coarsely crystalline calcite cement is not abundant. The most significant amounts are in association with large fenestral cavities, fractures, and stylolites. Anhydrite and celestite are sparce and occur as pore-filling and calcite replacement. Baroque dolomite occurs chiefly as mold and cavity fill and it sometimes partly replaces allochems.

Non-selective calcite dissolution has been recognized mainly where pore spaces are now lined with a hydrocarbon residue. For this reason, it has been assumed that this is related to hydrocarbon maturation. However, not all hydrocarbon-bearing porosity appears to have been enhanced by late-stage dissolution. Non-selective, late-stage dissolution is more common at bedding interfaces, where permeability differences related to grain size, sorting, and cementation were already present. In samples from near the top of the formation, dissolution appears also to have affected some sulfate minerals.

Source rocks for the hydrocarbon have been thought to be the lime muds of the lower Smackover, or the fine-grained basinward equivalent facies of the Smackover (Moore and

Figure 12. Thin-section photomicrographs of Smackover Formation lithologies.

a. Poorly-washed pelsparite, common in the upper part of the lower Smackover. Note crustacean fecal pellets (A) and finely crystalline calcite cement (B). Barret No. 1 Stricklin (No. 25) well, 6,245 feet (1,903 m).

b. Poorly-sorted allochems (A) and allochem molds (B) cemented by finely cyrstalline calcite coarsening towards pore centers. Sample from the laminated part of the cross-bedded and burrowed pelsparite-pelmicrite. Barret No. 2 Kuhn (No. 18) well, 6,168 feet (1,880 m).

c. Allochems now represented by molds of peloid-size (<0.2 mm) grains (arrow). Burrowed part of the cross-bedded and burrowed pelsparite-pelmicrite. Barret No. 1 Stricklin (No. 25) well, 6,223 feet (1,897 m), crossed nicols.

d. Portion of an oncolite. Note the irregularity of laminae accretion and the concentration of pyrite at some laminae interfaces (arrow). Gastropod mold nucleus is now filled by coarse sparry calcite. Grande No. 1 Yeager (No. 22) well, 6,339 feet (1,932 m).

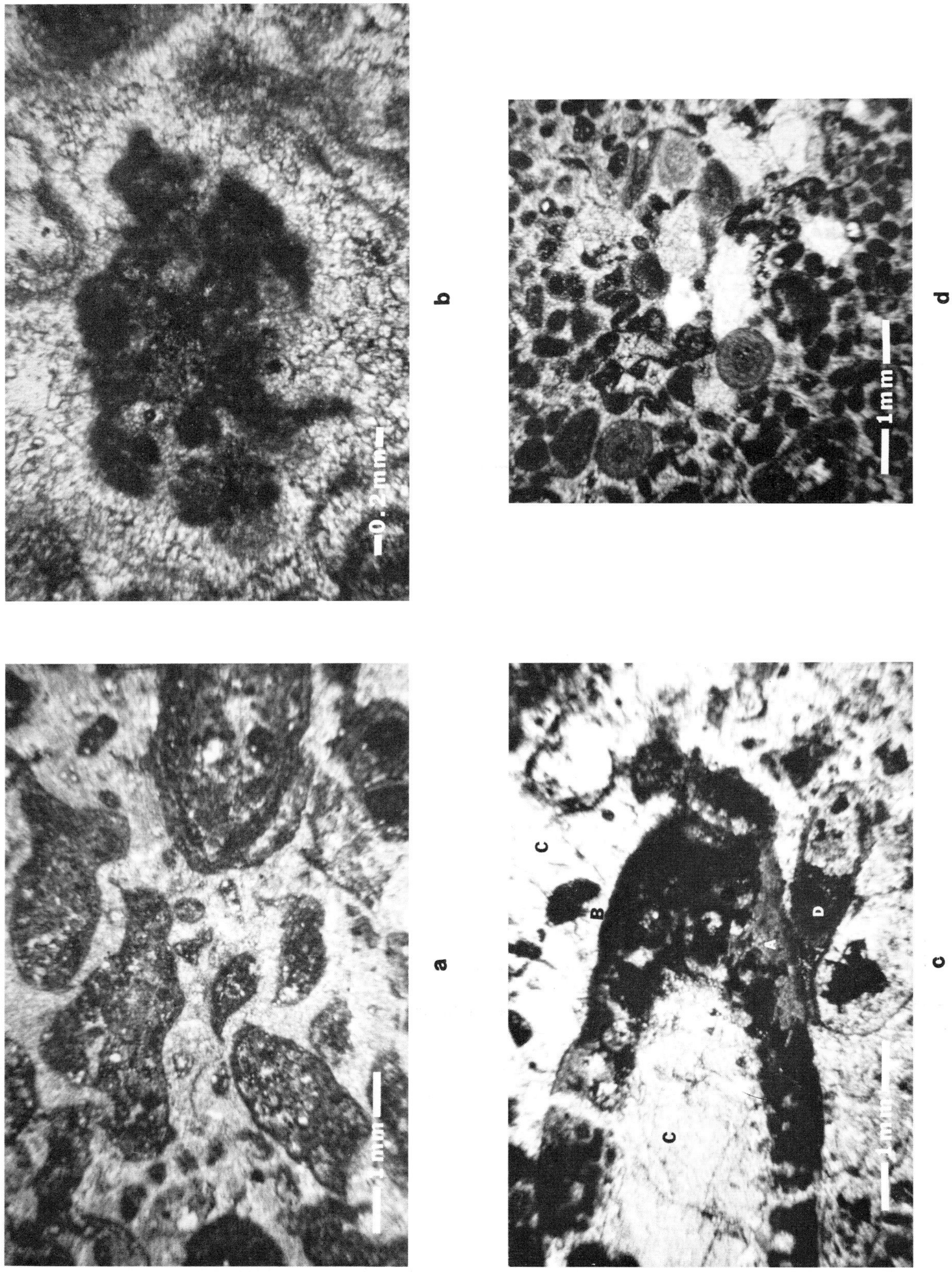

Figure 13

Druckman, 1981). In this updip area of Smackover deposition where permeability was greatly reduced during the early stages of burial, it is probable that hydrocarbons did not migrate far from their source. Therefore, the lower Smackover is favored as a hydrocarbon source rock. Also, faults associated with the development of PS-M Fields structures may have acted as conduits for hydrocarbon migration.

CONCLUSIONS

(1) The upper Smackover Formation of the PS-M Fields represents subtidal to intertidal carbonate-sand deposition under the influence of strong wave agitation. Blue-green algae played an important role in carbonate grain composition and in the modification of depositional environments.

(2) The upper Smackover Formation of the PS-M Fields was deposited north of the updip limit of Louann Salt deposition. The fields coincide with structural highs. There is reasonable evidence that these structures existed as low-relief features during Smackover deposition and had some influence on the geometry of upper Smackover sediments. Post-burial structural movement is responsible for the present configuration of the PS-M Fields.

(3) The bulk of the lithification of the upper Smackover sediments and secondary porosity development occurred at near-surface conditions by solutions of meteoric water. Porosity distribution mainly coincides with facies and grain types because most of the original interpartical porosity is now occluded by cements, and secondary porosity is determined by the original mineralogy and solubility of the allochemical constituents. Other minor phases of cementation and dissolution occurred later in the burial history of the formation.

ACKNOWLEDGMENTS

This investigation was originally undertaken in partial fulfillment of the requirements for the degree of Master of Arts from The University of Texas at Austin. Special thanks to R. L. Folk and D. G. Bebout for their help and guidance, and to R. O. Kehle and R. Sorenson for stimulating interest in studying the Smackover.

Special thanks to Bill Poynter of Texarkana, Arkansas, and to Ray Sorenson of Anadarko Production Company, Oklahoma City, Oklahoma for supplying core samples and well material and for helpful discussions; and to the Department of Geosciences, University of Arizona, Tucson, Arizona, for the use of their laboratory and photo equipment.

Figure 13. Thin-section photomicrographs of Smackover Formation lithologies.

a. Algal clasts. Clasts are bound by one or more algal laminae and have no well-defined nuclei. The irregularity of their shapes is sometimes further enhanced by incipient stylolitization. Barret No. 1 Stricklin (No. 25) well, 6,187 feet (1,886 m).

b. Intraclast. This grain was formed by the micritization of a former sedimentary grain and is similar to the "lumps" described by Illing (1954). Finely crystalline, equant calcite cement fills the interparticle pore space. Barret No. 2 Kuhn (No. 18) well, 6,150 feet (1,875 m).

c. Dasycladacean fragment (A) rimmed with finely crystalline calcite followed by scattered blades of calcite (B). Poikilotopic celestite (C) fills subsequent fractures and surrounding pore space. Remaining pores and some post-burial porosity is lined with hydrocarbon residue (D). Grande No. 1 Yeager (No. 22) well, 6,315 feet (1,925 m), crossed nicols.

d. Fine ooids interbedded with larger pisoids, superficial ooids, and fossil fragments. Most allochems are now represented by moldic pore space. Anhydrite fills molds along laminae made up of coarser allochems defining former pathways of increased permeability. Barret No. 1 Stricklin (No. 25) well, 6,172 feet (1,881 m), crossed nicols.

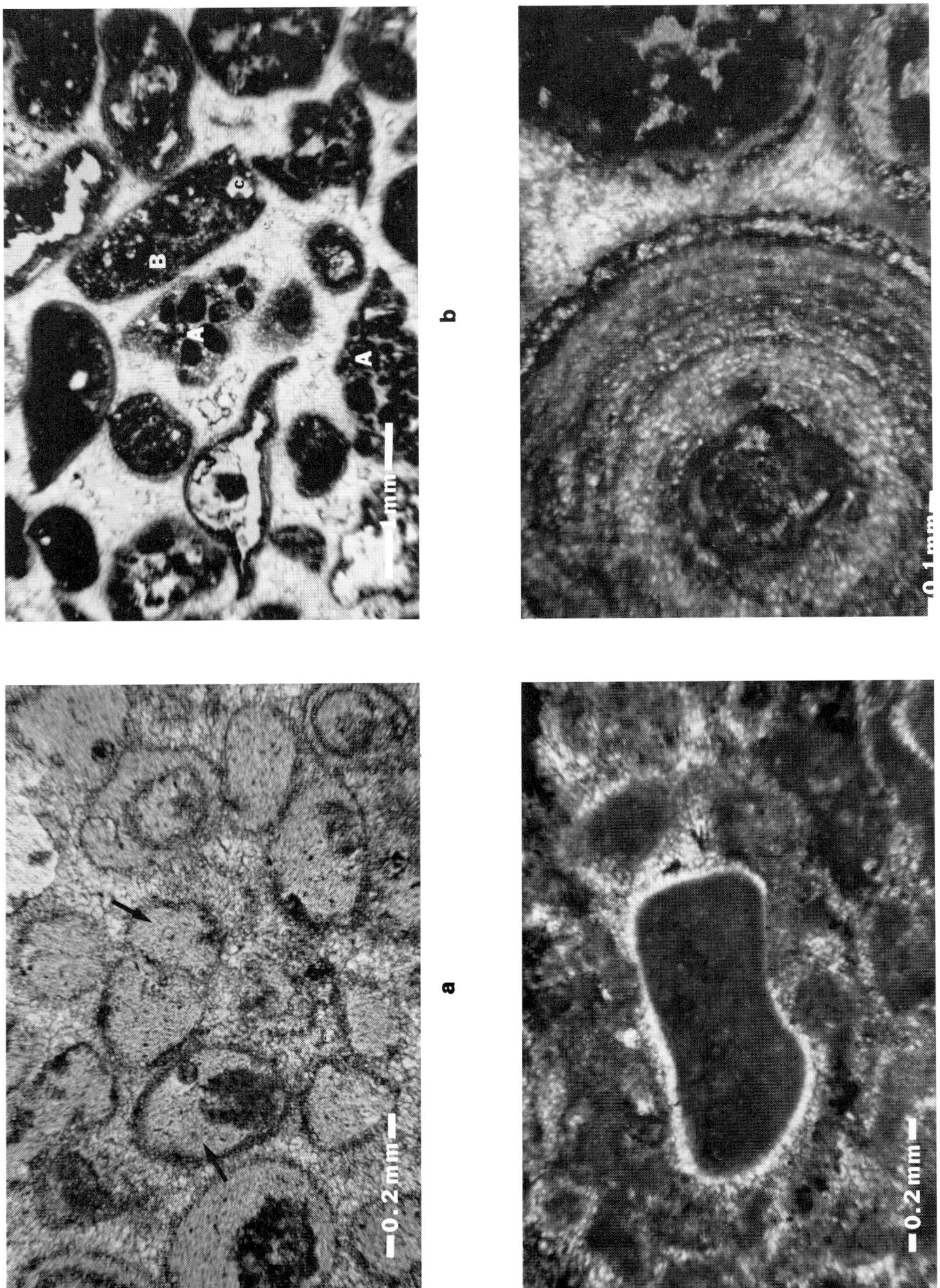

Figure 14

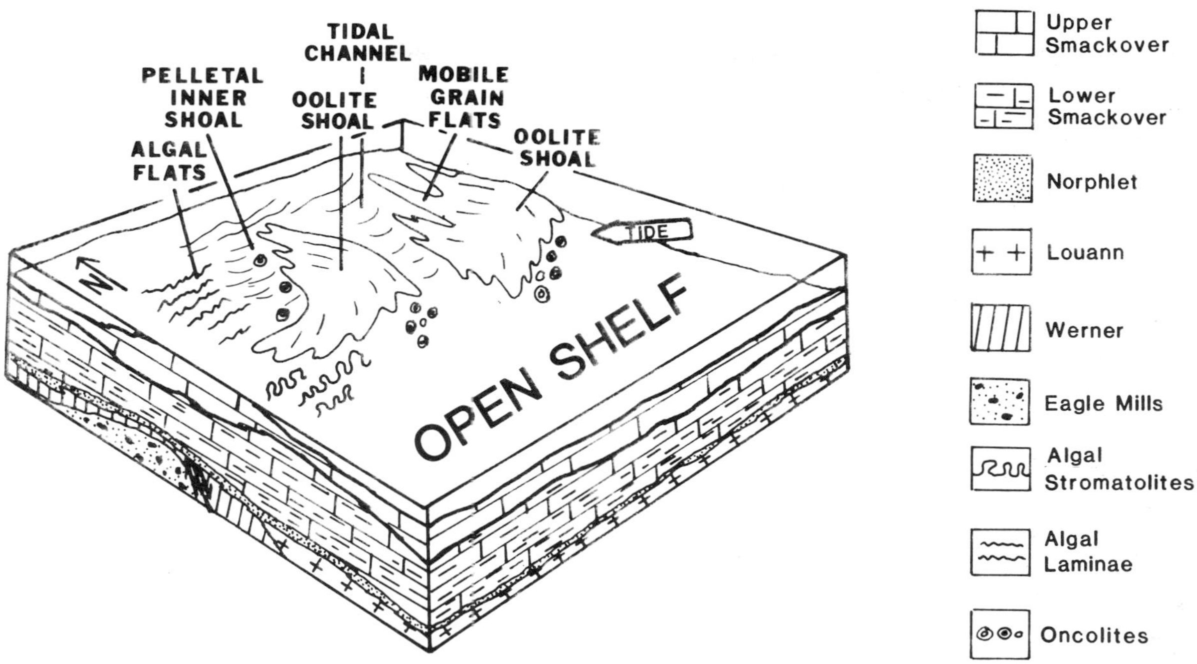

Figure 15. The depositional setting for the Smackover Formation of the PS-M Fields showing the formation of ooid shoals over pre-Smackover structural highs. Maximum wave energy was expended on the southeast flanks of these shoals. Peripheral environments show characteristics of lowered wave energy because of the protection offered by the oolitic barrier or by the presence of slightly deeper water. The pelletal inner shoal probably served as a source for most ooid nuclei. The algal flats and mobile grain flats developed in moderately protected areas where wind-driven tidal variations had more influence. More open circulation behind the eastern shoal may have inhibited stabilization of the mobile grain flats by algae and prevented the deposition of carbonate mud. Algal stromatolites grew where sediments were rapidly but intermittantly deposited.

Figure 14. Thin-section photomicrographs of Smackover Formation lithologies.

a. Oomoldic porosity (arrow). Some concentric laminae and pellet nuclei are preserved. Loosely packed ooids were cemented by a finely crystalline calcite mosaic. Anhydrite lath fills mold and replaces calcite cement in the top right corner. Barret No. 1 Stricklin (No. 25) well, 6,172 feet (1,881 m).

b. Superficial ooids and intraclasts. Some intraclasts (A) appear to be aggregates of peloids cemented by very finely crystalline equant calcite. The oblong grain (B) is a crustacean fecal pellet with a small replacement crystal of dolomite (C). Medium crystalline equant to bladed calcite crusts incompletely fill interparticle pore space. Fortune No. 1 Collins (No. 31) well, 6,198 feet (1,889 m).

c. Finely crystalline calcite which cement primary sediments throughout the PS-M Fields. That these cements were precipitated before significant compaction is illustrated by the fine calcite fringe (arrow) precipitated around a more coherent allochem. This would have occurred before the surrounding micrite was compacted enough to interfere with crystal growth. Barret No. 2 Kuhn (No. 18) well, 6,153 feet (1,876 m).

d. Dissolution of allochems which creates the bulk of the porosity found in the PS-M Fields. Here individual laminae and the nucleus of an ooid are selectively dissolved. Johnson No. 1 Stricklin (No. 23) well, 6,219 feet (1,896 m).

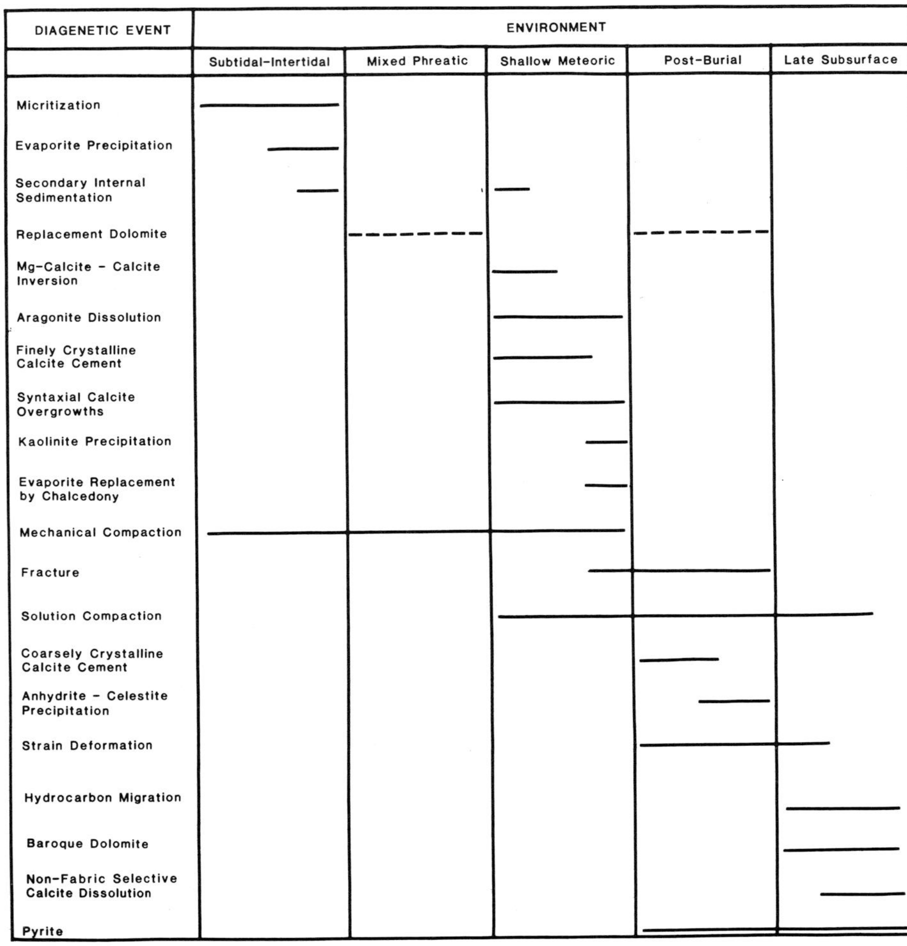

Figure 16. The diagenetic events identified in Smackover samples of the PS-M Fields, showing their relative timing as a function of burial environment. The sequence of events is based on textural associations and inferred environment of formation.

REFERENCES

Arper, W.B., 1953, The Smackover Formation in southern Arkansas and northern Louisiana, and adjacent areas of northeastern Texas and west-central Mississippi: Doctoral Dissertation, Univ. of Oklahoma, 74 p., appendices.

Ball, M.M., 1967, Carbonate sand bodies of Florida and the Bahamas: Jour. Sed. Petrology, v. 37, no. 2, p. 556-591.

Bishop, W.F., 1967, Age of pre-Smackover Formations, north Louisiana and south Arkansas: Am. Assoc. Petroleum Geologists Bull., v. 51, no. 2, p. 244-278.

__________, 1968, Petrology of upper Smackover limestone in North Haynesville field, Claiborne Parish, Louisiana: Am. Assoc. Petroleum Geologists Bull., v. 52, no. 1, p. 92-128.

Budd, D.A. and R.G. Loucks, 1981, Smackover and lower Buckner Formations, South Texas: Depositional systems on a Jurassic carbonate ramp: Univ. Texas at Austin, Bureau Econ. Geology, Rept. of Investigations no. 112, 38 p.

Dickinson, K.A., 1968, Upper Jurassic stratigraphy of some adjacent parts of Texas,

Louisiana, and Arkansas: U.S. Geological Survey Prof. Paper 594-E, p. E1-E25.
Folk, R.L., 1959, Practical petrographic classification of limestones: Am. Assoc. Petroleum Geologists Bull., v. 43, no. 1, p. 1-38.
____________, 1962, Spectral subdivision of limestone types: *in* Classification of Carbonate Rocks - a Symposium, Am. Assoc. Petroleum Geologists Memoir no. 1, p. 62-84.
Gebelein, C.D., 1969, Distribution, morphology, and accretion rate of Recent subtidal algal stromatolites, Bermuda: Jour. Sed. Petrology, v. 39, no. 1, p. 49-69.
Illing, L.V., 1954, Bahamian calcareous sands: Am. Assoc. Petroleum Geologists Bull., v. 38, no. 1, p. 1-95.
Logan, B.W., R. Rezak, and R.N. Ginsburg, 1964, Classification and environmental significance of algal stromatolites: Jour. Geology, v. 72, no. 1, p. 68-83.
Lokke, D.H. and M.D. Brent, 1966, Geologic conditions associated with nitrogen occurrences in Jurassic oil and gas fields, Northeast Texas and southwest Arkansas: *in* Natural Gases of North America, Am. Assoc. Petroleum Geologists Memoir No. 9, v. 1, p. 1074-1084.
Moore, C.H. and Y. Druckman, 1981, Burial diagenesis and porosity evolution, upper Jurassic Smackover, Arkansas and Louisiana: Am. Assoc. Petroleum Geologists Bull., v. 65, no. 4, p. 597-628.
Rosenkrans, R.R., and J.D. Marr, 1967, Modern seismic exploration of the Gulf Coast Smackover trend: Geophysics, v. 32, no. 2, p. 184-206.

WILLIAM C. WARD

Department of Earth Sciences
University of New Orleans
New Orleans, Louisiana 70148

ABSTRACT

A thick section of Zuloaga Limestone (Oxfordian) crops out in Astillero Canyon about 30 km south of Concepcion del Oro, Zacatecas, Mexico and about 50 km southeast of the type locality of the Zuloaga Formation. Here, the Zuloaga Formation is predominantly a carbonate mudstone-wackestone with minor amounts of grainstone. The section is divided into the following facies, in order of their volumetric importance: (1) skeletal mudstone and wackestone, (2) pellet mudstone and wackestone, (3) pellet packstone-grainstone, (4) ooid-peloid grainstone-packstone, (5) ooid grainstone, and (6) algal boundstone. In addition, there are a few thin dolomite beds. Northward toward the Coahuila Peninsula and eastward toward the Tamaulipas Arch the Zuloaga Limestone thins and contains more grainstone and evaporite.

The Zuloaga Limestone was deposited on a carbonate ramp which sloped gently southward from the Coahuila Peninsula and westward from the Tamaulipas Arch. On the inner part of the ramp, carbonate sediment accumulated in peritidal environments. The area of Astillero Canyon was on the "outer ramp", probably 150 km south of the southern shoreline of the Coahuila Peninsula. At this locality, Zuloaga this was a relatively rapidly subsiding part of the epicontinental basin, but carbonate accumulation kept up with subsidence and shallow-water regimes prevailed.

Measured sections of the Zuloaga Limestone on both limbs of the same anticline are difficult to correlate, indicating that lithic units are not laterally extensive. The vertical sequences suggest that shallow-water depositional environments frequently shifted from place to place as the epicontinental shelf continuously subsided. Thus, the Zuloaga strata at Astillero Canyon lack the predominately regressive characteristics of the age-equivalent Smackover Limestone in South Texas. Assuming that the grainstone and algal-stromatolite represent deposition in the shallowest water, the stratigraphic section at Astillero Canyon may be divided into at least eight major "shoaling-upward" sequences.

INTRODUCTION

The Zuloaga Formation is the Upper Jurassic (Oxfordian) limestone facies of northeastern Mexico (Imlay, 1938). These carbonate rocks accumulated on the gently sloping floor of an epicontinental sea (Weidie, 1968)

landmass to the west (Fig. 1). The few previous petrographic studies of the Zuloaga Formation concentrated on limestones that were deposited in shallow water southeast of the Coahuila Peninsula and on the flanks of Tamaulipas Arch paleohighs (Oivanki, 1974; and Sandstrom, 1982). The Zuloaga limestones described here accumulated on a relatively rapidly subsiding area of the epicontinental shelf south of the Coahuila Peninsula and west of the Tamaulipas Arch (Fig. 1).

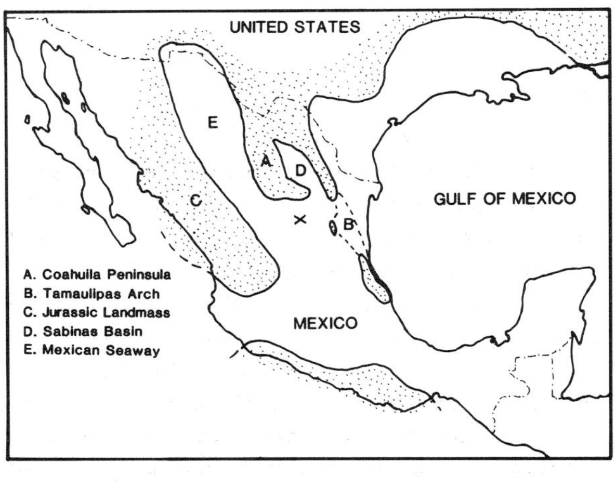

Figure 1. Jurassic paleogeography for the Gulf Coast and Mexico. Land areas are denoted by the stippled pattern. "X" marks approximate location of the study area. After Oivanki, 1974.

Our study focuses on a thick section of the Zuloaga Formation exposed in Astillero Canyon about 20 km, south of Concepcion del Oro, Zacatecas, Mexico (Fig. 2). Here, the formation crops out in the core of a northwest-trending isoclinal anticline, which is overturned toward the northeast. The limestone is exposed along a narrow arroyo which cuts through both limbs of the anticline. In Astillero Canyon, the Zuloaga Formation is about 450 m thick; the base is not exposed. Although the Zuloaga in the vicinity of the type locality about 50 km to the northwest has been described as "deep-water" pelagic limestone (Oivanki, 1974), we interpret the

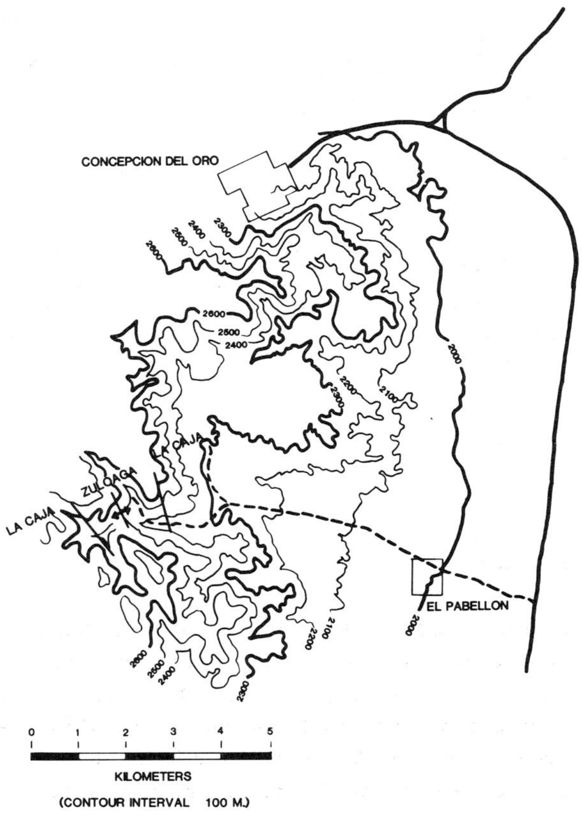

Figure 2. Location map for Astillero Canyon study area. Sections are accessible by car with the last kilometer reached on foot.

STRATIGRAPHIC FRAMEWORK

During Oxfordian time, terrigenous sediment was shed into the shallow sea around the Coahuila Peninsula and the islands of the Tamaulipas Arch. However, carbonate sediments accumulated over most of the epicontinental shelf.

Interfingering terrigenous and carbonate rocks around the southern flank of the Coahuila Platform are included in the La Gloria Formation (Imlay, 1936). This unit grades southward and eastward into limestones and thin evaporites of the Zuloaga Formation.

In much of northeastern Mexico, the Zuloaga Limestone overlies the Minas Viejas evaporite (Fig. 3). However, on the south-beds of the La Joya Formation (Huizachal Group), and in parts of northern Zacatecas it overlies the volcanic and volcaniclastic red beds of the Nazas Formation (Blickwede, 1982). Presumably the Nazas Formation underlies the Zuloaga in the vicinity of Astillero Canyon (Fig. 3). In the study area, the Zuloaga strata are conformably overlain by shale, fine-grained sandstone, and phosphatic limestone and sandstone of the La Caja Formation (Fig. 3). Farther northward the La Caja Formation interfingers with the coarser sandstones, shale, and thin limestones of the La Casita Formation (Fig. 3).

At Astillero Canyon, the Zuloaga Limestone is thicker than it is closer to the Coahuila Peninsula (Fig. 4). In addition, the Zuloaga Limestone here is predominantly a

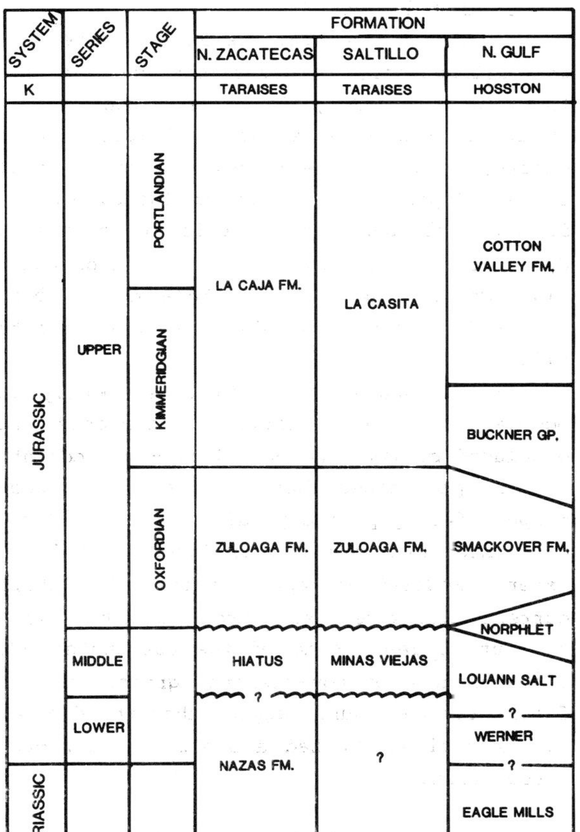

Figure 3. Stratigraphic nomenclature in northern Mexico and its correlation with the northern Gulf Coast of the United States.

carbonate mudstone-wackestone with only minor grainstones, whereas midway between Saltillo and Concepcion del Oro (Fig. 4) the formation contains abundant grainstones (Oivanki, 1974).

ROCK TYPES

The Zuloaga Formation exposed in Astillero Canyon is divided into the following interbedded lithofacies, in order of their volumetric abundance: (1) skeletal mudstone and wackestone, (2) pellet mudstone and wackestone, (3) pellet packstone-grainstone, (4) ooid-peloid grainstone-packstone, (5) ooid grainstone, and (6) algal boundstone. In addition, a few thin dolomite beds are exposed on the northeastern flank of the anticline.

Skeletal Mudstone and Wackestone

About 55 percent of the section is made up of dark-gray, medium- to massively-bedded (0.25-2m), burrowed skeletal mudstone and

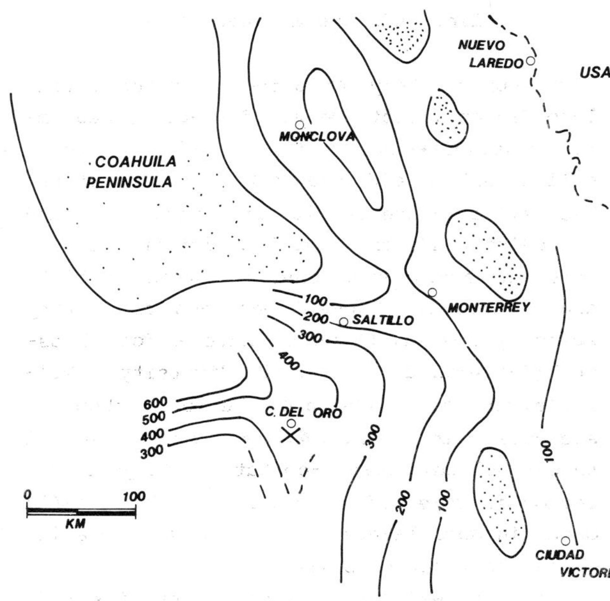

Figure 4. Isopach map of the Zuloaga Formation after Oivanki (1974). Contour interval is 100 meters. Stippled pattern locates area where the Zuloaga Formation was not deposited. "X" shows the location of Astillero Canyon.

wackestone. The most abundant fossils are mollusks, echinoids, ostracods, sponge spicules, and benthic foraminifers. Mollusks include thick- and thin-shelled pelecypods and the gastropod *Nerinea*. Some samples of the Zuloaga contain the green alga *Cayeuxia,* dasyclads, and a few blue-green-algal(?) oncoids. Some beds contain common to abundant "calcispheres", which are subspherical bodies of sparry calcite that lack wall structure and may be radiolaria that have been replaced by calcite.

Non-skeletal carbonate constituents are pellets and peloids. Most of the mud-supported limestones contain minor amounts of quartz silt, and many of the quartz grains have euhedral overgrowths. Finely disseminated pyrite and irregular chert nodules and layers are common.

These muddy limestones were deposited on a low-energy subtidal shelf. The green algae and probably the thick-shelled mollusks suggest shallow-water; the echinoids indicate approximately normal-marine salinity.

Pellet Mudstone and Wackestone

Approximately 25 percent of the Astillero Canyon limestones are dark-gray, medium- to massively-bedded (0.15-1.5m), burrowed pellet and peloid mudstone and wackestone. The pellets range in size from very fine sand to medium sand, and most lack internal structure; however, some *Favreina* fecal pellets are preserved. These limestones are either sparsely fossiliferous or have a fossil assemblage with a low species diversity. Mollusks and ostracods are the most abundant skeletal components. *Nerinea* gastropods and thick-shelled, coarse-costate pelecyods are common in some beds. Oncoids of *Cayeuxia(?)* occur in some layers. Echinoids and foraminifers are rarer constituents.

A burrowed lime mud which contains calcite pseudomorphs after evaporite (gypsum?) crystals occurs at the base of the section on the southwestern limb of the anticline. Some of the evaporite apparently grew within burrows. In the middle and lower part of both sections, many mudstones and wackestones contain scattered dolomite rhombs. Some beds contain irregular chert nodules.

These pelleted, sparsely fossiliferous limestones accumulated in a restricted, low-energy environment. The fossils indicate shallow water, and the low species diversity suggests abnormal salinity. Gypsum-crystal pseudomorphs indicate that some of the sediment occasionally was immersed in hypersaline waters. Budd and Loucks (1981) describe isolated evaporite nodules in basinal (not necessarily deep-water) rocks of the Smackover Formation in South Texas.

Pellet Packstone and Packstone-Grainstone

Medium- to dark-gray pellet packstone-grainstone makes up 11 percent of the Zuloaga Formation in Astillero Canyon. Bedding ranges in thickness from 0.1 to 2 m.

Favreina fecal pellets and the more common ellipsoidal, structureless pellets are the most abundant grain types. Other allochemical constituents are peloids, intraclasts, skeletal fragments, and superficial ooids. Typical fossils are mollusks, echinoderms, ostracods, and benthic foraminifers. Sponge spicules, "calcispheres", and dasyclad algae occur in a few beds. In addition, most of the beds contain small amounts of quartz silt.

Intergranular pores in these poorly to moderately sorted packstone-grainstones are occuluded by lime mud or calcite-spar cement. In some packstones there are calcite pseudomorphs after gypsum crystals.

These limestones accumulated in shallow-water, moderate-energy environments where currents and waves were strong enough to winnow out at least some of the carbonate mud. The abundance of non-skeletal grains and the fairly limited fauna suggest that the depositional environment had a somewhat restricted circulation.

Ooid-Peloid Grainstone-Packstone

Brown-gray, black, and light-gray ooid peloid grainstone-packstone constitutes five percent of rocks in the measured sections.

The grains are generally well rounded and well sorted, and a few samples from the middle of the section show rhythmic laminae composed of alternating layers of medium-sand-sized and very fine sand-sized grains.

Ooids, peloids, and pellets are the major grain types. Most of the ooids have large nuclei and cortices with radial structure. Their nuclei are peloids, pellets, mollusks, echinoderms, dasyclad algae, and quartz silt. *Favreina* pellets are common in some beds, but small structureless pellets are the most abundant type. Other common grains are peloids, intraclasts, skeletal fragments, and oncoids. Fossils include mollusks, echinoderms, dasyclad algae, ostracods, and foraminifers.

These grain-supported limestones were deposited in moderate- to high-energy shoal areas. The rhythmic-laminated grainstones may represent intertidal deposition. Loreau and Purser (1973) describe Persian Gulf tidal-delta sands which are compositionally and texturally similar to the Zuloaga rhythmites.

Ooid Grainstone

Medium- to dark-gray, well-cemented ooid grainstones [1 to 1.5 m thick] make up about one percent of the limestones. The thickly coated ooids are medium- to coarse-sand sized and are well sorted. Their cortices have radial structure and their nuclei are peloids, pellets, and quartz silt. Some quartz nuclei have euhedral overgrowths which replace part of the ooid. Some mollusk nuclei were dissolved before deeper burial and compaction, which is demonstrated by ooid rims which are collapsed toward the leached centers of the ooids. Following compaction, calcite spar filled the remaining void in the center. On the northeastern limb of the anticline some of the oolite layers are dolomitic.

These grainstones probably were deposited on high-energy shoals or beaches. Judging from the paucity of grainstone in the Astillero Canyon section, high-energy shoal-water areas were widely scatterd and short-lived. Evidence for early dissolution of aragonite in these rocks may indicate that the oolite deposits formed small islands underlain by lenses of fresh or brackish phreatic water.

Algal Boundstone

Medium- to dark-gray algal-laminated mudstone is exposed in a few places in Astillero Canyon and makes up about one percent of the total section. The thickest stromatolitic unit is about one meter thick. The stromatolites are composed of thin undulose laminae of brown-gray pelleted lime mud and organic-rich mud. The wavy laminae are generally flat lying with low domal structures and in some places are interrupted by calcite pseudomorphs after gypsum.

This facies was deposited in the low-energy intertidal zone, perhaps in the protection of a grainstone shoal or island. Evidence for supratidal exposure, such as desiccation fractures and fenestral fabric, is absent.

Dolomite

In the upper third of the Zuloaga Formation, 0.5-3m thick beds of brown, coarsely crystalline dolomite crop out on the northeastern limb of the anticline. Dolomite makes up about three percent of the exposed section. The dolomites are composed of interlocking subhedral crystals, and intercrystalline porosity is estimated to be about two percent. "Ghosts" of allochemical constituents in some samples indicate that the dolomitized limestone was peloid-ooid packstone and grainstone.

Small patches of saddle dolomite occur in the skeletal mudstone and wackestone and the pellet mudstone and wackestone near the middle of the section measured on the southwestern limb of the anticline. The coarsely crystalline saddle dolomite is in 3-4cm diameter milky-white "nodules" which replace the host rock. These "nodules" probably are late-diagenetic features. Estimates for the temperature of formation of saddle dolomite range from 60°C to 150°C, but saddle do-range from 60°C to 150°C, but saddle dolomite is most likely to form at temperatures higher than 100°C (Radke and Mathis, 1980).

Depositional Framework

Stratigraphy and regional paleography suggest that the Zuloaga Formation was deposited on a carbonate ramp (Ahr, 1973) which sloped gently southward from the Coahuila Peninsula and westward from the Tamaulipas Arch (Fig. 1). On the inner part of the ramp (Fig. 5), carbonate sediment accumulated in peritidal environments, including supratidal, low-energy intertidal, and shallow-water subtidal, and high-energy shoals and bars (Oivanki, 1974). In several places, evaporites were deposited in hypersaline ponds and lagoons that existed in the more restricted environments of the inner ramp.

The carbonates exposed in Astillero Canyon were deposited on the "outer ramp", probably 150 km south of the southern shoreline of the Coahuila Peninsula (Figs.1 and 5). The predominance of wackestones and mudstones is evidence that most of these Zuloaga limestones were deposited in low-energy subtidal environments. These carbonates form a complexly interbedded stratigraphic sequence. The thickness of the Zuloaga in this area (Fig. 4) indicates that this was a relatively rapidly subsiding part of the epicontinental basin, but carbonate accumulation kept up with subsidence, and shallow-water regimes prevailed.

The following constituents of the Zuloaga Formation at Astillero Canyon are evidence of shallow-water environments of deposition:

1. shallow-water algae such as *Cayeuxia*, dasyclads, and blue-green algal stromatolites and oncolites;
2. well-sorted oolite and peloid grainstones; and
3. peloid-ooid grainstone rhythmites,

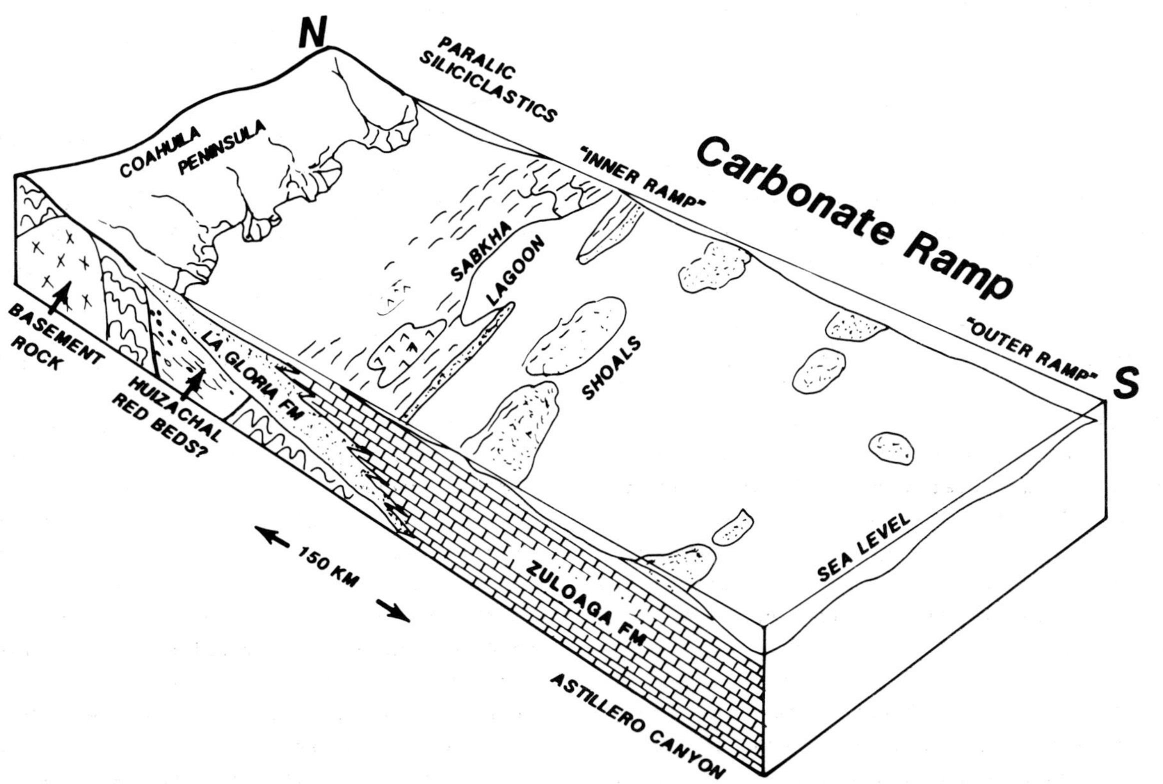

Figure 5. Conceptual model of the depositional setting from Coahuila Peninsula to Astillero Canyon area during Oxfordian time.

similar to modern carbonate (Loreau and Purser, 1973) and siliciclastic (Reineck and Singh, 1975) tidal deposits.

Much of the Zuloaga Limestone in Astillero Canyon is similar to the part of the South Texas Smackover Limestone interpreted by Budd and Loucks (1981) to be low-energy open-shelf ("top of a ramp") deposits. The grainy limestones of the Zuloaga Formation are like the Smackover limestones deposited in a near-shore "high-energy shoal system" (Budd and Loucks, 1981). The few unfossiliferous, poorly burrowed and wavy-laminated mudstones could be interpreted as "basinal" limestones. However, in general, the Zuloaga Formation in Astillero Canyon lacks the characteristics of "cratonic-basin" limestones (Wilson, 1975).

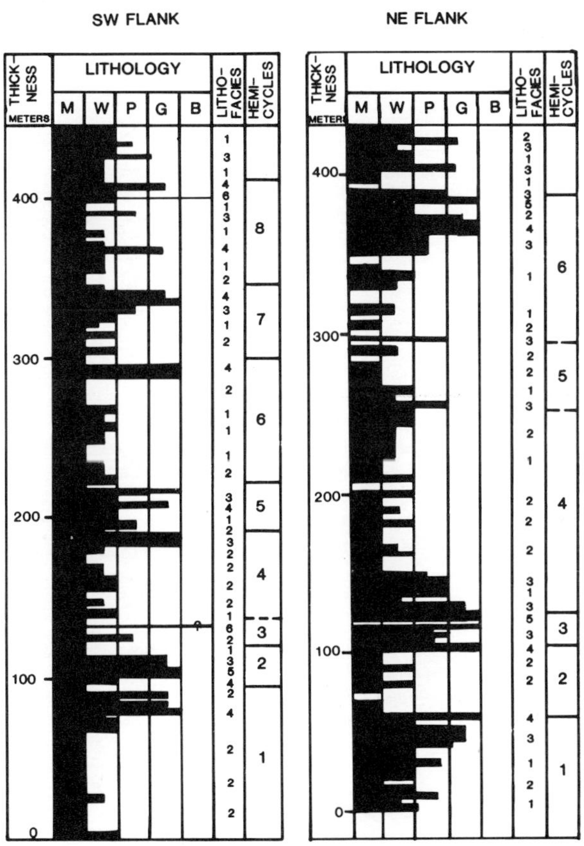

Figure 6. Zuloaga sections measured on the southwestern (SW) and northeastern (NE) flanks of the anticline cut by Astillero Canyon. Datum is the base of the La Caja Formation. The SW section may be divided into at least eight major shoaling-upward sequences, and the NE section, tentatively into six. M=limestone, W=wackestone, P=packstone, G=grainstone, B=algal boundstone. 1=skeletal mudstone and wackestone, 2=pellet mudstone and wackestone, 3=pellet packstone and packstone-grainstone, 4=ooid-peloid grainstone-packstone, 5=ooid grainstone, and 6=algal boundstone.

The measured sections of the Zuloaga Formation (Fig. 6) on both limbs of the same anticline are difficult to correlate, indicating that the lithic units are not laterally extensive. The vertical sequences (Fig. 6) suggest that shallow-water depositional environments shifted from place to place as the epicontinental shelf continuously subsided. The Zuloaga limestones lack the overall regressive or progradational vertical trend characteristic of the Smackover section deposited at the same time on the carbonate ramp in South Texas (Budd and Loucks, 1981).

Assuming that the grainstone and algal-stromatolite facies represent deposition in the shallowest water, the stratigraphic section may be divided into eight major "shoaling-upward" hemicycles plus the uppermost unit of the Zuloaga (Fig. 6).

SUMMARY

The Zuloaga limestone exposed in Astillero Canyon was deposited on a shallow-water, mostly low-energy epicontinental shelf (ramp) about 150 km south of the Coahuila Peninsula and 150 km west of the Tamaulipas Arch. Small, short-lived shoaling areas developed locally as carbonate buildups shifted from place to place on the gently undulating sea floor. In general, carbonate accumulation kept up with subsidence. In the Late Jurassic, carbonate deposition was terminated by an influx of siliciclastic sediment (La Caja-La Casita Formations) on to the epicontinental shelf.

ACKNOWLEDGMENTS

The research for this paper was completed as part of a M.S. thesis at the University of New Orleans. Financial aid was provided by grants from the Gulf Coast Association of Geological Societies, American Association of Petroleum Geologists, and University of New Orleans Department of Earth Sciences and Graduate School. Amoco Production Company, New Orleans, provided typing services. A.E. Weidie provided valuable discussion during preparation of this paper.

REFERENCES

Ahr, W. M., 1973, The carbonate ramp: An alternative to the shelf model: Gulf Coast Assoc. Geol. Soc. Trans., v. 23, p. 221-225.

Blickwede, J. F., 1981, Stratigraphy and petrology of Triassic(?) "Nazas Formation", Sierra de San Julain, Zacatecas, Mexico: Unpub. M.S. thesis, University of New Orleans, New Orleans, Louisiana.

Budd, D. A. and R. G. Loucks, 1981, Smackover and lower Buckner formations, south Texas - depositional systems on a Jurassic carbonate ramp: Bureau of Economic Geology, University of Texas at Austin, Report of Investigations 112, 38 p.

Imlay, R. W., 1936, Evolution of the Coahuila Peninsula, Mexico; Pt. 4, Geology of the western part of the Sierra de Parras: Geol. Soc. America Bull., v. 47, p. 1091-1152.

____________, 1938, Studies of the Mexican geosyncline: Geol. Soc. America Bull., v. 49, p. 1651-1694.

Loreau, J. P. and B. H. Purser, 1973, Distribution and ultrastructure of Holocene ooids in the Persian Gulf, *in:* Purser, B. H., (ed.), The Persian Gulf: New York, Springer-Verlag, p. 279-328.

Oivanki. S. M., 1974, Paleodepositional environments in the Upper Jurassic Zuloaga Formation (Smackover), northeastern Mexico: Gulf Coast Assoc. Geol. Socs. Trans., v. 24, p. 258-279.

Radke, B. M., and R. L. Mathis, 1980, On the formation and occurrence of saddle dolomite: Jour. Sed. Petrol., v. 50, p. 1149-1168.

Reineck, H. E. and I. B. Singh, 1975, Depositional sedimentary environments, with reference to terrigenous clastics: Springer-Verlag, New York, 439 p.

Sandstrom, M., 1982, Stratigraphy and environments of deposition of the Zuloaga Group, Victoria, Tamaulipas, Mexico: Program and Abstracts, The Jurassic of the Gulf Rim, Gulf Coast Section Soc. Econ. Paleontologists and Mineralogists, 3rd Annual Research Conf., p. 94-97.

Weidie, A. E., 1968, The myth of the Mesozoic Mexican geosyncline (abs.): Geol. Soc. Am. Abs. 1968, Spec. paper 121, p. 314-315.

Wilson, J. L., 1975, Carbonate facies in geologic history: New York, Springer-Verlag, 471 p.

THE UPPER SMACKOVER OF THE GULF RIM: DEPOSITIONAL SYSTEMS, DIAGENESIS, POROSITY EVOLUTION AND HYDROCARBON PRODUCTION

CLYDE H. MOORE

Applied Carbonate Research Program
Department of Geology
Louisiana State University
Baton Rouge, Louisiana 70803

ABSTRACT

A generalized stratigraphic framework for the upper Jurassic is suggested, in which the name Haynesville Formation is utilized, and the name Lower Cotton Valley Lime is suppressed in favor of the Gilmer, where appropriate. Eustatic sea-level fluctuations have resulted in patterns of sedimentation common to upper Jurassic sequences across the entire northern Gulf of Mexico region and hence indicate the wide applicability of this generalized stratigraphic framework.

The presently accepted model of the Smackover-Haynesville sedimentation must be modified to take into account sea-level fluctuations, subsidence, and sediment availability. The model developed here is simply one of lower Smackover basin fill during a rapid transgressive phase and upper Smackover regional shoaling during a sea-level stillstand in which sedimentation was in equilibrium with subsidence. The lower and upper Smackover are not necessarily time equivalent, but represent two separate sedimentologic sea-level regimens. The Haynesville Formation is thought to be a separate sedimentologic package that was deposited during the next sea-level rise; the Gilmer Limestone formed a shelf-margin barrier behind which the lagoonal Buckner evaporites were deposited. The evaporites graded landward into quartzose clastics.

Predictable regional porosity patterns have developed in the Smackover-Haynesville, in response to early diagenetic overprints, controlled largely by eustatic sea-level subsidence interactions. These patterns include: updip oomoldic porosity in a regional meteoric-water system developed during the upper Smackover sea-level stillstand; downdip porosity preservation under marine conditions along the shelf margin; regional dolomitization associated with reflux of evaporitive waters from the Buckner lagoon behind the Gilmer shelf-margin barrier.

Structural hydrocarbon traps associated with salt movement are the most common type of Late Jurassic trap. Buckner evaporites or Haynesville shales usually form the seals in the Late Jurassic reservoirs. Jurassic source rocks are probably lower Smackover limestones and Norphlet shales. Sourcing is generally local with migration into updip areas, particularly where regional dolomitization has occurred. The time of migration, which is a key factor in a viable Smackover exploration strategy, varies across the Gulf in response to the subsidence history of each individual basin.

Future Jurassic exploration will center on south Texas, the Gilmer shelf margin, and the updip Smackover along the bounding graben fault systems. Most production will be gas, but some oil should occur along the updip faulted fairways.

INTRODUCTION

The limestones and dolomites of the Upper Jurassic Smackover have been a target for explorationists for almost half a century. The active exploration fairway from the Rio Grande to the Florida panhandle (Fig. 1) is a mature petroleum province. Because of the large number of wells that have been drilled and cored and the time span of continued exploration, a considerable body of knowledge has accumulated, both in the literature and in confidential company files. The development of the Jurassic literature of the Gulf Coast has paralleled the economic development of the Smackover trend itself, because the U.S. Gulf Jurassic is a wholly subsurface sequence. Because of this dependence on drilling, the result is a geographic and temporal compartmentilization of the literature as the exploration of the trend was sequentially broadened to encompass ultimately the entire northern margin of the Gulf of Mexico. Exploration started in southern Arkansas in the 1930's and has continued with Alabama, Florida, and south Texas in the 1970's and 1980's.

The early works of Weeks (1938); Hazzard (1939); Hazzard *et al.* (1947); and Philpott and Hazzard (1949) concentrated on the trend in Arkansas and easternmost Texas. Dickinson (1963) first published the Jurassic stratigraphy of Mississippi and Alabama and later extended his work into east Texas and parts of Arkansas and Louisiana (Dickinson, 1968a, b). As the active fairway was extended across Mississippi, Oxley *et al.* (1967) and Badon (1974) incorporated new subsurface data into a more comprehensive view of the Mississippi segment of the subcrop. Ottmann *et al.* (1973) and Sigsby (1976) gave us our first views of the Alabama-Florida Smackover which have been enhanced by the recent work of Mancini and Benson(1980). Aiken and Graves (1969) outlined the critical facies relationships of the upper Smackover in southern Arkansas, while Bishop illuminated the complex structure and stratigraphy of the Arkansas-Louisiana State Line Fault Trend (1971a, b, c, and 1973). Ahr (1973) established the ramp model for the Smackover, and finally, Budd and Loucks (1981) outlined, for the first time, the stratigraphic and diagenetic setting of the Smackover in south Texas. Of course, the works of Imlay, concentrating primarily on the paleontologic framework of the Jurassic, have spanned this entire period (1943, 1952, and 1980).

Most of the aforementioned works were regional-stratigraphic in nature; however, during the 1970's and 1980's increased attention was focused on the petrographic and diagenetic aspects of the Smackover both in individual fields (Bishop, 1971a, b, c, Becher and Moore, 1976), and regionally (Moore and Druckman, 1981 and Loucks and Budd, 1981).

The present paper is first an attempt to emphasize certain patterns of sedimentation common to Late Jurassic sequences from the Rio Grande to Florida; to relate these patterns to regional events, such as salt tectonics and subsidence rates, and world-wide eustatic sea level changes; and to use these common sedimentation patterns to refine a depositional model for the Smackover and associated units that more closely satisfies the constraints of regional subsurface data. Secondly, this paper outlines some striking regional patterns of common porosity types that are the result of diagenetic processes related to these same Gulf-wide sedimentation, structural, and sea-level patterns. Finally, in a most general way, the regional patterns of oil versus gas and trap type are discussed as a function of these same events.

Stephen Collins (1980) provided the initial stimulus for much of the present work and his publication has been extensively utilized.

JURASSIC TECTONIC SETTING OF THE GULF RIM

The structural elements that affect Late Jurassic sediments from south Texas to panhandle Florida (Fig. 1) generally reflect the Late Triassic and Early Jurassic rift tectonics that were associated with the opening of the Gulf of Mexico. The South Texas, East Texas, North Louisiana, and Mississippi salt basins are separated from the main Gulf Basin by the Angelina, LaSalle, and Wiggins Arches. This series of major positive features represents continental blocks which

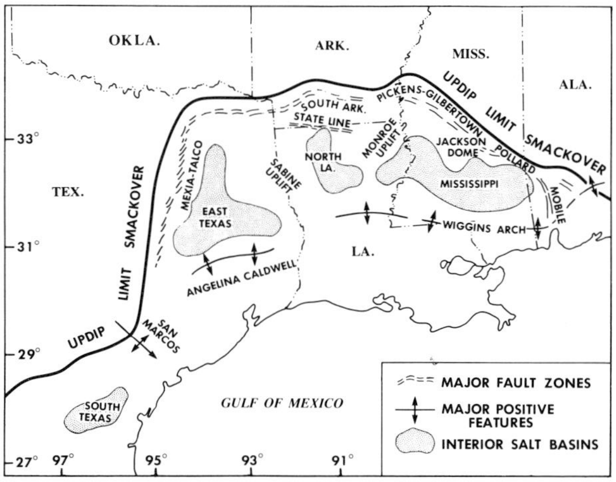

Figure 1. Main structural elements of the Gulf Rim affecting Late Jurassic sequences.

	STAGE		MEMBER	FORMATION	GROUP
UPPER JURASSIC	TITHONIAN	0-2000+'		SCHULER	COTTON VALLEY
		0-600'		BOSSIER	
	KIMMERIDGIAN	0-1000'	GILMER LIMESTONE	HAYNESVILLE FORMATION	LOUARK
		0-800'	UPPER BUCKNER MEMBER		
		0-500'	BUCKNER EVAPORITE MEMBER		
	OXFORDIAN	0-400'	UPPER GRAINSTONE MEMBER	SMACKOVER FORMATION	
		0-500'	MIDDLE MEMBER		
		0-500'	LOWER MEMBER		
		0-700'		NORPHLET FORMATION	
MIDDLE JURASSIC	CALLOVIAN	0-5000'+		LOUANN FORMATION	LOUANN

MODIFIED IN PART FROM IMLAY, 1980.

Figure 2. Middle-Upper Jurassic stratigraphic nomenclature of the Gulf Rim.

were stranded during the early opening phases of the Gulf (Walper, 1980; Buffler *et al.*, 1980). Each of these discrete salt basins are separated by positive elements, the San Marcos, Sabine, and Monroe uplifts, which generally trend normal to the axis of the main Gulf Basin. These features reflect high heat flux associated with adjacent basin subsidence (Nunn, 1982). Each salt basin has high-rise salt structures which indicate thick Louann salt, although a thin Louann sequence may extend across the Sabine, Monroe, and San Marcos uplifts. A major graben fault system bounds the inner limits of the Late Jurassic subcrop. In Texas, the fault system is called the Mexia-Talco; in Arkansas, the South Arkansas fault system; and in Mississippi, the Pickens-Gilbertown-Pollard system (Murray, 1961). Although some fault movement can be detected in the Smackover strata, most activity along these faults occurred in Haynesville-Cotton Valley strata, and subsequently in Cretaceous rocks. Initial fault movement probably reflects gravity sliding of the Louann salt toward the basin (Bishop, 1973). The fault zone does mark the updip limit of thick Louann Salt, although a thinner Louann-Smackover sequence may extend several tens of miles landward of the fault system. Subtle salt anticlinal ridges delineate the area just basinward of the fault zone to the area of high-rise salt structures in the basin proper.

A well-defined graben fault zone which is sub-parallel to, and basinward of, the main fault trend occurs at the Arkansas- Louisiana border. Most movement of this system took place during Smackover-Haynesville time, perhaps in response to salt-related gravity sliding toward the basin (Bishop, 1973).

In the western Alabama-Florida extremities of the trend, two smaller basins, termed the Manila and Conecuh embayments, apparently developed during Smackover-Haynesville deposition. These basins are separated from the Mississippi Salt Basin by a distinct salt ridge, and from each other by the Conecuh-Wiggins Arch (Benson and Mancini, 1982).

In summary, during the Late Jurassic, the basic tectonic pattern of the northern Gulf Rim is one of isolated interior salt basins which contain high-rise salt structures and which are rimmed by relatively stable platforms with subtle salt anticlines. The platforms terminated updip against a major gravity-related graben fault system.

MIDDLE-UPPER JURASSIC STRATIGRAPHIC FRAMEWORK OF THE GULF RIM

The use of the term Haynesville Formation, with the Buckner and Gilmer reduced to member status, is designed to emphasize the basic sedimentologic differences between the Smackover and Haynesville Formations, as well as to standardize the nomenclature on a regional basis (Fig. 2). The inclusion of the Bossier in the Cotton Valley follows Todd and Mitchum's seismic stratigraphic analysis of the Jurassic of East Texas (1977), in which they demonstrate Bossier onlap onto the upper surface of the Gilmer member of the Haynesville. The term Lower Cotton Valley Lime, generally used for the limestones developed in the upper Haynesville, is therefore suppressed and the term Gilmer is substituted where appropriate.

The following discusson is sequentially centered about each of the interior salt basins, and is designed to emphasize gross Gulf-wide similarities and differences in the Haynesville-Smackover sequence that may reflect important regional sedimentologic controls. The cross sections utilized in these discussions are generally schematic, and are based on the available literature and the author's present understanding of stratigraphic relationships in the various areas.

Central Gulf
(Southern Arkansas-North Louisiana)

A schematic north-to-south cross section across southern Arkansas into north Louisiana (Fig. 3) combines elements of similar sections taken from Moore and Druckman (1981), Aiken and Graves (1969), Dickinson (1968b), and Bishop (1971a, b, c). The Smackover Formation is represented as a separate entity relative to the overlying Haynesville Formation, and the upper Smackover is a blanket carbonate grainstone from central Arkansas to northern Louisiana. This interpretation differs significantly from that held by many workers, who believe that the Smackover-Buckner (Buckner member of the Haynesville) interfingers regionally from north to south (Budd and Loucks, 1981). However, Moore and Druckman (1981) found no evidence of significant regional interfingering between the Smackover and Buckner in this region. Both Aiken and Graves (1969) and Dickinson (1968a) recognized that the Buckner anhydrite of Arkansas generally postdates the Smackover in response to post-Smackover barrier formation. They both stated that the salt anticlines of south Arkansas formed the barrier that resulted in massive anhydrite deposition in the Buckner. It is recognized, however, that supratidal dolomites associated with nodular anhydrites may occur locally in the upper 20 feet of the Smackover in southern Arkansas. This zone seems to be related to local Smackover shorelines rather than to the massive evaporites of the Buckner member of the overlying Haynesville Formation, and may well be responsible for much of the confusion surrounding Smackover-Buckner stratigraphic relationships. I also feel, that although the subtle salt structures called upon by Aiken and Graves (1969) and Dickinson (1968a) may have contributed to a general evaporite lagoonal environment during Buckner deposition, a more well-defined Haynesville barrier was localized by the tectonic activity along the State Line Fault Trend as outlined by Bishop (1971a, b, and c). This Haynesville barrier consists of quartzose clastics and locally well-defined carbonate sand buildups. These carbonate sands along the State Line Fault Trend have generally been termed the "Smackover A" because of their oolitic composition. However, I strongly recommend the suppression of the term "Smackover A" in favor of the term Haynesville Lime, a term that more clearly suggests the true affinities of these buildups. The Buckner Anhydrite Member is restricted to a narrow subcrop through southern Arkansas just north of the Dorcheat Macedonia Field and south of the South Arkansas Graben. In this area, the Buckner Anhydrite contains significant amounts of intercalated clay.

In those areas where the Buckner Anhydrite is present, the anhydrite acts as an effective seal, and hydrocarbon production is generally confined to the Smackover Formation. In other areas, however, where the Buckner Anhydrite is missing, significant Haynesville production has been established.

The Bossier Shale thickens rapidly basinward of the State Line Fault Trend-Haynesville barrier at the expense of the Haynesville-upper Smackover. As stated earlier, seismic evidence (Todd and Mitchum, 1977) indicates Bossier onlap with the strong possibility of an unconformity between the Bossier and the Haynesville-Smackover.

East Texas

The schematic cross section of the Upper Jurassic across the western margin of the East Texas Salt Basin in Navarro County (Fig. 4) is characteristic of East Texas from the San Marcos Arch on the south, to Cass county on the east. This section shows an increase in limestone in East Texas in the Haynesville (the Gilmer Limestone) as compared to that of the central Gulf, where the limestones in the Haynesville occur in restricted areas fronting the North Louisiana Salt Basin and are essentially encased in quartzose clastics. The Gilmer in East Texas, however, becomes quite massive, is oolitic, and comprises thick limestone along the basin margin; this limestone cannot be differentiated from the Smackover, where the Buckner is absent, throughout East Texas. The Buckner Anhydrite member of the Haynesville Formation becomes considerably thicker and more massive in Texas, and contains isolated occurrences of halite (Dickinson, 1968 a). Again, there is no significant evidence of regional interfingering of the massive Buckner Anhydrite and the underlying Smackover. The Bossier Shale onlaps the Gilmer and extends northward and westward, almost to the Mexia-Talco Fault Zone. As in the central Gulf, the Gilmer and upper Smackover of East Texas are replaced basinward by the Bossier. The area of relatively stable shelf (basin margin to updip limit) is much narrower in the southern limits of East Texas than in the Arkansas-Louisiana region, but it does widen toward Arkansas and Louisiana where the basin margin trend is in an east-west direction. The Sabine Uplift area, which separates the North Louisiana Salt Basin from the East Texas Salt Basin, was probably a positive feature during Louann deposition because of the lack of high-rise salt structures over the uplift. However, the status of the Sabine Uplift is uncertain during Smackover-Haynesville deposition. Where the Buckner Anhydrite is present hydrocarbon production is from the Smackover; and where the anhydrite is absent, production is from the massive Gilmer-Smackover carbonates, a pattern similar to that to the east in the Arkansas-Louisiana area.

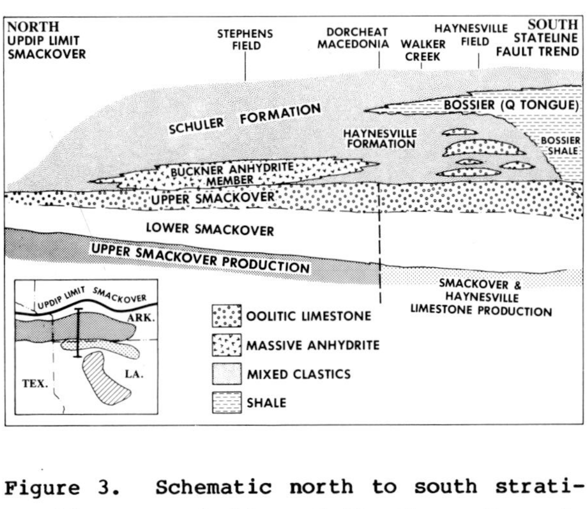

Figure 3. Schematic north to south stratigraphic cross section of the Upper Jurassic through south Arkansas into north Louisiana.

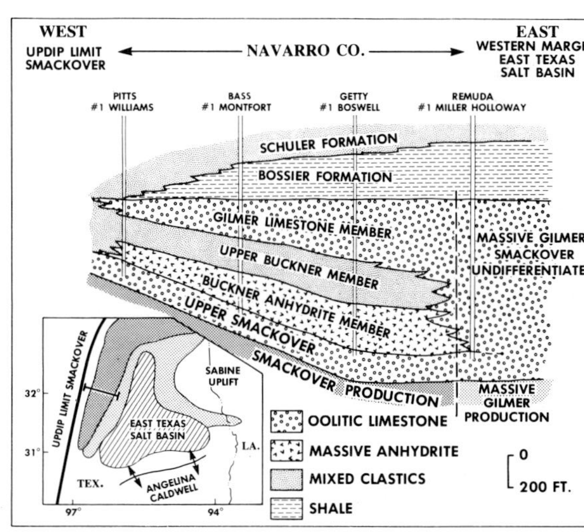

Figure 4. Schematic west to east stratigraphic cross section of the Upper Jurassic, East Texas.

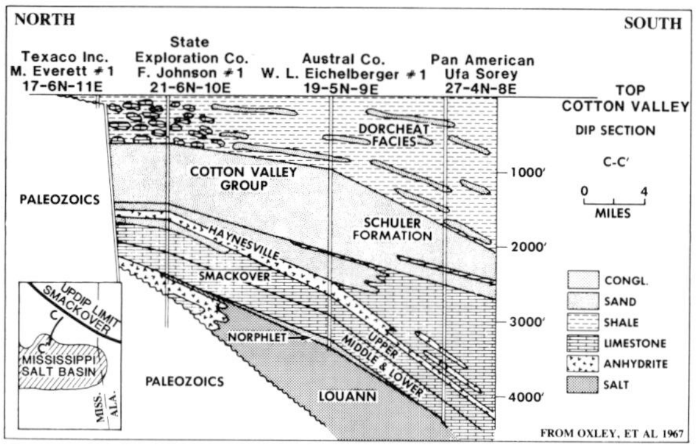

Figure 5. Schematic north to south stratigraphic cross section of the Upper Jurassic, south Texas.

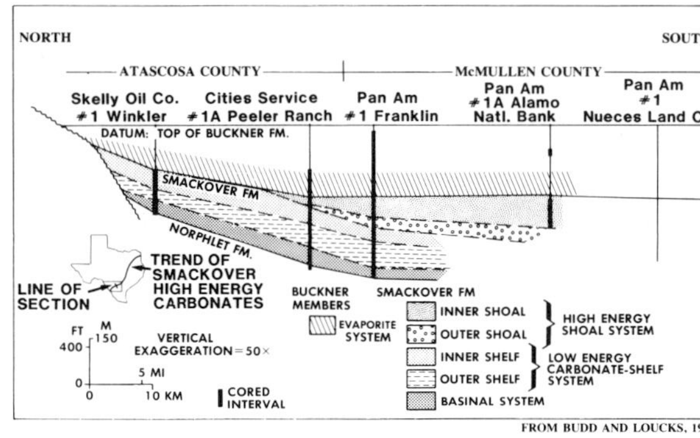

Figure 6. Schematic north to south stratigraphic cross section of the Upper Jurassic, Mississippi.

South Texas

Figure 5 is a schematic cross section by Budd and Loucks (1981) across the stable Smackover shelf into the South Texas Salt Basin. Because there are so few wells in the area, the actual stratigraphic relationships are still very much in doubt. Data from cores and mechanical logs from the Budd and Loucks report indicate that the Smackover Formation in South Texas is very similar to the Smackover throughout the rest of the Gulf Rim. The Buckner Anhydrite Member of the Haynesville grades into lime grainstones basinward. These grainstones are assumed to be the South Texas equivalent of the Gilmer, indicating that there is probably a massive limestone buildup along the Gilmer-Smackover basin margin in South Texas similar to that present in East Texas. Budd (personal communication) believes that the Buckner Anhydrite Member pinches out basinward near the location of the Pan Am #1A Alamo National Bank well (Fig. 5). No information is available about Bossier equivalents in South Texas, but it probably occurs basinward of the Pan Am well (Fig. 5). Stabler (1982) indicated that the Middle-Late Jurassic of northeastern Mexico is very similar to the sequences developed in East and South Texas.

Mississippi Salt Basin

Figure 6 is a schematic stratigraphic cross section of the Jurassic from basin edge to the updip limit of the Jurassic in the eastern extent of the salt basin as reported by Oxley *et al*. (1967). Although most of the exploitation of the Mississippi Jurassic has been later than 1965, Oxley's understanding of the stratigraphy seems to be still generally valid today. As to the west, the upper Smackover is a regressive sequence that in stable areas formed a blanket carbonate sand out to the shelf margin. However, the upper Smackover in the western reaches of the basin contains significant quartzose clastics, presumably derived from the Monroe uplift. The Haynesville Formation consists of a basal Buckner Evaporite Member overlain by quartzose clastics which change to carbonates at the basin margin; a seemingly identical stratigraphic setting as in Louisiana, Arkansas, and Texas. The Dorcheat Shale in Mississippi may be equivalent to the Bossier. Hydrocarbon production in the Haynesville Formation appears to be controlled by the presence of the Buckner Evaporite Member.

Alabama-Florida

The Alabama-Florida part of the trend is outside the main Mississippi Salt Basin (Mancini and Benson, 1980), and contains only occasional high-rise salt structures. This region is characterized by two sub-basins probably developed during Smackover-Haynesville time, which are separated from each other and the Mississippi Salt Basin to the west by basement and/or salt highs. The northernmost basin, the Manila, is separated from the Mississippi Salt Basin by a basement, or salt ridge, which trends in a northeast-southwest direction. A schematic east-west cross section from Oxley *et al.* (1967) (Fig. 7), illustrates a Gilmer Limestone buildup on the western margin of the Manila Embayment and a thick Buckner evaporite in the center of the embayment behind the Gilmer barrier. Significant thicknesses of Haynesville salt occur in the center of the Manila Embayment, similar to that in East Texas. The Conecuh Embayment to the south is separated from the Manila Embayment by the Conecuh Ridge, which is the northeastern extension of the Wiggins Uplift (Mancini and Benson, 1980). The barrier across at least part of the Conecuh Ridge seems to be formed by quartzose clastics. The lack of a Gilmer Limestone buildup is perhaps due to isolation from more normal marine conditions. The upper Smackover in the Conecuh Embayment contains a large quantity of high-energy carbonate grainstones which are associated with subtle salt structures, as contrasted with the low-energy muddy limestones (Mancini and Benson, 1980) of the Manila Embayment. These relationships may indicate that the Manila Embayment developed earlier (during Smackover time) than did the Conecuh Embayment which did not fully develop until Haynesville time.

Wells drilled in the vicinity of Pensacola, Florida, along the southern margins of the Conecuh Embayment, encountered a limestone buildup in the Haynesville Formation, that probably represents a seaward-facing Gilmer barrier similar to that developed elsewhere in the Gulf.

SMACKOVER-HAYNESVILLE DEPOSITIONAL SYSTEMS

The discussion of the stratigraphic framework of the Smackover-Haynesville outlined above emphasizes the similarity of the depositional patterns and facies relationships of these units from Texas to Florida. These patterns are: the Smackover Formation across the entire northern Gulf is a thick carbonate wedge that probably was deposited on a relatively featureless ramp; however, the overlying Haynesville Formation formed a basin-edge carbonate barrier (Gilmer), which created a restricted evaporite lagoon domi-

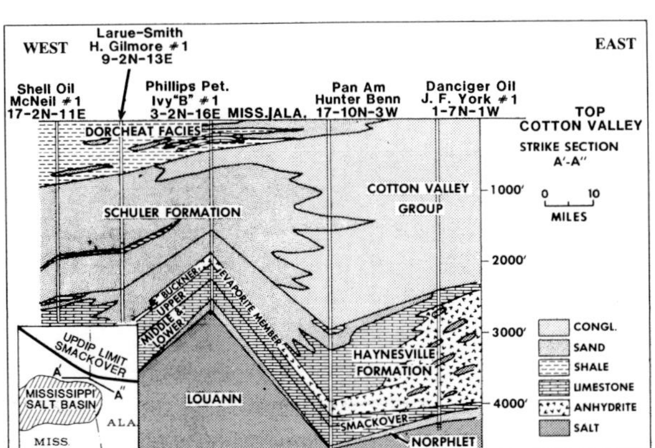

Figure 7. Schematic west to east stratigraphic cross section of the Upper Jurassic, Mississippi Salt Basin to Manila Embayment, Mississippi and Alabama.

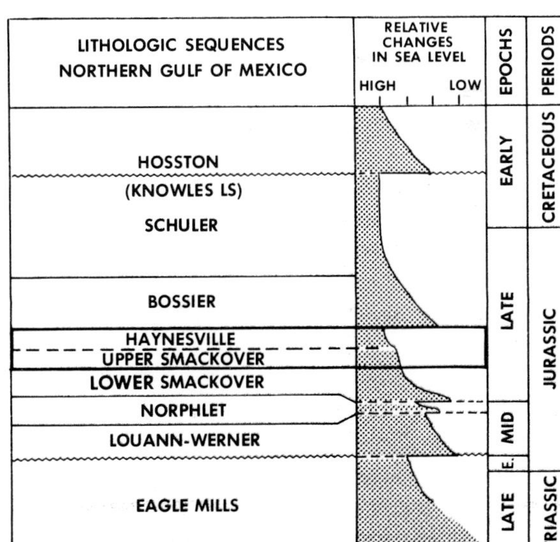

Figure 8. Mesozoic sea level curve for the northern Gulf of Mexico developed from the seismic stratigraphic studies of Todd and Mitchum (1977).

nated by anhydrite and halite (the Buckner Member). This lagoon was ultimately filled by terrigenous clastics (upper clastic member of the Haynesville Formation). Because of their Gulf-wide distribution, I believe that these similar depositional patterns are the result of eustatic sea-level changes that have affected the entire Gulf, modified by the local subsidence of each individual basin. A seismic stratigraphic study of the Jurassic of the East Texas Basin (Todd and Mitchum, 1977) provides a Jurassic sea-level curve for the Gulf (Fig. 8). The lower Smackover was deposited under a rising sea-level regimen, the upper Smackover was deposited under conditions of a sea-level stillstand, or slight fall, and the overlying Haynesville was deposited during a rapid rise in sea level (Fig. 8). The depositional history of the Jurassic Smackover-Haynesville sequences has not been correlated to eustatic sea-level changes and differential subsidence patterns across the Gulf, and as a result, the present day models of these sequences are inadequate. The following discussion is an attempt to correlate eustatic sea level, subsidence, and depositional patterns across the Gulf, in order to clarify our understanding of the depositional history of these important rock sequences (Fig. 11).

Smackover Formation

Budd and Loucks (1981) published a depositional model for the Smackover in South Texas which seems to represent most workers' views of the cirum-Gulf Smackover depositional system (Fig. 9). This is basically a simple, stacked, progradational model, which makes the shoal facies of the upper Smackover the time equivalent of the deep-basin facies at the base of the sequence (Fig. 10). The fact that this sequence attains thicknesses of more than 1,000 feet indicates that the model is an oversimplification. If this is a simple progratational system, the upper Smackover should be no more than 50 feet thick, and the middle Smackover should be perhaps 100+ feet thick. Thus, the total relief from sea level to the commencement of the basinal facies would be somewhat over 150 feet. The upper Smackover, however, is commonly 300 to 400 feet thick, with the entire sequence attaining thicknesses of over 1,000 feet.

The first three panels of Figure 11 represent an idealized developmental sequence of the Smackover during the initial post-Louann transgression and the stillstand (or slight drop) at the top of the Smackover. The depositional environments and characteristics of the Smackover during this time were modified from Budd and Loucks (1981).

The initial sea-level rise was probably rapid, far outstripping carbonate deposition, with the transgression surface represented by a thin sandy zone. The first deposition on this surface would be relatively deep-water pelagic anoxic laminated carbonate muds (lower Smackover). Sediment bypass and nondeposition are common during the initial stages of any rapid transgression. Sediment bypass would result in the accumulation of a wedge of lower Smackover mudstones in the basin center, with non-deposition upslope in those basin margin areas affected by the rapid movement of a high-energy shoreline across the margin during transgression.

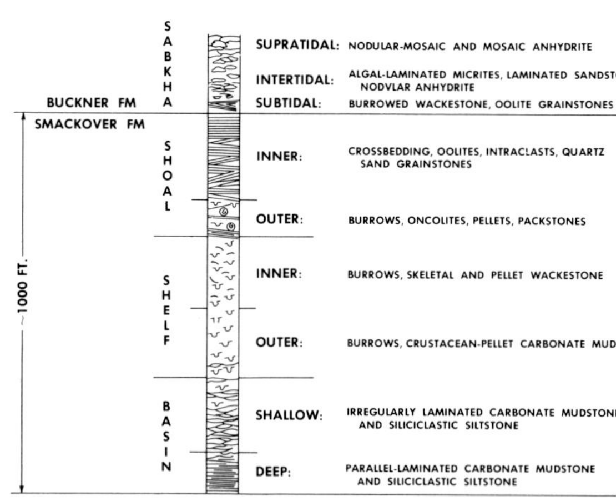

Figure 9. Vertical sequence of facies and interpreted depositional environments for the Upper Jurassic Smackover and Buckner, as seen by Budd and Loucks (1981) in South Texas. This sequence is generally characteristic of these units across the Gulf rim.

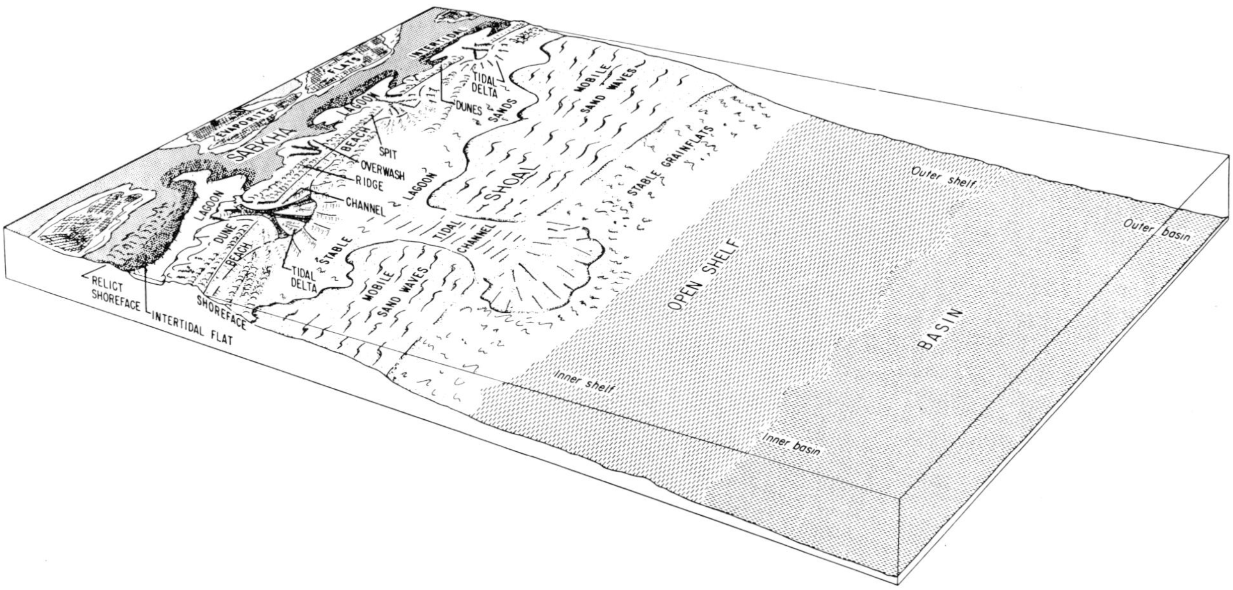

Figure 10. Schematic block diagram developed by Budd and Loucks (1981) from Figure 9 (this paper) to represent the depositional model for the Smackover in South Texas. This model is the one generally accepted by geologists across the Gulf and implies chronostratigraphic equivalency of the Buckner-upper Smackover, as well as upper Smackover with lower Smackover.

As the sea-level rise slowed, carbonate sedimentation began to overtake sea level and the sediment-water interface would emerge into oxic environments allowing moderate biologic activity, which is shown by burrowing, bioclasts and pellets in the middle Smackover. Maximum reef growth may have occurred in the Smackover at this time (Harris and Crevello, 1983). Updip, along the basin margins, little deposition took place because sea level was still rising across the shelf, causing a high-energy bypass zone as above.

As the sea-level rise ceased and maintained a high stand and the sediment-water interface built above wave base, a widespread high-energy shoal system was established, dominated by ooids, pellets, and oncolites. This system was maintained for a relatively long period of time as an equilibrium between subsidence and sedimentation was attained, during which more than 400 feet of high-energy carbonate sands were deposited in some areas. Although not entirely progradational, the vertical sequence of the upper Smackover is dominated by numerous stacked, coarsening-upward sequences (Fontana and Harwood, 1982) that represent localized base-level sedimentation and result in a complex mosaic of high-energy shallow-shoal environments. The upper Smackover high-energy deposits form a wedge that thickens in a basinward direction and probably reaches its maximum thickness along the margins of the salt basins because of increased rates of subsidence toward the basin. If this is a true ramp at this point, the sands should give way basinward to burrowed, pellet-dominated carbonate muds with little change in slope or relief. I believe that shelf-margin relief had already been established by the time of deposition of the uppermost Smackover because of incipient basin-margin faulting caused by the initiation of salt movement.

As Smackover deposition began to draw to a close, local shorelines, especially those associated with salt structures, were established with the development of sabkha condi-

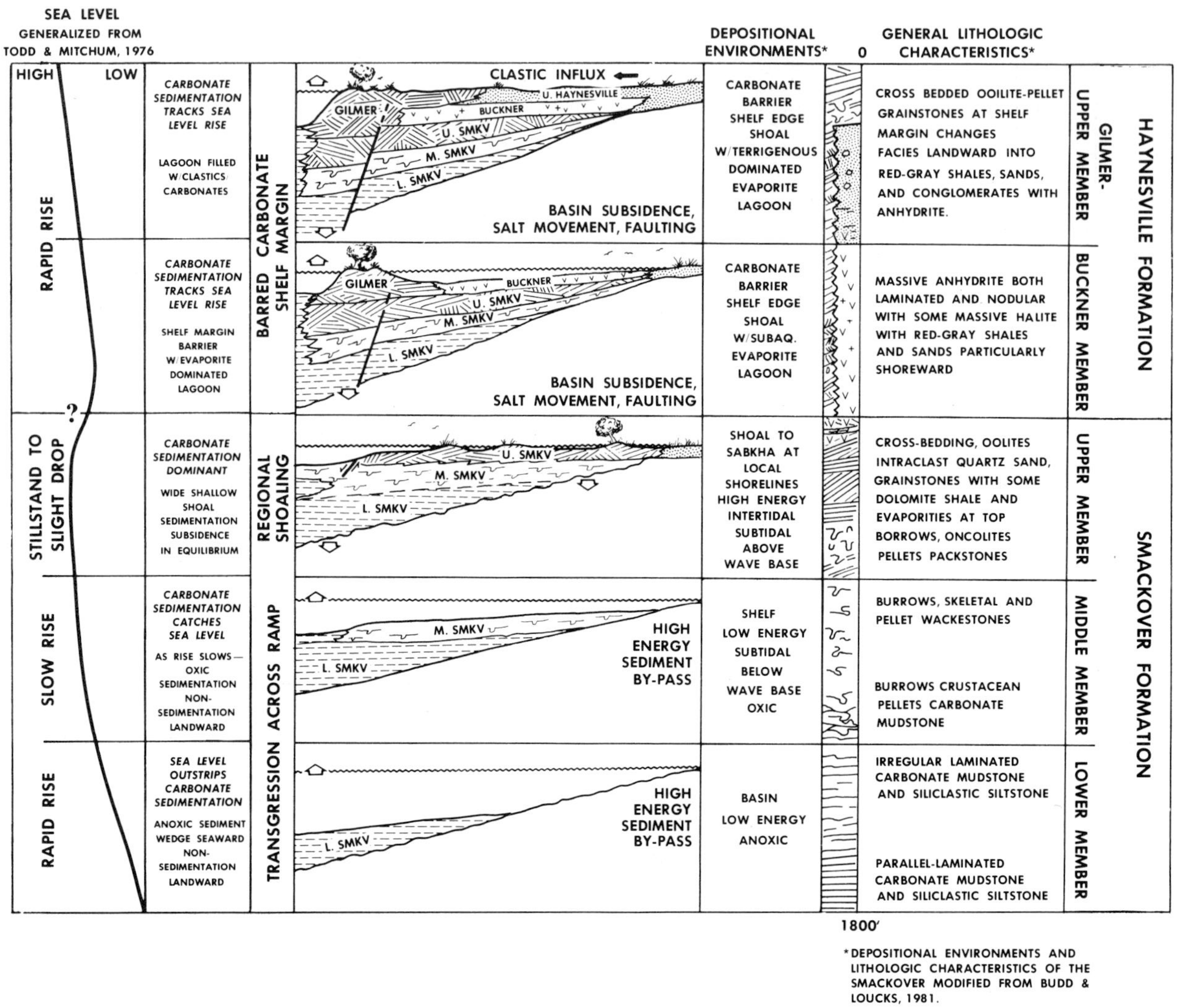

Figure 11. Idealized developmental sequence of the Late Jurassic Oxfordian-Kimmeridgian, relating sea level and subsidence to lithologic development and depositional environment.

tions on their shoreward flanks. Dolomite, evaporites, and some shale were deposited on these sabkhas in the uppermost Smackover. As sea level dropped, or as subsidence slowed, the updip portions of the Smackover facies tract was subaerially exposed and local evaporite-pan and sabkha conditions prevailed.

Haynesville Formation

During the deposition of the Haynesville Formation, sea level maintained a slow but steady rise (Figs. 8 and 11). As salt movement began in most of the salt basins, faulting and increased subsidence associated with that movement caused basin-margin relief. Carbonate deposition was localized by that basin-margin relief and the Gilmer carbonate barrier was formed. Broad lagoons developed behind the Gilmer barriers, and the massive anhydrite deposition of the Buckner Member commenced. In the deepest parts of these evaporite lagoons, halite deposition was common (Cass and Bowie Counties, Texas; Manila and Conecuh Embayments in Alabama-Florida). This interpretation of a Buckner lagoon formed behind a limestone barrier was proposed by Dickinson (1968b) (Fig. 17, this paper) but subsequent workers favored the popular sabkha model instead. This writer revives Dickinson's original concept, however, since most of the subsurface data used in this study fit the Dickinson model. Terrigenous clastics in Arkansas and Mississippi began to rapidly fill the lagoons from their landward sides to such an extent that evaporite deposition was suppressed or even eliminated. Depositional environments ranged from true evaporitic sub-aqueous lagoons to sabkha, marsh and tidal flat.

As Haynesville deposition continued, the Gilmer carbonate barriers were maintained and the terrigenous clastic influx covered most of the Gulf rim, obliterating most evaporite deposition. These clastics, which are the red shales to massive conglomerates and sandstones of the upper Haynesville, filled elongate depocenters which paralleled the Gilmer carbonate shelf-edge barriers. These elongate depocenters generally overlie the Buckner evaporite depocenters proposed by Dickinson (1968b) (Fig. 17, this paper). In some areas, such as East Texas, clastic influx waned, subsidence increased, and the limestones of the Gilmer Member extended landward as far as the Mexia-Talco Fault Zone.

According to Todd and Mitchum (1977), there was a significant drop of sea level at the end of Haynesville deposition (Fig. 8). The deep-water shales of the Bossier onlapped the Haynesville-Gilmer surface during the subsequent sea-level rise.

In summary, this interpretation of the depositional history of the Smackover-Haynesville formations differs from other accepted models in that: lower and middle Smackover wedges may not be time equivalents of the upper Smackover, but may be separate depositional systems representing sedimentary fill to base level during a major rapid sea-level rise. Also, the upper Smackover is not a simple, prograding basin-shoal-sabkha sequence, but is instead a widespread, complex, shallowshoal environment which developed during a sea-level stillstand on a slowly subsiding shelf; although evaporites and sabkha dolomites locally are developed in the upper Smackover, the updip facies equivalents of the ooid shoals are generally terrigenous clastics, and not the massive Buckner evaporites. In addition, a distinct shelf margin, with relief into the basin, developed during the Haynesville; and finally, Buckner evaporites are generally lagoonal deposits rather than dominantly sabkha deposits, which developed behind a shelf-edge carbonate barrier, the Gilmer, and represent a separate depositional system from the underlying Smackover Formation.

POROSITY TYPES OF THE UPPER SMACKOVER

The upper Smackover, the major reservoir unit of the Louark Group, occurs as a widespread blanket-type deposit that throughout most of its extent consists of ooid-dominated lime grainstones with only minimal organic components. Porosity, permeability, and pore types do vary widely, however, across the upper Smackover subcrop. Much of the following discussion is based on a regional diagenetic study in Arkansas and Louisiana by Moore and Druckman (1981).

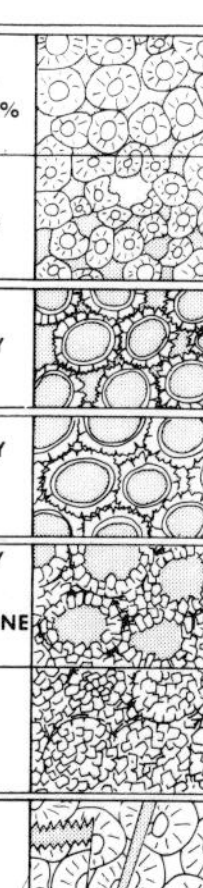

POROSITY TYPE	ASSOCIATED PERMEABILITY	FIELD EXAMPLES
PRIMARY PRESERVED INTERGRANULAR 1%-15%	HIGH TO MODERATELY LOW	WALKER CREEK, ARK. CHALYBEAT SPRINGS, ARK. EAST NANCY, MISS.
PRIMARY PRESERVED INTERGRANULAR SOLUTION ENHANCED 1%-20%	HIGH TO MODERATELY LOW	NORTH HAYNESVILLE, LA. WALKER CREEK, ARK.
PRIMARY PRESERVED OOMOLDIC SECONDARY 5%-50%	VERY HIGH	MT. VERNON, ARK.
OOMOLDIC SECONDARY CALCITE CEMENTED 5%-40%	VERY LOW TO TIGHT	MIDWAY, ARK. PAUP SPUR, ARK.
OOMOLDIC SECONDARY WITH DOLOMITE REPLACEMENT WITH DOLOMITE INTERCRYSTALLINE 15%-40%	MODERATE TO LOW	BRYAN'S MILL, TEX. JAY, ALA. FLA.
INTERCRYSTALLINE DOLOMITE 1%-10%	MODERATE TO LOW	FOUKE, ARK.
FRACTURE-STYLOLITE 1%-5%	LOW	GROGAN, LA.

Figure 12. Basic porosity types developed in the Jurassic upper Smackover across the Gulf rim.

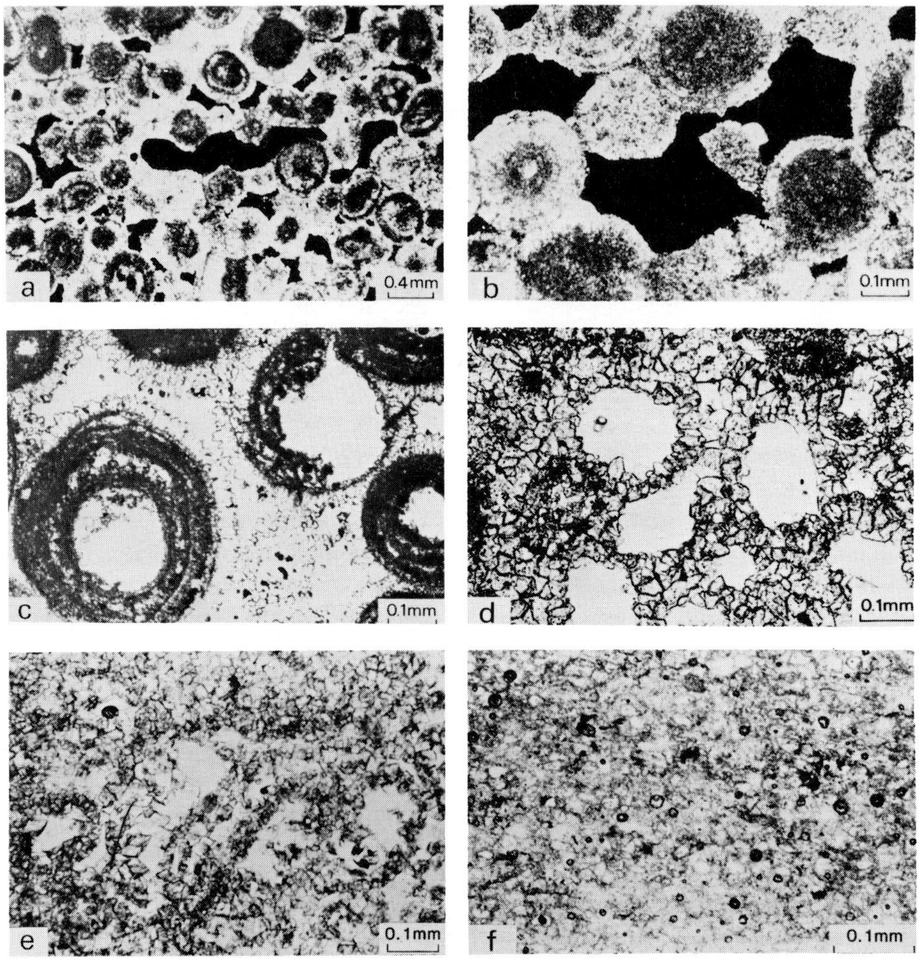

Figure 13. Thin-section photomicrographs of basic porosity types developed in the Jurassic upper Smackover.

a. Primary preserved porosity without an early diagenetic overprint. Elongate pore fabrics suggestive of late secondary leaching. Burial depth 10,000+ feet, Walker Creek Field, Columbia County, Arkansas. Crossed polars.

b. Primary preserved porosity with well-developed late-stage secondary porosity enhancing primary porosity, same well as a. Crossed polars.

c. Primary preserved, with oomoldic secondary porosity. Note early first generation bladed circumgranular cement, selective leaching of ooids and partial ooid collapse. Murphy Giffco, Miller County, Arkansas, 7,762 feet. Plane light.

d. Oomoldic secondary porosity with primary porosity totally occluded by early fine crystalline calcite cement and high porosity, low permeability. Mount Vernon Field, 7,998 feet Columbia County, Arkansas. Plane light.

e. Dolomitized oomoldic secondary porosity. Dolomite replacing an original calcite cement and enhancing interconnection of oomoldic porosity. Bryan's Mill Field, Cass County, 9,000+ feet burial. Plane light.

f. Intercrystalline dolomite porosity. An original ooid grainstone totally dolomitized. Low porosity, moderate permeability. East Texas, Cass County, 9,000 feet burial. Plane light.

Primary Preserved Intergranular Porosity (Figs. 12, 13)

Primary pore preservation is common in southern Arkansas, north Louisiana, eastern Mississippi, along the southern margins of the Mississippi Salt Basin, and western Alabama (Moore and Druckman, 1981; Badon, 1974) (Fig. 15). It occurs in rocks buried to depths in excess of 15,000 feet and may extend westward along the northern margins of the East Texas Salt Basin as a dominant pore type. Original porosity is commonly occluded by grain compaction, pressure solution, and late-stage calcite and dolomite cementation (Moore and Druckman, 1981). Rocks with primary porosity commonly do not have early diagenetic cements, other than occasional early marine circumgranular crust cements. Original porosities that may have exceeded 40 percent (Halley and Harris, 1979) are commonly reduced from 1 to 20 percent. Walker Creek and Chalybeat Springs Fields in Arkansas and East Nancy Field in Mississippi are examples of fields with reservoirs that are dominated by preserved primary porosity.

Deeply buried upper Smackover rocks containing preserved primary porosity commonly have significant late-stage subsurface solution that enhances porosity and permeability (Moore and Druckman, 1981). This type of late solution is well developed at North Haynesville Field in Louisiana and Walker Creek Field in Arkansas (Figs. 12, 13, 15).

Oomoldic Secondary Porosity (Figs. 12, 13, and 15)

Oomoldic secondary porosity is one of the most spectacular pore types in the upper Smackover. Oomoldic porosity is formed by the preferential dissolution of ooids along with concurrent precipitation of intergranular calcite cements, which may later be replaced by medium to coarse-crystalline dolomite (Figs. 12 and 13). Ooid grainstones with some preserved primary pore space not filled with early calcite cement, in addition to oomoldic porosity, have extremely high porosity and permeability (Fig. 13). This type of porosity development is common at the Mt. Vernon Field in central Arkansas. The primary pore space may, however, be completely occluded by early calcite cements, which effectively reduce permeability below economic levels while unaffecting the high porosity values (Fig. 13). This pore type occurs at Midway and Paup Spur Fields in central Arkansas. Reservoir rocks at Bryan's Mill Field, Texas, and Jay Field, Alabama-Florida have resulted from dolomitization of oomoldic rocks, re-establishing interconnection between oomolds (Fig. 13).

Oomoldic fabrics occur in the updip reaches of the upper Smackover subcrop from South Texas to Florida (Fig. 15), probably the result of diagenesis accompanying a regional fresh-water system. In general, calcite-cemented oomoldic rocks with poor permeability occur farther updip, while those oomoldic rocks with preserved primary porosity occur farther downdip toward the basin (Moore and Druckman, 1981).

However, dolomitized oomoldic fabrics (perhaps after an original calcite fabric) have a more restricted distribution and only occur as a dominant pore type in Texas, eastern Mississippi, Alabama, and Florida (Fig. 15)

Intercrystalline Dolomite Porosity (Figs. 12, 14, and 15)

Totally dolomitized reservoir rocks without significant oomoldic fabrics are not common, but do occur such as at Fouke Field in southwestern Arkansas. Although porosity values are low (5-10 percent), intercrystalline permeability may be moderate (40 md), and the reservoir may produce significant hydrocarbons.

Fracture and Stylolite Porosity (Figs. 12, 14, and 15)

Fractures and stylolites may act as open pores in the subsurface, because abundant petrographic evidence shows that both of these features may act as fluid conduits (Mazzullo, 1981). However, the Grogan Field

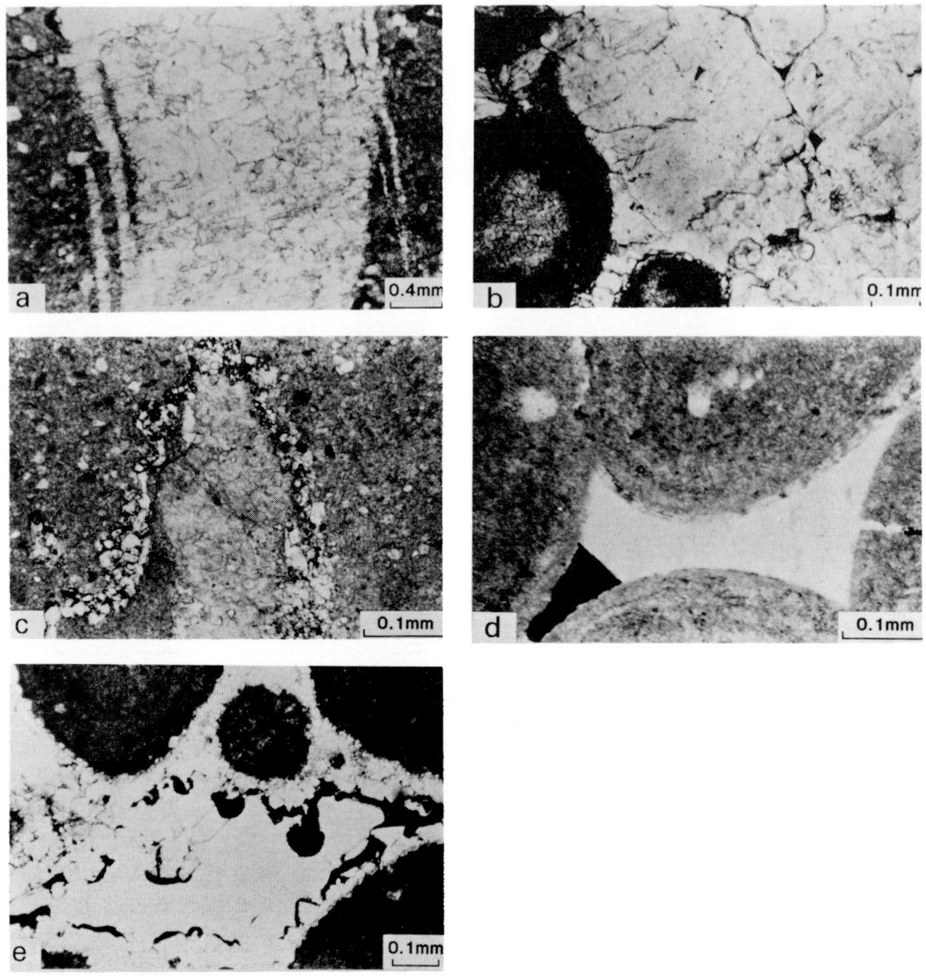

Figure 14. Thin-section photomicrographs of basic porosity types in the Jurassic upper Smackover and relationship of hydrocarbons to cementation in Arkansas and Texas.

a. Fracture in Jurassic upper Smackover ooid grainstones filled with sparry calcite cement Limestone County, Texas 14,000+ feet burial. Plane light.

b. Close up of a, above, showing hydrocarbons trapped in areas of incomplete cementation.

c. Stylolite in ooid lime grainstone from East Texas at 12,000 feet burial. Note insolubles along stylolite, relatively small amplitudes. Plane light.

d. Compacted ooid grainstone with post-compaction calcspar cement in the center. Dark area is hydrocarbon showing late arrival (post-compaction, post-cementation) southern Arkansas, 10,000 feet burial. Plane light.

e. Ooid grainstone from East Texas Bryan's Mill field showing relatively early timing of hydrocarbon, after initial cementation, dolomitization and shortly before late-stage calcspar cements. Compare with d. Plane light.

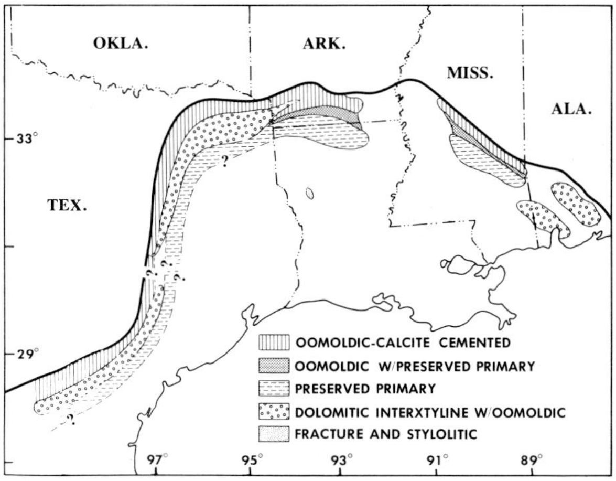

Figure 15. Regional porosity type distribution, Upper Jurassic Smackover.

in north central Louisiana may be the only example of production from open fractures in the upper Smackover (Fig. 15).

Regional Controls Over Porosity Development

In summary, the most common Late Jurassic Smackover porosity types are oomoldic and primary preserved intergranular. Their distributional patterns across the northern Gulf Coast area are predictable; oomoldic porosity occurs along the updip margins of the Smackover and primary preserved porosity is found in downdip areas near the salt-basin margins. Such a distributional pattern indicates that a regional meteoric water system was established in the upper Smackover shortly after deposition, during the Late Jurassic sea-level still stand. The aragonitic ooids were dissolved and calcite precipitated which led to the generation of the dominant oomoldic fabrics in updip directions. Downdip, the meteoric water system passed into a marine environment where the only early diagenetic event was the local precipitation of circumgranular marine cements (Moore and Druckman, 1981) (Fig. 16, phase 1, this paper). With later burial, oomoldic porosity was retained updip, while downdip, porosity occlusion and preservation were functions of burial compaction (Moore and Druckman, 1981).

But the question remains regarding the oomoldic dolomite that is an important reservoir rock in the western and eastern reaches of the upper Smackover subcrop. Two plausible explanations for these regional dolomitization events have been proposed: dolomitization in the mixing zone between the updip meteoric water system and downdip marine water system (Stamatedes, 1982; Vinet, 1982; Loucks and Budd, 1981); dolomitization associated with reflux from evaporites of the overlying Buckner Formation (Dickinson, 1968 a; Harris and Dodman, 1982). The mixing zone model, in its strictest sense, is not viable, because the mixing zone of fresh and marine waters occurs also in Arkansas and Mississippi on the shelf, and there the upper Smackover is not dolomitized. On the other hand, the direct relationship between the Buckner evaporites and the dolomitization in the upper Smackover (Dickinson, 1968 a), (Fig. 17, this paper) in East Texas has been clearly demonstrated by later drilling in Texas, Mississippi, Alabama, and Florida. No significant amount of dolomite has been found in the upper Smackover in eastern Arkansas, Louisiana, and Mississippi, and the interval that usually contains the evaporites of the Buckner Formation has been filled in these areas with terrigenous clastics derived in part from the Monroe uplift. In the Manila and Conecuh Embayments, thick sequences of Buckner evaporites that contain halite directly overlie oomoldic upper Smackover dolomite.

The upper Smackover dolomitization model favored by this author is schematically outlined in Fig. 16, phase 2, and discussed below.

At the end of Smackover deposition, as sea level began to rise during the deposition of the Haynesville Formation, the barrier formed by the Gilmer Limestone was established and the evaporite lagoon developed in which the evaporites of the Buckner Formation were deposited over the porous, upper Smackover oomoldic limestones. These limestones may still have been under the influence of fresh meteoric waters from the north. Reflux of highly saline fluids from the lagoon, which mixed with either fresh meteoric or ma-

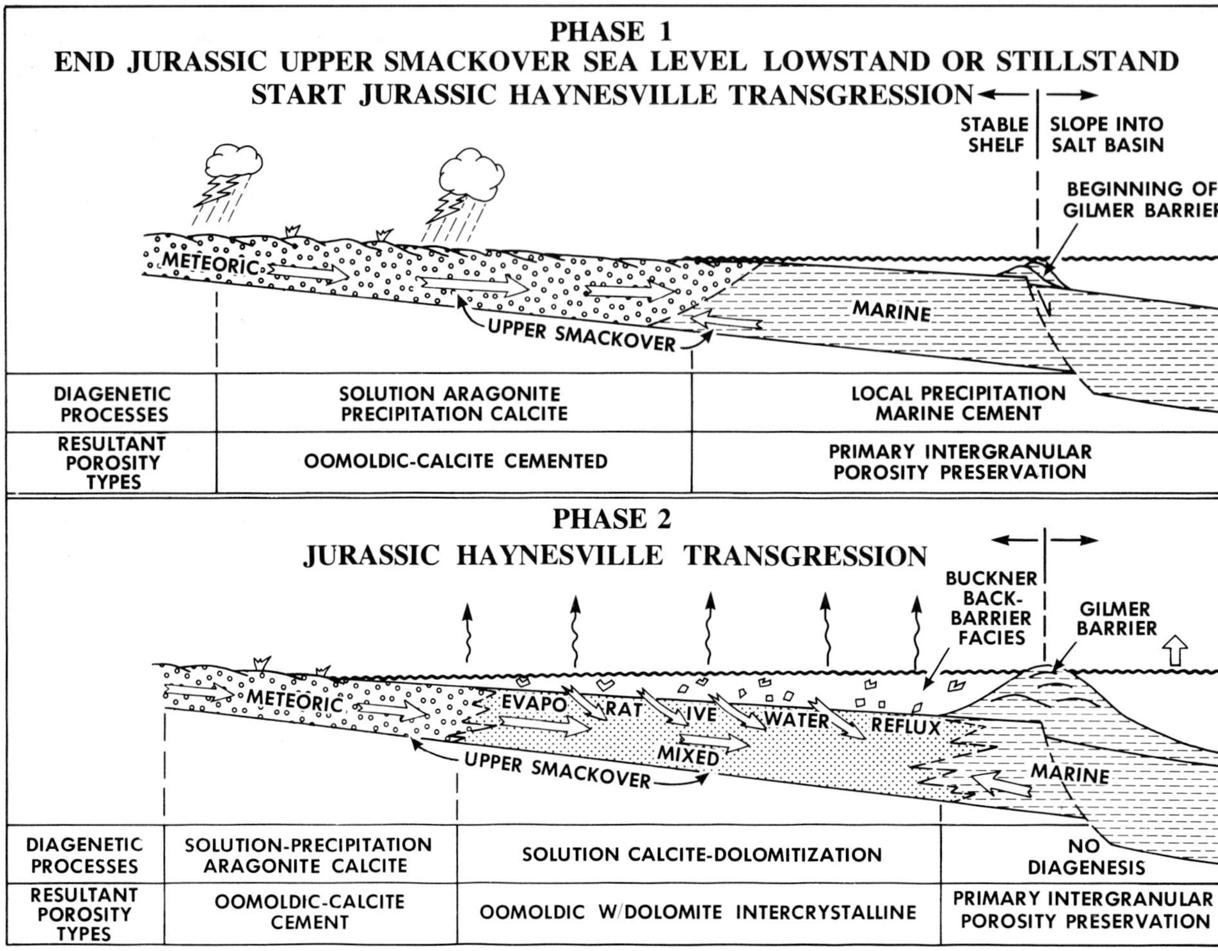

Figure 16. Conceptual model of the early diagenetic environments and processes responsible for the striking regional distributional patterns of porosity types in the upper Smackover across the Gulf Rim.

rine waters in the underlying porous limestones, resulted in massive dolomitization of the upper Smackover. Extensive upper Smackover dolomitization is almost always coincident with oomoldic porosity, hence not only are the Buckner evaporites necessary for dolomitization, but a fresh-water system is also a necessary ingredient. Geochemical and petrographic investigations currently under way should help to further refine the actual dolomitization process, or processes, and their timing.

UPPER JURASSIC SMACKOVER HYDROCARBON SOURCE ROCKS AND MIGRATION

For a geologic unit to have the potential to produce commercial quantities of hydrocarbon, the unit must have porous and permeable reservoir rocks and a mature hydrocarbon source rock nearby. In addition, a viable trapping mechanism and seal must be present at the time of hydrocarbon migration. In the preceding sections, the basic controls over the type and regional distribution of upper Jurassic Smackover reservoir rocks have been discussed. The next sections will cover the final two variables of the equation for the Smackover: source and trap.

Source Rock

Few published accounts of Gulf Jurassic source rocks exist, so that little specific data are available for discussion. However, well-defined geologic constraints limit the number of sources of Smackover-Haynesville oil and gas. Throughout most of its extent, the Smackover-Haynesville limestone wedge lies directly on the Louann Salt or Werner Anhydrite, Triassic-aged Eagle Mills coarse continental clastics, or highly metamorphosed upper Paleozoic rocks of the Ouachita Orogenic Belt to the west, and the Appalachians to the east. With the possible exception of a few isolated areas where Smackover limestones may directly overlie undeformed Palezoic marine sedimentary sequences (possibly East Texas), the Smackover-Haynesville sequence itself must be considered as a hydrocarbon source rock. The most logical source rocks within these strata are the dark limestones of the lower Smackover, which occur within and around the basin margins and which extend partly across the shelf in an updip direction, attaining thicknesses in excess of 500 feet. Basinward, the dark-colored organic-rich Bossier Shale, which in Texas and Louisiana onlaps the HaynesvilleSmackover along the basin margin and partway across the shelf, may be a source rock.

To the east, in eastern Mississippi, Alabama, and Florida, the quartzose clastic Norphlet Formation which underlies the Smackover reaches several hundred feet in thickness and contains black shales which could, in part, be the source of the hydrocarbons found in the Smackover Formation (Badon,

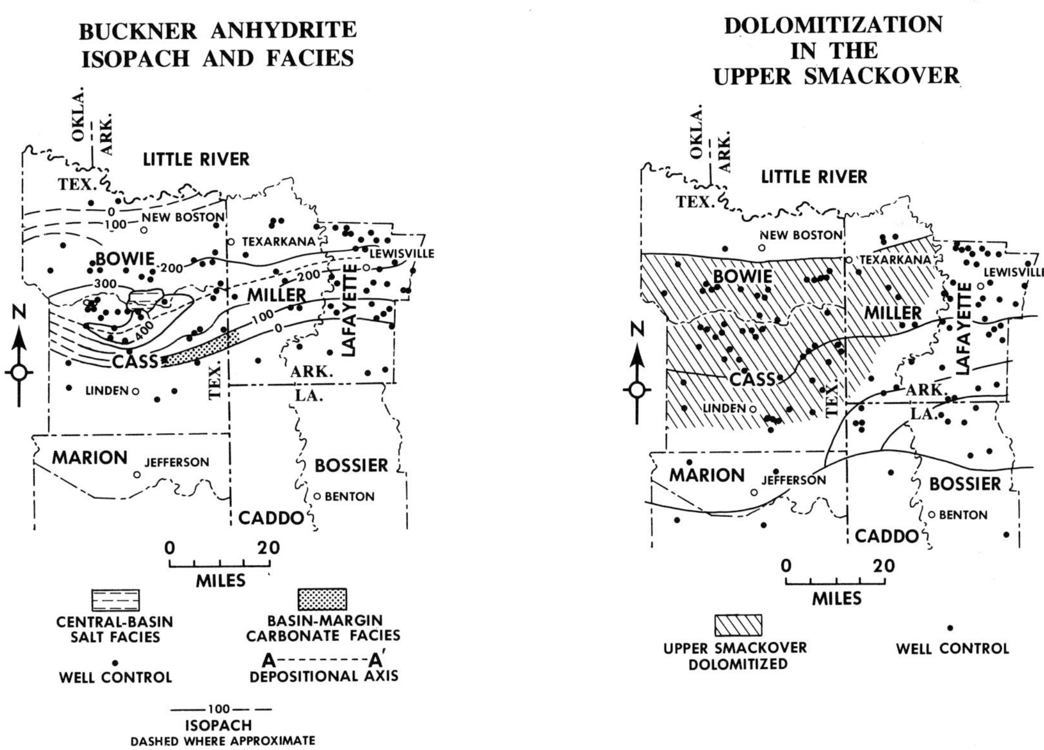

Figure 17. Relationship of Buckner Anhydrite and dolomitization in the upper Smackover, East Texas and western Arkansas-Louisiana as determined by Dickinson (1968). Both diagrams modified from Dickinson (1968a, b).

1974). The Norphlet Formation in Arkansas, Louisiana, and East Texas is too thin and coarse to represent a reasonable source for hydrocarbons. The Norphlet could, however, thicken at the basin margins and be a significant hydrocarbon source. There are no published data relative to the type of organic material in any of these potential source beds, nor any data about the level of maturation that has been reached. However, the geothermal gradient during the Late to Middle Jurassic was higher than it is today (Nunn, 1982), and many of the upper Smackover rocks across the Gulf have been exposed to temperatures that would place them well within the oil-generation window (Moore and Druckman, 1981).

Migration

Because most of the potential Smackover source rocks occur as wedges that thin onto the shelf, a local hydrocarbon source becomes unlikely in an updip direction. Thus, regional fluid migration from the basin margin across the shelf is very important in the Smackover (Moore and Druckman, 1981). Blanket carbonate sands of the upper Smackover are an ideal conduit for such regional migration, particularly in areas where regional dolomitization is common.

Oil emplacement in an area can be timed relative to the diagenetic history, or to the paragenetic sequence of the carbonate unit at that site. As an example, it can be observed that hydrocarbons arrived after one generation of cement, but before the next (Fig. 14). Petrography, stable isotopes, and two-phase fluid inclusions allow the geologist to estimate depth of burial, and hence, time of formation of these cements; and, by association, the time of arrival of the hydrocarbons can be estimated (Moore and Druckman, 1981). It can be documented that hydrocarbons migrated into south Arkansas reservoirs relatively late, after the percipitation of calcite cements at 90°C (Moore and Druckman, 1981). To the west, in East Texas, hydrocarbons arrived in reservoirs much earlier, after an early, precompaction cementation event, but before the later, deeper subsurface cements were precipitated (Fig. 14). This would imply that East Texas has a different subsidence history than does north Louisiana-south Arkansas, and indicates that the source rocks in East Texas reached thermal maturity earlier than those in Louisiana. Because the Smackover shelf in east Texas is much narrower than the equivalent shelf in Arkansas, hydrocarbons would arrive at the trap considerably earlier. These preliminary data parallel subsidence data being developed by J. Nunn by thermal modeling techniques (J. Nunn, personal communication, L.S.U., Baton Rouge, La.).

The question of timing of migration is critical, relative to mechanical traps, because the trap must be present during migration for entrapment to take place. This relationship has played a major role in hydrocarbon distribution within the upper Smackover, particularly in East Texas. Unfortunately at this time, the relative timing of hydrocarbon migration in Mississippi, Alabama, and Florida is unknown. Because two major salt basins seem to have different subsidence histories, there is no reason to believe that the subsidence rates of those eastern basins, and the South Texas basin to the west, will be identical to the rates in East Texas or Louisiana.

UPPER JURASSIC SMACKOVER TRAPS

Five major trap types are common in upper Smackover sequences across the northern Gulf (Fig. 18). The mechanical trap, associated either with faulting or folding is the dominant type; true stratigraphic traps are rare. In the majority of cases, the upper seal on Smackover traps is either Buckner anhydrites, or Haynesville shales resting on porous upper Smackover limestones. In the absence of these seals, hydrocarbon production is absent in the Smackover, but is commonly found in the overlying Gilmer Limestone. The regional distribution of trap types is shown on Fig. 19.

Updip Graben Fault Systems

Graben-fault systems near the updip limit of the Smackover, such as the Mexia-Talco System, form a regional system of potential mechanical traps that almost spans the entire northern Gulf. The time of activity of individual faults along the system may vary considerably, extending from the time of deposition of the lower Haynesville through the Cotton Valley and into the Cretaceous. The systems are discontinuous and each block may move independently of the next. In areas where porosity development is regional and migration is early (East Texas regional dolomitization), these fault systems form important traps. The typical trap occurs in the first upthrown block to the basin with the seal formed by the downthrown Haynesville shales or anhydrites in the adjacent graben (Fig. 18). Brantly-Jackson and Tawakoni are fields that produce from the Smackover along the Mexia-Talco Fault System (Collins, 1980).

TRAP TYPES	FIELD EXAMPLES
1. UPDIP FAULT TRAPS 	TAWAKONI, TX. BRANTLEY JACKSON, TX
2. SALT ANTICLINES 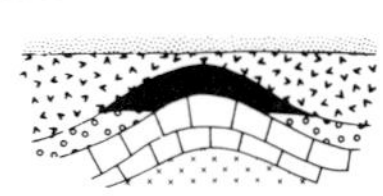	EUSTACE, TX. DORCHEAT MACEDONIA, ARK. NANCY, MISS. JAY-ESCAMBIA, ALA, FLA. HATTERS POND, ALA.
3. BASEMENT STRUCTURES (NO SALT, OR THIN SALT) 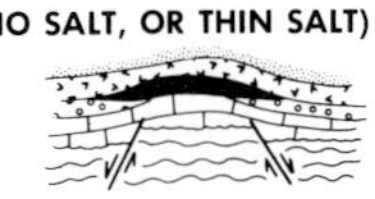	JOHN W., TX. BRYAN'S MILL, TX. MIDWAY, ARK. VOCATION, ALA.
4. COMPLEX GRABEN FAULT STRUCTURES 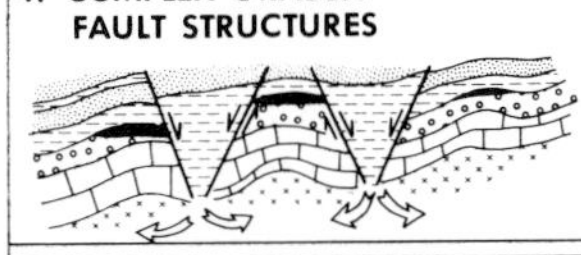	EAST HAYNESVILLE, LA. BAYOU MIDDLEFORK, LA
5. STRATIGRAPHIC TRAPS 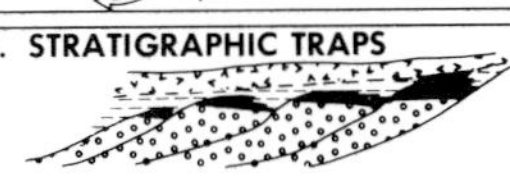	MT. VERNON, ARK. OAKS, LA. HARMONY, MISS.

Figure 18. Major hydrocarbon trap types, Jurassic Smackover and Gilmer carbonates.

Little production occurs along the Arkansas Graben Trend because migration was later and the majority of the hydrocarbons were trapped downdip, and/or regional migration barriers were established by later diagenetic events (Moore and Druckman, 1981).

Salt Anticlines

Subtle, salt-cored anticlines are perhaps the most common, and most important trap type in the Smackover-Haynesville sequence (Collins, 1980). Downdip from the bounding graben system (Mexia-Talco, etc.), where the Smackover wedge and the Louann Salt begin to thicken, subtle low-amplitude salt structures began to form in response to loading from the overlying Smackover-Haynesville, from increased density differential during compaction, and from increased basinward tilt through time. These anticlines are generally elongate and parallel the shelf edge, and occur in well-defined bands between the Gilmer shelf margin and the updip bounding graben-fault system in the East Texas Basin, south Arkansas, and Mississippi Salt Basin. Small tensional fault systems are commonly associated with the crests of the anticlinal structures. Eustice Field in East Texas, Dorcheat Macedonia Field in southern Arkansas, East

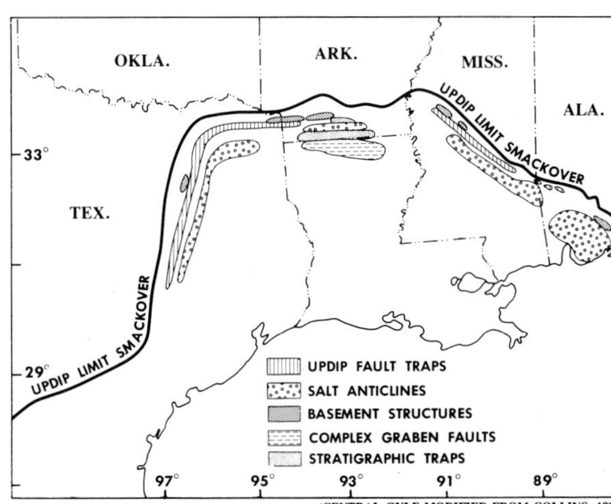

Figure 19. Regional distribution of major hydrocarbon trap types, Jurassic upper Smackover and Gilmer carbonates.

Nancy field in Mississippi, and probably Jay field in Alabama-Florida are examples of this most common trap type.

Basement Structures

Traps associated with discrete basement structures generally occur updip of the bounding graben-fault system in areas of thin salt, where horst blocks, Paleozoic cuestas, and erosional remnants formed topography at the time of the Late Jurassic transgression. This topography modified the Late Jurassic depositional environments, and resulted in local high-energy conditions over the structures. With subsequent burial, differential compaction of contrasting facies (ooids versus pellets, for example) can form hydrocarbon traps. Rejuvenation of basement faulting by loading leads to further enhancement of subtle basement-related traps. Production from such traps is limited and is always closely associated with major graben fault trends, such as the John W. and Bryan's Mill Fields in East Texas and the Midway Field in Arkansas (Fig. 19).

Complex Graben Fault Traps

The State Line Fault Trend along the Arkansas-Louisiana border is perhaps the site of the most complex trapping environment of the entire Jurassic trend. This fault system marks the northern border of the North Louisiana Salt Basin and consists of a series of east-southeast trending graben faults that were active during the deposition of upper Smackover carbonate sands, as well as throughout Haynesville deposition. The fault system probably formed in response to salt movement into the fronting North Louisiana Salt Basin, and developed a tectonically active shelf margin during Smackover deposition. Thick Smackover ooid deposits were generated on adjacent horst blocks; the deposition of ooids extended into the overlying Haynesville and formed a Haynesville barrier. Traps were often created by structural bowing of ooid-covered horst blocks by salt movement in response to graben formation, and bounding faults which juxtapose fine-grained Haynesville quartzose clastics against clean ooid sands (Croft, 1980; Bishop, 1973). The Haynesville and Bayou Middle Fork Fields in Louisiana are examples of this type of trap.

Stratigraphic Traps

Many upper Smackover fields have strong stratigraphic components to their traps. There is often evidence of updip porosity pinchout and a lack of structural closure. These traps are generally of two types in the upper Smackover. The first type exhibits loss of porosity due to facies changes from ooid shoal and shoreline conditions updip into sabkha environments (perhaps the Mount Vernon Field). This facies change only involves the upper 10 to 20 feet of the Smackover, and the resulting trap is long and narrow, is independent of structure, and has a thin oil column. These types of traps generally occur updip on the stable shelf part of the trend, and just downdip from the bounding graben fault trend.

The second stratigraphic trap type is associated with very subtle salt structures (as opposed to the well-defined salt anticlines). These traps do not cause structural closure, but because of salt movement during the upper Smackover, tend to concentrate

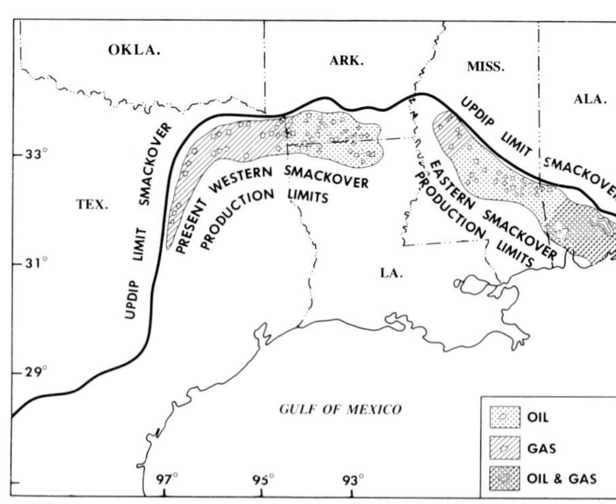

Figure 20. Hydrocarbon production, and type from the Jurassic upper Smackover, across the Gulf rim.

ooids over the axis of structural growth with lower energy sands (pellet and oncolites) deposited on the flanks. Thick sequences of ooid sands encased in pellet sands are localized in this manner. Subsequent burial and compaction results in porosity preservation in the ooid facies, and porosity destruction by compaction in the pellet facies (Brock and Moore, 1982). The resulting trap has no structural closure, but often exhibits intricately stacked multiple reservoirs such as at Walker Creek field.

CONTROLS OVER OIL AND GAS PRODUCTION UPPER JURASSIC SMACKOVER-HAYNESVILLE

Figure 20 illustrates the general distribution of oil and gas production in the Late Jurassic Smackover-Haynesville across the northern Gulf Rim. No production, to date, has been established in South Texas. East Texas is generally a gas province, in which the only oil production is concentrated in the fault traps against the Mexia-Talco Fault Trend, and in the occasional basement structure west of the Mexia-Talco System, such as the John W. Field in Henderson County. The oil along and west of the Mexia-Talco Fault Trend migrated early, used the porosity developed from regional dolomitization of the upper Smackover as a conduit, and filled the available traps on the updip side of the Mexia-Talco Trend before faulting. The oil was then trapped along the basinward side of the Mexia-Talco system as major faulting commenced. Little oil has been found in the salt structures east of the Mexia-Talco Fault Trend, because the traps were not present during early oil migration. The salt anticlines are, however, the site of major gas fields that produce from the Smackover where the Buckner Anhydrite is present, and from the Gilmer along the shelf margin where the Buckner is absent and where the Gilmer directly overlies the upper Smackover.

The Arkansas, Louisiana, and Mississippi region is primarily an oil province. It is not clear whether this west to east change from gas to oil is a function of source type (gas-prone sources in East Texas?), or burial history (deeper and more rapid burial in east Texas), or both. It is clear, however, that oil migration took place later in Arkansas and Louisiana, and that many traps were present by the time major oil migration commenced. Because of very early movement on the State Line Fault Trend, the area north of that trend may have a predominantly local source hydrocarbon. The shelf in southern Arkansas is very wide and the Smackover wedge thickens along the southern margins of the shelf, just north of the fault trend. Both of these features indicate some major sourcing from the lower Smackover north of the State Line Fault Trend. Regional diagenetic studies do, however, indicate large-scale fluid migration from south to north across this area (Moore and Druckman, 1981). It is uncertain whether this hydrologic system was bounded on the south by the State Line Fault System. Discrete areas of high H_2S and sour oil production in south Arkansas may be related to a separate Arkansas shelf hydrologic system in contact with interbedded Louann Salt and Werner Anhydrite.

Patterns of hydrocarbon production in Mississippi seem to parallel that found in Arkansas. However, the western part of the Mississippi Salt Basin is dominated by overpressured reservoirs that contain significant quantities of gas trapped in reservoir rocks composed of quartzose clastics which were presumably transported from the Monroe uplift (McCunn, *et al.*, 1982).

The Alabama-Florida part of the trend, particularly in the Conecuh Embayment, produces mixed oil and gas from salt-related structures and dolomitized reservoirs such as those in the Jay and Hatters Pond Fields. Here, dolomitization is related to the distribution of porous facies (Vinet, 1982; Bradford, 1982), which, in turn, can be related to early salt movement. The timing of hydrocarbon migration is uncertain, but if the source is local (the lower Smackover-Norphlet), it can be surmised that these rocks did not pass through the hydrocarbon-generation window until the Early Cretaceous, so that migration was probably Cretaceous in age (Bradford, 1982). The localization of dolomitization and the porosity ultimately preserved over early structural features would preclude regional hydrocarbon migration in this area.

FUTURE DEVELOPMENT

The frontier of Jurassic hydrocarbon exploration is no doubt in South Texas. Drilling and sedimentologic patterns in Mexico indicate a depositional model that is similar in most respects to that of East Texas, except in South Texas there is much deeper burial. Wells drilled to date have been well onto the shelf, and perhaps behind the Gilmer shelf edge, and updip from significant salt anticlinal development. The early regional dolomitization of the upper Smackover in South Texas is certainly favorable for porosity preservation, lateral migration, and significant accumulations, particularly along the Gilmer shelf margin. A South Texas equivalent of the Mexia-Talco Fault Trend has not been discovered. However, if one is present, it could play a significant role in future hydrocarbon exploration in South Texas. All accumulations would be gas.

In East Texas, the Gilmer shelf margin is a prime target that is relatively unexplored. Traps will be both structural as well as stratigraphic because of rapid lateral facies changes in a shelfward direction, and reefing may occur along the trend. The low sea-level stand before the Bossier transgression could lead to major secondary porosity development along the Gilmer trend across the entire northern Gulf.

Gilmer equivalents on the basinward side of the Conecuh Embayment in Florida also appear to be promising prospects, but they will probably produce only gas.

The updip bounding graben system across the northern Gulf will continue to be a promising exploration target in those areas associated with regional dolomitization, which allowed migration into, and in some cases across the trend into updip basement structures. South Texas, East Texas and parts of the trend in eastern Mississippi, Alabama, and Florida are most promising. Production should be oil, but the geologic constraints will make exploration difficult and success will require skillful subsurface geologic work in combination with close seismic control.

South Arkansas and East Texas will surely be the sites of numerous small stratigraphic discoveries related to shorelines and localized by subtle salt and basement structures.

Locally, high-rise structures in and along the margins of all the interior salt basins should be legitimate targets for exploration. Reefing, such as at Hico-Knolls field in Claiborne Parish, Louisiana, should be common even in the centers of the interior salt basins. Although porosity will probably be low, fracturing due to periodic salt rejuvenation, salt-related faulting, and late secondary porosity enhancement by solution could result in economic reservoir development.

CONCLUSIONS

1. Eustatic sea-level fluctuations, modified by local subsidence and sediment availability, have resulted in patterns of sedimentation common to Upper Jurassic sequences from the Rio Grande to Florida, and hence indicate the wide applicability of a generalized stratigraphic framework.
2. The sedimentologic model of the Smackover-Haynesville must be modified to take into account sea-level fluctuations, subsidence, and sediment availability. The new model presented here is one of basin fill during a long-term transgression in lower Smackover time, followed by regional shoaling during the deposition of the upper Smackover during a sea-level stillstand when sedimentation was in equilibrium with subsidence. The lower and upper Smackover are not necessarily time equivalent, but represent two separate sedimentologic-sea-level regimens. The Haynesville is thought to be a

separate sedimentologic package deposited during the next sea-level rise. The Gilmer Limestone formed a shelf-edge barrier behind which the Buckner evaporites were deposited under lagoonal conditions; these limestones graded into quartzose clastics in a landward direction.

3. Predictable regional porosity patterns have been developed in the Smackover-Haynesville in response to early diagenetic overprints, which were controlled primarily by eustatic sea-level subsidence interactions. These patterns include updip oomoldic porosity in a regional meteoric water system developed during the Oxforidan sea-level still-stand; downdip porosity preservation under marine conditions along the shelf margin; and regional dolomitization associated with reflux of evaporitive waters from the Buckner lagoon, formed behind the Gilmer shelf-margin barrier.
4. The mechanical hydrocarbon trap associated with salt movement is the most common Late Jurassic trapping mechanism. Buckner evaporites or Haynesville shales are common seals in Late Jurassic reservoirs.
5. Jurassic source rocks are thought to be lower Smackover limestones and Norphlet shales. Sourcing is generally local with migration into updip areas, particularly where regional dolomitization has occurred.
6. The timing of hydrocarbon migration varies across the Gulf Rim in response to the subsidence history of each individual basin. Time of migration is a key factor in a viable exploration strategy.
7. Future Jurassic exploration will center on South Texas, the Gilmer shelf margin, and the updip Smackover along the bounding graben fault systems. Most production will be gas, but some oil will be found along the updip faulted fairways.

ACKNOWLEDGMENTS

It is impossible to acknowledge all of the individuals who have unselfishly shared their knowledge of the Jurassic with me over the past several years, and who helped to shape my own limited view of this most interesting rock sequence. I would like, however, to single out; Emily and David Stoudt for furnishing an enormous amount of data, cores and encouragement; George Herman for his ever helpful nudges; Jim Sullivan, Jan van Sant and the Pennzoil Company for their encouragement and help during the formative stages of this paper; the industrial associates of the Applied Carbonate Research Program for sustained funding of our Jurassic program; Barbara Delaville, Marlene Moore, Danny Peace, Dupe Duplechin and Mary Eggart for help in manuscript preparation. Finally, I would like to acknowledge a very special group that over the past months has helped crystallize many of these ideas by acting as a collective, patient sounding board; Gill Harwood, Susan Cooper, Kathy McGillis, Sara Stewart and Scott Wilkinson - the "East Texas Gang."

REFERENCES

Ahr, W.M., 1973, The Carbonate Ramp: an alternative to the shelf model: Trans. Gulf Coast Assoc. Geol. Soc., v. 23, p. 221-225.

Aiken, R.H., and R.W. Graves, 1969, Reynolds oolite of southern Arkansas: Am. Assoc. Petroleum Geologists Bull., v. 53, p. 1909-1922.

Badon, C.L., 1974, Petrology and reservoir potential of the upper member of the Smackover Formation Clarke County, Mississippi: Trans. Gulf Coast Assoc. Geol. Soc., v. 24, p. 163-174.

Becher, J., and C.H. Moore, 1976, the Walker Creek Field: a Smackover diagenetic trap: Trans. Gulf Coast Assoc. Geol. Soc., v. 26, p. 34-56.

Benson, D.J., and E.A. Mancini, 1982, Porosity development and reservoir characteristics of the Smackover Formatiom in south Alabama: Program and Abstracts; the Jurassic of the Gulf Rim; 3rd Annual Research Conference, Gulf Coast Section Soc. Econ. Paleontologists Mineralogists, p. 11-12.

Bishop, W.F., 1969, Environmental control of porosity in the upper Smackover Limestone,

North Haynesville Field, Claiborne Parish, Louisiana: Trans. Gulf Coast Assoc. Geol. Soc., v. 19, p. 155-169.

__________, 1971a, Stratigraphic control of production from Jurassic calcarenites, Red Rock Field, Webster Parish, Louisiana, Trans. Gulf Coast Assoc. Geol. Soc., v. 21, p. 125-137.

__________, 1971b, Geology of upper member of Buckner Formation, Haynesville Field area, Claiborne Parish, Louisiana: Am. Assoc. Petroleum Geologists Bull., v. 55, p. 566-580.

__________, 1971c, Geology of a Smackover stratigraphic trap, Am. Assoc. Petroleum Geologists Bull., v. 55, p. 51-63.

__________, 1973, Late Jurassic contemporaneous faults in north Louisiana and south Arkansas: Am. Assoc. Petroleum Geologists Bull., v. 57, p. 858-877.

Bradford, C.A., 1982, Transgressive-regressive carbonate facies of the Jurassic Smackover formation, Big Escambia Creek and Flomaton fields, Escambia County Alabama: Program and Abstracts; The Jurassic of the Gulf Rim; 3rd Annual Research Conference, Gulf Coast Section Soc. Econ. Paleontologists Mineralogists, p. 15-17.

Brock, F.C., and C.H. Moore, 1982, Walker Creek revisited: a re-interpretation of the diagenesis of the Smackover Formation of Walker Creek Field, Arkansas: Trans. Gulf Coast Assoc. Geol. Soc., v. 31, p. 49-58.

Budd, D.A., and R.G. Loucks, 1981, Smackover and lower Buckner formations, south Texas: Depositional systems on a Jurassic carbonate ramp: Bureau of Economic Geology, the Univ. of Texas, Austin. Report of Investigations No. 112, 38 p.

Buffler, R.T., J.S. Watkins, F.J. Schaub, and J.L. Worzel, 1980, Structure and early geologic history of the deep central Gulf of Mexico basin; in The Origin of the Gulf of Mexico, proceedings of a symposium held at Baton Rouge, La, p. 3-16.

Collins. S.E., 1980, Jurassic Cotton Valley and Smackover reservoir trends, east Texas, north Louisiana, and south Arkansas: Am. Assoc. Petroleum Geologists Bull., v. 64, no. 7, p. 1004-1013

Croft, W.S., 1980, Deposition and diagenesis of the Jurassic upper Smackover Formation North Haynesville Field, northern Louisiana: Unpublished M.S. Thesis Louisiana State University, 96 p.

Dickinson, K.A., 1963, The upper Jurassic, stratigraphy of Mississippi and southwestern Alabama (Abs.): Dissertation abstract, v. 24, p. 248-249.

__________, 1968a, Petrology of the Buckner Formation in adjacent parts of Texas, Louisiana, Arkansas: Jour. Sed. Petrology, v. 38, p. 555-567.

__________, 1968b, Upper Jurassic stratigraphy of some adjacent parts of Texas, Louisiana, and Arkansas: U. S. Geological Survey Professional Paper 594-E, 25 p.

Fontana, C., and G.M. Harwood, 1982, Deposition and diagenesis of upper Smackover carbonates, Bryan's Mill Field, east Texas: Program and abstracts: The Jurassic of the Gulf Rim; 3rd Annual Research Conference Gulf Coast Section Soc. Econ. Paleontologists Mineralogists, p. 45-46.

Halley, R.B., and P.M. Harris, 1979, Fresh water cementation of a 1000-year-old oolite: Jour. Sed. Petrology, v. 49, p. 969-988.

Harris, P.M., and P.D. Crevello, 1982, Smackover Reefs - an example from Walker Creek Field, Arkansas: Program and abstracts; the Jurassic of the Gulf Rim; 3rd Annual Research Conference, Gulf Coast Section Soc. Econ. Paleontologists Mineralogists, p. 11-12.

__________, 1983, Upper Jurassic Smackover Reefs - an example from Walker Creek Field, Arkansas: Soc. Econ. Paleontologists Mineralogists Core Workshop No. 4, Dallas, p. 366-379.

Harris, P.M., and C.A. Dodman, 1982, Jurassic evaporites of the U. S. Gulf Coast: The Smackover-Buckner contact: Depositional and Diagenetic Spectra of Evaporites - a core workshop, Soc. Econ. Paleontologists Mineralogists core workshop no. 3, p. 174-192.

Hazzard, R.T., 1939, Notes on the Comanche and pre-Comanche? Mesozoic formations of the Arkansas-Louisiana-Texas area, and a suggested correlation with northern Mexico, in Shreveport Geological Society Guidebook, 14th Annual Field Trip: p. 155-165.

__________, W.C. Spooner, and B.W. Blanpied, 1947, Notes on the stratigraphy of the formations which underlie the Smackover Limestone in south Arkansas, northeast Texas and north Louisiana: Shreveport Geol. Society Reference Rept., v. 2, p. 483-503.

Imlay, R.W., 1943, Jurassic formations of the Gulf region: Am. Assoc. Petroleum Geologists Bull., v. 27, no. 11, p. 1407-1533.

__________, 1952, Correlations of the Jurassic formations of North America, exclusive of Canada: Geol. Soc. America Bull., v. no. 9, p. 953-992.

__________, 1980, Jurassic paleobiogeography of the conterminous United States in its continental setting: U. S. Geological Survey Professional Paper 1062, 134 p.

Loucks, R.G., and D. A. Budd, 1981, Diagenesis and reservoir potential of the Upper Jurassic Smackover Formation of South Texas: Trans. Gulf Coast Assoc. Geol. Soc., v. 31, p. 339-346.

Mancini, E.A., and D. J. Benson, 1980, Regional stratigraphy of Upper Jurassic Smackover carbonates of southwest Alabama: Trans. Gulf Coast Assoc. Geol. Soc., v. 30, p. 151-163.

Mazzullo, S.J., 1981, Facies and burial diagenesis of a carbonate reservoir: Chapman Deep (Atoka) Field, Delaware Basin, Texas, Am. Assoc. Petroleum geologists Bull., v. 65, p. 850-865.

McCunn, H.J., G.C. Ward, and C. Cox, 1982, Johns Field, Upper Smackover, Mississippi: Program and Abstracts: the Jurassic of the Gulf Rim; 3rd Annual Research Conference, Gulf Coast Section Soc. Econ. Paleontologists Mineralogists, p. 45-46.

Moore, C.H., and Y. Druckman, 1981, Burial diagenesis and porosity evolution, Upper Jurassic Smackover, Arkansas and Louisiana: Am. Assoc. Petroleum Geologists Bull., v. 65, no. 4, p. 597-628.

Murray, G.E., 1961, Geology of the Atlantic and Gulf coastal province of North America: Harper & Bros., New York, 692 p.

Nunn, J.A., 1982, Subsidence and temperature histories for Jurassic sediments in the northern Gulf Coast: A thermal-mechanical model: Program and Abstracts; The Jurassic of the Gulf Rim; 3rd Annual Research Conference, Gulf Coast Section Soc. Econ. Paleontologists Mineralogists, p. 45-46.

Ottmann, R.D., P.L. Keyes, and M.A. Ziegler, 1973, Jay Field, a Jurassic stratigraphic trap: Trans. Gulf Coast Assoc. Geol. Soc., v. 23, p. 146-157.

Oxley, M.L., E. Minihan, and J.M. Ridgeway, 1967, A study of the Jurassic sediments in portions of Mississippi and Alabama: Trans Gulf Coast Ass. Geol. Soc., v.17, p. 24-48

Philpot, T.H., and R.T. Hazzard, 1949, Preliminary correlation chart of upper Gulf Coast: Shreveport Geological Society 17th annual field trip guide book (in pocket).

Rosenkrans, R.R., and J.D. Marr, 1967, Modern seismic exploration of the Smackover Trend: Jour. Geophysics, v. 32, no. 2, p. 184-206.

Sigsby, R.J., 1976, Paleoenvironmental analysis of the Big Escambia Creek - Jay - Blackjack Creek Field area: Trans. Gulf Coast Assoc. Geol. Soc., v. 26, p. 258-270.

Stabler, C.L., 1982, Influence of the Tamaulipas Arch on Jurassic sedimentation in eastern Mexico: Program and abstracts; The Jurassic of the Gulf Rim; Third Annual Research Conference, Gulf Coast Section, Soc. Econ. Paleontologists Mineralogists, p. 98.

Stamatedes, M.R., 1982, Dolomitization of the upper Smackover in Miller County, Arkansas and adjacent areas: unpublished M.S. thesis, Louisiana State University, 155 p.

Todd, R.G., and R.M. Mitchum, 1977, Seismic stratigraphy and global changes of sea level, Part 8: identification of upper Triassic, Jurassic, and lower Cretaceous seismic sequences in Gulf of Mexico and offshore West Africa: in, Seismic stratigraphy applications to hydrocarbon exploration; AAPG Memoir 26, p. 145-163.

Vinet, M.J., 1982, Stratigraphy and dolomitization of the Smackover and Buckner formations (Upper Jurassic) in the Jay-Big Escambia Creek Field area, southern Alabama - western Florida: Program and abstracts; The Jurassic of the Gulf Rim; Third Annual Research Conference, Gulf Coast Section Soc. Econ. Paleontologists Mineralogists, p. 110-113.

Walper, J.L., 1980, Tectonic evolution of the Gulf of Mexico: in, The opening of the Gulf of Mexico; proceedings of a symposium held in Baton Rouge, La., p. 87-98.

Weeks, W.B., 1938, South Arkansas stratigraphy with emphasis on the older coastal plain beds: Am. Assoc. Petroleum Geologists Bull., v. 22, p. 953-983.

SUBSIDENCE AND TEMPERATURE HISTORIES FOR JURASSIC SEDIMENTS IN THE NORTHERN GULF COAST: A THERMAL-MECHANICAL MODEL

JEFFREY A. NUNN

Department of Geology
Louisiana State University
Baton Rouge, Louisiana 70803

ABSTRACT

The northern coast of the Gulf of Mexico contains major depositional basins in East Texas, North Louisiana, and Central Mississippi. Basin subsidence can be interpreted as resulting from the extension of continental crust by rifting related to the opening of the Gulf of Mexico during Late Triassic to Early Jurassic times. For example, subsidence of the North Louisiana Salt Basin, determined from well data, is consistent with crustal extension by a factor of 1.5 to 2. A seismic survey across the Central Mississippi Salt Basin also indicates crustal extension by about a factor of 1.6. The thermal-mechanical model presented here describes the subsidence and temperature histories of a continental margin, assuming that the lithosphere is stretched and thinned during rifting. The effects of thermal blanketing by overlying sediments are included. The observed position of strata and current thermal regime are used as model constraints. Subsidence and temperature histories for Jurassic-age sediments predicted from this model are compared with observational data. The results of this comparison indicate that geothermal gradients were higher in the past and that thermal conditions in Jurassic-age sediments were favorable for hydrocarbon generation.

INTRODUCTION

Many recent studies have shown that the evolution of rifted, Atlantic-type continental margins can be explained by thinning of the crust and lithosphere as a result of lateral extension prior to final continental breakup and onset of sea-floor spreading (Watts and Steckler, 1979; Royden *et al.*, 1980; Royden and Keen, 1980; Bradley, 1982; Sawyer *et al.*, 1982). It is the intent of this paper to demonstrate the applicability of the extension model to the northern Gulf Coast.

The northern coast of the Gulf of Mexico contains a thick sequence of predominantly shallow-water Mesozoic and Cenozoic sediments (Martin, 1978). Most of the Mesozoic sediments are concentrated in major depositional basins in East Texas, North Louisiana, and Central Mississippi (Fig. 1). They are bounded on the north and west by the continental platform and are separated by several uplifts (arches or posiments).

The East Texas, North Louisiana, and Central Mississippi Salt Basins have been interpreted as resulting from opening of the Gulf of Mexico during Late Triassic to Early Jurassic times (Fig. 1; Pilger, 1981). The presence of thick Jurassic salt deposits indicates that the basins formed before or during the earliest phases of opening of the Gulf. The basins also contain Eagle Mills Formation clastics which are similar to the Newark Series of the East Coast Triassic rift basins. In addition, Jurassic-Triassic age mafic rocks, presumably associated with rifting, have been reported from numerous wells around the margins of the Gulf (Edick,1982).

This study concentrates on the North

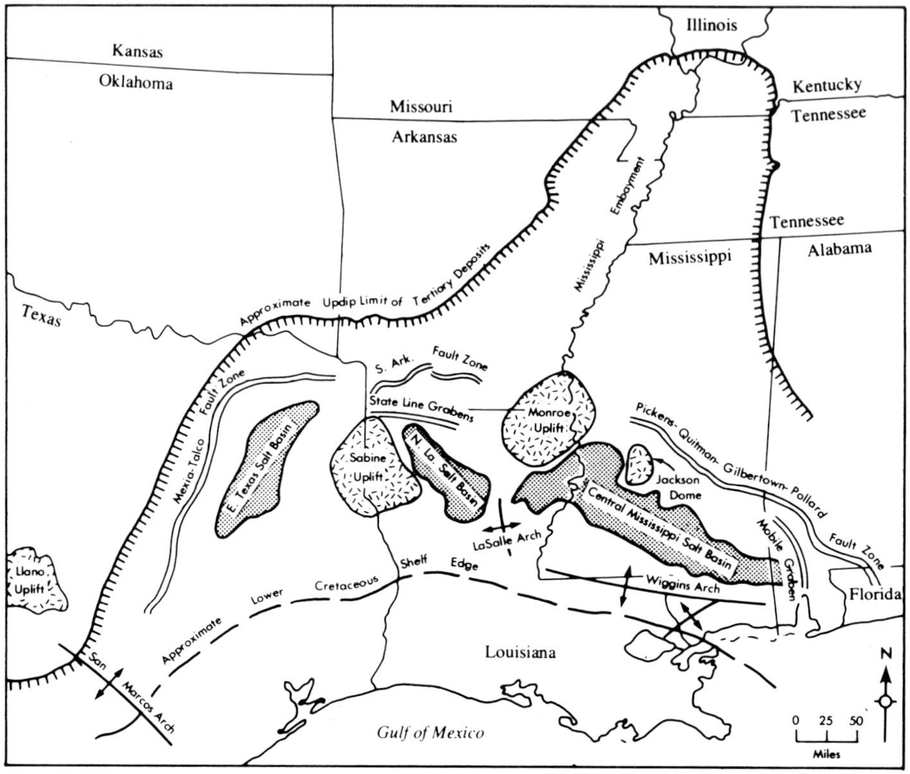

Figure 1. Index map of the major tectonic features of the northern coast of the Gulf of Mexico (after Scardina, 1982). The Mississippi Embayment is outlined by the updip limit of Tertiary deposits.

Louisiana Salt Basin because of the availability of subsurface data (Scardina, 1982). Subsidence curves based on well data are compared with curves predicted from the extension model. Once the age and intensity of the rifting process have been established from the subsidence history, the thermal evolution of the northern Gulf Coast can be computed using a thermal-mechanical model.

EXTENSION

The two models used here which describe the rifting process are the uniform extension model (McKenzie, 1978) and the non-uniform or depth dependent extension model (Sclater *et al.*, 1980; Royden and Keen, 1980). In both models, thinning of the crust and upper mantle as a result of lateral extension is the dominant process during rifting (Fig. 3). Although these models have been discussed in detail elsewhere (McKenzie, 1978, Sclater *et al.*, 1980; Royden and Keen, 1980), a brief description of each model is given below to maintain continuity and to illustrate how the models were constructed.

Uniform Extension Model

The uniform extension model, first described by McKenzie (1978), assumes constant extension of both the crust and mantle by a factor β (Fig. 3). Thinning of the lithosphere is compensated by passive upwelling of hot asthenospheric material. Elevation changes predicted by the extension model can be divided into two phases: initial and thermal. The initial elevation change, which occurs during rifting, is rapid, and occurs in isostatic response to net density changes resulting from the replacement by thinning of continental-crust material with passively upwelled mantle material. Density changes due to this compositional difference are partially offset by heating and thermal expansion as a result of the higher temperature of the passively upwelled asthenospheric material. The combined effect of these two types of density changes results in uplift for initial crustal thicknesses less than 20 km and initial subsidence for crustal thicknesses initially greater than 20 km (Royden and Keen, 1980). Following rifting, cooling and thermal contraction of hot mantle material which was brought abnormally close to the surface during extension produces a long period (approximately 100 million years) of gradual thermal subsidence (McKenzie, 1978). The sum of initial and thermal subsidence is the tectonic subsidence. The magnitude of tectonic subsidence is only dependent on the original thickness and amount of extension of continental crust (Jarvis and McKenzie, 1980). Model calculations are made with the assumption that basement subsidence takes place in the absence of sediment. In general, tectonic subsidence is greatly amplified by the weight of accumulated sediment.

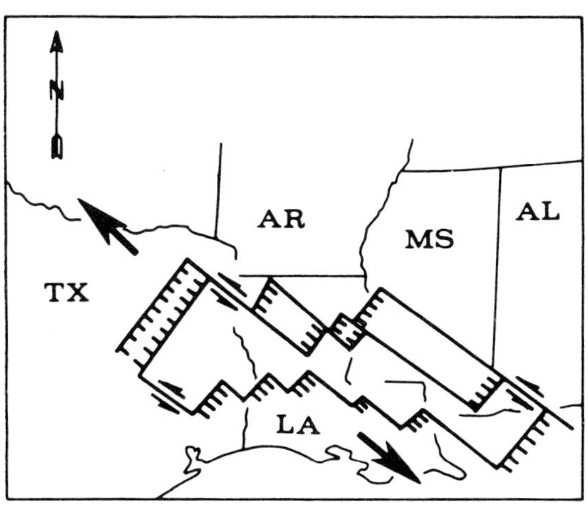

Figure 2. Spreading and transform model of the northern Gulf Coast consistent with the early opening direction of North Atlantic Ocean (after Pilger, 1981).

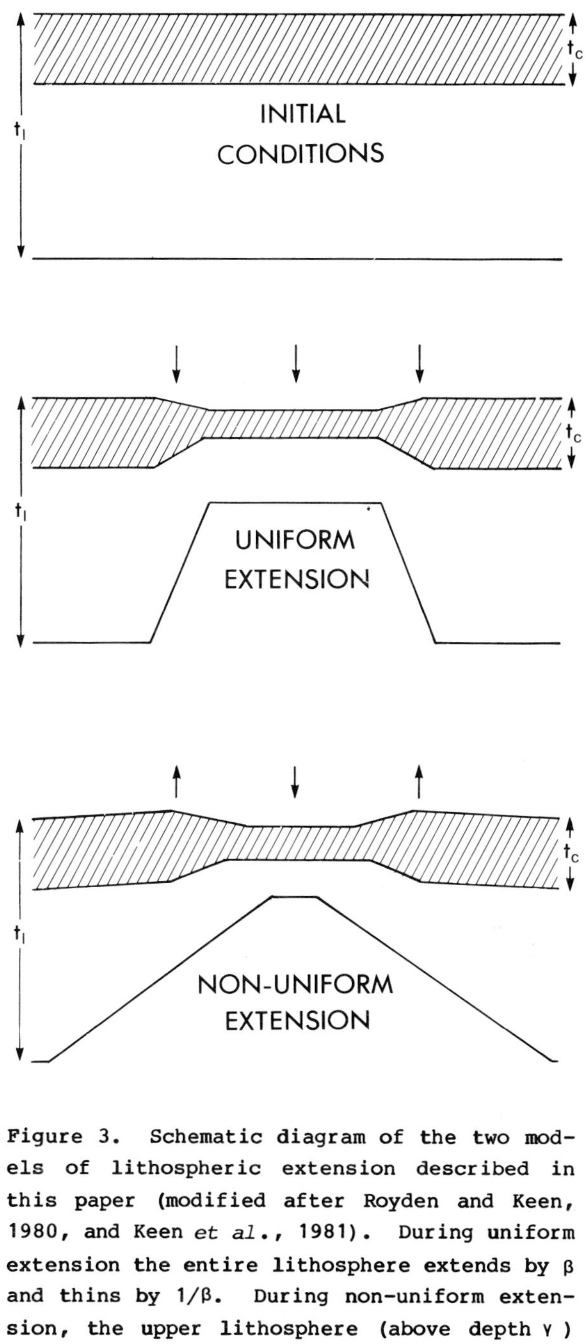

Figure 3. Schematic diagram of the two models of lithospheric extension described in this paper (modified after Royden and Keen, 1980, and Keen *et al.*, 1981). During uniform extension the entire lithosphere extends by β and thins by $1/\beta$. During non-uniform extension, the upper lithosphere (above depth γ) extends by δ and thins by $1/\delta$. The lower lithosphere extends by β and thins by $1/\beta$. The arrows above each model indicate whether initial uplift or subsidence is predicted.

Non-Uniform Extension Model

The non-uniform or depth-dependent extension model differs from the uniform extension model in that it assumes different amounts of extension in the upper and lower

Table 1. Physical Constants Used in Model Calculations

Meaning	Value
Thermal Expansion Coefficient	3.3×10^{-5} °C^{-1}
Thermal Conductivity of Lithosphere	8×10^{-3} cal/cm°C sec
Heat Capacity	.3 cal/gm°C
Density of Lower Lithosphere	3.3 gm/cm^3
Density of Crust	2.9 gm/cm^3
Density of Water	1.0 gm/cm^3
Initial Crust Thickness	32 km
Initial Lithosphere Thickness	120 km
Initial Temperature of Asthenosphere	1350°C

lithosphere, decoupled at a depth, γ (Fig. 3). For example, the crust generally deforms by brittle failure, whereas the lower lithosphere probably extends by ductile flow. In this paper, the upper and lower lithosphere are decoupled at the crust-mantle boundary. Continental crust is thinned by an amount δ, while the lithosphere beneath the crust is independently thinned by an amount β. The initial elevation change caused by non-uniform extension is a complex function of the original crustal thickness and relative amounts of crustal and subcrustal extension. Thus, the non-uniform extension model can predict either initial uplift or initial subsidence during rifting. Because the upper mantle is usually presumed to extend more than the crust, the net effect of non-uniform extension is to increase heating and thermal expansion by introducing more hot, passively upwelled mantle material into the lithosphere. Consequently, a non-uniform extension model has less initial (early) subsidence and more thermal (late) subsidence.

Other Factors

All calculations made in this paper assume instantaneous extension, where no heat is lost by the lithosphere during active rifting. This assumption is only valid when rifting occurs over a geologically small time span (5-10 million years) (Cochran, 1981). If extension occurred over a longer period of time, the model results will underestimate rift-stage (initial) subsidence and overestimate thermal subsidence. Although the duration of rifting in the northern Gulf Coast

Table 2. Physical Properties of Sediments

Meaning	Value
Thermal Conductivity of Sediment	3×10^{-3} cal/cm°C sec
Porosity Parameter, ϕ_o	.4
Porosity Parameter, C	10^{-5} cm^{-1}
Density of Sediment Matrix	2.6 gm/cm^3
Initial Temperature of Sediment	0°C

has not been constrained by geologic evidence, preliminary results from a finite extension rate model indicate that the area was actively extending for a long period of time (50 million years) (Nunn, in preparation).

One-dimensional models discussed in the next section also implicitly assume that only vertical heat conduction is important. Because of the relatively small lateral dimensions of these basins, this assumption is probably not valid. Lateral heat conduction will increase the rate of heat loss from extended lithosphere and, therefore, will increase the rate of thermal subsidence (Steckler, 1983). However, it will not change the total amount of tectonic subsidence. Despite these limitations, the model results are very good.

SUBSIDENCE HISTORIES

Subsidence histories were calculated at six locations in the North Louisiana Salt Basin (Fig. 4). The "backstripping" technique for calculating basin subsidence through time used here was first described by Steckler and Watts (1978) and modified by Sclater and Christie (1980). Results for tectonic subsidence (effects of sediment loading removed) are compared with theoretical subsidence curves predicted from the uniform extension model (Fig. 5).

Analysis

The depth and absolute ages of time horizons were taken from Scardina (1982). In his study, subsurface data from 893 well logs were used to make structural trend surface maps. These maps were then filtered to re-

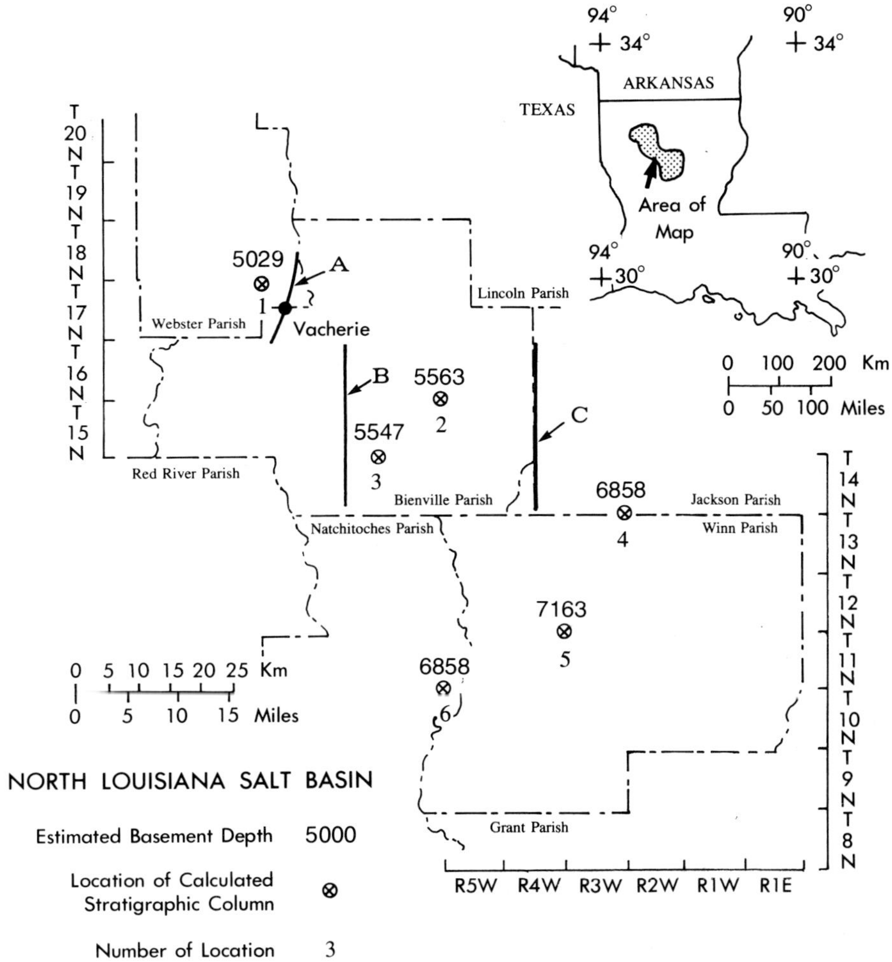

Figure 4. Location map showing the six computed stratigraphic columns and the three seismic lines used to estimate depth to basement (modified after Scardina, 1982). Seismic line A is from Newchurch and Romberg (1976) and lines B and C are from Reese (1977).

move the effects of local salt movement. The six locations shown in Figure 4 represent points of maximum stratigraphic coverage. The depth to basement at these six locations was estimated using seismic data. Stage classifications of the formations in the North Louisiana Salt Basin are from Berryhill *et al.*, (1968), Todd and Mitchum (1977), Hardenbol and Berggren (1978), Salvador (1979), Anderson (1979) and Imlay (1980). The relationships between biostratigraphic and absolute ages are those given by Van Hinte (1976 a, b) and Hardenbol and Berggren (1978). A complete discussion of the subsurface data analysis is given by Scardina (1982).

The sediment thickness and average density at various times in the past were calculated by decompacting the sediments according to a simple porosity-depth relationship:

$$\phi = \phi_o \exp(-cz)$$

where ϕ is porosity, z is depth below the top of the sediment, and ϕ_o and c are parameters which must be determined empirically (Table 2). This was done using a generalized density versus depth curve for the region (Dobrin, 1976). Porosity is related to

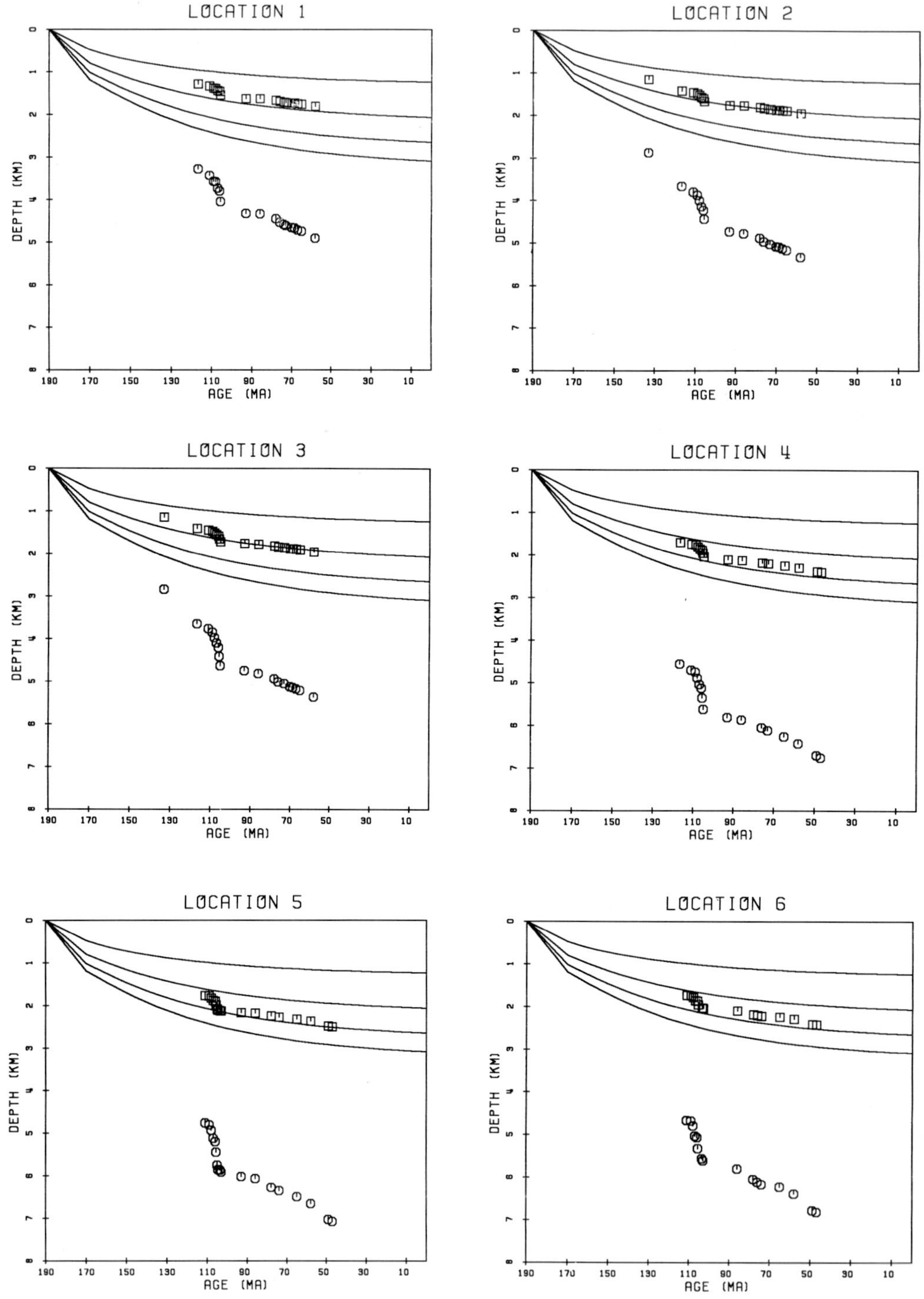

Figure 5. Subsidence for six locations in the North Louisiana Salt Basin. The circles and boxes indicate subsidence with and without the weight of sediment overburden, respectively. Subsidence was calculated from biostratigraphic data and filtered for salt diapirism. Solid lines delineate theoretical tectonic subsidence for uniform extension by factors of 1.25, 1.5, 1.75, and 2. The period of initial subsidence (extension) is 20 My (190-170 Ma).

density by the relationship:

$$p = p_s(1-\phi) + p_w\phi$$

where p is density and p_s, p_w are the densities of sediment matrix and water, respectively. A detailed account of decompaction techniques is given by Sclater and Christie (1980).

Paleowater depths and eustatic sea-level corrections have not been included in calculating depth to basement, and I have assumed that the sediments were deposited in a shallow-water environment. These corrections were not made because they are generally small and the inaccuracies in determining paleowater depths may be greater than the correction itself (Sleep, 1971). Estimates of the water depth in North Louisiana during the Early Cretaceous range from 0-60 meters (Scardina 1982).

Corrections for sediment loading were made using pointwise, or Airy-type isostasy. Neither the flexural rigidity of the plate nor its viscosity have been considered (Watts and Ryan, 1976; Walcott, 1972). The proper computation of sediment loading on a plate with finite flexural rigidity depends on the thickness, which varies with time for a cooling lithosphere, and position within the basin (Watts *et al.*, 1982; Steckler, 1983). Before flexural effects can be computed, an initial approximation for the temperature distribution must be derived using simplified assumptions about plate loading. Such a thermal model is developed in this paper using Airy isostasy. Its consequences in terms of more complex rheological models is the object of current work (Nunn, 1983). Other studies suggest that more complex models may change the total amount of tectonic subsidence by 10-20 percent but do not change the overall shape of the curves (Beaumont *et al.*, 1982; Steckler, 1983).

Model Parameters

Tectonic subsidence curves computed using Airy isostasy for each location were then compared with results from the extension models discussed above. Values for physical constants used in all model calculations are listed in Table 1. Physical and thermal properties of the lithosphere were derived from studies of mid-ocean ridges (Parsons and Sclater, 1977).

The pre-rift crustal thickness was fixed at 32 km, which corresponds to the present crustal thickness in northern Mississippi and South Texas (Worzel and Watkins, 1973). For the non-uniform extension model, the detachment horizon between the upper and lower lithosphere was also set at 32 km, the crust-mantle boundary.

The age of 190 Ma was assigned to the northern Gulf Coast rifting based on dated igneous rocks. Age dates from Mississippi and Arkansas average about 190 Ma (Scardina, 1982). Although substantial igneous activity can occur before and/or after continental breakup (± 15-20 million years), peak activity usually occurs slightly before the end of rifting and the onset of seafloor spreading (Scrutton, 1973). Active rifting was assumed to be completed and the onset of cooling to begin 20 million years after the start of rifting. A rifting period from 190-170 Ma is compatible with place reconstructions of the Gulf of Mexico (Buffler *et al.*, 1980; Klitgord and Schouten, 1982; Pilger, 1982).

Results

Figure 5 shows subsidence curves for six locations in the North Louisiana Salt Basin. The solid lines are theoretical subsidence curves for uniform extension by factors of 1.25, 1.5, 1.75, and 2 (or extension of 25, 50, 75 and 100 percent respectively). None of these locations had stratigraphic coverage down to basement and the calculations of earliest subsidence is poorly constrained. The fact that total thickness of sediment estimated from seismic lines (Scardina, 1982) may be in error by as much as 1/2 km adds further uncertainty to estimates of early subsidence, but should only affect tectonic subsidence by 0.125 km or less.

For locations 1 to 3 in the northern and central portions of the basin, subsidence profiles are in reasonable agreement with uniform extension by a factor of 1.4 to 1.5.

In the southern part of the basin (locations 4 to 6), observed tectonic subsidence is consistent with the uniform extension by a factor of 1.65 to 1.75. The lack of information about early subsidence makes it difficult to distinguish between the uniform and non-uniform extension models. Thus, until deeper wells drilled in this region provide data to the contrary, the subsidence may be said to be compatible with, but not necessarily the result of, uniform extension of the lithosphere.

Because the northern Gulf Coast margin is fairly old and therefore close to thermal equilibrium, the total amount of crustal thinning can also be constrained by the requirement that the present depth be compatible with the theoretical crustal stretching factor (δ). On a cooled margin, present depth (corrected for sediment loading) is very sensitive to the magnitude of crustal extension and very insensitive to the magnitude of subcrustal extension. The extension values of 1.4 to 1.5 for the northern and central portion and 1.65 to 1.75 for the southern portion of the basin, which have been determined from the shape of the subsidence curves, satisfactorily account for the total sediment thickness since rifting.

Discussion

During the 170 Ma which have elapsed since the end of rifting in the northern Gulf Coast, the thermal anomaly which accompanied lithospheric extension has completely dissipated. Thus, there is little additional evidence which can be used to support or contradict the subsidence histories described in the previous section. One comparison which can be made is between the present crustal thickness predicted by the extension model and the crustal thickness measured by seismic refraction. Unfortunately, there are no published, deep seismic refraction data for North Louisiana. However, a north-south seismic line across the adjacent Central Mississippi Salt Basin indicates that the crust has thinned from 32 km in the north to as little as 20 km underneath the basin (Fig. 6; Worzel and Watkins, 1973). Farther south, the crust thickens to about 30 km beneath the Wiggins Arch before it thins again in the continent-ocean transition zone within the Gulf of Mexico. Crustal thinning from 32 km to 20 km corresponds to an extension factor of 1.6, which agrees with subsidence profiles for the North Louisiana Salt Basin.

The major deviation from theoretical subsidence curves is a period of rapid subsidence in the Cretaceous from 110 to 105 Ma. Possible explanations for this phase of rapid subsidence are: poor chronostratigraphic control, flexural effects caused by rapid loading on the shelf edge and deflation due to salt movement. Absolute age-stage correlations for this time period (Aptian-Albian boundary) are poorly constrained. If the duration of this period were increased, the apparent rapid subsidence would be substantially reduced. Another explanation is that the local loading (Airy) model used to compute the subsidence histories failed to account for extrabasinal loading during the period. The Louisiana shelf edge has migrated about 150 km southward since the Cretaceous because of massive deposition of clastics (Fig. 1). It is possible that extensive deposition of sediments on the shelf edge depressed the surrounding area and thus generated the additional subsidence (Scardina, 1982). A preliminary study indicates that such a situation could produce the observed subsidence pattern (Nunn *et al.*, 1983). However, lack of data on Cretaceous sedimentation rates south of the Lower Cretaceous shelf edge makes it difficult to prove that this situation existed. Finally, rapid subsidence may be due to salt deflation. A similar phase of rapid sediment accumulation occurred in the East Texas Salt Basin (Jackson and Seni, 1983; Nunn, 1983), which is synchronous with the period of maximum rate of gross diapiric growth in that basin (112 to 104 Ma)(Jackson and Seni, 1983).

PALEOTEMPERATURES

Once the age and intensity of the rifting process have been established from the subsidence history, the thermal evolution of the northern Gulf Coast can be computed for selected times during the development of the margin.

Computational Methods

Computation of temperatures within the sediments is complicated by diferences in conductivity between basement rocks and sediments which vary with lithology, porosity, and temperature. In order to include this factor, a finite-difference model is used to simulate the thermal and mechanical evolution of the northern Gulf Coast.

The thermal-mechanical model is designed to predict temperature and subsidence through time during the thermal subsidence (post-rifting) phase of basin development. Extension parameters were specified and were used to compute the initial subsidence and temperature structure of the lithosphere. The temperature distribution within the sediments at a subsequent point in time was obtained by solving the finite difference equivalent of the heat conduction equation (Carslaw and Jaeger, 1959; Richtmyer and Morton, 1967). In parallel to this conduction calculation, effects of thermal contraction, sediment deposition, and sediment compaction during this time interval are determined. After subsidence and sedimentation effects have been computed, the model returns to the conduction equation to calculate the change in temperature for the next time interval. The entire history of the basin is evolved in this manner.

The volume of sediment deposited during each time interval is computed assuming a continually-filled basin. New sediment is added to the top of the column of material, its initial temperature is set to 0°C and the underlying sediment is compacted according to the porosity-depth relationship given in a previous section. The porosity parameters used are the same as those used to "backstrip" the six locations in the North Louisiana Salt Basin (Table 2). Isostatic equilibrium is maintained throughout basin development. As before, the effects of sea-level changes, paleobathemetry, lateral heat conduction, and radioactive heat production are ignored.

A thermal conductivity of 3×10^{-3} cal/cm°C sec was assigned to the basin sediments as a best fit to the observed temperature gradient and heat flow. This relatively low value for thermal conductivity is required to produce the present high thermal gradient of about 33°C/km on an essentially cooled margin. Changes in conductivity due to porosity were computed using the Maxwell model (Beck, 1976). The effect of temperature on conductivity was computed using the data of Birch and Clark (1940).

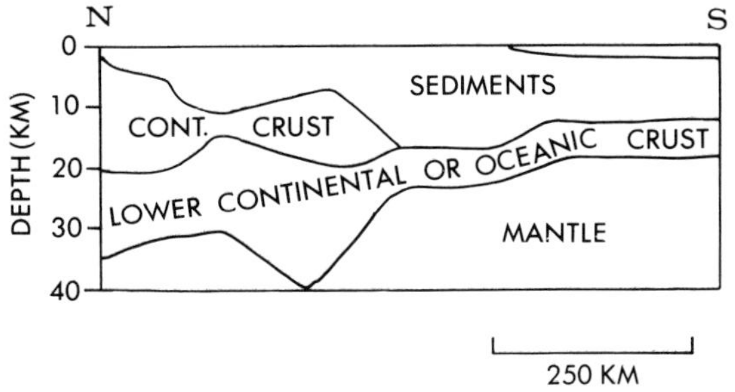

Figure 6. North-south structure cross section of the crust and upper mantle in Mississippi (after Worzel and Watkins,1973). Area of thinned continental crust is the Central Mississippi Salt Basin.

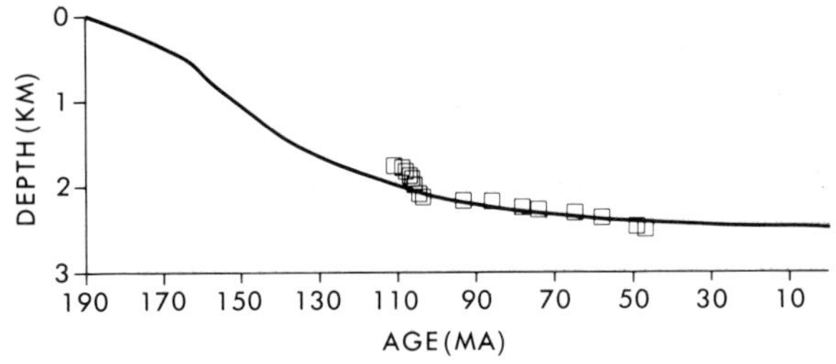

Figure 7. Tectonic subsidence versus time for location 5 (boxes). The solid curve is the theoretical tectonic subsidence versus time for crustal extension by a factor of 2 and subcrustal extension by a factor of 4. The period of initial subsidence (extension) is 25 My (190-165 Ma).

Results

The theoretical subsidence curve that gives the best fit for models that include the thermal effects of overlying sediments is crustal extension by a factor of 2 and subcrustal extension by a factor of 4 (Fig. 7). Duration of rifting was also extended to 25 million years (from 190 Ma to 165 Ma). A non-uniform extension model and an increase in crustal extension were required to counterbalance a decrease in thermal (post-rift) subsidence caused by the thermal insulation of overlying sediments. Although these extension parameters are within observational limits because of poor control on early subsidence and total sediment thickness, they should probably be considered close to maximum values. As vertical heat conduction becomes more difficult due to thermal blanketing by overlying sediments, the error in predicted subsidence introduced by neglecting lateral heat conduction becomes progressively larger because proportionally more heat flows out the sides than out the top.

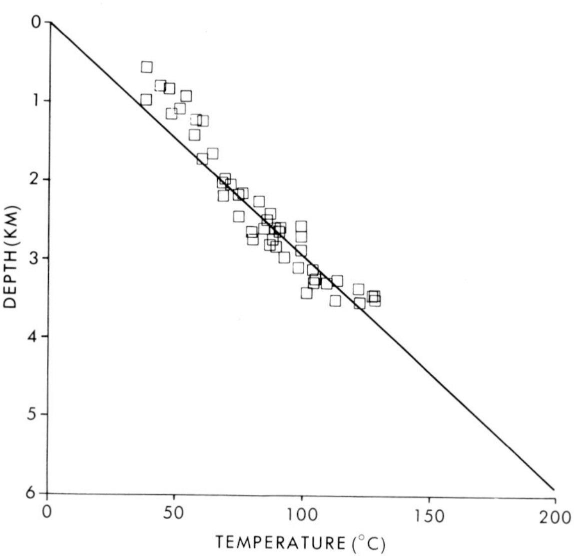

Figure 8. Observed temperature plotted as a function of depth for 52 wells in North Louisiana (boxes). Temperatures are from the AAPG Basic Data File. The solid line is the temperature gradient predicted from the non-uniform extension model (Fig. 7). Deviations from observed temperatures at shallow depths may be due to model constraint of 0°C at the surface.

Present temperatures predicted by the model are compared with measured bottom-hole temperatures from approximately 50 wells in the North Louisiana Salt Basin (Fig. 8; AAPG Basic Data File). The agreement between measured and predicted values is excellent given the errors in observational data and the uncertainties in the models. Deviations from observed temperatures at shallow depths are primarily due to the model constraint of a temperature of 0°C at the surface. Available heat-flow data in North Louisiana range from 1.1 to 2 μcal/cm^2-s with an average value of 1.5 μcal/cm^2-s (Smith *et al.*, 1981), which compares with a predicted value of 1.0 μcal/cm^2-s. Areas of high heat flow are attributed either to exceptionally high conductivity of buried salt deposits or rich abundances of heat producing radioactive elements in buried alkalic igneous rocks. Ten heat-flow values in adjacent areas of Mississippi average 0.95 μcal/cm^2-s (Smith *et al.*, 1981). Although present-day temperature

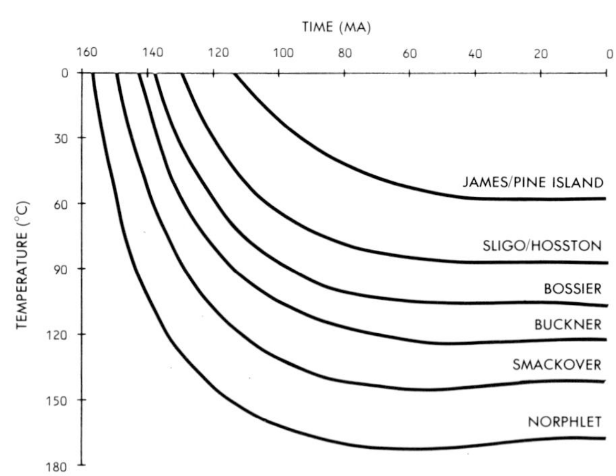

Figure 9. Temperature histories for selected Jurassic and Cretaceous sedimentary horizons in the North Louisiana Salt Basin calculated from the non-uniform extension model (Fig. 7). Absolute age correlation is for the base of the unit.

and heat flow are not restrictive enough to uniquely determine a model, the consistency between observed and predicted values does lend credence to the results.

Temperatures within Jurassic and Cretaceous sediments vary with time according to two competing effects - elapsed time since rifting and depth of burial (Fig. 9). Temperature at a given depth decreases with elapsed time since rifting as the margin cools. At the same time, the temperature of sediments of a given age tends to increase as the basement subsides, sediments are deposited, and the depth of burial increases. For the northern Gulf Coast, depth of burial is the dominant effect. Thus, the present-day temperature of a given unit is also close to its maximum paleotemperature. This correlation is consistent with results of thermal-mechanical model studies of other rifted continental margins such as the eastern seaboard of North America (Royden and Keen, 1980; Sawyer *et al.*, 1982).

For the northern Gulf Coast margin, model results predict that the most rapid change in the thermal regime occurred during the Jurassic and Early Cretaceous and that a constant temperature distribution has existed within the sediments for about the last 60 million years. As a result, Jurassic-age sediments were heated very early and have remained near their present temperature (maximum paleotemperature) for the last 100 million years (Fig. 9). Theoretical thermal histories for Jurassic-age sediments are conducive to the generation of mature hydrocarbons, if all other factors are favorable. Model results show that these units (Norphlet-Bossier) have been in the oil generation window (90 to 150°C) throughout most of their histories.

Although the temperature of sediments of a given age primarily reflects depth of burial, there has been a dramatic decrease in thermal gradient since rifting. Predicted temperature gradients at the end of the deposition of the Smackover Formation (143 Ma) and Sligo-Hosston Formations (113 Ma) and the present are shown in Figure 10. The change in slope on each thermal gradient curve (indicated by arrows) marks the sediment/basement boundary at that time and is caused by the difference in their thermal conductivities. The temperature gradient within basin sediments has decreased from 72°C/km at the end of deposition of the Smackover Formation (143 Ma), to 58°C/km at the end of deposition of the Sligo-Hosston Formations and finally down to its present level of 33°C/km. The much higher geothermal gradient in the past also helped to produce the rapid early heating of Jurassic-age sediments (Fig. 9). The non-linear relationship between temperature and depth for basin sediments during the Jurassic (Fig. 10, Smackover Curve) indicates that they were not in conductive equilibrium (*i.e.*, they were still actively heating up).

Unfortunately, quantities which could assess the validity of the predicted paleotemperatures are for the most part lacking. Vitrinite reflectance, which is a measure of thermal alternation of the organic material

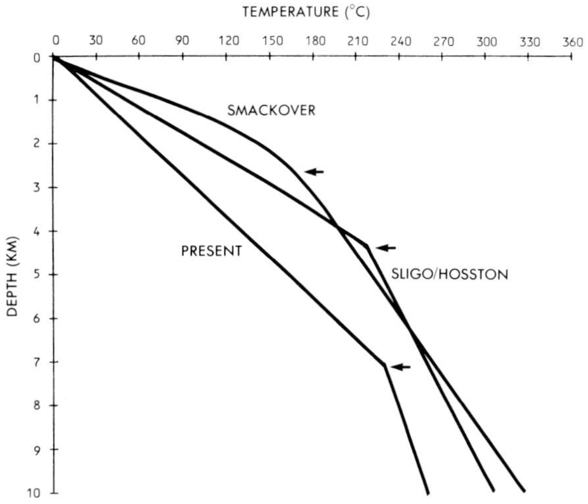

Figure 10. Temperature versus depth predicted from the non-uniform extension model (Fig. 7) at end of Smackover (143 (Ma), end of Sligo/Hosston (113 Ma), and present. Arrows indicate the computed depth to basement at that time. The dramatic change in the temperature gradient at these depths is due to the difference in thermal conductivity between the sediments and the crustal rocks.

within the sediments and depends on the temperature history, could, in theory, be used to test model results, but these data were not available. However, recent work on dolomite diagenesis and two-phase fluid inclusions within the Jurassic-age sediments are consistent with a much higher geothermal gradient in the past (C. H. Moore, Jr. personal communication, 1982).

CONCLUSIONS

1. The subsidence of the northern Gulf Coast, calculated from deep exploratory well data in the North Louisiana Salt Basin, appears to be the result of extension of the lithosphere during rifting.
2. The onset of subsidence is compatible with the opening times for the Gulf of Mexico during Late Triassic to Early Jurassic.
3. Subsidence of the North Louisiana Salt Basin is consistent with crustal extension by a factor of 1.5 to 2. Lack of information on the early subsidence history makes it difficult to constrain the amount of subcrustal extension.
4. Assuming an initial crustal thickness of 32 km, the magnitude of crustal thinning predicted from subsidence data alone is in good agreement with crustal thicknesses inferred from a seismic refraction profile across the adjacent Central Mississippi Salt Basin.
5. The present day temperature profile, computed from a similar modeling procedure, agrees with bottom-hole temperatures from wells in the region. Basal heat-flow measurements are also in good agreement with those predicted by the model.
6. Paleotemperature estimates indicate that Jurassic-age sediments have been close to their current temperatures (90-170°C) for the last 100 million years. Thus, they should be good sources for mature hydrocarbons.
7. Model results also suggest that Jurassic temperature gradients were more than twice the present value (33°C/km).

ACKNOWLEDGMENTS

I am grateful for encouragement and numerous suggestions from A.D. Scardina and Rex H. Pilger, Jr., who initiated tectonic subsidence analysis of the North Louisiana Salt Basin. C.H. Moore, Jr., M. Vinet and M.P.A. Jackson provided helpful discussions. This work was supported by the Louisiana Geological Survey Deep Basin Analysis Project and the L.S.U. Geology Department.

REFERENCES

Anderson, E.G., 1979, Basic Mesozoic study in Louisiana, the northern coastal region, and the Gulf basin province; LA. Geol. Survey Folio Series 3, v. 1, 58 p.

Beaumont, C., C. E. Keen, and R. Boutilier, in press, On the evolution of rifted continental margins: comparisons of models and observations for the Nova Scotia margin: Geophys. J. Roy Astron. Soc.

Beck, A.E., 1976, An improved method of computing the thermal conductivity of fluid-filled sedimentary rocks: Geophysics. v. 41, p. 133-144.

Berryhill, R.A., *et al.*, 1968, Stratigraphy and selected gas field studies of north Louisiana; Part A, stratigraphy, *in* Beebe, B.W., and G.B. Curtis, (eds.); Natural Gases of North America: Am. Assoc. Petrol. Geol. Mem. 9, v.1, p. 1099-1137.

Birch, F. and H. Clark, 1940, The thermal conductivity of rocks and its dependence on temperature and composition: Am. J. Sci., v. 238, p. 529-558.

Bradley, D.C., 1982, Subsidence in Late Paleozoic basins in the northern Applachians: Tectonics, v. 1, p. 107-123.

Buffler, R.T., J.S. Watkins, F.J. Shaub, and J.L. Worzel, 1980, Structure and early geologic history of the deep central Gulf of Mexico Basin: *in* Pilger, R.H., Jr., ed., Proceedings of a Symposium, the origin of the Gulf of Mexico and the early opening of the central North Atlantic Ocean: School of Geoscience, Louisiana State University, Baton Rouge, p. 3-16.

Carslaw, H.S. and J.C.Jaeger, 1959, Conduction of Heat in Solids, 2nd Ed.: Oxford at the Clarendon Press, London.

Cochran, J.R., 1981, Simple models of diffuse extension and the pre-seafloor spreading development of the continental margin of the northeast Gulf of Aden, *in* Blanchet, R., and L. Montadert (eds.), Geology of the Continental Margins: Oceanologica, Acta, v. 4, p. 155-166.

Dobrin, M.B., 1976, Introduction to Geophysical Prospecting: McGraw-Hill, Inc., New York, 3rd ed., 630 p.

Edick, M., 1982, The nature and timing of Mesozoic igneous activity: northern Gulf of Mexico basin: unpublished M.S. thesis, Louisiana State University, Baton Rouge.

Hardenbol, J. and W.A. Berggren, 1978, A new Paleogene numerical time scale in Cohee, G.V., M.F. Glaessner, and H.D. Hedberg (eds.), Contributions to the Geologic Time Scale: Am. Assoc. Petrol. Geol. Studies, v. 6, p. 213-234.

Imlay, R.W., 1980, Jurassic paleobiogeography of the conterminous United States in its continental setting: U.S. Geol. Survey Prof. Paper 1062, 134 p.

Jackson, M.P.A., and S.J. Seni, 1983, Geometry and evolution of salt structures in a marginal rift basin of the Gulf of Mexico, east Texas: Geology, v. 11, p. 131-135.

Jarvis, G.T. and D.P. McKenzie, 1980, Sedimentary basin formation with finite extension rates, Earth Planet. Sci. Lett., v. 48, p. 42-52.

Klitgord, K.D. and H. Schouten, 1980, Mesozoic evolution of the Atlantic, Caribbean and Gulf of Mexico: (Abs.), *in* Pilger, R.H.,Jr., (ed.), Proceedings of a Symposium, The origin of the Gulf of Mexico and the early opening ofthe Central North Atlantic Ocean: School of Geoscience, Louisiana State University, Baton Rouge, p. 100-101.

Martin, R.G., 1978, Northern and eastern Gulf of Mexico continental margin - Stratigraphic and structural framework, Facies and Oil-trapping Characteristics of the Upper Continental Margin: Am. Assoc. Petrol. Geol. Studies, v. 7, p. 21-42.

McKenzie, D.P., 1978, Some remarks on the development of sedimentary basins: Earth Planet. Sci. Lett., v. 40, p. 25-32.

Newchurch, E.M., and M. Romberg, 1976, IV. Tectonic Stability, Part E. Seismic studies of Vacherie Salt Dome: *in* An investigation of the utility of Gulf Coast Salt domes for the storage of disposal of radioactive wastes: Institute for Environmental Studies, Louisiana State University, Baton Rouge, p. 193-227.

Nunn, J.A., 1983, Subsidence and heat flow across a sedimentary basin-uplift boundary - a thermal-mechanical model (Abs.): Amer. Assoc. Petrol. Geol. Bull., in press.

__________, A.D. Scardina and R.H. Pilger, Jr., in preparation. Thermal Evolution of the North-Central Gulf Coast.

Parsons, B. and J.G. Sclater, 1977, An analysis of the variation of ocean floor bathymetry and heat flow with age: J. Geophys. Res., v. 82, p. 803-827.

Pilger, R.J., Jr., 1981, The opening of the Gulf of Mexico - implications for the tectonic evolution of the northern Gulf Coast: Gulf Coast Assoc. Geol. Soc. Trans., v. 31, p. 377-381.

Reese, R.J., 1977, Salt kinematics of southern Bienville Parish, Louisiana: unpublished master's thesis, Louisiana State University, Baton Rouge, 136 p.

Richtmyer, R.D. and K.W. Morton, 1967, Difference methods for initial value problems, 2nd ed.: Interscience Pub., New York.

Royden, L. and C.E. Keen, 1980, Rifting processes and thermal evolution of the continental margin of eastern Canada determined from subsidence curves: Earth Planet. Sci. Lett., v. 51, p. 343-361.

__________, J.G. Sclater, and R.P. von Herzen, 1980, Continental margin subsidence and heat flow: Important parameters in formation of petroleum hydrocarbons: Amer. Assoc. Petrol. Geol. Bull., v. 64, p. 173-187.

Salvador, A., 1979, Late Triassic-Jurassic paleogeography and origin of Gulf of Mexico (Abs.): Am. Assoc. Petrol. Geol. Bull., v. 63, p. 520.

Sawyer, D.S., B.A. Swift, J.G. Sclater and M.N. Toksoz, 1982, Extensional model for the subsidence of the northern United States Atlantic continental margin: Geology, v. 10, p. 134-140.

Scardina, A.D., 1982, Tectonic subsidence history of the North Louisiana Salt Basin: LSU Publications in Geology and Geophysics, Gulf Coast Studies, v. 2, 34 p.

Sclater, J.G. and P.A.F. Christie, 1980, Continental stretching: an explanation of the post Mid-Cretaceous subsidence of the Central North Sea Basin: J. Geophys. Res., v. 85, p. 3711-3739.

__________, L. Royden, F. Horvath, B.C. Birchfiel, S. Semkin, and L. Stegena, 1980, The formation of the intra-Carpathian basins as determined from subsidence data: Earth Planet. Sci. Lett., v. 51, p. 139162.

Scrutton, R.A., 1973, The age relationship of igneous activity and continental break-up: Geol. Mag., v. 110, p. 227-234.

Sleep, N. H., 1971, Thermal effects of the formation of Atlantic continental margins by continentl break-up: Geophys. Jour. Roy. Astr. Soc. v. 24, p. 325-350.

Smith, D.L., W.T. Dees, and D.W. Harrelson, 1981, Geothermal conditions and their implications for basement tectonics in the Gulf Coast Margin: GCAGS Trans., v. 31, p. 181-190.

Steckler, M.S., 1983, Two-dimensional considerations in the subsidence of continental margins: J. Geophys. Res., in press.

__________ and A.B. Watts, 1978, Subsidence of the Atlantic-type continental margin off New York: Earth Planet. Sci. Lett., v. 41, p.1-13.

Todd, R.G., and R.M.Mitchum, Jr., 1977, Seismic stratigraphy and global changes of sea level, part 8 - Identification of Upper Triassic, Jurassic and Lower Cretaceous seismic sequences in Gulf of Mexico and offshore west Africa, *in* Payton, E.D., (ed.), Seismic Stratigraphy - Application to Hydrocarbon Exploration: Am. Assoc. Petrol. Geol. Mem. 26, p. 145-163.

Van Hinte, J.E., 1976a, A Cretaceous time scale: Am. Assoc. Petrol. Geol., Bull., v. 60, p. 498-516.

__________, 1976b, A Jurassic time scale: Am. Assoc. Petrol. Geol., Bull., v. 60, p. 489-497.

Walcott, R.I., 1972, Gravity, flexure and the growth of sedimentary basins at a continental edge: Geol. Soc. Amer. Bull., v. 83, p. 1845-1848.

Watts, A.B., G.D. Karner and M.S. Steckler, 1982, Lithospheric flexure and the evolution of sedimentary basins: Phil. Trans. R. Soc. Lond., v. 305A, p. 249-281.

__________ and W.F.B. Ryan, 1976, Flexure of the lithosphere and continental margin basins: Tectonophysics, v. 36, p. 25-44.

__________, and M.S. Steckler, 1979, Subsidence and eustacy at the continental margin of eastern North America, *in* Talwani, M., W.W. Hey, and W.B.F. Ryan, (eds.): Deep Drilling Results in the Atlantic Ocean: Continental Margins and Paleoenvironments.

Worzel, J.L. and J.S. Watkins, 1973, Evolution of the northern Gulf Coast deduced from geophysical data: Gulf Coast Assoc. Geol. Soc. Trans., v. 23, p. 84-91.

COMPUTER-GENERATED INTERPRETATIONS OF STRATIGRAPHY AND DEPOSITIONAL ENVIRONMENTS FROM A LOGGING/CORE STUDY: COTTON VALLEY GROUP, EAST TEXAS

GORDON PIRIE

Schlumberger-Doll Research
P.O. Box 307
Ridgefield, Connecticut 06877

INTRODUCTION

Schlumberger-Doll Research, Schlumberger Well Services (Shreveport, LA and Houston, TX), and Delta Drilling Company (Tyler, TX) cooperatively examined a whole core and a conventional (DIT*/BHC, GR/CNL/FDC*, HDT*/FIL) and experimental/prototype (Digital Sonic, CNT-G, EPT*, GST*-A, NGT*, NML*) suite of logs from the Middle Cotton Valley sandstones (Upper Jurassic) in a 10,200 foot well (Alice Snider No. 1). Refer to the Appendix for a brief explanation of the tools and their geological application in gas-bearing, shaly sandstone units. The well is located in the Carthage tight gas sand field (Rogers, 1968) in northwestern Panola County, about 60 miles east of Tyler, Texas. The primary purpose of this study was to provide an environment in which to examine the conventional and prototype tool responses against a detailed examination of the rock core. However, there also was considerable interest in interpreting the sedimentary depositional environments from the logs and core and to relate these environments to a variety of new, computer-generated interpretations - the major theme of this paper. Additional papers on related studies of this well include: Brown *et al* (1980), Flaum and Pirie (1981a, 1981b), Hunt *et al* (1981), Pirie and Hunt (1981), and Pirie (1982).

GEOLOGICAL BACKGROUND

The Cotton Valley Group is made up of the Bossier, Schuler, and Knowles Formations (Todd and Mitchum, 1977). The Schuler Formation is a nearshore to continental sandstone, conglomerate, and shale unit. The Bossier Formation is a deep-water shale and is the facies equivalent of the Schuler Formation. Further seaward, the upper part of the Cotton Valley Group contains a low-energy, fine-grained carbonate mud with some carbonate grains. This unit is the Knowles Limestone (Mann and Thomas, 1964).

A total of 390 feet were cored and 340 feet of oriented core were recovered. Schlumberger-Doll Research examined 139 feet from the "Middle Cotton Valley Sands". These cores were selected to provide a comparison of the two gas-producing zones: Zone 1 = 9,354-9,638 feet which was cored and logged, and Zone 2 = 9,700-10,027 feet, which was logged but not cored. Zone 2 is located primarily in the "Taylor Sands" - the lowest sandstone reservoir unit in the Cotton Valley Group.

In the Middle Cotton Valley Sands examined, the lithology is transitional upwards from a shale-dominant to a sandstone-dominant sequence. This trend is interrupted by calcite-cemented sandstones and/or fossiliferous zones. There are also minor occurrences of

*Mark of Schlumberger

pebble conglomerates and coals interbedded throughout the section. In general, however, the lithology of this portion of the well is laminated-bedded sandstone and shale sequence of very fine-grained, bioturbated, quartzose-lithic sandstones with low porosity (5%) and permeability (3 to 30μd) interlaminated and interbedded with very thinly to thinly-bedded, bioturbated, black shales with low porosity (4%) and permeability (<10 md).

BEDDING

The most obvious features of the 139 feet of rock core examined are the sedimentary bedding and laminations. In addition to direct observation and analysis of sedimentary bedding, the High Resolution Dipmeter Tool (HDT) provides a more generalized, yet useful, evaluation of bedding. The upper portion of Figure 1 shows the standard presentation of the four-arm dipmeter with azimuth, deviation, and two calipers. The logged section (9,545-9,560 feet) is, in part, enlarged below which, in turn, is compared to the core or core photographs. From these presentations and other data, the depositional environments are be determined (Fig. 1, far right). Additional processing of the HDT data also provides further insight into the dynamics of the operating processes of deposition (Fig. 4).

Generally, the four arms track each other and the curves demonstrate clearly the thicker bedded shale and sandstone units and highlight zones where shale beds may have been washed out during the rock-coring process. The resolution of the standard HDT allows only a gross evaluation of the laminae, particularly in thick sandstone units which are laminated but have no lithological contrast. The new Stratigraphic High-Resolution Dipmeter Tool (SHDT), which samples every 0.1 inch, will provide better resolution.

Ideally, once the lithologies are characterized by inspection of the whole or sidewall cores and a depositional model is established, the dipmeter curves assist in setting the boundaries between sedimentological

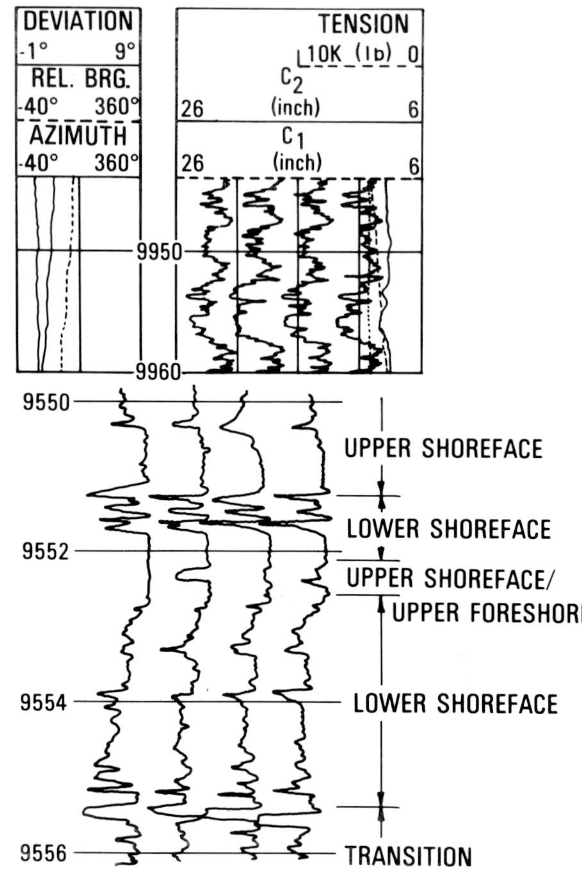

Figure 1. Standard High-Resolution Dipmeter (HDT) presentation (top) (9545-9560 feet) and cored section (9550-9556 feet) in log depths (lower) rescaled to core photographs (not shown). With rock core study and dipmeter evaluation, sedimentary environments (shown on right) are identified.

(genetic) units in the context of the depositional model.

Another important observation regarding bedding is the effect of diagenesis in masking laminations and, to a lesser extent, bedding. Petrographic study of the shale laminae indicates an abundance of coarse, well-crystallized, micaceous minerals dominated by muscovite. In contrast, shale beds are not so obviously affected. It also appears that at least part of the diagenetic clays in the sandstone is derived from the effects of diagenetically-altered laminations. A similar observation is made on thicker shale units

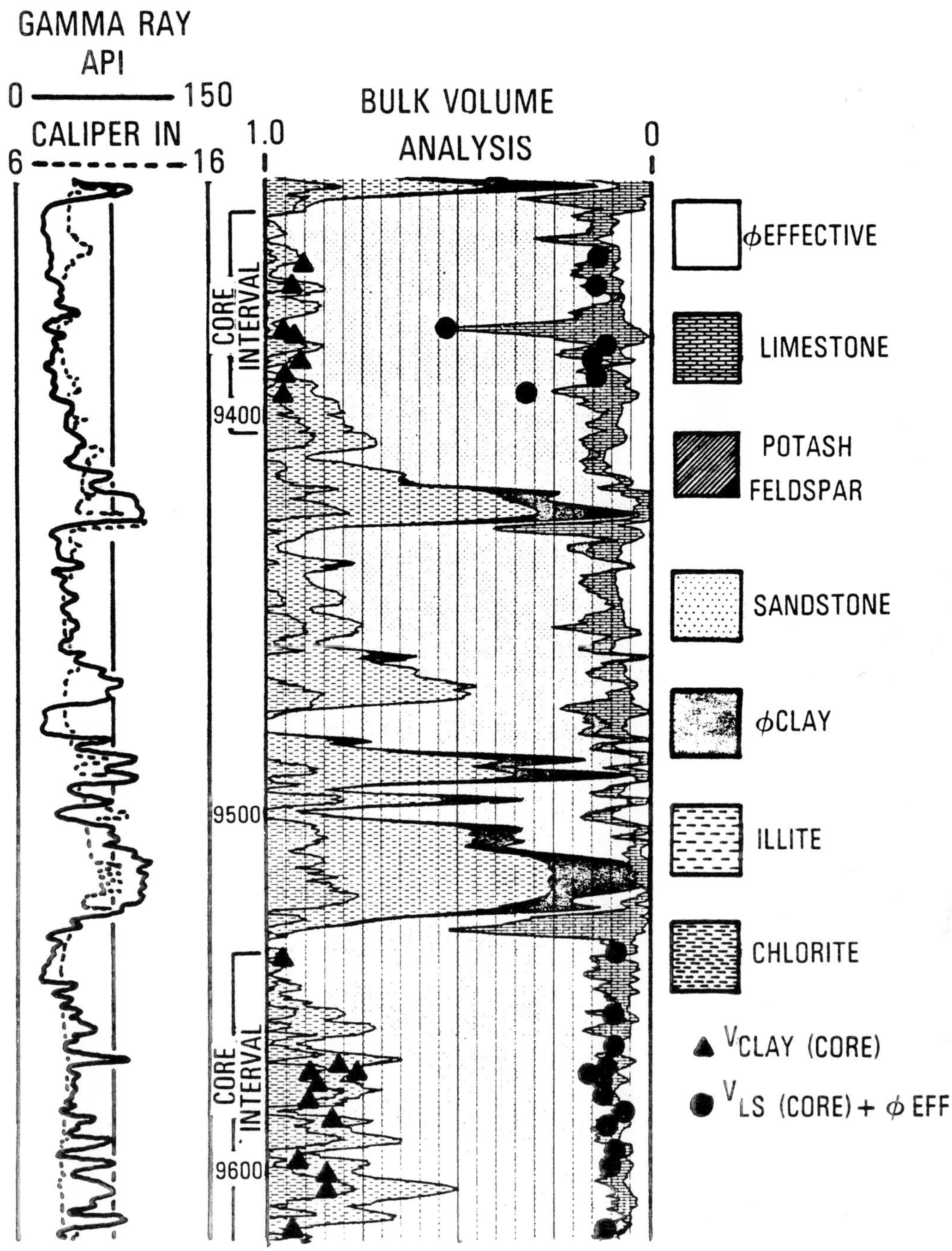

Figure 2. GST-NGT mineralogy/lithology log over cored and non-cored intervals, Cotton Valley Group, DDC Alice Snider well. Note how core data are used to calibrate log data in non-cored section to identify thick water-saturated shales and calcite-cemented sandstones which may act as vertical barriers to hydraulic fracturing.

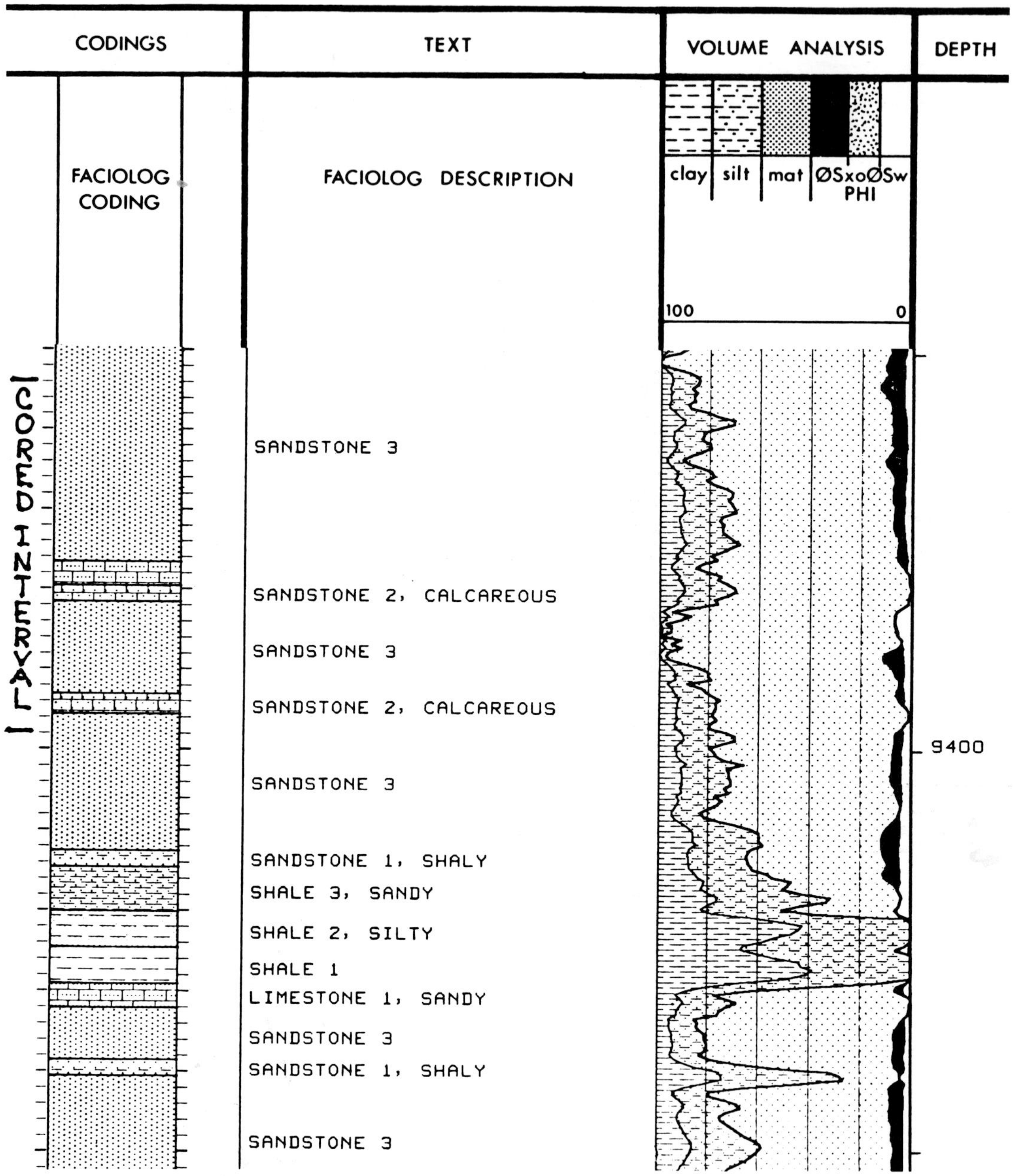

Figure 3. FACIOLOG coding and text with volume analysis from a VOLAN* presentation over cored and non-cored section of Cotton Valley Group, DDC Alice Snider well. See text for further explanation.

*Mark of Schlumberger

where the clay minerals authigenically precipitate in the adjacent sandstones as products of clay transformation, *i.e.*, continued compaction *after* burial, compaction and water release, clay dehydration, and clay destruction (Foscolos and Powell, 1979). Also, the commonly-observed effects of bioturbation destroy both laminations and bedding and tend to increase the porosity of the sandstones - a logical zone for gas storage.

Other sedimentary structures common in this rock sequence include: planar crossbedding, usually at low angles; shallow-water interference ripple marks; and convoluted bedding due to penecontemporaneous loading.

In spite of the complexities in defining sedimentary bedding, sedimentary structures, etc., from rock study and the dipmeter, these observations and conclusions regarding laminations/bedding play a major role in designing better fracturing programs to develop tight gas sands (Tindell *et al*, 1981).

Stratigraphy

A unique presentation of the stratigraphic section is obtained using nuclear spectroscopy logging, specifically with the induced gamma-ray tool (GST) which measures the relative yields of gamma rays resulting from interactions of neutrons with different elements (isotopes) present in the downhole formation. Elements detected and quantitatively measured include: hydrogen, silicon, sulfur, chlorine, calcium, and iron. In combination with additional information (SARABAND*, NGT, whole or sidewall core data), a computerized presentation (Fig. 2) yields volume fractions of effective porosity, sandstone, limestone, feldspar, and total clay ("shale") which in the Middle Cotton Valley Sands is further divided into volumes of clay porosity, illite, and chlorite in the cored *and* non-cored sections. In the upper section of the Alice Snider well, the Cretaceous Glen Rose Group (evaporite sequence of interbedded limestone and anhydrite), a computerized interpretation of the GST log was prepared *without* core data.

For comparison, another presentation, the FACIOLOG* (Fig. 3), is shown over the same section of the Middle Cotton Valley Sands. Here the facies or "electrofacies" are determined by an automatic clustering in n-dimensional space of *only* the log responses (Wolff and Pelissier-Combescure, 1982). The annotations under the FACIOLOG coding and description are determined by inspection of the resulting principal component analyses, logs, and crossplots. This computerized presentation displayed alongside edited logs, a processed dipmeter, a bulk volume analysis from a VOLAN* (far right), or other presentations provides a valuable visual comparison of the stratigraphic section with or without core data.

ROCK HISTORY AND DEPOSITIONAL MODEL

The primary source rocks from which the sediments for the Middle Cotton Valley tight gas sandstones were derived were sedimentary rocks dominated by sandstone, shale, and chert. Originally, the potash feldspars were derived from granites and the calcium-rich plagioclase feldspars were derived from gabbros.

In the arid environment of the Jurassic (Habicht, 1979), mechanical weathering was dominant over chemical weathering, *i.e.*, grain breakage and abrasion were common. The presence of unstable calcium-rich plagioclases suggests a short distance of transportation - perhaps ten miles or less. The uniformly fine-grained sandstones throughout the barrier bars suggest low relief and a source area dominated by pre-existing sandstones. The sediments were transported through a flash flood-dominated, low relief area via rivers and through a lagoonal/tidal (deltaic?) shallow marine system where the sediments were deposited in a shoreface environment. After deposition in moderate/high

*Mark of Schlumberger

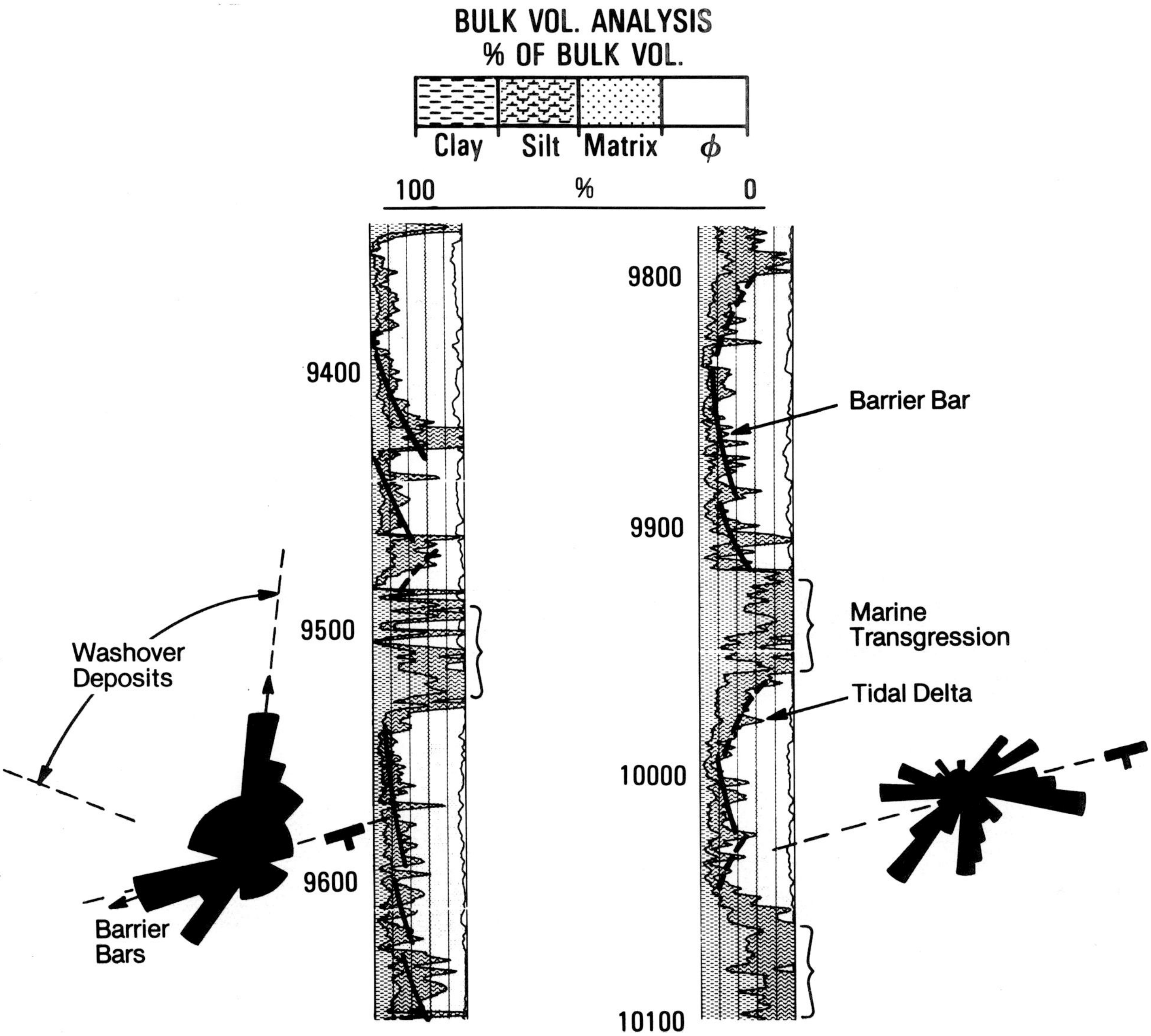

Figure 4. Bulk volume analysis from SARABAND, dipmeter (HDT), and core study assist identification of major depositional units: barrier bar and tidal delta. These sand units are interrupted by marine transgressive sequences of shaly sandstones. Zoned azimuth frequency diagrams from dipmeter suggest dominant southwest direction of longshore current (single arrow). Also note position of washover deposits of barrier bars and flood- and ebb-tide components of tidal deltas.

(sands) and low (muds) energy levels, some bedding was destroyed by bioturbation followed by the usual complement of processes of lithification (compaction, cementation, crystallization, etc.) to produce the sandstone/shale sequence. There are two stages of calcite cementation: the first posibly related to marine transgressions and the redeposition of calcite dissolved from carbonate shell fragments, and the second perhaps related to dewatering and dissolution of calcite from the overlying shales, a mechanism proposed by Foscolos and Powell (1979). The first stage of calcite cementation is followed by quartz cementation which may also be a product of shale dewatering.

The sedimentary rock sequence of stacked or layered barrier bars in the Alice Snider well was deposited on a nearly-flat depositional slope in a shallow (3-35 feet) marine coastal environment - perhaps the reworking by surf waves of nearby deltaic sediments (Heward, 1981). Frank (1978) classified these sandstone units with the High Resolution Dipmeter as "channel-like" or "barlike". The depositional strike of the barrier bar deposits trends northeast-southwest, and the barrier bar units are thin and marginal to the sedimentary basin. When the discrete barrier bar beds are generalized in a SARABAND presentation (Fig. 4), numerous barrier bar units are identified as coarsening upwards with thicknesses of 30 to 35 feet. Using zoned azimuth frequency diagrams from the dipmeter, the dominant southwest direction of the longshore current is noted. Also, there is a strong suggestion of inclined bedding to the northwest which is interpreted as washover deposits. In the same SARABAND presentation, there are less frequent and thinner (20-25 feet or less) sedimentary rock sequences which fine upwards and are interpreted as flood and ebb tidal delta units which breach the barrier bar system. These sequences are interrupted by thick ($\simeq$40 feet) sandy/shale units which represent major marine transgressions over the area.

The majority of the sandstones in the Middle Cotton Valley unit represent sediments deposited in the upper and lower shoreface. Transitional (lower shoreface-offshore) and lower foreshore deposition bracket the shoreface deposits. There is a striking similarity in the bedding type and lithology of these ancient barrier bars and a modern-day analog, Galveston Island of the Texas coastal plain (Bernard *et al*, 1959, 1962). The subaerial beach and dune deposits seen in Galveston Island are not preserved in the core. The bedding types also have a strong parallelism with the summer sequence of middle shoreface and transitional zones in the modern beach-shelf deposits of Italy's Gulf of Gaeta in the Tyrrhenian Sea (Reineck and Singh, 1973). In summary, the entire Cotton Valley Group represents a *major* transgressive sequence; however, in the Alice Snider well, the Middle Cotton Valley Sands sequence is punctuated by several transgressive/regressive events.

REFERENCES

Bernard, H.A., D.F. Major, Jr., and B. S. Parrott, 1959, The Galveston barrier island and environs; a model for predicting reservoir occurrence and trend, Trans., Gulf Coast Assoc. Geol. Socs., v.9, p. 221-224.

______________, R.J. LeBlanc, and C.F. Major, 1962, Recent and Pleistocene Geology of Southeast Texas *in* geology of the Gulf Coast and Central Texas and Guidebook of Excursions, Houston Geol. Soc., p.175-224.

Brown, R.O., J.M. Forgotson, and J.M. Forgotson Jr., 1980, Predicting the orientation of hydraulically created fractures in the Cotton Valley Formation of East Texas, Annual Meeting, Soc. Petroleum Eng., Dallas, SPE 9269, 12 p.

Collins, S.E., 1980, Jurassic Cotton Valley and Smackover reservoir trends, East Texas, North Louisiana and South Arkansas, Am. Assoc. Petroleum Geologists Bull., v.64, No.7, p. 1007-1013.

Flaum, C. and G. Pirie, 1981a, Induced spectroscopy: a mineralogy-lithology well log (Abs.), Amer. Assoc. Petroleum Geologists, Annual Meeting, San Francisco.

_________ and _________, 1981b, Determination of lithology from induced gamma-ray spectroscopy, Symposium, Soc. Prof. Well Log Analysts, Mexico City, 16 p. Also presented at Formation Evaluation Symposium, Canadian Well Logging Society, Calgary.

Foscolos, A.E. and T.G. Powell, 1979, Catagenesis in shales and occurrence of authigenic clays in sandstones, North Sabine H-49 Well, Canadian Arctic Island, Can. Jour. Earth Sci., v 16, p. 1309-1314.

Frank, R.W., 1978, Formation evaluation with logs in the Ark-La-Tex Cotton Valley, Trans., Gulf Coast Assoc. of Geol. Socs., v.28, p.131-141.

Habicht, J.K.A., 1979, Paleoclimate, paleomagnetism, and continental drift, AAPG Studies in Geology No.9, Am. Assoc. Petroleum Geologists, Tulsa, OK., 31 p.

Heward, A.P., 1981, A Review of wave-dominated clastic shoreline deposits, Earth-Science Reviews, v.17, p.223-276.

Hunt, E.R., D.H. Raymond, C.E. Haskett, and R.G. Pirie, 1981, Application of new well logs and geology to fracturing and producibility in tight gas sands, Cotton Valley Group, Proc., Soc. Petroleum Eng./Dept. of Energy Symposium on Low Permeability Gas Reservoirs, Denver, SPE/DOE 9832, 10 p.

Mann, C.J. and W.A. Thomas, 1964, Cotton Valley (Jurassic) nomenclature, Louisiana and Arkansas, Trans., Gulf Coast Assoc. Geol. Soc., v.14, p.143-152.

Pirie, G., 1982, Geology and log study of tight gas sandstones: Cotton Valley Group, Trans., Gulf Coast Assoc. of Geol. Socs., V. 32, p. 77-88.

__________, and E. Hunt, 1981, Sedimentology of tight gas sandstones: well log core evaluation (Abs.), Am. Assoc. Petroleum Geologists, Annual Meeting, San Francisco.

Reineck, H.E. and I.B. Singh, 1973, Depositional sedimentary environments - with reference to terrigenous clastics, Springer-Verlag, New York, 412 p.

Rogers, R.W., 1968, Carthage Field, Panola County, Texas *in* Natural Gases of North America - A Symposium v. 1, Amer. Assoc. of Petroleum Geologists, p. 1020-1059.

Tindell, W.A., J.K. Neal, and J.C. Hunter, 1981, Evolution of fracturing the Cotton Valley Sands in Oak Hill Field, Jour. Petroleum Tech., v.33, p.799-807.

Todd, R.G. and R.M. Mitchum, Jr., 1977, Seismic stratigraphy and global changes of sea level, Part 8: Identification of Upper Triassic, Jurassic, and Lower Cretaceous seismic sequences in Gulf of Mexico and Offshore West Africa, *in* Seismic Stratigraphy - Applications to hydrocarbon exploration (Payton, C.E., Ed.), Am. Assoc. Petroleum Geologists Mem. 26, p. 145-163.

Wolff, M. and J. Pelissier-Combescure, 1982, FACIOLOG-Automatic electrofacies determination, Soc. Prof. Well Log Analysts, Corpus Christi, 24 p.

APPENDIX

The following is a brief explanation of the tools and their geological application to gas-bearing, shaly sandstones.

BHC (Borehole Compensated Sonic Log): Determines porosity in liquid-filled boreholes. In complex and/or shaly sandstone lithologies, must be used in conjunction with density and neutron logs to determine accurate porosity and lithology.

CNT-G (Compensated Neutron Tool): Measures thermal and epithermal neutrons released from formation when bombarded by high-energy neutron source. In complex and/or shaly sandstone lithologies, must be used in conjunction with density and sonic logs to determine accurate porosity and lithology.

Digital Sonic Tool: Measures values of shear and compressional velocities from which Poisson's ratio is determined. These velocities in combination with bulk density yield Young's Modulus which is commonly used to calculate fracture properties. By combining Poisson's ratio with a shaly sand analysis, vertical containment boundaries of a hydraulic fracture may be determined.

DIL (Dual Induction-Laterolog 8 or Dual Induction - Spherically Focused Log (DISF)): Records SP and formation resistivities in boreholes filled with water or oil-base mud, or air, with best results obtained in medium resistivities and fresh mud.

EPT (Electromagnetic Propagation Tool): Measures water-filled porosity and is unaffected by fresh water. In combination with CNL/FDC logs, identifies oil, gas, and water zones.

FACIOLOG: See text for explanation.

FDC (Formation Density Log): For porosity analysis in boreholes drilled with gas or liquid. In complex and/or shaly sandstone lithologies, must be used in conjunction with neutron and sonic logs to determine accurate porosity and lithology.

GST-A (Gamma Spectrometry Tool): When run in capture mode, detects and quantitatively measures hydrogen, silicon, sulfur, chlorine, calcium, and iron, which may be combined to produce a mineralogy/lithology log. In addition, carbon and oxygen may be detected when run in inelastic mode for calculation of oil saturation.

HDT (High-Resolution Dipmeter Tool): For formation dip and direction, stratigraphic interpretation, structural identification and location of faults and unconformities. With Fracture Identification Log (FIL) can be used to locate and evaluate fractures.

LSS (Long-Spaced Sonic Log): Primarily used for same purposes as BHC sonic logs but provides more accurate data for seismic applications in areas where shales are altered during drilling process, and in zones with enlarged boreholes.

NGS (Natural Gamma Ray Spectrometry): Detects and quantitatively measures total Gamma Ray (GR), and thorium, uranium, and potassium spectra to determine clay content in presence of radioactive non-clay minerals. Also determines zones of constant clay properties.

NML (Nuclear Magnetic Log): In gas-filled formations determines free fluid index which is essentially the water which is free to move and therefore producible.

SARABAND: Recently replaced by VOLAN, is a computer-processed shaly sand analysis which uses resistivity log with SP, gamma ray, density, neutron, sonic, and EPT logs. VOLAN (Volumetric Analysis) is a computer program which incorporates concept of bound water in shales and free water in sands.

SIGNIFICANT FOSSILS OF THE KNOWLES LIMESTONE, LOWER CRETACEOUS, TEXAS

R.W. SCOTT

Amoco Production Company
P.O. Box 591
Tulsa, Oklahoma 74102

ABSTRACT

Cores of the Knowles Limestone in Robertson and Milam Counties, Texas, East Texas Embayment, have yielded 25 taxa of algae, tintinnids, foraminifers, sponges, corals, stromatoporoids, and *Incertae Sedis*. Many of these Tethyan taxa are reported for the first time from the Gulf Coast region. The tintinnids demonstrate correlation with upper Berriasian to Valanginian assemblages of the Mediterranean area. The other fossils are longer ranging Upper Jurassic and/or Lower Cretaceous taxa. This Knowles assemblage represents a shelf-to-basin suite of paleocommunities, including coral-algal reefs.

INTRODUCTION

The Knowles Limestone is the uppermost unit of the Cotton Valley Group in the Texas-Louisiana part of the Gulf Coast (Finneran *et al.*, this volume). Cores from four downdip wells have been studied by means of acetate peels and thin-sections. At this point in the analysis, 25 taxa have been identified and are described and illustrated here. This is the first documentation of these fossils from post-Jurassic and pre-Barremian carbonates in the Gulf Coast of the United States. Additional taxa remain to be studied from these and other available cores. However, many specimens are not readily identifiable because they are poorly preserved. Lower Cretaceous tintinnids were most recently illustrated from Mexico by Trejo (1975) and from Cuba by Pop (1976). Pre-Aptian corals and mollusks were described from Venezuela by von der Osten (1957) and Neocomian corals and mollusks from Mexico were described by Felix (1891). Many of these corals from Venezuela and Mexico need to be studied in thin-section and compared with Mediterranean faunas in order to understand their affinities.

The *biota* is briefly described and illustrated in order to demonstrate the age and to permit their identification in future studies of carbonate cores. An exhaustive bibliography of each species was not compiled, but the original reference and, where possible, a reference to a current description providing more complete documentation are given. Generally, the latter reference is more readily accessible than the former. All specimens and thin-sections used are cataloged in the Amoco Paleo File collections under the appropriate well locality or taxon.

SYSTEMATICS

Phylum Rhodophycophyta Papenfuss, 1946
Affinities Uncertain
Genus Marinella Pfender, 1939
Marinella lugeoni Pfender, 1939
Pl. 1, figs. 1-3

Marinella lugeoni Pfender, 1939, p. 3, Pl. 2; Johnson, 1961, p. 147-148, Pl. 31, figs. 1, 2.

Diagnosis: Nodular structure comprised of distinct columnar lobes or digitate branches with concentric growth laminae; thallus consists of narrow filaments that intertwine and diverge distally; cross partitions in filaments indistinct.

Discussion: The Knowles specimen is 26 mm high and 22 mm wide; thalli are 270-710 μm wide, and the filaments are 4-7 μm wide. *Marinella lugeoni* differs from the Lower Cretaceous *M. yugoslavica* Maslow by its less distinct cross partitions in the filaments. *Marinella lugeoni* was found in the Upper Jurassic Cotton Valley limestone (probably Haynesville Formation) in Navarro County, Texas (Johnson, 1961). Johnson (1961) noted the lack of cross partitions but the presence of local constrictions; close examination of our specimens shows that locally these constrictions join across the filaments to form indistinct partitions. This taxon ranges in age from Lias to Early Cretaceous in the Mediterranean region (Johnson, 1961) and lived in shallow shelf, restricted environments (Persoz & Remane, 1973).

Phylum Chlorophycophyta Papenfuss, 1946
Family Codiaceae
Genus Cayeuxia Frollo, 1938
Cayeuxia piae Frollo, 1938
Pl. 1, figs. 4, 5

Cayeuxia piae Frollo, 1938, p. 269, Pl. 8; Johnson, 1969, p. 37, Pl. 25, fig. 3, Pl. 26 fig. 2.

Diagnosis: Nodular thallus with subparallel circular filaments radial to slightly diverging, branching at angle of 40-45 degrees, filament diameter 70-125 μm with local constrictions; wall dark-brown-gray microspar with fine, black granules, wall 9-21 μm wide.

Discussion: These nodules are from 6 to 13 mm tall and 5 to 7.5 mm wide. This species differs from other species of *Cayeuxia* by the large filament diameter and the dark wall structure. *Cayeuxia americana* Johnson (1961) has a filament diameter of 65-95μm and a dark micrite wall and probably is a junior synonym of *C. piae*. *Cayeuxia piae* is a Tethyan Middle Jurassic to Early Cretaceous species. *Cayeuxia americana* was found in Upper Jurassic rocks in Texas (Johnson, 1961).

Genus Lithocodium Elliott, 1956
Lithocodium aggregatum Elliott, 1956
Pl. 1, fig. 6; Pl. 3, fig. 6

Lithocodium aggregatum Elliott, 1956, p. 331,

PLATE 1

Figures 1-3. ***Marinella lugeoni*** **Pfender, Amoco No. 1 Seale, 14,914 ft; 1, thallus 5.2X; 2, longitudinal section, 84X; 3, transverse section, 84X.**

Figures 4-5. ***Cayeuxia piae*** **Frollo; 4, thallus, Inexco No. 1 Seale, 15,107.5 ft, 13X; 5, thallus, Amoco No. 1 Seale, 14,967 ft, 13X.**

Figure 6. ***Lithocodium aggregatum*** **Elliott and** ***Micritosphaera ovalis*** **Scott, Amoco No. 1 Seale, 14,905 ft, thallus encrusting** ***Solenpora*** **with** ***M. ovalis*** **in-situ, 33X.**

Figures 7-16. Tintinnids, Inexco No. 1 Seale at 15,593 ft, 336X:

Figures 7-8. ***Tintinnopsella carpathica*** **(Murg. & Filip.);**

Figures 9-10. ***Lorenziella hungarica*** **Knauer & Nagy;**

Figures 11-12. ***Calpionella alpina*** **Lorenz;**

Figure 13. ***Calpionellopsis oblonga*** **Cadisch;**

Figures 14-15. ***Calpionellopsis simplex*** **(Colom);**

Figure 16. ***Calpionellites darderi*** **(Colom).**

Figures 17-19. ***Pseudocyclammina*** **sp., 33X; 17, Inexco No. 1 Seale, 14,964 ft; 18, 19, Amoco No. 1 Seale, 14,955 ft.**

Figures 20-21. ***Pseudocyclammina lituus*** **(Yokoyama), Amoco No. 1 Seale, 15,074 ft; 20, 21X; 21, 33X.**

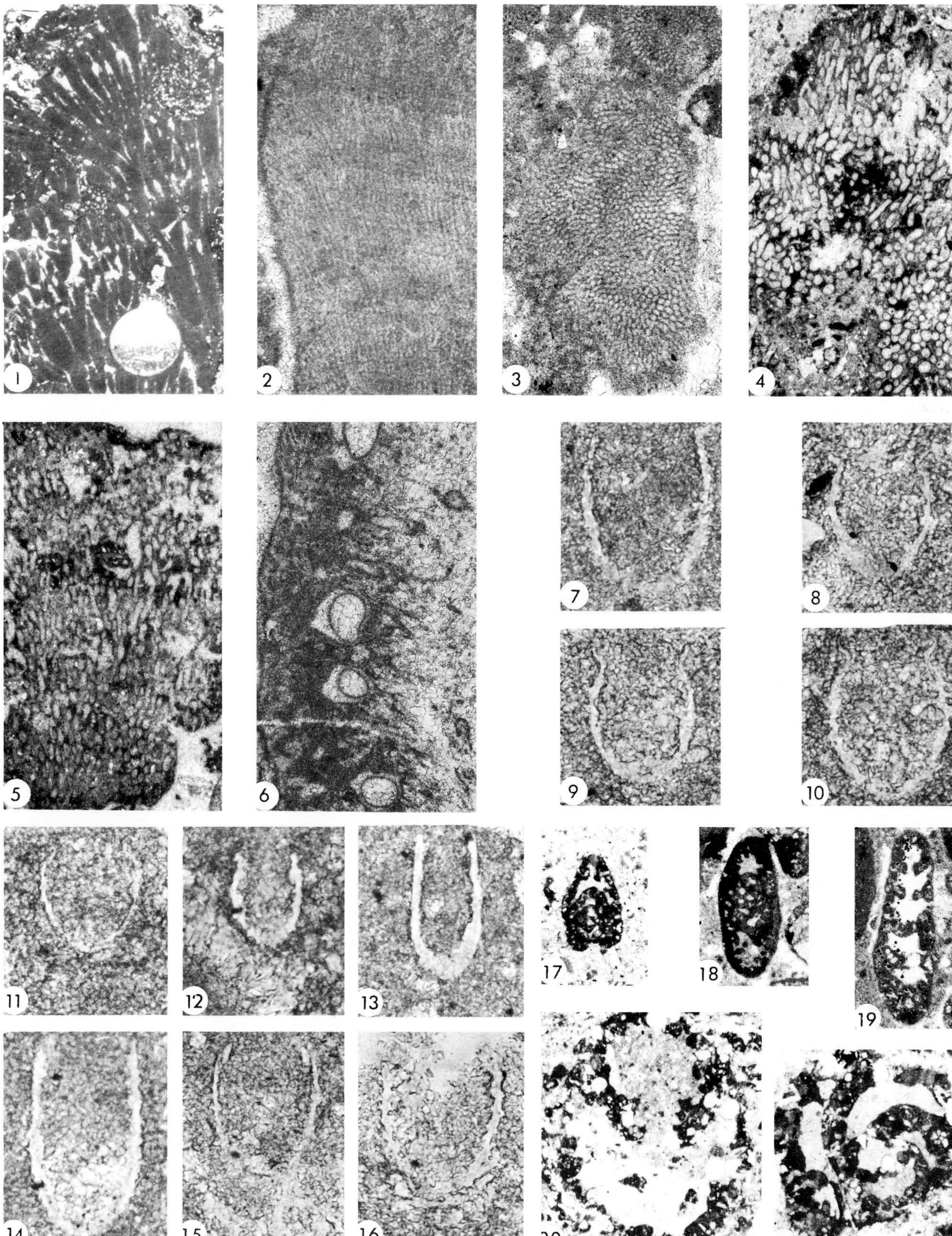
1
2
3
4
5
6
7
8
9
10
11
12
13
14
15
16
17
18
19
20
21

Pl. 1, figs. 2, 4, 5; Johnson, 1969, p. 38-39, Pl. 27.

Diagnosis: Thallus irregular, encrusting; dark micritic wall; inner layer of irregularly spaced large filaments, outer layer of irregularly radial, branching filaments; wall thickness variable, about 250 µm.

Discussion: The thallus encrusts a variety of objects resulting in the irregular exterior shape and wall thickness. Locally, the inner wall consists of a coarse reticulate structure of curved or straight partitions. Tangential sections of the outer wall show a densely perforate pattern resulting from the intersection of the numerous, closely spaced filaments. In the Knowles specimens, the diameter of the large filaments in the inner wall is about 80-90 µm; and the small filaments of the outer wall are about 25-35µm in diameter; wall thickness ranges up to 750µm. This is an Early Cretaceous, Tethyan species.

Genus Micritosphaera Scott, 1978
Micritosphaera ovalis Scott, 1978
Pl. 1, fig. 6

Micritosphaera ovalis Scott, 1978, p. 508-510, text-fig. 1.

Diagnosis: Test spherical to oval; wall structure a single layer of subtranslucent micrite, 5-25µm thick; diameter 490 to 1,180 µm.

Discussion: Originally this species was considered *Incertae Sedis*. However, the discovery of *M. ovalis* in the conceptacle chambers of *Lithocodium aggregatum* indicates that it is a cyst or conceptacle of this green alga. Because *M. ovalis* is commonly found separated and different species of *Lithocodium* may have *M. ovalis* as their cyst, it is recommended that this species be retained as a parataxon. The *in situ* Knowles specimens range from 145 to 260 m in long diameter, and the wall is about 4-7 m thick. This occurence extends the range to Berriasian-Albian.

Phylum Protozoa Goldfuss, 1818
Class Ciliata Perty, 1852
Order Tintinnida Corliss, 1955
Family Codonellidae Kent, 1881
Genus Tintinnopsella Colom, 1948
Tintinnopsella carpathica (Murgeanu and Filipescu, 1933)
Pl. 1, figs. 7, 8

Calpionella carpathica Murgeanu & Filipescu, 1933, p. 63-64, text-figs. 1a-c.

Tintinnopsella carpathica (Murgeanu & Filipescu). Le Hegarat & Remane, 1968, p. 32-33, Pl. 6, figs. 14-16; Pl. 7, figs. 4-11; Pl. 8, figs. 1, 6, 16, 19-23; Pl. 9, figs. 10, 11, 13-15; Pl. 10, figs. 4, 11, 13-15.

Diagnosis: Lorica subcylindrical to bell-shaped; collar flares out at an angle of 45-90° to long axis of lorica; aperture about as wide as width of lorica; base pointed with caudal spine; lorica 60-100 µm long, 46-53 µm wide.

Discussion: A distinct caudal spine is not seen on these specimens, presumably because of the orientation of the cuts. The specimen on Pl. 1, fig. 7, is 104µm long by 59µm wide, a ratio of 1.76. *Tintinnopsella longa* (Colom) has no caudal spine and differs from these specimens by its greater length and narrower width. *Tintinnopsella longa* has a length-width ratio of about 3.2, based on ranges of published measurements. The age of *T. carpathica* is Late Tithonian to Barremian (Remane, 1969).

Genus Lorenziella Knauer and Nagy, 1963
Lorenziella hungarica Knauer and Nagy, 1963
Pl. 1, figs. 9, 10

Lorenziella hungarica Knauer and Nagy, 1963, p. 144, Pl. 1, figs. 3, 5, 9-13; Le Hégarat & Remane, 1968, p. 35, Pl. 10, figs. 6, 8-10.

Diagnosis: Lorica ovoid, distinct shoulders; collar arcuate, convex to aperture, arc radius short; basal portion sharply curved to nearly pointed; length 52-88µm, width 41-68µm

Discussion: Our specimens possess a distinctive inward deflection of the collar above the outward arc. This inward deflection also occurs on specimens illustrated by Trejo (1975, p. 373-375, Pl. 12, figs. 19, 28). Age is Berriasian-Valanginian (Remane, 1969).

Genus Remaniella Catalano, 1965
Remaniella cadischiana (Colom, 1948)

Tintinnopsella cadischiana Colom, 1948, p. 247, Pl. 33, fig. 17.

Remaniella cadischiana (Colom). Le Hegarat & Remane, 1968, p. 36-41, Pl. 7, figs. 12, 17; Pl. 8, fig. 25; Pl. 10, figs. 17, 18.

Diagnosis: Lorica subcylindrical; collar topped by a broadly v-shaped or u-shaped ring separated from wall by a suture; base sharply curved to pointed; length 90-160 μm, width about 70μm.

Discussion: The Knowles specimen is 97 μm long and 64μm wide (Inexco No. 1 Seale well, 15,558.5') The age is Berriasian to Hauterivian (Remane, 1969).

Family Codonellopsidae Kofoid and Campbell, 1929
Genus Calpionella Lorenz, 1902
Calpionella alpina Lorenz, 1902
Pl. 1, figs. 11, 12

Calpionella alpina Lorenz, 1902, p. 60-61, Pl. 9, fig. 1; Le Hégarat & Remane, 1968, p. 31, 32, Pl. 6, figs. 4-10, 18-21; Pl. 8, figs. 2-5; Pl. 9, figs. 21-23; Pl. 10, figs. 2, 3.

Diagnosis: Lorica ovoid; collar with short segment subparallel with long axis of lorica; aperture narrower than lorica width; base smoothly rounded, lorica 50-110μm long, 40-60 μm wide.

Discussion: These specimens are recrystallized and some have secondary overgrowths that alter the typical appearance. These specimens range from 45-56 μm long and 34-50 μm wide. *Calpionella alpina* ranges in age from Late Tithonian to Valanginian (Remane, 1969).

Genus Calpionellopsis Colom, 1948
Calpionellopsis oblonga Cadisch, 1932
Pl. 1, fig. 13

Calpionellopsis oblonga Cadisch, 1932, p. 252, figs. 20, 21; Le Hégarat & Remane, 1968, p. 34-35, Pl. 8, figs. 17, 18; Pl. 9, figs. 1-6; Pl. 10, figs. 1, 7, 16.

Diagnosis: Lorica subcylindrical, walls tend to converge towards aperture; simple neck set into inner side of collar; basal end angular to pointed; up to 50μm wide.

Discussion: The Knowles specimen is 81 μm long and 36 μm wide. *Calpionellopsis oblonga* ranges in age from Late Berriasian to Valanginian (Remane, 1969).

Calpionella simplex (Colom, 1939)
Pl. 1, figs. 14, 15

Calpionella simplex Colom, 1939, p. 820, Pl. 2, fig. 11; Pl. 3, fig. 6, 9.

Calpionellopsis simplex (Colom). Le Hégarat & Remane, 1968, p. 34, Pl. 7, fig. 13; Pl. 8, figs. 7-12; Pl. 9, fig. 8.

Diagnosis: Lorica subcylindrical; neck a simple ring set flush into inner side of collar, parallel with wall; basal end rounded; length about 90μm, width about 40 μm.

Discussion: The inner ring of the collar has a different extinction angle than the wall. The Knowles specimens are 70-101 μm long and 53-56μm wide. Geologic range is Berriasian to possibly Hauterivian (Remane, 1969).

Genus Calpionellites Colom, 1948
Calpionellites darderi (Colom, 1934)
Pl. 1, fig. 16

Calpionella darderi Colom, 1934, p. 381, Pl. 31, fig. 3.

Calpionellites darderi (Colom). Le Hégarat & Remane, 1969, p. 36-41, Pl. 10, figs. 19-20.

Diagnosis: Lorica subspherical to subcylindrical; wall curves toward aperture; neck a separate ring set into inner side of collar and appearing to bifurcate from wall; base smoothly rounded to slightly pointed; length 80-130μm, width 65-80μm.

Discussion: The inner ring of the collar has a different extinction angle than the wall. A single specimen was found, 78μm long and 59 μm wide. The geologic range of *C. darderi* is Berriasian to Hauterivian (Remane, 1969), although its common occurrence begins in the Valanginian.

Class Rhizopodea von Siebold, 1845
Order Foraminiferida Eichwald, 1830
Suborder Textulariina Delage and Hérouard, 1896
Superfamily Lituolacea de Blainville, 1825
Family Lituolidae de Blainville, 1825
Subfamily Cyclammininae Marie, 1941
Genus Pseudocyclammina Yabe and Hanzawa, 1926

Pseudocyclammina lituus Yokoyama, 1890)
Pl. 1, figs. 20, 21

Cyclammina lituus Yokoyama, 1890, p. 26, Pl. V, fig. 7.

Pseudocyclammina lituus (Yokoyama). Maync, 1959, p. 158-164, Pl. 1-6.

Diagnosis: Test large, inflated, ovate in axial section, periphery broadly rounded, chambers tall and wide; sutures distinctly impressed; test tightly coiled, later stage uniserial; wall thick; up to 10 chambers in first whorl.

Discussion: Unfortunately none of the four specimens in this study have perfect axial or equatorial sections. The Knowles specimens possess the labyrinthic, microgranular wall with some coarse agglutinated particles. The chambers are broadly rounded and closely appressed. This Tethyan species ranges in age from Bathonian to Hauterivian (Maync, 1959).

Pseudocyclammina sp.
Pl. 1, figs. 17-19

PLATE 2

Figures 1-3. *Nautiloculina oolithica* **Mohler, 53X; 1, transverse section, Inexco No. 1 Seale, 14,981 ft; 2, 3, Amoco No. 1 Seale, 14,955 ft and 14959 ft; transverse and axial sections.**

Figure 4. *Pfenderina* **sp., 53X; basal section, Amoco No. 1 Seale, 14,955 ft.**

Figure 5. *Trocholina* **cf.** *T. elevata* **Paalzow, 53X; Amoco No. 1 Seale, 14,959 ft.**

Figures 6-7. *Corynella mexicana* **Rigby & Scott, Amoco No. 1 Seale; 6, transverse section, 14,924.5 ft, 5.2X; 7, details of spicules, 15,094, 33X.**

Figure 8. *Styosmilia* **sp., transverse section of calyces, Inexco No. 1 Seale, 14,964 ft,13X**

Figures 9-10. *Microsolena distefanoi* **(Prever); transverse sections, Inexco No. 1 Seale, 5.2X; 9, 14,974 ft; 10, 14,972.5 ft.**

Figure 11. *Microsolena* **sp. cf.** *M. guttata* **Koby, Inexco No. 1 Seale, 14,974.5 ft, 5.2X.**

Figures 12-13. *Actinaraea* **sp. aff.** *A. tenuis* **Morycowa, Inexco No. 1 Seale, 15,006 ft, 5.2X**

Figure 14. *Isastraea* **sp., Shell No. 2 Adoue, acetate peel at 12,148 ft, 5.2X.**

Figure 15. *Shuqraia arabica* **Hudson; Amoco No. 1 Seale, 14,979.5 ft, crossed nicols, fibrous structure, 84X.**

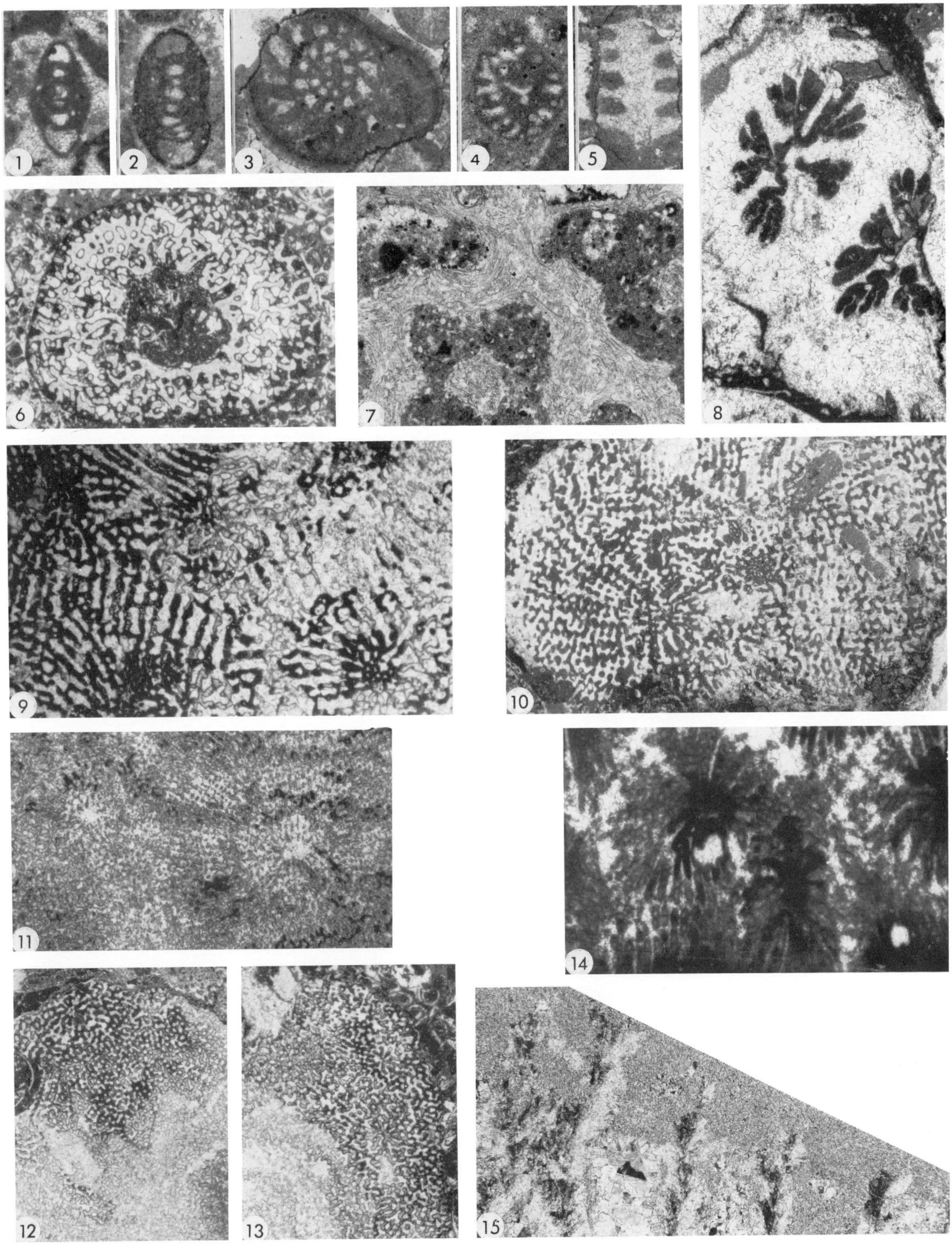
1
2
3
4
5
6
7
8
9
10
11
12
13
14
15

Description: Test planispiral, involute, tending to become uniserial, lenticular, umbilicate for at least three whorls; periphery subacute, narrowly rounded; wall agglutinated calcareous microgranular, thin, imperforate outer layer, thick, reticulate to labyrinthic inner layer, tending to become denser on later chambers; septal walls with few perforations. The axial diameter ranges from 251-322 µm and equatorial diameter is at least 537-662µm. Proloculus diameter is 21-28 µm.

Discussion: Only axial and tangential sections were found so that the complete coiling style is unknown. This taxon is distinctive and common in the back-reef grainstones of the Knowles. It is much smaller than any other species of *Pseudocyclammina* and its test wall is denser, less reticulate than most other species of this genus. It closely resembles *Cyclammina greigi* Henson (1948), but its periphery is more angular, the test wall is labyrinthic, and the test is consistently smaller. *Cyclammina greigi* occurs above *P. lituus* in Arabia.

Family Pavonitinidae Loeblich and Tappan, 1961
Subfamily Pfenderininae Smout and Sugden, 1962
Genus Pfenderina Henson, 1948
Pfenderina sp.
Pl. 2, fig. 4

Description: Test wall calcareous microgranular, imperforate; numerous chambers arranged radially around a solid axis; septa about normal to axis and inclined to test wall; diameter 322-394 µm.

Discussion: The single specimen is not adequate for a specific assignment. The genus ranges in age from Late Jurassic to Cenomanian (Henson, 1948, p. 610).

Suborder Miliolina Delage and Hérouard, 1896
Superfamily Miliolacea Ehrenberg, 1839
Family Fischerinidae Millett, 1898
Subfamily Fischerininae Millett, 1899
Genus Nautiloculina Möhler, 1938
Nautiloculina oolithica Möhler, 1938
Pl. 2, figs. 1-3

Nautiloculina oolithica Möhler, 1938, p. 19, Pl. 4, figs. 1-3, 6, text-fig. 6.

Diagnosis: Periphery of test broadly rounded in axial section; in equatorial section septa about normal to exterior wall; axial diameter about 280-350 µm, equatorial diameter about 450-600µm; 4-5 whorls, up to 14 chambers in last whorl.

Discussion: The Knowles specimens range from 394-626µm in equatorial diameter and 215-519 µm in axial diameter. This species is generally smaller than *Nautiloculina cretacea* Peybernes (1976), which has a more sharply rounded periphery and inclined septa in the last whorl.

Nautiloculina circularis (Said and Baraket, 1959 *in* Maync, 1966, p. 9) also has a subacute periphery, inclined septa and up to 17 chambers in the last whorl. *Nautiloculina oolithica* ranges in age from Oxfordian to Early Cretaceous. *Nautiloculina cretacea* ranges in age from Upper Valanginian to Upper Albian, and *N. circularis* ranges in age from Aalenian to Oxfordian in Israel (Maync, 1966).

Suborder Rotaliina Delage and Herouard, 1896
Superfamily Cassidulinacea d'Orbigny, 1839
Family Cassidulinidae d'Orbigny, 1839
Genus Trocholina Paalzow, 1932
Trocholina cf. *T. elevata* Paalzow, 1932
Pl. 2, fig. 5

Trocholina elevata Paalzow, 1932, p. 140, Pl. 11, fig. 4.

Diagnosis: Test conical, trochospiral, height greater than axial diameter; base plate, without sculpture; height about 300µm, diameter about 200µm.

Discussion: Numerous species of *Trocholina* have been based on exterior features of free specimens, and these cannot be recognized in thin-section. The Knowles specimens are sectioned off center and are about 286 by 465μm in size. They can be compared to *T. elevata,* which has a similar size, shape and stratigraphic distribution. Comparable specimens from the Neocomian of southern France were called *Trochlina alpina-elongata* Leupold by Peybernes (1976, p. 176), which are Late Jurassic to Early Cretaceous-aged taxa erected in 1935.

Phylum Porifera Grant, 1872
Class Calcispongia De Blainville, 1834
Order Pharetronida Zittel, 1878
(?)Suborder Charlarina De Laubenfels, 1955
Family Lelapiidae Dendy and Row, 1913
Genus Corynella Zittel, 1878
Corynella mexicana Rigby and Scott, 1981
Pl. 2, figs. 6, 7

Corynella mexicana Rigby and Scott, 1981, p. 552-559, Pl. 1, figs. 1, 2, 4-6, text-fig. 2a.

Description: Small subcylindrical to turbinate skeleton, wide osculum extending length of skeleton; prominent radial subhorizontal canals cross connected by concentric canals and small vertical canals; skeletal tracts meandriform with numerous straight to curved spicules and few large triacts.

Discussion. The Knowles specimens range in diameter from 7 to 28 mm, the walls of larger specimens are 6-12 mm thick and the osculum is 8 by 11 mm in diameter. The diact spicules are 17 to 21μm in diameter and up to 120 m in length. The triaxons are about 35μm by 280 μm in size. These specimens match the descriptions of *C. mexicana,* so that its range is extended from Albian to Berriasian in the Gulf Basin area. The species also occurs in the Shell No. 1 Adoue well at a depth of 12,167 feet.

Phylum Coelenterata Frey & Leuckart, 1847
Class Anthozoa Ehrenberg, 1834
Order Scleractinia Bourne, 1900
Suborder Astrocoeniina Vaughan & Wells, 1943
Family Stylinidae d'Orbigny, 1851
Subfamily Stylininae d'Orbigny, 1851
Genus Stylosmilia M.-Edwards & Haime 1848
Stylosmilia sp.
Pl. 2, fig. 8

Description: Corallum dendroid to phaceloid; exterior of branch costate; calyces oval, not interconnected by septa; 3 cycles of septa, 5-6 tending to join styliform columella, 6 septa of second cycle, 12 septa of third cycle; septal margins with costae and auriculae appearing as spines and projections on inner edges of septa; coenosteum dense, recrystalized spar. Branches 2.7 by 4.6 mm in diameter, calyx diameter 1.9 by 3.2 mm, intercalycal distance about 2 mm.

Discussion: The phaceloid clumps commonly are overgrown by other corals and encrusted by algae. The exterior ribbing is nearly obscured by these symbionts. Numerous species are assigned to this Middle Jurassic-Early Cretaceous genus, which are known in Europe and South America.

Suborder Fungiina Verrill, 1865
Superfamily Agariciicae Gray, 1847
Family Calamophylliidae Vaughan & Wells, 1943
Genus Isastrea M.-Edwards & Haime, 1851
Isastrea sp.
Pl. 2, fig. 14

Description: Massive ceroid corallum; corallite walls septothecate, septa confluent; trabecular columella lacking or poorly developed; common dissepiments; corallites about 6-7 mm diameter, 30 or more septa.

Discussion: Poorly preserved corals with polygonal corallites and virtually no columella appear to belong to this genus. This specimen is similar to those of *Isastrea whitneyi* Wells (1932, p. 245, Pl. 39, figs. 4, 5), which comes from the Lower Albian Glen Rose Formation of Texas. The species is also reported from the pre-Aptian Barranquin Formation in Venezuela (von der Osten, 1957, p. 576, Pl. 63, fig. 10).

Superfamily Poriticae Gray, 1842
Family Microsolenidae Koby, 1890
Genus Microsolena Lamouroux, 1821
Microsolena distefanoi (Prever, 1909)
Pl. 2, figs. 9, 10

Microsarea distefanoi Prever, 1909, p. 71, Pl. 2, figs. 6, 6a.

Microsolena distefanoi (Prever); Turnšek and Buser, 1974, p. 101-102, 117, Pl. 11, fig. 2.

Diagnosis: Corallum massive to laminar; corallite surface flat, calyces shallow, randomly arranged, columella projecting slightly; columella formed by anastomosing palli; septa more or less regularly perforated, straight, arcuate or undulating, synapticulae quite numerous, dissepiments rare; calycal diameter about 4-5 mm, distance between calyces 5-9 mm, 25-48 septa, 6-8 septa per 2 mm (data from Morycowa, 1971; Turnšek and Buser, 1974).

Discussion: The Knowles specimens have regularly perforated septa; but in the calyx, perforations are widely spaced and between calyces they are closely spaced. Some septa tend to be thickened by secondary calcite. The dimensions are similar to European specimens: calycal diameter 3-6 mm, intercalycal distance 5-9 mm, 20 to 36 septa, 12-14 septa per 5 mm, septa 0.11-0.31 mm thick. However, the columella appears better developed than in the Yugoslavian specimens (see Pl. 2, fig. 2, Turnšek & Buser, 1974). *Microsolena distefanoi* differs from other Lower Cretaceous species of this genus by its thick, widely spaced septa, moderate number of septa, and moderate calycal diameter. Its septa are of the same size and density as *M. thurmanni* Koby, but *M. thurmanni* has very widely spaced septal performations and a larger number of septa per calyx, which tend to be arranged in rows; *M. thurmanni* also has more dissepiments than *M. distefanoi*. *Microsolena thurmanni* occurs in Oxfordian and Kimmeridgian strata; *M. disteranoi* is known mainly from the Barremian to the Aptian but also may occur in the Senonian (Turnšek & Buser, 1974).

Microsolena sp. cf. *M. guttata* Koby, 1898
Pl. 2, fig. 11

Microsolena guttata Koby, 1898, p. 83, 84, Pl. 21, figs. 1, 2; Turnšek and Mihajlovic, 1981, p. 35, Pl. 41, figs. 1-5.

Description: Massive to laminar; calyces randomly arranged; septa straight to arcuate, perforated sparsely at widely spaced intervals; synapticulae numerous, closely spaced; calycal diameter about 3-5 mm, intercalycal distance about 7-8 mm, 24 or more septa, 8-10 septa per 2 mm.

Discussion: This poorly preserved specimen is characterized by having thinner, more closely spaced septa, than *M. distefanoi*.

PLATE 3

Figures 1-5. *Shuqraia arabica* **Hudson; 1-3, Inexco No. 1 Seale, 14,991 ft; 1, transverse section of branch, 5.2X; 2, closeup of periphery, 13X; 3, recrystallized microstructure, 33X; 4-5, Amoco No. 1 Seale, 14,979.5 ft; 4, longitudinal section of nodule with three branches, 5.2X; 5, closeup of axial area of distal end of left-hand branch, 13X.**

Figures 6-7. *Koskinobullina socialis* **Cherchi & Schroeder, Amoco No. 1 Seale; 6, encrusting** *Lithocodium* **and overgrown by same, 14,994 ft, 33X; 7, closeup of wall structure, 15,035 ft, 84X.**

Figures 8-9. *Tubiphytes morronensis* **Crescenti; 8, transverse sections showing keyhole outline of tube, Inexco No. 1 Seale, 15,593 ft, 21X; oblique section of tube, Amoco No. 1 Seale, 15,094 ft, 33X.**

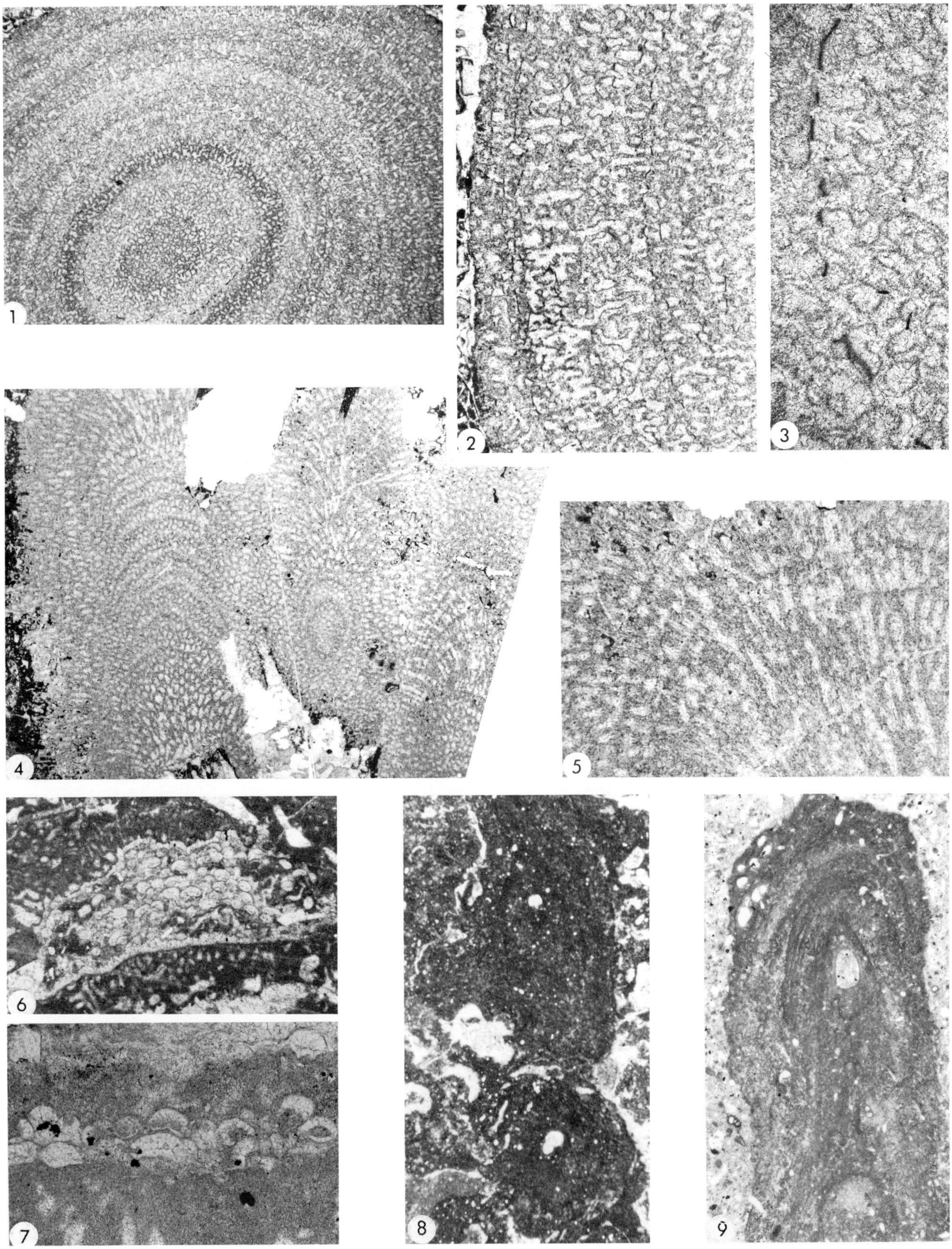

Also, *M. guttata* has larger calyces, about in 5-9 in mm diameter and 40-48 septa; the intercalycal distance is 5-6 mm, but septal thickness and density are close to this specimen (7-10 per 2 mm). *Microsolena guttata* is a Barremian to Early Aptian taxon in the European Tethys.

Family Actinacididae Vaughan & Wells, 1943
Genus Actinaraea d'Orbigny, 1849
Actinaraea sp. aff. *A. tenuis* Morycowa, 1971
Pl. 2, figs. 12, 13

Actinaraea tenuis Morycowa, 1971, p. 128-130, Pl. 35, fig. 1: Pl. 36, fig.1; text-fig. 37.

Description: Corallum nodular; corallites small, 1.4-1.8 mm, calycal margins poorly defined, 9-10 septa at the center; coenosteum irregularly trabeculate, intercalycal distance 2.3-2.6 mm, 11-13 septa per 2 mm; columella trabeculate, palli poorly developed.

Discussion: These poorly preserved fragments of colonies show few definite corallites. They differ from Morycowa's species by the more densely perforated septa, which are also developed in the Serbian specimens (Turnšek and Mihajlovic, 1981). *Actinaraea tenuis* has calyces 1.2-1.5 mm diameter, distance between calyces is 3-5 mm, 8-9 septa per 2 mm, and 12-24 septa, 8-10 of which reach the center of the calyx. Palli do not appear to be developed. This species is Early Aptian in age in Romania.

Class Hydrozoa, Owen, 1843
Order Stromatoporoidea Nicholson & Murie, 1878
Family Milleporidiidae Yabe & Sugiyama, 1935
Genus Shuqraia Hudson, 1954
Shuqraia arabica Hudson, 1954
Pl. 2, fig. 15; Pl. 3, figs. 1-5

Shuqraia arabica Hudson, 1954, p. 216-217, Pl. 7, figs. 3a-c.

Diagnosis: Dendroid colonies with axial and peripheral reticulum; branches about 30-35 mm diameter; axial reticulum small, about 5 mm diameter; oval axial tubules continuous, coenosteal laminae rare; peripheral tubules diverging, coenosteal laminae common; lateral tubules connect major tubules. Tubule diameter 60-190µm, tubule wall thickness 80-100µm.

Discussion: This genus differs from *Milleporidium* Steinmann by the approximately constant size of the axial and peripheral tubules. In *Milleporidium* the axial tubules are smaller than those in the peripheral cortex (Lecompte, 1956). Wall microstructure of *Shuqraia* is fibrous and fasiculate (Hudson, 1954), however, one of the Knowles specimens has been altered to sparry calcite. The fibrous wall structure is well preserved in a second specimen. The calcite fibers diverge distally from a central axis in the wall, which has a sinuous or corrugated exterior surface. Tubule diameter ranges from 88 to 226µm, and wall thickness is 18 to 54µm in the axial area and 44 to 95µm in the peripheral region. Branch diameter is 20 by 35 mm and the axial region is about 5-7 mm across. Hudson (1954) reported this species from the Late Jurassic of Arabia. The Knowles occurrence extends the range into the Early Cretaceous.

One specimen is a basal nodule with three small, contiguous branches 5-8 mm in diameter, which would be characteristic of a young colony. As the colony grows, the peripheral reticulum expands as the tubules grow outwards. Thus, the relative size of the two areas should not be a species character. Likewise, the walls are thicker in this specimen than in the other Knowles specimen, 88-100µm in the axial area and 88-132µm in the peripheral region. This thicker wall may also be a feature of a youthful colony. *Shuqraia arabica* has shorter, less distinct axial tubules, closer spaced tabulae, and more distinct latilamellae than *S. heybroeki* Hudson, also from the Late Jurassic of Arabia.

Incertae Sedis
Family Wetheredellidae Vachard, 1977
Genus Koskinobullina Cherchi and Schroeder, 1979

Koskinobullina socialis
Cherchi and Schroeder, 1979
Pl. 3, figs. 6, 7

Koskinobullina socialis Cherchi and Schroeder, 1979, p. 520-523, Pl. 1, figs. 1-3.

Diagnosis: Encrusting stratiform or irregularly rounded aggregates of hemispherical chambers averaging 120µm in diameter; calcite wall of radial fibers, porous, convex side of wall about 20µm thick, basal wall about 10µm thick.

Discussion: The Knowles specimens are 89-125 µm in diameter, the convex wall is 14-28µm thick, and the basal wall is 4-14µm thick. It commonly encrusts *Lithocodium* and is overgrown in turn by skeletal algae. *Koskinobullina socialis* is present in Tethyan reefal facies of Bathonian to Cenomanian age (Cherchi and Schroeder, 1979). In the Pyrennes Mountains, *K. socialis* was identified as *Incertae Sedis* (Aurouze et al., 1956 *in* Peybernes, 1976, Pl. IV, fig. 6, Pl. X, fig. 12), and occurred in Late Bathonian to Late Oxfordian-aged platform facies. This distinctive fossil may be either an alga or a foraminiferid (Cherchi and Schroeder, 1979).

Family Indeterminate
Genus Tubiphytes Maslov, 1956

Tubiphytes morronensis Crescenti, 1969
Pl. 3, figs 8, 9

Tubiphytes morronesis Crescenti, 1969, p. 35, Pl. 1, figs. 20-22.

Diagnosis: Tubular to irregular masses of micrite; distinct dark concentric laminae encircling a central tube filled with spar; key-hole-like restrictions in central tube at irregular intervals.

Discussion: These micritic masses are nodular or encrust other particles in the forereef/basin facies of the Knowles; rare fragments occur in the backreef grainstones. In the European Tethys, this species is in reefal facies from Bajocian to Aptian in age (Flügel, 1981). Although the biologic affinities are uncertain, Flügel (1981) suggests that these were cyanophycean(?) algae.

ACKNOWLEDGMENTS

Amoco Production Company has permitted the release of this information and has supported this study by preparation of the manuscript and plates. Many colleagues shared their experiences and references on many of these fossils: P. L. Brenckle, J. A. Babcock, B. A. Masters, R. J. White, and D. Turnšek.

REFERENCES

Cadisch, J., 1932, Ein beitrag zum Calpionellen-problem: Geol. Rundsch., v. 23, p. 241-257.

Cherchi, A. and R. Schroeder, 1979, *Koskinobullina* n. gen., Microorganism en colonie Incertae Sedis (Algues?) du Jurassique-Crétacé de la région Méditerranéene: Note preliminaire: Bull. Cent. Rech. Explor.-Prod. Elf-Aquitaine, v. 3, no. 2, p. 519-523, 1 pl.

Colom, G., 1934, Estudios sobre las Calpionelas: Bol. R. Soc. Esp. Hist. Nat., v. 34, p. 379-389.

________ 1939, Tintinnidos fosiles (Infusorios Oligotricos): Las Ciencias, v. 4, p. 815-825.

________ 1948, Fossil Tintinnids: Loricated Infusoria of the order of the Oliogotricha: Jour. Paleont., v. 22, p. 233-263.

Crescenti, U., 1969, Biostratigrafia delle facies mesozoiche dell' Appenino centrale: Correlazioni: Geol. Romana, v.8, p. 15-40, 1 pl.

Elliott, G. F., 1956, Further records of fossil calcareous algae from the Middle East: Micropaleontology, v. 2, p. 327-334.

Felix, J., 1891, Versteinerungen aus der mexicanischen Jura- und Kreide-Formation: Palaeontographica, v. 37, p. 140-194, 30 pl.

Flügel, E., 1981, "Tubiphyten" aus dem frankischen malm: Geol. Bl. NO-Bayern, v. 31, no. 14, p. 126-142.

Frollo, M. M., 1938, Sur un nouveau genre de Codiacee du Jurassique Superieur des Carpates Orientales: Soc. geol. France Bull., ser. 5, v. 8, p. 269-271.

Henson, F. R. S., 1948, New Trochamminidae and Verneuilinidae from the Middle East: Ann. Mag. Nat. Hist., ser. II, v. 14, p. 605-630, pl. XIV-XVIII.

Hudson, R. G. S., 1954, Jurassic Stromatoporoids from southern Arabia: Paris Mus. Nat. Hist. Nat., Notes Memo. Moyen-orient, v. 5, p. 208-220, pl. 6-8.

Johnson, J. H., 1961, Jurassic algae from the subsurface of the Gulf Coast. Jour. Paleont., v. 35, p. 147-151.

____________1969, A review of the Lower Cretaceous Algae. Colo. School Mines, Prof. Cont. no. 6, 180 p., 68 pl.

Knauer, J., and I. Nagy, 1963, *Lorenziella* nov. gen. uj Calpionellidea nemzetseg: Foldt. Int. Evi. Jel. 1961-röl II, p. 143-153, 1 pl.

Koby, F., 1896-1898, Monographie des Polypiers Crétacés de la Suisse, Pt. I-III: Mem. Soc. Paleont. Suisse, v. 22-24, 100 p., 21 pl.

Le Hégarat, G. and J. Remane, 1968, Tithonique Superieur et Berriasien de l'Ardeche et de l'Herault correlation des ammonitea et des calpionelles: Geobios, no. 1, p. 7-70, 10 pl.

Lecompte, M., 1956, Stromatoporoidea *in* R. C. Moore, ed., Treatise on Invertebrate Paleontology, Pt. F., Coelenterata, p. F107-F144, Geol. Soc. Amer. and Univ. Kansas Press, Lawrence, Kansas.

Lorenz, T., 1902, Geologische studien im grenzgebiete zwischen helvetischer und Ostalpiner fazies II. Der südliche Rhätikon: Ber. Nat. Ges. Freiburg, v. 12, p. 34-95, 9 pl.

Maync, W., 1959, Biocaractéres et analyse morphometrique d' especes Jurassique de genre Pseudocyclammina (Foraminiféres). I. Pseudocyclammina lituus (Yokoyama): Rev. Micropal., v. 2, p. 153-172, 6 pl.

__________1966, Microbiostratigraphy of the Jurassic of Israel: Israel Geol. Sur. Bull. no. 40, 56 p., 10 pl.

Möhler, W., 1938, Micropaläontologische untersuchungen in der nordschweizerischen Juraformation: Schweiz. Pal. Ges., Abh., v. 60, p.1-53, pl. 1-4.

Morycowa, E., 1971, Hexacorallia et Octocorallia du Crétacé Inferieur de Rarau (Carpathes Orientales Roumaines): Acta Palaeont. Polonica, v. 16, p. 1-149, 40 pl.

Murgeanu, G., and M. G. Filipescu, 1933, *Calpionella carpathica* n. sp. dans les Carpathes roumaines: Notat. Biol., v. 1, p. 63-64.

Paalzow, R., 1932, Die Foraminiferen aus den Transversarius-Schichten und Impressa-Tonen der nördostlichen Schwäbischen Alb: Ver. Vaterl. Naturk, Wurttemberg, Jahrehsefte, v. 88, Abh. 3, p. 81-142, pl. 4-11.

Persoz, F. and J. Remane, 1973, Evolution des milieux de dépôt au Dogger supérieur et au Malm dans le Jûra neuchâtelois méridional. Eclogae geol. Helv. v. 66, p. 41-70.

Peybernes, B., 1976, Le Jurassique et le Cretace Inferieur des Pyrenées Franco-Espagnoles, entre la Gargonne et la Méditerranee: These Doctorat Sciences Naturelles, Toulouse, 459 p., XLII pl.

Pfender, J., 1939, Sur un calcaire phytogene du lias inférieur d'Espagne et l'extension de ce facies en quelque autres régions: Univ. Lausanne Lab. geologie Bull. 66, 16 p., 5 pls.

Pop, G., 1976, Tithonian-Valanginian Calpionellid zones from Cuba: Dari de seama ale sedintelor, Inst. Geol. Geofizica, Bucarest, v. 62, 3. Paleont., p. 237-266, 10 pl.

Prever, L., 1909, La fauna coralligena cretacio dei Monti d'Ocre nell Abruzzo Aquillano: Mem. Carte geol. Italia, v. 5, p. 1-145.

Remane, J., 1969, Les possibilités actuelles pour une utilisation stratigraphique des calpionelles (Protozoa Incertae Sedis, Ciliata?): Proc. First Int. Conf. Plank. Microfossils, E. J. Brill, Leiden, v. 2, p. 559-573.

Rigby, J. K. and R. W. Scott, 1981, Sponges from the Lower Cretaceous Mural Limestone in Arizona and northern Mexico: Jour. Paleont., v. 55, p. 552-562, 1 pl.

Scott, R. W., 1978, *Micritosphaera,* a new Cretaceous microfossil: Jour. Paleont., v. 52, p. 508-511.

Trejo, M., 1975, Tintinidos Mesozoicos de Mexico: Bol. Asoc. Mex. Geol. Petrol., v. 27, p. 329-449, 25 pl.

Turnšek, D. and S. Buser, 1974, The Lower Cretaceous corals, Hyrdozoans and Chaetetids of Banjska Planota and Trnovski Gozd: Acad. Sci. Art. Slovenica, Razprave Dissert., v. 17, no. 2, p. 79-124, 16 pl.

___________and M. Mihajlovic, 1981, Lower Cretaceous Cindarians from eastern Serbia: Acad. Sci. Art. Slovenica, Razprave Dissert., v. 23, no. 1, p. 1-54, 50 pl.

Von der Osten, E., 1957, A fauna from the Lower Cretaceous Barranquin Formation of Venezuela: Jour. Paleont., v. 31, p. 571-590, pl. 63-65.

Wells, J. W., 1932, Corals of the Trinity Group of the Comanchean of central Texas: Jour. Paleont., v. 6, p. 225-256, pl. 30-39.

Yokoyama, M., 1890, Foraminiferen aus dem Kalksteine von Torinosu und Kompira: Denk. Akad. Wissen, math. nat. Cl., v. 57, p. 26-27, pl. 5.

JURASSIC UNCONFORMITIES, CHRONOSTRATIGRAPHY AND SEA-LEVEL CHANGES FROM SEISMIC STRATIGRAPHY AND BIOSTRATIGRAPHY

P.R. VAIL

J. HARDENBOL

Exxon Production Research Company
P.O. Box 2189
Houston, Texas 77001

R.G. TODD

Esso Exploration and Production Norway
Stavanger, Norway

ABSTRACT

Seventeen global unconformities and their correlative conformities (sequence boundaries) subdivide the strata of the Jurassic and earliest Cretaceous into genetic sequences produced by 16 eustatic cycles. These 16 cycles make up the Jurassic supercycle. Eight of the global unformities are both subaerial and submarine (Type 1), and are believed to have been caused by rapid eustatic falls of sea level. Nine of the unconformities are subaerial only (Type 2), and are believed to be related to slow eustatic falls of sea level. In addition, 16 marine condensed sections (starved intervals) have been identified. These condensed sections are interpreted to be related to rapid eustatic rises of sea level. Unconformity recognition is locally or regionally enhanced by periodic truncation of folded and faulted strata during sea-level lowstands and onlap onto topographic highs during sea-level highstands, but we find no evidence that the tectonics caused the global unconformities. The 16 eustatic cycles that make up the Jurassic supercycle correspond to 16 global chronostratigraphic intervals that subdivide Jurassic strata into a series of genetic sequences, which are recognizable from seismic, well and outcrop data.

The Jurassic unconformities and the stratal and facies patterns between them are caused by the interaction of basement subsidence, eustatic changes of sea level, and differing sediment supply. Detailed analyses of the sediments with seismic stratigraphy and well data permit quantification of the subsidence history and reconstruction of paleoenvironment and sea-level changes through time. The intergrated use of seismic stratigraphy and biostratigraphy provides a better geologic age history than could be obtained by either method alone. Paleobathymetry, sediment facies, and relative changes of sea level can be interpreted from seismic data and confirmed or improved on by well control. Geohistory analysis based on geologic time-depth diagrams provides a quantitative analysis of total basin subsidence. When this subsidence is corrected for compaction and sediment loading, the tectonic subsidence and long-term eustatic changes may be determined. Short-term, rapid changes of sea

Reprinted from AAPG Memoir 36, Interregional Unconformities and Hydrocarbon Accumulation

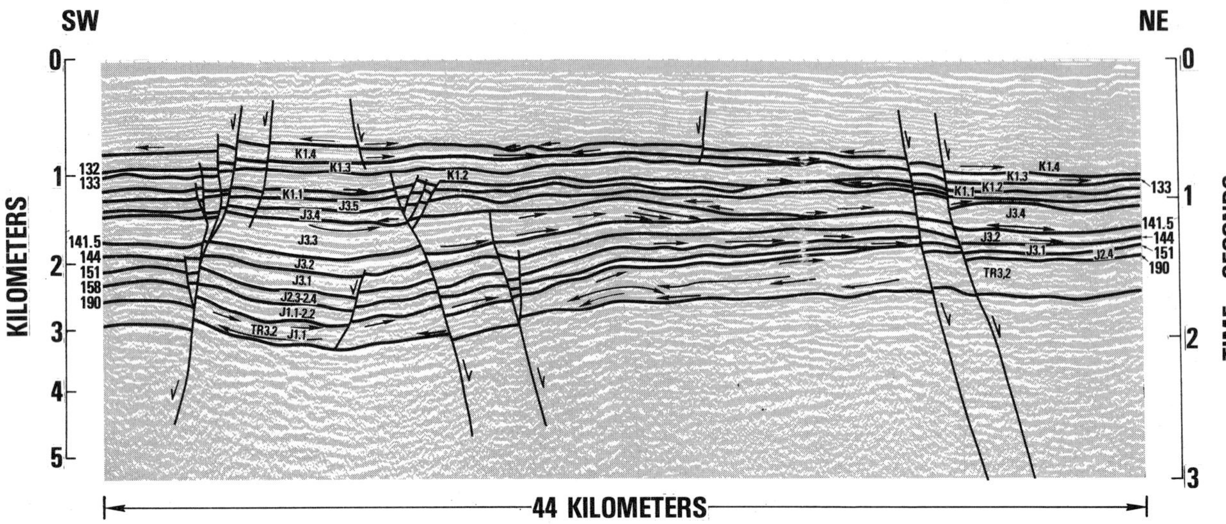

Figure 1. Seismic section northeast of Beatrice Field, Inner Moray Firth, North Sea, United Kingdom, showing seismic sequences of the Jurassic supersequence. Numbers (e.g., 132) refer to the ages of the sequence boundaries where they are comformable. Symbols (e.g., J3.2) identify the individual sequences. Seismic section courtesy of Geophysical Company of Norway (U.K.) Ltd. (GECO).

level can be demonstrated from seismic, well and outcrop data. The stratigraphic resolution of these changes rarely allows exact quantification of their magnitude, but a minimum rate of change of sea level often can be determined.

INTRODUCTION

A representative seismic line from the Inner Moray Firth in the North Sea (Fig. 1) illustrates a seismic sequence interpretation of the Jurassic. Black lines indicate seismic sequence boundaries (unconformities or their equivalent conformities), arrows mark representative reflection terminations, numbers (*e.g.*, 132) indicate the age of the sequence boundaries determined where they are conformable, and symbols (*e.g.*, J3.2) identify the sequence. Seismic stratigraphic studies tied to well control in more than 60 areas located all over the globe indicate that any regional grid of reflection seismic profiles with reasonable data quality can be interpreted in a similar manner. Not only are sequence boundaries always present, but their ages dated in areas where they become conformable are the same in different basins around the world.

This paper discusses the Jurassic supercycle and shows how eustatic changes of sea level cause the global similarities of unconformity age, coastal onlap, and marine-condensed sections. It also shows how transgressions and regressions of the shoreline are related to these criteria and why transgressions and regressions are not necessarily globally synchronous.

The material for this paper is derived from a series of previously published papers: Vail *et al.* (1977), Vail *et al.* (1980), Vail and Mitchem (1979), Vail and Hardenbol (1979), Vail and Todd (1981), and Hardenbol, Vail and Ferrer (1981).

AGE, TYPES, AND CAUSES OF JURASSIC GLOBAL UNCONFORMITIES AND CONDENSED SECTIONS

Seventeen global unconformities and

their correlative conformities (sequence boundaries) subdivide the strata of the Jurassic and earliest Cretaceous into genetic sequences produced by 16 eustatic cycles. These 16 cycles make up the Jurassic supercycle (Vail *et al.*, 1977, Part 8)(Table 1). Eight of the global unconformities are both subaerial and submarine (Type 1) and are believed to have been caused by rapid eustatic falls of sea level. Nine of the unconformities are subaerial only (Type 2) and are believed to be related to slow eustatic falls of sea level. In addition, 16 marine condensed sections (starved intervals) have been identified. These condensed sections are interpreted to be related to rapid eustatic rises of sea level (Vail and Todd, 1981). Unconformity recognition is enhanced locally or regionally by periodic truncation of folded and faulted strata during sea-level lowstands and onlap onto topographic highs during sea-level highstands, but we find no evidence that the tectonics caused the global unconformities. Table 1 is our latest summary of the ages of Type 1 and Type 2 unconformities and condensed sections. Of special significance is the fact that by more careful age dating we have discovered that the unconformities we had previously placed at the base of the Sinemurian, Callovian, Oxfordian, Kimmeridgian, and Valanginian (Vail and Todd, 1981) are all older and occur within or at the base of the late portion of the preceding age. Figure 2 shows the relation of unconformity age and type and condensed interval age to the relative changes of coastal onlap and eustatic changes of sea level in the Jurassic supercycle.

Unconformities, marine condensed sections, coastal onlap, and marine transgressions-regressions are controlled by the interaction of the variables: subsidence, eustatics, and sediment supply. Empirical observations from seismic stratigraphic studies indicate that unconformities, condensed sections, coastal onlap, and marine transgression-regression are related as shown diagrammatically on Figure 3 and discussed below.

Unconformity is defined (Bates and Jackson, 1980, p. 675, def. b) as the structural relationship between rock strata in contact characterized by a lack of continuity in deposition, and corresponding to a period of nondeposition, weathering, or especially erosion (either subaerial or subaqueous) prior to the deposition of the younger beds, and often (but not always) marked by absence of parallelism between strata. We use the term unconformity for a surface representing a significant time gap with erosional truncation (subaerial or subaqueous) and/or subaerial exposure. Marine surfaces with significant hiatuses but without evidence of erosion are not unconformities according to this usage. This definition also restricts the usage of unconformity as used in Vail *et al.*, 1977. Angularity of the underlying strata commonly is due to erosion and truncation of tilted strata, although sedimentary bypass may produce a similar effect referred to as *toplap* (Vail *et al.*, 1977, Part 2). Discordance of the overlying rocks with an unconformity is commonly caused by terminations of overlying strata by sedimentary lapouts. These lapouts are termed *onlap* and *downlap*, depending upon whether they lapout in an updip or downdip direction, respectively, at the time of deposition (Fig. 3).

Global unconformities are unconformities that are present at one place or another in all sedimentary basins with a sea-level baselevel at the time of deposition, and become conformable at the same geologic age. They are not necessarily unconformable surfaces everywhere throughout each basin. In other words, although an unconformity may be present within a basin, it may be characterized by extensive conformable areas in that basin, as well.

Type 1 (subaerial-submarine) unconformities are characterized by downward shifts of coastal onlap (defined below), commonly below the shelf edge, typically producing subaerial exposure of the shelf, valley entrenchment, and initiation of canyon cutting along the shelf edge (Fig. 4A, 5). Fluvial sediments and lowstand deltas commonly fill the entrenched valleys, and submarine fans and slope-front fill deposits accumulate in deepwater basins. Lowstand evaporite deposits and regional truncation of folded or faulted strata are characteristic in the proper geological setting. Type 1 unconformities are typified by both subaerial and submarine erosion and are formed when the rate of eustatic fall of sea level exceeds the rate of subsidence at the shelf edge.

Type 2 (subaerial) unconformities are characterized by downward shifts of coastal onlap to a position at or landward of the shelf edge, subaerial exposure of the landward portion of the shelf, and no evidence of canyon cutting (Fig. 4B, 5). Lowstand evaporite deposits develop in the prop-

Table 1. Jurassic Supercycle Global Unconformities (Sequence Boundaries) and Condensed Sections

Type		Age	Sequence Identification	Van Hinte (1976) Time Scale in Ma	Relative Magnitude Maj.	Med.	Min.
Subaerial/Submarine	1	latest Berriasian	-	132	X		
Condensed Section	CS	mid-Late Berriasian	K1.3	132.5			X
Subaerial	2	early Late Berriasian	-	133			X
Condensed Section	CS	basal Late Berriasian	K1.2	133.5			X
Subaerial	2	mid-Early Berriasian	-	134			X
Condensed Section	CS	basal Berriasian	K1.1	135		X	
Subaerial/Submarine	1	within Tithonian	-	138		X	
Condensed Section	CS	within Tithonian	J3.5	138.5		X	
Subaerial/Submarine	1	within Tithonian	-	139		X	
Condensed Section	CS	within Tithonian	J3.4	139.5		X	
Subaerial/Submarine	1	within Tithonian	-	140	X		
Condensed Section	CS	basal Tithonian	J3.3	141		X	
Subaerial/Submarine	1	mid-Late Kimmeridgian	-	141.5		X	
Condensed Section	CS	basal Kimmeridgian	J3.2	143	X		
Subaerial	2	mid-Late Oxfordian	-	144		X	
Condensed Section	CS	basal Middle Oxfordian	J3.1	147	X		
Subaerial	2	basal Late Callovian	-	151		X	
Condensed Section	CS	mid-Callovian	J2.4	153	X		
Subaerial/Submarine	2	mid-Early Callovian	-	155			X
Condensed Section	CS	basal Callovian	J2.3	156			X
Subaerial	2	basal Late Bathonian	-	158	X		
Condensed Section	CS	basal Bathonian	J2.2	165		X	
Subaerial	2	basal Late Bajocian	-	168		X	
Condensed Section	CS	basal Bajocian	J2.1	171	X		
Subaerial/Submarine	1	mid-Early Aalenian	-	173		X	
Condensed Section	CS	mid-Toarcian	J1.5	175		X	
Subaerial	2	mid-Early Toarcian	-	177			X
Condensed Section	CS	basal Toarcian	J1.4	178			X
Subaerial/Submarine	1	mid-Late Pliensbachian	-	179		X	
Condensed Section	CS	basal Late Pliensbachian	J1.3	180		X	
Subaerial	2	basal Pliensbachian	-	184			X
Condensed Section	CS	mid-Late Sinemurian	J1.2	185		X	
Subaerial/Submarine	1	latest Hettangian	-	190		X	

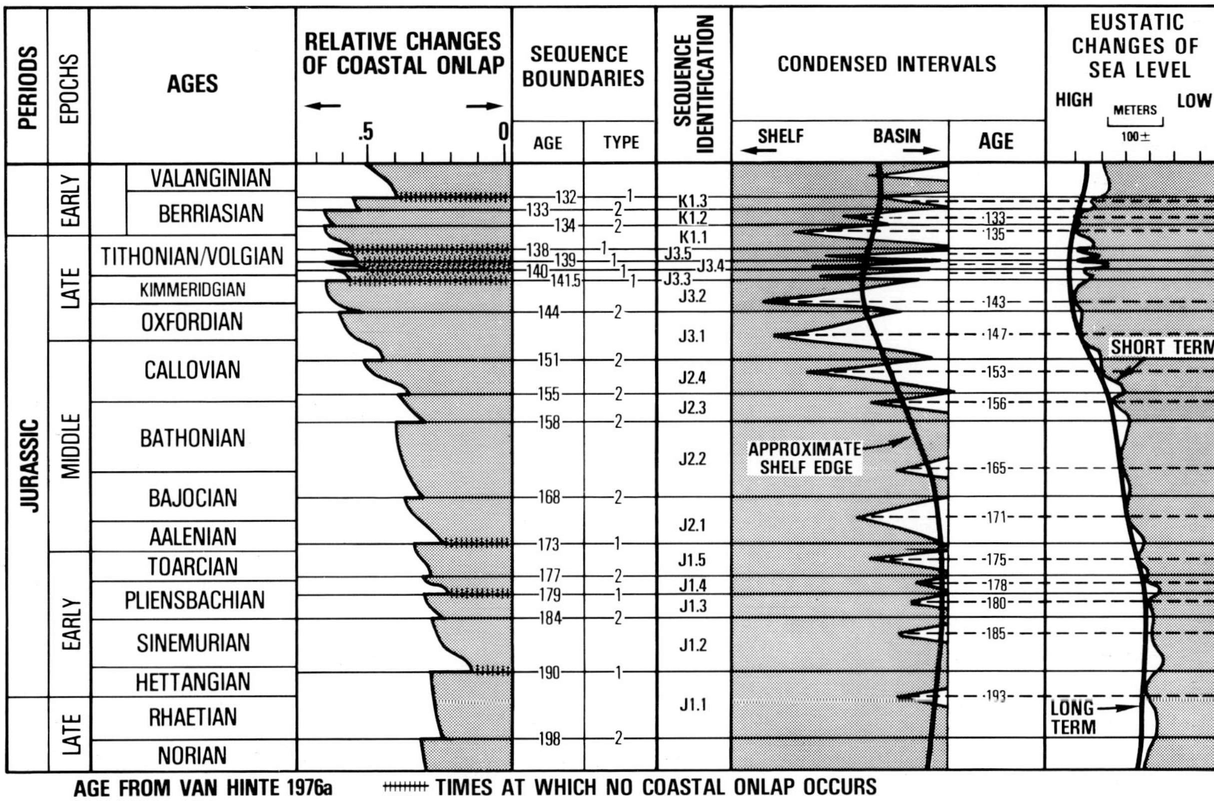

Figure 2. Global cycle chart of the Jurassic supercycle showing age and type of sequence boundaries (unconformities and their correlative conformities), sequence identification, age of condensed intervals and charts showing relative changes of coastal onlap and eustatic changes of sea level.

er geologic setting. Type 2 unconformities have only subaerial erosion, but their equivalent submarine conformities can commonly be recognized by submarine onlap resulting from shifts in the sites of deposition. Type 2 unconformities are formed when the rate of fall of eustatic sea level is less than the rate of subsidence at the shelf edge, but exceeds the rate of subsidence on the inner portion of the shelf.

The **chronostratigraphic significance** of an unconformity is that all the rocks below the unconformity are older than the rocks above it. The ages of the strata immediately above and below the unconformity differ geographically according to the areal extent of erosion or nondeposition. The duration of the hiatus associated with an unconformity differs correspondingly, but the unconformity itself is a chronostratigraphic boundary because it separates rocks of different ages, and no chronostratigraphic surfaces cross it. Although many chronostratigraphic surfaces may merge along an unconformity, none actually cross the unconformity. For these reasons, **unconformities are not diachronous** but are time boundaries and may be assigned a specific geologic age dated in those areas where the hiatus is least and/or they become conformable. By careful correlation of unconformities and their correlative conformities, a sedimentary interval can be divided into genetic sequences bounded by these unconformities. A sequence is a chronostratigraphic interval because it contains all the rocks deposited during a given interval of geologic time limited by the ages of the sequence boundaries where they are conformi-

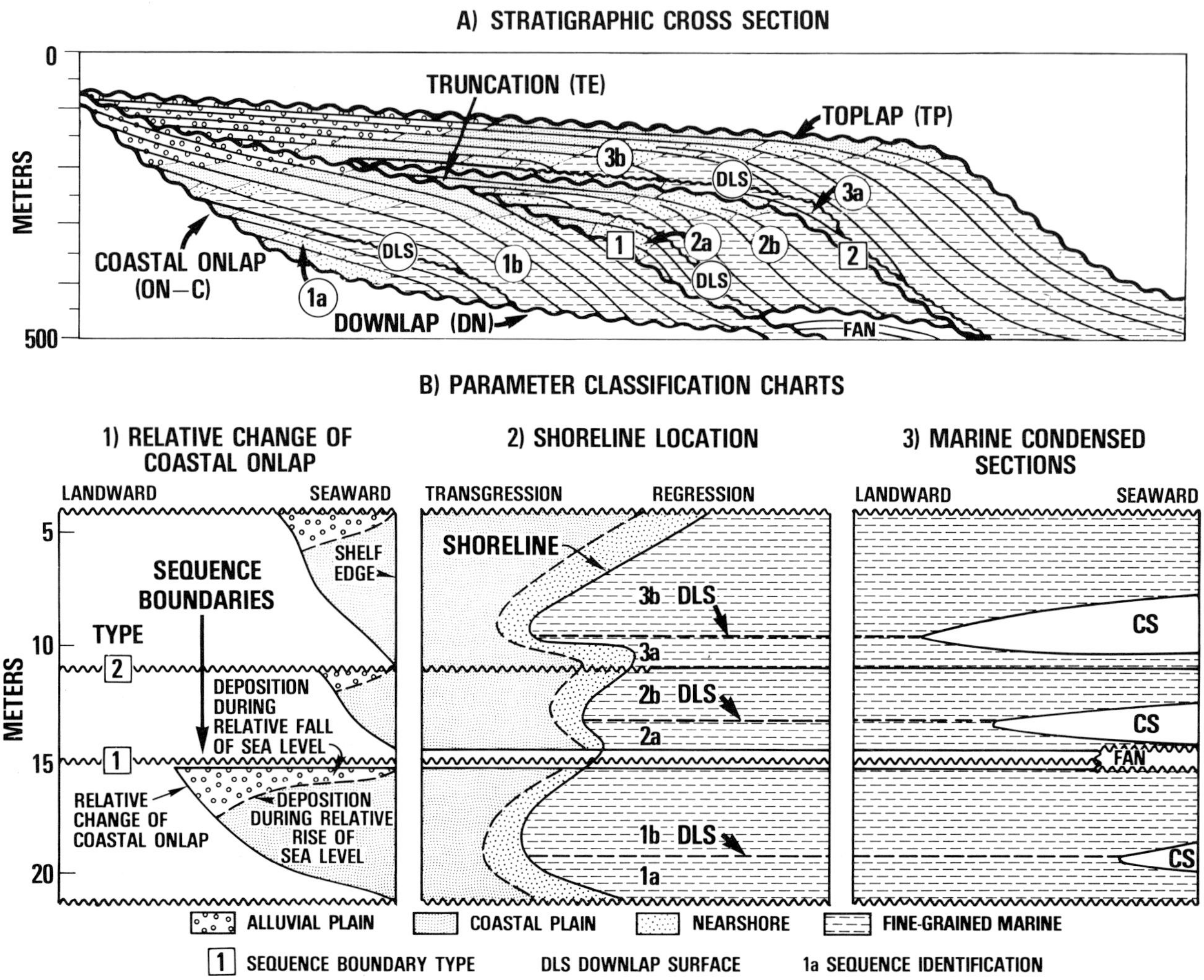

Figure 3. Diagrammatic stratigraphic cross section and chart showing parameters used to make the global cycle chart of the Jurassic supercycle. The stratigraphic cross section (A) shows the distribution of boundary types, downlap surfaces (condensed sections), and facies of three idealized sequences. The parameter classification charts show the relation between the parameters of the three sequences in time.

ties. The global Jurassic unconformities and their correlative conformities subdivide the Jurassic supersequence (all sediments deposited within the Jurassic supercycle) (Fig. 2) into a series of genetic sequences with global chronostratigraphic significance. Our studies indicate that the most widespread chronostratigraphic unit at the time of deposition is near the top of each sequence, as shown on Figure 5. The upper portions of clastic sequences characteristically show the greatest landward onlap of the alluvial plain and at the same time they exhibit major seaward progradation.

In most basins of the world, Type 1 unconformities are characterized by both subaerial and submarine erosion. However, there are occasional exceptions where the shelf margin is locally subsiding so rapidly that its rate exceeds that of the fall of sea level, and the unconformity has the characteristics of a Type 2. An example is a shelf mar-

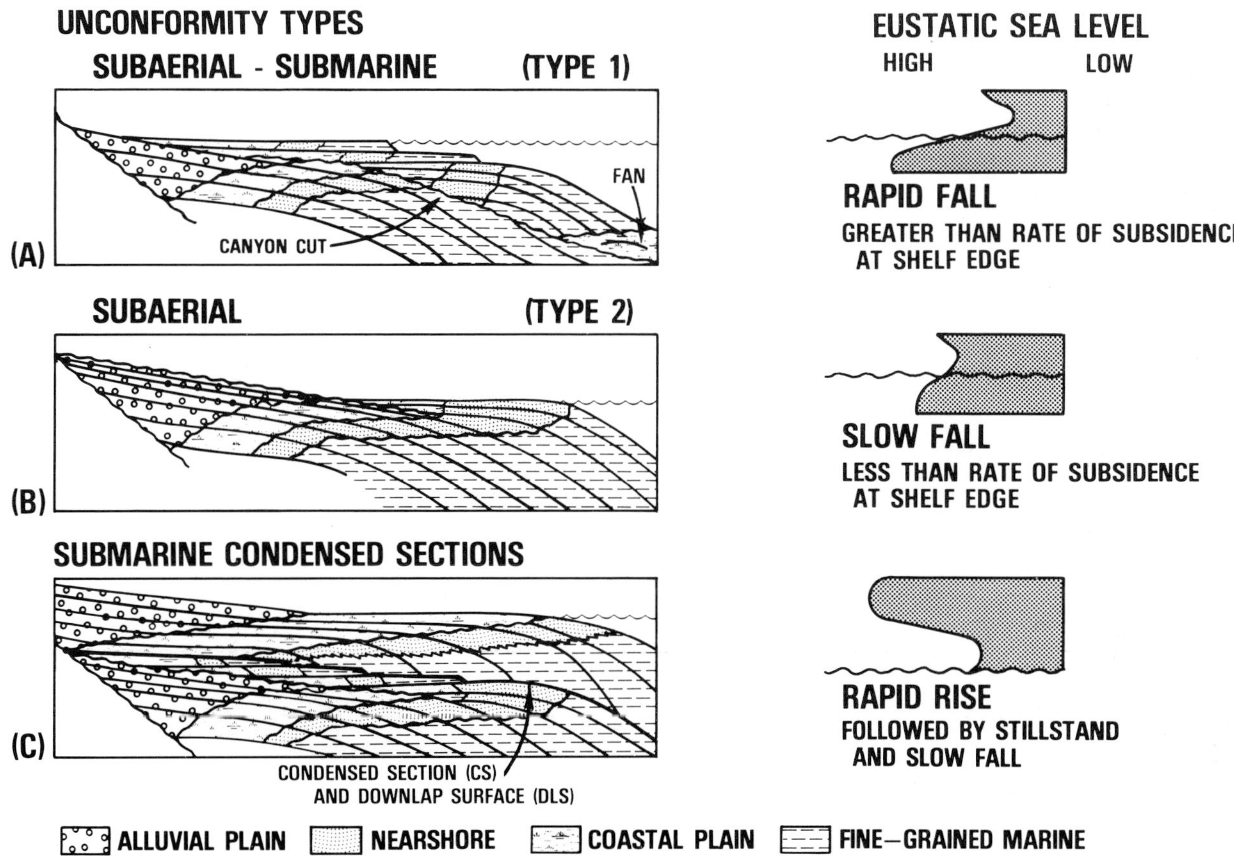

Figure 4. Diagrammatic charts showing the relation between Type 1 and Type 2 unconformities and submarine condensed sections to eustatic changes of sea level.

gin controlled by a rapidly subsiding growth fault, such as in the Neogene of the Gulf of Mexico. Similarly, Type 2 unconformities may locally show Type 1 characteristics in areas where structural activity causes stability or uplift so that sea level may locally fall below the shelf edge.

Condensed sections are thin marine stratigraphic intervals characterized by very slow depositional rates (less than 1 cm/1000 years). They are commonly associated with thin but continuous zones of burrowed or lightly lithified beds (hardgrounds, omission surfaces). They are conspicuous by the presence of scattered pebbles or high concentrations of volcanic ash, glauconite, phosphate, or radioactive minerals, and are commonly linked to marine hiatuses or faunal mixing horizons (Fig. 4C, 5). In vertical deepening-upward sections, a condensed section commonly marks the greatest water depth. However, in areas of rapid subsidence the greatest water depths may be above the condensed section. Condensed sections should not be confused with a winnowed interval associated with high-energy removal of fine sediment.

A condensed section develops when the rate of relative rise (eustatic rise plus subsidence) of sea level is significantly greater than the rate of accumulation, causing the depositional site to shift landward and resulting in low sedimentation rates (starved conditions) seaward of the depositional site. The marine hiatuses associated with condensed sections tend to disappear landward into essentially conformable successions developed at the site of deposition (Figs. 3, 5). The age of a condensed section within a given depositional sequence tends to be synchronous globally but may differ slightly from basin to basin with changes in rates of deposition and subsidence.

Downlap surfaces are submarine surfaces characterized seismically by a downlap over a concordant pattern and are commonly associated with a marine hiatus (Figs. 4C, 5). Downlap surfaces associated with condensed sections mark the change from the end of transgression to the start of regression as the rate of eustatic rise decreases, and sediments begin to prograde out over the old starved surface. The downlap represents deep-marine clinoform toes developed as prograding progresses seaward. Downlap surfaces may also be present above submarine fans.

Coastal onlap refers to the progressive landward encroachment of the coastal (littoral, coastal plain, or alluvial plain) deposits of a given sequence (Vail *et al*., 1977, Part 3, Figs. 3, 5). All sequences exhibit almost continuous coastal onlap throughout the time of their deposition. Periodically, there is a major downward (basinward) shift in the coastal onlap pattern at the times of Type 1 or Type 2 unconformities. This abrupt downward shift in coastal

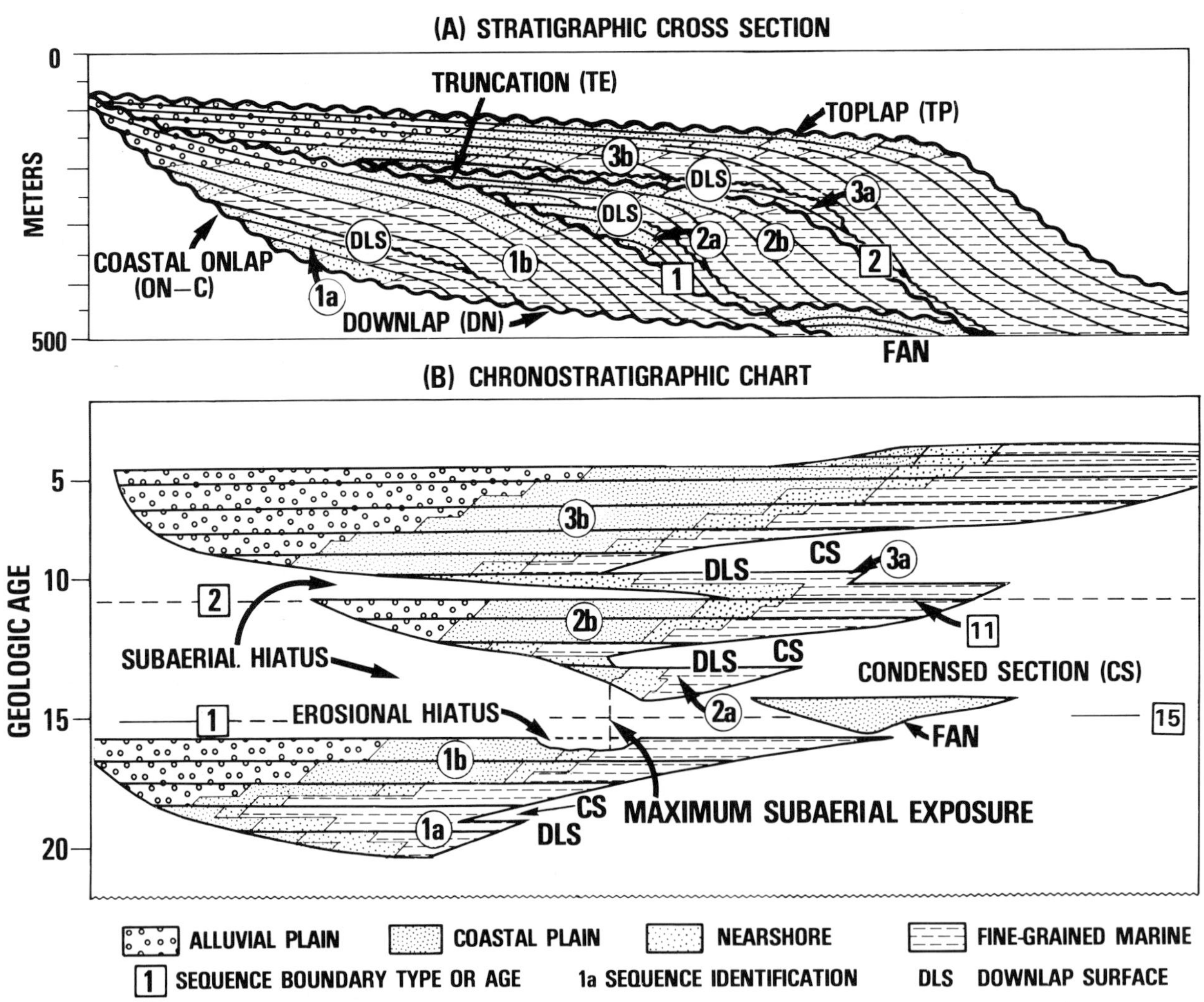

Figure 5. Diagrammatic stratigraphic cross section and chronostratigraphic chart showing the distribution of sequence boundary types, downlap surfaces (condensed sections) and facies of three idealized sequences in depth and time.

onlap gives our chart of relative changes of coastal onlap the sawtooth character observed in Figure 2. The almost continuous presence of coastal onlap indicates that most sediments are deposited during a relative rise of sea level. Either eustatic sea level is rising or the depositional surface is subsiding or the two are combining to create space for the accomodation of sediments. Type 1 and Type 2 unconformities mark those exceptional periods of time when the rate of eustatic sea-level fall is greater than that of subsidence, producing relative falls of sea level and globally synchronous downward shifts of coastal onlap. During the periods of time involved in Type 1 relative falls, and prior to the beginning of the following relative rises, the only sediments deposited in significant volumes are submarine fans, which have no corresponding coastal onlap. Thus, the downward shift in coastal onlap appears only as a gap or discontinuity in the coastal onlap record and is instantaneous in terms of geologic time (Fig. 3, 5).

In previous papers (Vail *et al*, 1977, 1980), we equated directly relative changes of coastal onlap with relative changes of sea level. Pitman (1978) and M. T. Jervey (personal communication) have demonstrated that changes observed in coastal onlap cannot be related directly to relative changes of sea level in the upper part of the cycle, as we have done in the past. We now refer to our sawtoothed global charts as relative changes of coastal onlap. Ideally, a chart of coastal onlap should be constructed by plotting the successive positions of the upper limit of the coastal plain. When an alluvial plain is asosciated with the coastal plain, the relative sea-level boundary approximates the facies change between the coastal plain and the alluvial plain (M. T. Jervey, pers. comm., Figure 3). Unfortunately, seismic stratigraphic techniques do not always permit identification of this facies change between coastal and alluvial plains. Consequently, the charts developed from studies of seismic sections include data from both the coastal plain and the alluvial plain with its slightly greater alluvial dips. This greater dip is the source of the differences between the coastal onlap chart and the relative change of sea-level chart. We recognized this factor in Vail *et al.* (1977, Part III), but did not take it into account when interpreting relative changes of sea level. The two charts, however, are similar. Charts of relative changes of coastal onlap versus distance typically show abrupt sawtooth shifts from widespread to restricted, whereas charts of relative changes of sea level versus distance commonly show more gradual shifts.

Transgression is defined as the landward displacement of the shoreline indicated by landward migration of the littoral facies in a given stratigraphic unit. **Regression** is the seaward displacement of the shoreline (Vail *et al.*, 1977, Part 3, Figs. 3,4). Our studies indicate that transgressions and regressions are **not** necessarily globally synchronous. Similar conclusions have been reached by other writers (Yanshin, 1983). There is, however, a general tendency for certain periods of time to be dominated by transgressive deposits, while in others the prevalent trend is regressive. Times of maximum transgression tend to be more globally synchronous than times of regression, because of the cumulative effects of sea-level rise and subsidence. In general, transgression results when eustatic sea level rises and regression results when eustatic sea level falls at a slower rate than subsidence (Fig. 6). We consider an interval transgressive if the overall tendency is to deepen upward, even though there may be a series of stacked, retrogradational regressive units.

We believe that cycles of relative changes of coastal onlap are more useful than transgressive-regressive cycles for defining stratigraphic intervals. The lows in the sawtooth pattern mark times of global unconformities that subdivide the section into sequences genetically related to cycles of coastal onlap. The cycles define chronostratigraphic intervals, which include all strata deposited during the time period between the points where the unconformities become conformable. Stratigraphic units bounded by un-

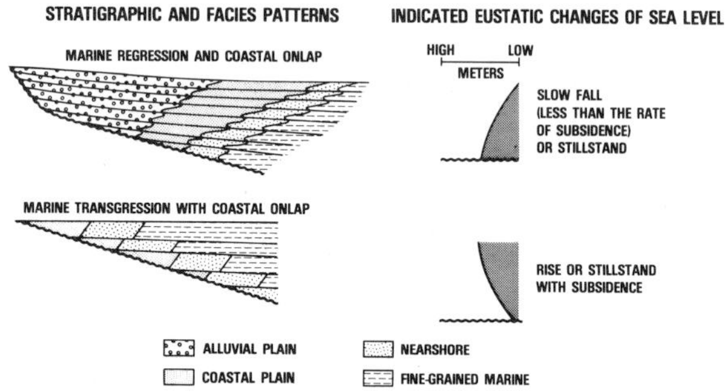

Figure 6. Diagrammatic charts showing the relation between transgression-regression to eustatic changes of sea level.

conformities or their equivalent conformities are defined as sequences (Vail *et al.*, 1977, Part 2). They have the advantage of being genetic units and are ideal for facies analysis and paleogeographic reconstructions. Marine condensed sections occur within sequences, and can be used to subdivide sequences.

Cycles of transgression and regression do **not** necessarily coincide with cycles of relative change of coastal onlap. In general, the maximum point of transgression occurs in the middle of a cycle of coastal onlap, and the maximum point of regression occurs after the downward shift of coastal onlap, and thus above the associated global unconformity (Figs. 3, 5). As stated previously, this relationship does not always exist because of the variations of sediment supply. In general, transgressions are more rapid than regressions.

The **stratigraphic relations** among global unconformities, condensed sections, coastal onlap, relative changes of sea level, marine transgressions and regressions, subsidence, and eustatic changs of sea level are diagrammed in Figure 7. In general, as we discuss later in this report, the tectonic subsidence along most passive margins is long term and gradually decreasing in rate, related to a thermal decay curve. It does not change rapidly enough to cause regional unconformities. Tectonic subsidence patterns differ from region to region, and are not globally synchronous. We therefore conclude that the many synchronous unconformities and abrupt changes in stratigraphy observed in basins globally are caused by eustatic changes of sea level, superimposed on regional tectonic and sedimentary regimes that change at much slower rates.

Type 1 and Type 2 unconformities (Fig. 7) are globally synchronous because when they are dated precisely they coincide in time with the inflection points of curves showing eustatic falls of sea level (Vail and Todd, 1981; M. T. Jervey, pers. comm.). This inflection point also marks the time of downward shift of coastal onlap. Global cycles of coastal onlap are chronostratigraphic intervals that can be dated precisely because they include all the rocks deposited during the interval of time between the inflection points of eustatic falls of sea level.

Truncation below unconformities caused by erosion during a relative fall of sea level has a very different pattern depending on whether or not tectonic activity (folding, faulting, uplift) is taking place during the time of the fall. If no tectonic activity other than subsidence is taking place, the truncation should be greatest at the shelf edge and should progress shelfward by headward erosion and downcutting of river valleys (Fig. 8a). In general, valley and canyon truncation dies out landward away from the shelf edge, the overlying and underlying strata are generally parallel, and very detailed stratigraphic correlation is necessary to identify the truncation and onlap patterns associated with the canyon and valley fills. Farther shelfward, as truncation effects die out, the unconformities must be identified by coastal onlap and related facies patterns (Fig. 8a). If folding, faulting, uplift, or large differential subsidence are taking place during the relative fall, erosion of the tilted beds produces an easily identifiable truncation unconformity (Fig. 8b). In general, most global unconformities are not associated with tectonic activity.

The **causes of global eustatic changes of sea level** in the Jurassic (Fig. 2) are unknown. Most workers agree, however, that the long-term changes of sea

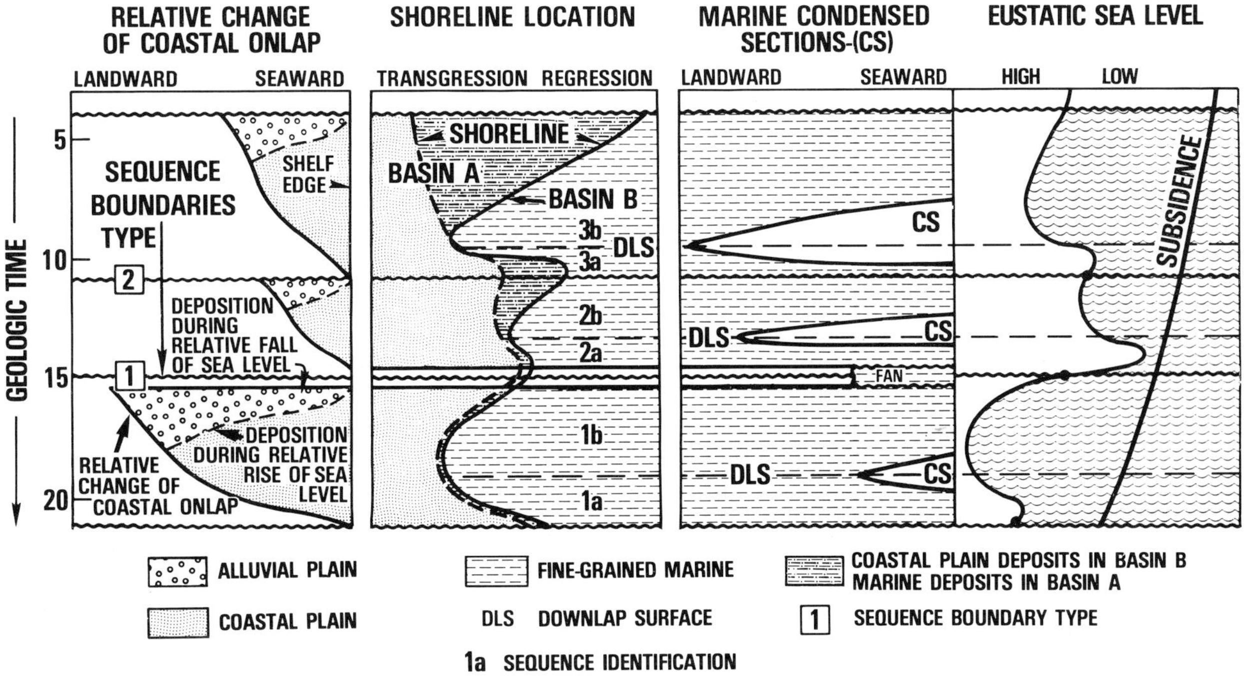

Figure 7. Summary chart showing relation of sequence boundary types, relative changes of coastal onlap, relative changes of sea level, transgression-regression, and marine condensed sections to eustatic changes of sea level and subsidence.

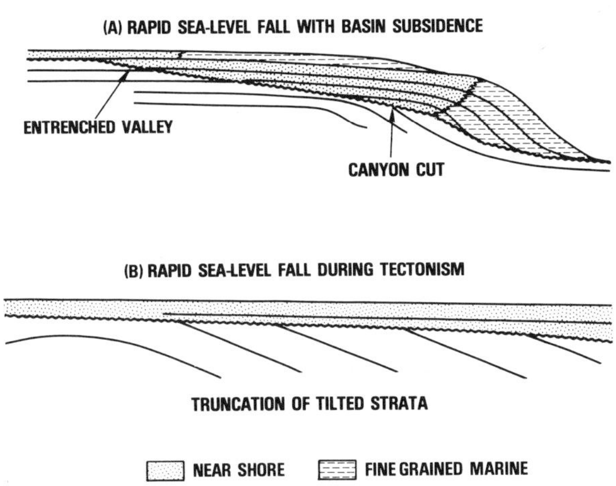

Figure 8. Diagrammatic sections showing stratal patterns associated with unconformities. Section A shows a typical headward erosion pattern associated with a Type 1 unconformity formed during basin subsidence. Section B shows the expression of a Type 1 unconformity formed during a period of tectonic uplift and folding.

level, such as the overall rise in the Jurassic, result from changes in volume of the ocean basins due to such processes as changes in rates of sea floor spreading (Hallam, 1963; Pitman, 1978). The causes of rapid changes of sea level (greater than 10 cm/1000 years), however, are problematical. The only known mechanism that produces rates of this magnitude is glaciation (Donovan and Jones, 1979), but conclusive evidence is lacking for glaciers of sufficient size to have caused the large and rapid changes of sea level we observe in the Jurassic.

In **summary**, eustatic changes of sea level are the primary cause of major abrupt vertical changes in stratigraphy at the time of deposition and are major factors controlling lateral changes. Eustatic changes of sea level produce interregional or "global" unconformities not only for continental margins, but for interior basins and deep-oceanic sites. As such, they provide a means to make exact correlations from the deep-sea basins across the shelf into interior basins. Figure 2 shows our best estimate of eustatic changes of sea level during the Jurassic Supercycle and how they relate to the ages of Type 1 and Type 2 regional unconformities, relative changes of coastal onlap, and age and relative magnitude of condensed sections. Transgressions and regressions are not included because of their global variability. Our studies have shown that Type 1 and Type 2 regional unconformities are globally synchronous when dated precisely at the age they become conformable. The age of conformity corresponds to the time of the most rapid rate of eustatic fall of sea level (inflection point on the sea-level curve) as shown in Figures 4 and 7 (M.T. Jervey, pers. comm.).

Ages of condensed sections may differ slightly with changes in rate of deposition and subsidence. Downward shifts of coastal onlap, however, are synchronous globally, because they occur at inflection points on the sea-level curve at the times of highest rates of eustatic fall. The inflection points also mark the most accurate dates to establish the synchrony of the global unconformities.

When the coastal onlap patterns of restricted and widespread sequences of the same age are compared in several areas globally, the patterns tend to be similar in areas where the subsidence pattern is continuous and uninterrupted over the interval studied. If stability or uplift occurs, the onlap patterns are distorted. Such distorted patterns are useful for indicating periods of tectonism. However, the ages of the sequence boundaries remain consistent in the distorted sections. Transgressions and regressions of the shoreline are the most variable of the parameters discussed; however, sequences tend to be mostly transgressive in the lower part of the coastal onlap cycle below the downlap surface, and mostly regressive in the upper part of the cycle above the downlap surface.

STRATIGRAPHIC NOMENCLATURE

A major operational problem is how to name and refer to unconformities (sequence boundaries) and sequences once they have been identified. The experience of the authors is that the sequence boundaries can be marked with colored pencils to facilitate identification as a grid of seismic or well and outcrop sections is correlated. After the areal continuity of the sequence boundaries has been verified by means of seismic-loop ties or well-log correlations, the geological ages of the strata are determined by means of paleontology. At this point, we believe that information is conveyed precisely by naming the unconformities after the oldest strata known to occur above the unconformity (*e.g.* the latest Berriasian unconformity) and by informally naming the sequences between the unconformities with symbols representative of age (*e.g.* Kl.3). If precise age dating cannot be achieved because a significant hiatus is present throughout the study area or because the paleontologic data are poor or lacking, the age of the closest global unconformity is used, and the error range is indicated on a summary chronostratigraphic chart. For reference purposes, we designate unconformities on seismic or well-log sections by the age of their correlative conformities in millions of years (Ma.). It is important to note which time scale is used, since the age in millions of years of a particular faunal zone may change in the future. The Jurassic global chart (Fig. 3 and Table I) shows the symbols we use for the global cycles and the ages of the global unconformities we have identified in the Jurassic supercycle. The numerical age is also useful for input into quantified stratigraphic studies, such as geohistory analysis discussed later in this paper.

Formal formation names for stratigraphic units serve a very useful descriptive purpose, but it must be remembered that formations are lithostratigraphic units and may transgress time. For proper facies and paleogeographic interpretation, the lithostratigraphy must be placed in the proper chronostratigraphic framework or serious interpretation errors will result. We also believe that naming unconformities after orogenic events is misleading. In the North Sea, for example, two major unconformities are called the Middle and Late Cimmerian. We prefer to designate the respective unconformities as latest Bathonian (158 Ma) and latest Berriasian (132 Ma) believing that more accurate stratigraphic information is thereby provided. Our studies indicate that these particular unconformities were caused by eustatic falls of sea level superimposed on longer term tectonic movements and were not caused directly by orogenic events.

PROCEDURE FOR INTERPRETATION OF UNCONFORMITIES, CHRONOSTRATIGRAPHY, COASTAL ONLAP, TECTONIC SUBSIDENCE, AND EUSTATIC CHANGES OF SEA LEVEL

The procedure used for charting the Jurassic unconformities, chronostratigraphy, coastal onlap, tectonic subsidence, and eustatic changes of sea level consists of the following five steps:

(1) Developing chronostratigraphic framework by determining sequence boundaries for a study region.
(2) Dating sequence boundaries and making a representative chronostratigraphic chart for a study region.
(3) Making regional chart of relative changes of coastal onlap for study regions and combining them to make a global onlap chart.
(4) Determining tectonic subsidence and long-term eustatic changes of sea level with geohistory analysis corrected for loading and compaction.
(5) Estimating short term eustatic changes of sea level.

Each of the preceding five steps is discussed in order.

(1) **Developing Chronostratigraphic Framework by Determining Sequence Boundaries for a Study Region.** Sequence boundaries are interpreted most accurately by using a combination of reflection seismic data, well control, and outcrop information. Figure 5A is a diagrammatic example of a geologic cross section showing chronostratigraphic surfaces, unconformities, and facies. In order to interpret unconformities and sea-level changes, it is critical that we make stratal correlations in sufficient detail to recognize truncation, lapout, prograding, and other patterns associated with different depositional environments. Seismic data tied to well control or closely spaced well logs with detailed marker-bed correlations are ideal for this purpose.

Figure 4 illustrates the three types of surfaces discussed previously, Type 1 and Type 2 unconformities, and condensed sections. In general, we find that if we can trace these types of surfaces over a regional grid of seismic and well data, they will correspond to one of the global unconformities or condensed sections listed in Table I. The 15-million-year unconformity is a Type 1; the 11-million-year a Type 2. Condensed sections are present at 9, 13, and 18-million years (Fig. 5B).

The condensed sections (CS) or downlap surfaces (DLS) shown schematically within sequences 1, 2, and 3 in Figure 5 overlie transgressive deposits and, thus, occur within a sequence between the bounding global unconformities. In deep water depositional sites such as ocean basins, condensed intervals or downlap surfaces commonly overlie directly both types of unconformities in areas where lowstand deposits are absent.

We have found that unconformities and condensed sections can be recognized best on reflection seismic sections by identifying onlap, downlap, and truncation patterns and dating them with well control (Vail *et al.*, 1977, Parts 2, 6; 1980). Because seismic reflections are produced largely by stratal surfaces and unconformities with significant velocity-density (impedance) contrasts (Vail *et al.*, 1977, 1980, Part 5), reflection patterns portray stratal configurations within the limits of seismic resolution.

Unconformities characterized by large hiatuses, such as those along basin margins, commonly extend laterally into areas where deposition has been more nearly continuous and where hiatuses gradually decrease as the

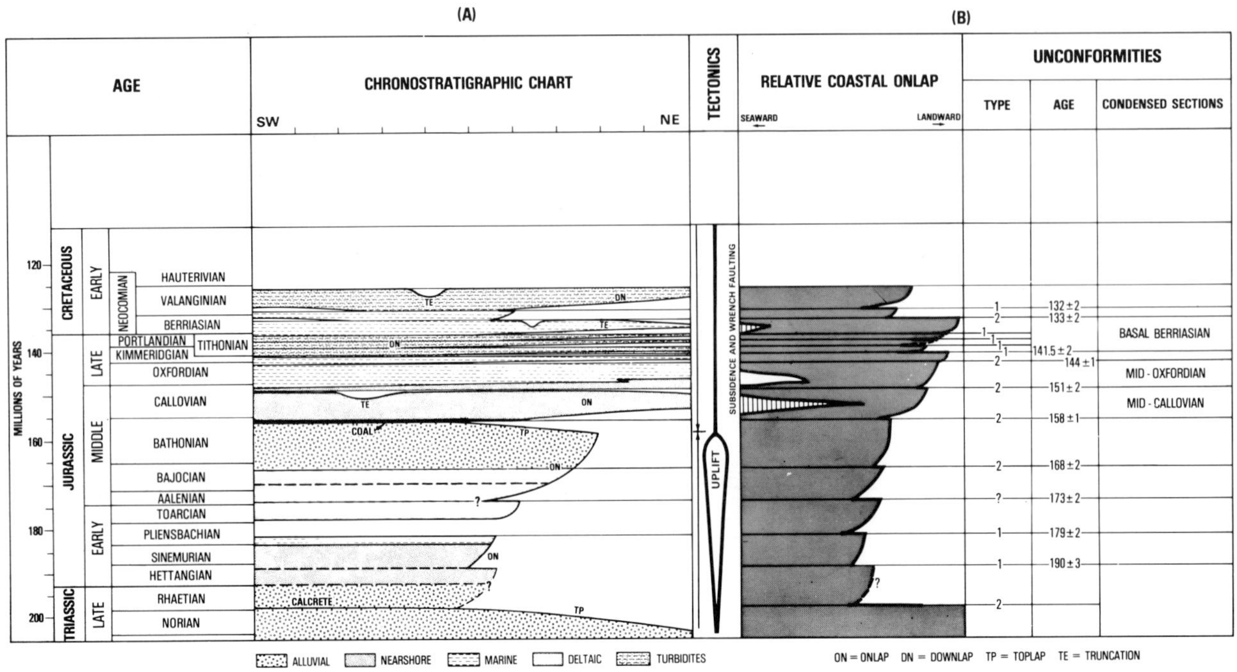

Figure 9. Charts showing chromostratigraphy, relative changes of coastal onlap, age and type of unconformities and condensed sections, from the Inner Moray Firth in the North sea. (This chart is discussed in detail in Vail and Todd, 1981).

unconformities approach conformity (Fig. 5). A conformable surface is a chronostratigraphic horizon and must be traced with its correlative unconformity in order to define completely the sequence it bounds. In this way, the three-dimensional sequence framework bounded by unconformities and their correlative conformities is completely defined for subsequent seismic facies and structural analysis (Vail *et al*., 1977, Parts 2 and 7). As shown on Figure 5, Type 1 and Type 2 unconformities become most conformable at the point of downward shift of coastal onlap. Locally, Type 1 unconformities have the least hiatal gap on the upper slope landward of the submarine fan. We give the submarine fan the same age as the unconformity, since we believe it formed during the time of most rapid eustatic fall of sea level, but we carry the base of the fan as our principal mapping surface.

(2) **Dating Sequence Boundaries and Making a Representative Chronostratigraphic Chart for a Study Region.** An unconformity is dated most accurately by establishing the ages of overlying and underlying strata at points along the unconformity where the hiatus is least. The ideal point is where the surface becomes a conformity. In many cases, however, conformity is not reached and the point of minimum hiatus is used. Dating is accomplished using biostratigraphy from wells and outcrop, and comparing with local and global onlap charts.

Figure 5B is an example of a chronostratigraphic chart which has been constructed from a geologic cross section. The chart portrays the sequences in geological time and thus emphasizes the genetic nature of the facies within the sequences and the abrupt discontinuities between sequences. Chronostratigraphic charts also emphasize the periods of time when no sediments are present because of nondeposition or erosion. Figure 9 is a chronostratigraphic chart for the Inner Moray Firth seismic section shown on Figure 1. This chart is discussed in detail in Vail and Todd (1981).

(3) **Making Regional Charts of Relative Changes of Coastal Onlap for Study Regions and Combining Them to Make a Global Onlap Chart.** Charts of relative changes of coastal onlap are derived from geological cross sections or interpreted seismic sections (such as Fig. 1) and from the chronostratigraphic charts (such as Fig. 9). The procedure for constructing charts of relative changes of coastal onlap is discussed in Vail *et al.*, (1977, Parts 3 and 4).

Figure 10 shows how we combine charts of relative changes of coastal onlap to produce the global chart. In this example, unconformity age and relative patterns of coastal onlap are compared among the North Viking Graben, the Inner Moray Firth, and the global chart. [See Vail and Todd (1981)]. Unconformities of the same age and type are added to the global chart when they have been identified in sedimentary basins on at least three different continents. Patterns of relative changes of coastal onlap are included on the global chart only from areas undergoing uniform subsidence, because the coastal onlap pattern in a given area will be influenced by a variable tectonic and subsidence history. Figure 11 shows the areas in which we have made seismic-stratigraphic and well-log studies used in the construction of the global Jurassic coastal onlap chart (Fig. 2).

(4) **Determining Tectonic Subsidence and Long-Term Eustatic Changes of Sea Level with Geohistory Analysis (Van Hinte, 1978) Corrected for Loading and Compaction (Hardenbol, Vail and Ferrer, 1981).** Sea floor subsidence, eustatic changes of sea level, and sediment supply are the principal interacting variables in a marine basin. This interaction is recorded in

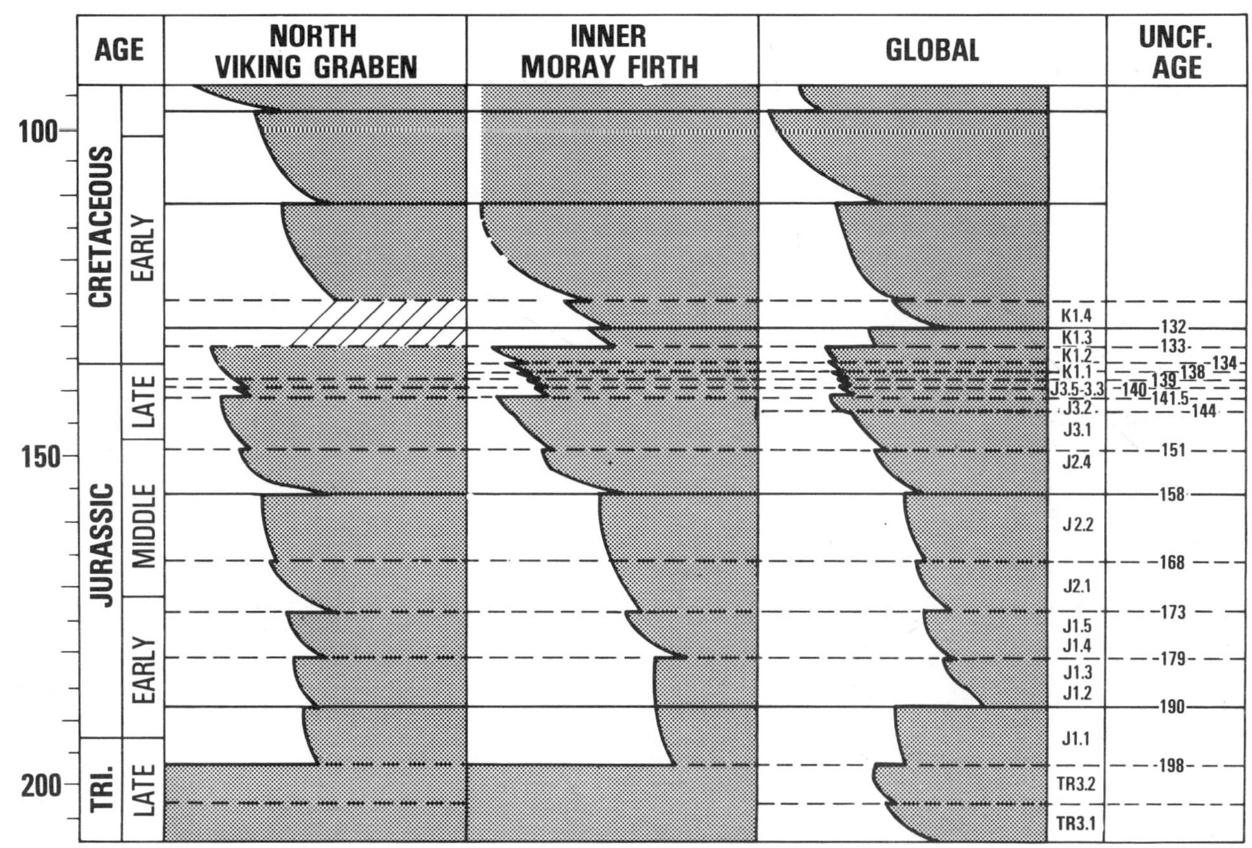

Figure 10. Charts comparing the Global Chart with relative changes of coastal onlap from the North Viking Graben and Inner Moray Firth, in the North Sea.

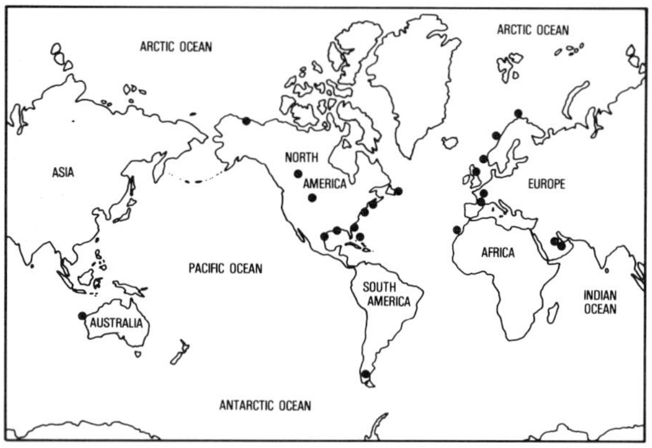

Figure 11. Map showing distribution of seismic stratigraphic and/or well studies used to develop the Jurassic supercycle.

the sedimentary section in the form of sequences filling the basin. Geohistory analysis provides a technique for depicting the interaction of eustatic changes of sea-level and sea-floor (tectonic) subsidence and relating these factors to the sedimentary record. This technique separates the total subsidence of an area into individual components due to the effcts of sediment and water-column loading, compaction, and tectonic activity (*e.g.*, thermal cooling, basement faulting, tectonic loading). Also separated are the apparent vertical movement effects of eustatic changes of sea level. By quantification of the individual effects of sediment compaction, sediment loading, and subsidence,

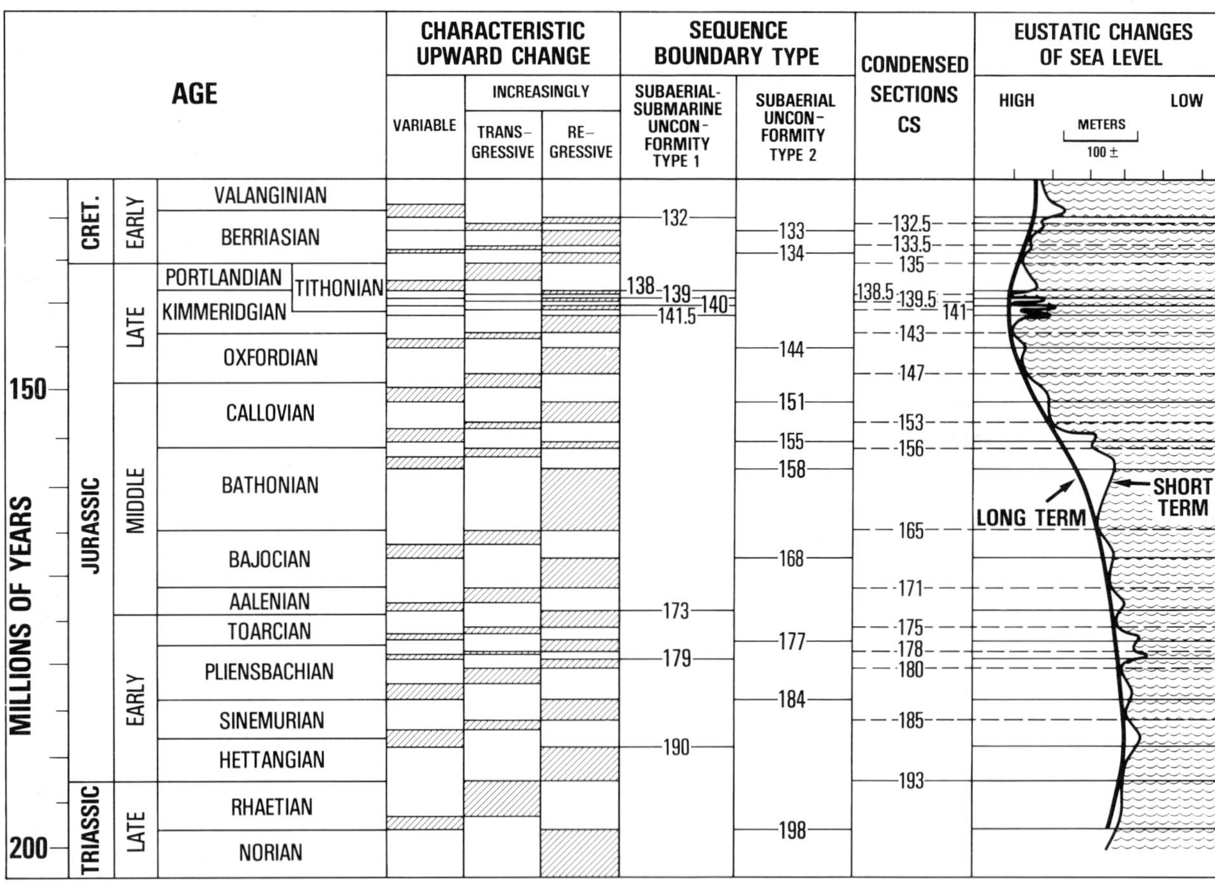

Figure 12. Diagram illustrating methodology used to construct the chart of eustatic changes of sea level during the Jurassic supercycle. The ages of Type 1 and Type 2 unconformities and condensed sections are plotted along with the intervals of time when the Jurassic sediments tend to be increasingly transgressive or regressive upward.

the magnitudes of the long-term eustatic changes can be established.

An example of this geohistory analysis from offshore northwestern Africa is described in Hardenbol, Vail, and Ferrer (1981). It shows that in the Early Cenomanian, sea level could have been nearly 300 meters higher than the present sea level.

The eustatic sea-level values obtained from geohistory analysis show close agreement with previous results obtained by different methods by Vail *et al*. (1977, Fig. 8, Part 4), Hayes and Pitman (1973), Pitman (1978), and Kominz (in press). A significant difference exists, however, with the much smaller values obtained by Watts and Steckler (1979), who used a similar method based on the subsidence history of the Atlantic margin of the North American east coast. Possibly their assumption of a single thermal contraction event for the thermo-tectonic subsidence of the Atlantic margin is not valid. Thermo-tectonic subsidence histories for Georges Bank and Baltimore Canyon C.O.S.T. wells suggest increased tectonic activity in the Lower Cretaceous.

(5) **Estimating Short-Term Eustatic Changes of Sea Level.** The pattern of short-term eustatic changes of sea level is derived from a combination of several data sets (Fig. 12). The ages of Type 1 and Type 2 unconformities and condensed sections are plotted, along with the intervals of time when Jurassic sediments tend to be increasingly transgressive or regressive upward. From these are derived the direction and slope of the short-term eustatic curve, based on the patterns shown in Figures 4 and 6. The final eustatic sea-level curve is then derived from the configuration of the short-term changes superposed upon the independently determined envelope of the long-term eustatic changes.

The absolute magnitudes of the rapid falls of sea level are difficult to calculate, but studies indicate that the 132 Ma and 140 Ma drops are the greatest and may approximate the magnitudes of the Pleistocene changes. The other changes appear to be less, as shown on Figures 2 and 12. These magnitudes are estimated from the downward shift of coastal onlap below the shelf edge and from the depth of subaerial erosion obtained from geohistory reconstructions.

CONCLUSIONS

Our experience indicates that analysis of stratigraphy based on depositional sequences is the most accurate approach for interpreting geologic age, depositional environment, lithofacies, and paleogeography. Not only does the approach subdivide strata into genetic depositional units, but it places them in a global context controlled by short-term variations in eustatic sea level. Sequences can be recognized readily from seismic interpretation, well-log and outcrop correlations, and facies studies. Interpretations based on lithofacies and biostratigraphy could be misleading unless they are placed within a context of detailed stratal chronostratigraphic correlations. This paper illustrates how sequence concepts may be applied to the Jurassic.

ACKNOWLEDGMENT

This paper represents the joint effort of many Exxon scientists who contributed to the development, documentation, and testing of the concepts presented in this report. We are especially indebted to R. M. Mitchum and M. A. Uliana, whose critical review of the manuscript provided a great many important improvements. In addition, we thank Exxon Production Research for permission to release the information included in this paper.

SELECTED REFERENCES

Bates, R.E. and J.A. Jackson (1980), Editors, Glossary of Geology, second edition, American Geological Institute, Falls Church, Virginia.

Donovan, D.T., E.J.W. Jones, M.F. Ridd, and J.A.E.B. Hubbard, (1979), Rates of marine transgressions and regressions, Jour. Geol. Soc. of London, Vol. 136, p. 187-193.

Hallam, A., 1963, Eustatic control of major cyclic changes in Jurassic sedimentation: Geol. Mag., v. 100, p. 444-450.

Hardenbol, J., P.R. Vail, and J. Ferrer, 1981, Interpreting paleoenvironments, subsidence history and sea-level changes of passive margins from seismic and biostratigraphy: Oceanologica Acta No. SP, p. 3-44.

Hayes, J.D., W.C. Pitman,III, 1973. LIthospheric plate motion, sea-level changes and climatic and ecological consequences: Nature, v. 246, p. 18-22.

Kominz, M.A., (In press 1983), Oceanic Ridge Volumes and Sea level change - An error Analysis: AAPG Memoir Interregional Unconformities and Hydrocarbon Accumulation.

Pitman, W.D., 1978, Relationship between sea-level change and stratigraphic sequences: Geol. Soc. Am. Bull., v. 89, p. 1389-1403.

Vail, P.R. and J. Hardenbol, 1979, Sea-level changes during the Tertiary: Oceanus, v. 22, p. 71-79.

__________ and R.M. Mitchum, 1979, Global cycles and sea-level change and their role in exploration: Proceedings of the Tenth World Petroleum Congress, v. 2, Exploration Supply and Demand, Bucharest, Romania, p. 95-104.

__________, R.M. Mitchum, T.H. Shipley, and R.T. Buffler, 1980, Unconformities of the North Atlantic: Phil. Trans. R. Soc. London, A 294, p. 137-155.

__________, R.M. Mitchum, R.G. Todd, J.M. Widmier, S. Thompson, J.B. Sangree, J.N. Budd, and W.G. Hatlelid, 1977, Seismic stratigraphy and global changes of sea level. In Seismic stratigraphy -- applications to hydrocarbon exploration: AAPG Memoir 26, p. 49-212.

__________ and R.G. Todd, 1981, North Sea Jurassic unconformities, chronostratigraphy and sea-level changes from seismic stratigraphy: Proc. Petrol. Geol. Cont. Shelf, N.W. Europe, p. 216-235.

Van Hinte, J.E., 1976a, A Jurassic time scale, AAPG Bull., 60, p. 489-497.

__________, 1976b, A Cretaceous time scale, AAPG Bull., 60, p. 498-516.

__________, 1978, Geohistory analysis - application of micropaleontology in exploration geology, AAPG Bull., 62, p. 201-222.

Watts, A.B. and M.S. Steckler, 1979, Subsidence and eustacy at the continental margin of eastern North America, *in:* Deep drilling results in the Atlantic ocean: continental margins and paleoenvironment, edited by M. Talwani, W.F. Hay and W.B.F. Ryan, Maurice Ewing, Ser. 3, p. 218-234.

Yanshin, A., 1973, On so-called worldwide transgressions and regressions, Bulletin M.O-VA Isp. Prirody Otd. Geologic, v. XLVIII, No. 1, p. 9-44.

GEOCHEMISTRY AND ORIGIN OF SMACKOVER AND BUCKNER DOLOMITES (UPPER JURASSIC), JAY FIELD AREA, ALABAMA - FLORIDA

MARSHALL J. VINET

Davis Oil Company
601 Poydras Street, Suite 2091
New Orleans, Louisiana 70130

ABSTRACT

The Smackover-Buckner sequence in the Jay-Big Escambia Creek area is divisible into three major lithostratigraphic units: lower Smackover, upper Smackover and Buckner with an uppermost Smackover transitional sequence. The lower Smackover is regionally extensive, mostly high intertidal, dolomitized packstones with abundant displacive anhydrite nodules. The upper Smackover is characterized both by areas of thick oolite grainstone buildups which form the primary reservoir rock and by areas of low energy, subtidal mudstones-wackestones. The Buckner is mostly nodular mosaic anhydrite with minor amounts of dolomite mudstones-packstones. The Smackover-Buckner contact is a transitional sequence (referred to here as uppermost Smackover) of high intertidal-low supratidal dolomitized mudstones-grainstones interbedded with nodular mosaic anhydrite layers.

A combination of stratigraphic and geochemical evidence indicate dolomites of four separate origins, three of which are early diagenetic. The dolomites of the Buckner-uppermost Smackover have supratidal stratigraphy and heavy oxygen isotopic geochemistry (means $\delta^{18}O$=+1.5‰PDB) and formed penecontemporaneously with deposition. Immediately below and separated by a sharp discontinuity in chemistry are the upper Smackover dolomitized oomoldic grainstones which were dolomitized prior to uppermost Smackover deposition. Stratigraphy, petrography and light oxygen isotopic geochemistry (mean $\delta^{18}O$= -2.0‰PDB) favor a mixed-water origin for the upper Smackover dolomites.

The lower Smackover dolomites formed in arid, high intertidal-low supratidal environments penecontemporaneously with deposition based upon petrographic similarity to those of the Buckner and trace element and oxygen isotopic values intermediate between upper Smackover and Buckner dolomite values.

Late, pore-filling saddle (baroque) dolomites precipitated under deep burial conditions.

INTRODUCTION

It is common practice to compare ancient dolomite geochemistry to Holocene counterparts in order to help determine an origin. In many studies, the comparisons are weak and the stratigraphy is not conclusive, but this uniformitarianistic approach is all that can be done considering the present state of knowledge. A step further can be taken, however, if within an ancient sequence, a dolomite of known origin exists with its original chemistry preserved. The dolomite can then be used as an "internal standard." Such is the case with the Smackover-Buckner sequence in the Jay Field area. In this study, Smackover dolomites are compared to Buckner dolomites which are encased in impermeable anhydrite and consequently can be used as an internal standard.

Another common problem is that doubts

are cast upon the geochemistry of ancient dolomites which occur in outcrop. What are the effects, if any, of exposure to present day, Tertiary or Mesozoic meteoric waters? In the Jay area, the Smackover Formation has a simple tectonic history of progressive burial without major unconformities in the overlying Mesozoic and Cenozoic section. This simple burial history and the sealing effect of the regionally continuous Buckner anhydrite minimize the chances of water coming from above which would change the Smackover water chemistry and cause rock-water reequilibration.

For this study, wells were selected where continuous core was taken through the entire Smackover and lower Buckner sequence. A total of 103 dolomite samples from the lower Smackover (40 samples from 5 wells), upper Smackover (56 samples from 3 wells) and Buckner (7 samples from 3 wells) were analyzed for Sr^{++}, Na^{+}, Mg^{++}, and Ca^{++} cation concentrations by atomic absorption spectroscopy. The magnesium-calcium ratios were double checked by X-ray diffraction (XRD) techniques (after Katz, 1971), and 70 samples were analyzed for the oxygen and carbon stable isotopes. The new geochemical evidence helps define the various dolomite types and their origins.

STRATIGRAPHY

The 270 to 370 foot-thick sequence of Smackover carbonates in the Jay Field area can be divided into three major lithofacies (Figs. 1 and 2). The lower Smackover member comprises a pellet-peloid packstone facies whereas in the upper Smackover member, both ooid grainstone facies and lime pellet mudstone-wackestone facies are present (Figs. 1 and 2). The lower Smackover member, 120 to 160 feet thick, is a regionally continuous, intertidal pellet-peloid packstone which is gray, moderately bioturbated, and contains assorted skeletal material.

The upper Smackover, high-energy, ooid grainstones are thickest in the middle of Jay Field and are interbedded with lime pellet mudstones-wackstones near the field periphery. Away from Jay and Big Escambia Creek Fields, the entire upper Smackover member is commonly composed of the low-energy, subtidal lime pellet mudstone-wackestone facies, which contrasts with the less extensive grainstone areas (Fig. 2). The ooid grainstones are dolomitized and composed of poorly defined, massive 20 to 40 foot-thick shoal cycles which are capped by low-energy mudstones with white anhydrite nodules. The grainstones possibly accumulated over highs formed from halokinetics of the Louann salt while the lime mudstones-wackestones were deposited in the intervening low areas.

The uppermost Smackover in Jay Field consists of interbedded nodular mosaic anhydrite layers and high intertidal-low supratidal dolomitized mudstones to grainstones. This transition sequence grades upward into supratidal nodular mosaic anhydrites of the Buckner Formation.

EARLY DIAGENETIC FABRICS

In the lower Smackover, early diagenetic products include gray, finely crystalline dolomite; 0.5 to 1 inch diameter, displacive, white, felted crystal anhydrite nodules; and a few intervals of porous, leached pellet fabric in an otherwise nonporous, impermeable unit. Early diagenesis of the upper Smackover produced breccias, coarsely crystalline dolomite, and extensive oomoldic porosity. The lime mudstones, where proximal to oomoldic grainstones, are recrystallized to microspar and contain various amounts of scattered dolomite rhombs.

DOLOMITE GEOCHEMISTRY

Four types of dolomite are present in the Jay area: Buckner dolomite, upper Smackover dolomite, lower Smackover dolomite, and saddle (baroque) dolomite. The dolomites of the Buckner and uppermost Smackover are finely crystalline (1 to 40μm) and slightly calcian (mean 2.0 mole % excess Ca^{++}) with diffuse XRD peak shapes. Bulk oxygen isotop-

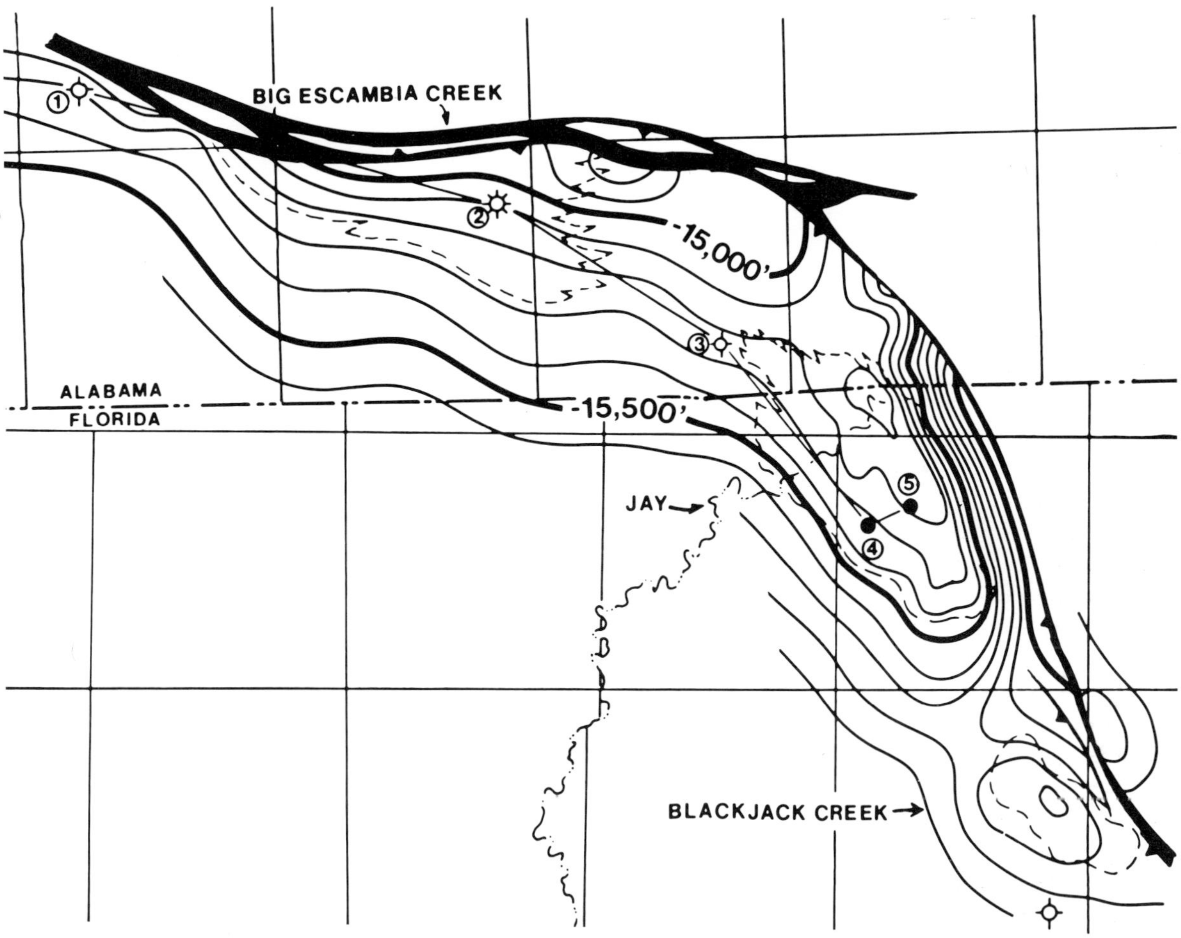

Figure 1. Structure map top Smackover of Jay area (modified after Ottman and others, 1973; and Sigsby, 1976). Wells 1-5 mark location of cross section in Figure 2. The detailed chemical analyses in this report are exclusively from Wells 1-5 and the Exxon #15 W. St. Regis well marked by the dry-hole symbol south of Blackjack Creek Field.

ic compositions are heavy (mean $\delta^{18}O$ = +1.5‰PDB, Fig. 3). Trace-element compositions (mean Sr^{++}=202ppm and Na^{+}=320ppm) are low relative to Recent dolomites.

Immediately below the uppermost Smackover and Buckner dolomites are the coarsely cyrstalline (80-300μm) upper Smackover dolomites. Restricted mainly to the grainstone facies, these dolomites are more stoichiometric (mean 1.4 mole % excess Ca^{++}) with sharp XRD peak shapes. Oxygen isotopic composition is relatively light (mean $\delta^{18}O$= -2.0‰PDB; Fig. 3) and trace-element contents are low (mean Sr^{++}=73ppm, Na^{+}=228ppm). The minor occurrences of dolomitized, anhydritic mudstones in the upper Smackover have isotopic and trace-element composition similar to Buckner values.

The lower Smackover member is approximately 80% aphanocrystalline (1-40μm) dolomite and 20% limestone. These dolomites are the least stoichiometric (mean 5.0 mole % excess Ca^{++}) and have diffuse XRD peak shapes. Oxygen isotopic compositions (mean $\delta^{18}O$=-0.5‰ PDB; Fig. 3) and trace element compositions (mean Sr^{++}=82ppm, Na^{+}=280 ppm) are intermediate between Buckner and upper Smackover values.

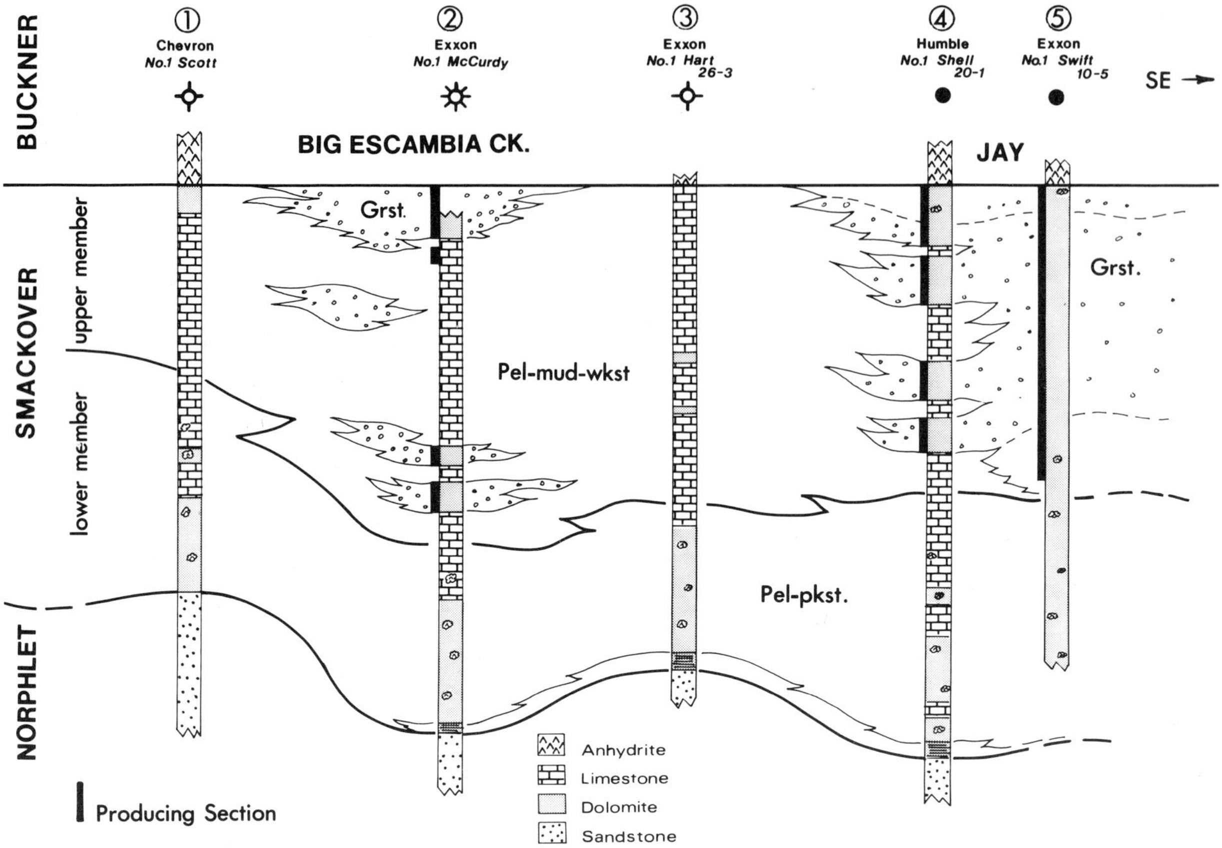

Figure 2. Cross section showing general stratigraphy of the area. The lower Smackover is mostly dolomite. The upper Smackover grainstones are nearly completely dolomitized. See Figure 1 for locations. The four lithologic patterns also apply to lithologic columns in Figures 4-6.

Sparse pore-filling saddle dolomite with characteristic curved crystal faces, undulose extinction, and coarse crystal size is the fourth dolomite type. One $\delta^{18}O$ analysis of -4.9‰ PDB indicates late precipitation at the elevated temperatures of deep burial conditions.

CHEMICAL DISCONTINUITY

Continuous sampling was possible across the Smackover-Buckner transition in three of the wells studied. In each case, a marked chemical discontinuity is present which separates the coarsely crystalline upper Smackover dolomite and the finely crystalline uppermost Smackover and Buckner dolomites (Figs. 4, 5, and 6). The discontinuity (dashed line) in each well is marked by a change in crystal size, trace-element composition, and oxygen-isotopic composition. The interval between the discontinuity and Smackover-Buckner contact (base of the massive Buckner anhydrite) was observed to vary in thickness from 6 to 30 feet (Figs. 4, 5, and 6).

Within the uppermost Smackover in the Humble No. 1 Shell 20-1 well (Fig. 4) are two thin anhydrite layers at 15,604 and 15,611 feet. Below the discontinuity (15,616 feet) is a breccia (15,620 feet) which probably represents a soil horizon or solution collapse of an anhydrite layer; in either case the breccia indicates meteoric water influence. The minor salt tectonism characteristic of the area precludes tectonic bedding plane slippage as a possible origin for the

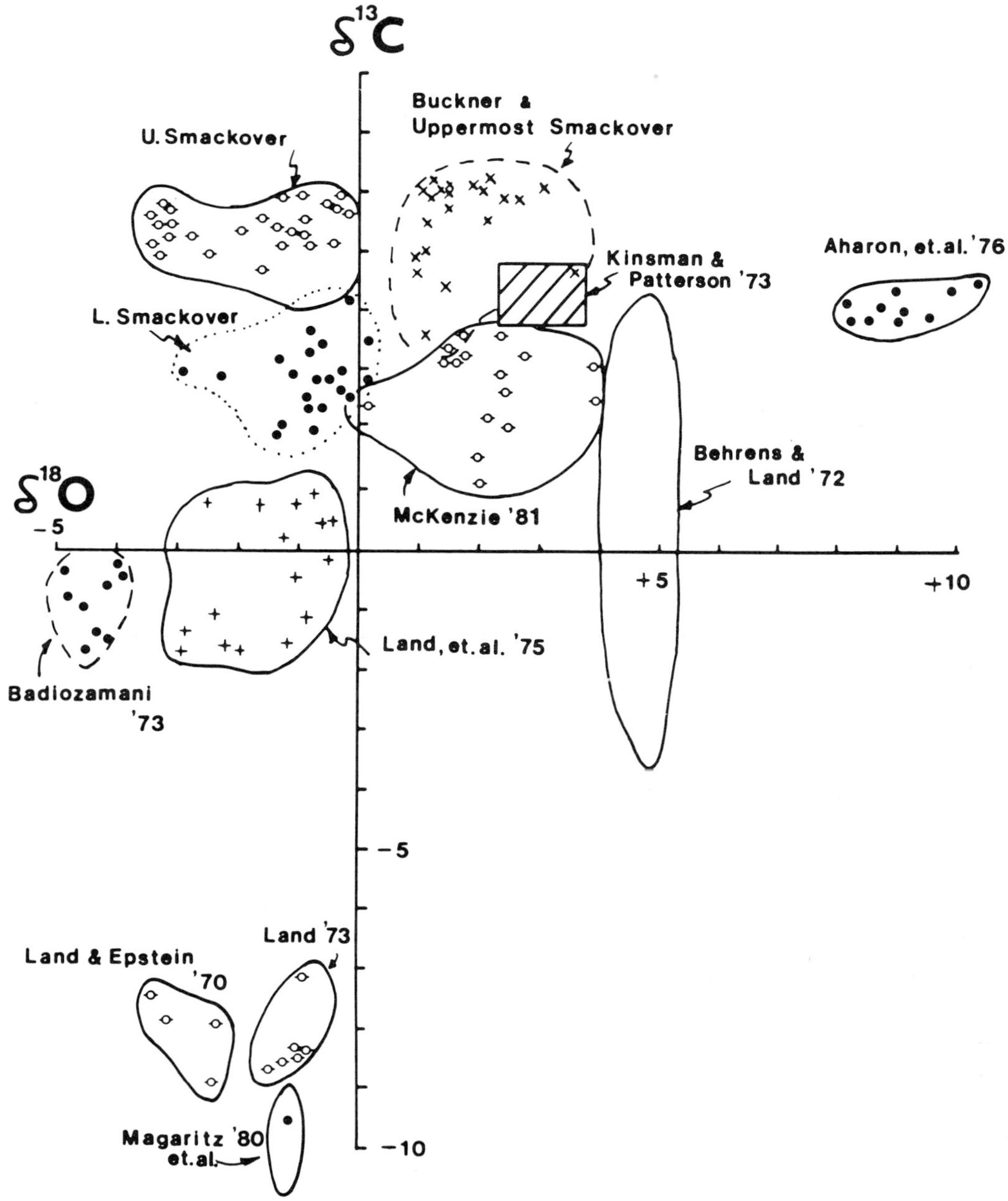

Figure 3. Cross plot of carbon and oxygen stable-isotope data for comparison of the Jay area dolomites to other studies.

breccia. The Exxon No. 1 Swift 10-5 well (Fig. 5) has 20 feet of uppermost Smackover section in which the aphanocrystalline dolomites are noticeably more calcian than the coarsely crystalline dolomites below. In the Exxon #15 W. St. Regis well (Fig. 6) the chemical discontinuity occurs at 16,041 feet, 6 feet below the Smackover-Buckner contact. This abrupt change in chemistry is coincident with the occurrence of a breccia similar to the one described in the Humber No. 1 Shell 20-1 well.

Anhydrite stringers (Figs. 4, 5, and 6) are present above the discontinuity and breccias (Figs. 4 and 6) occur only at or below the discontinuity. Hence, the discontinuity

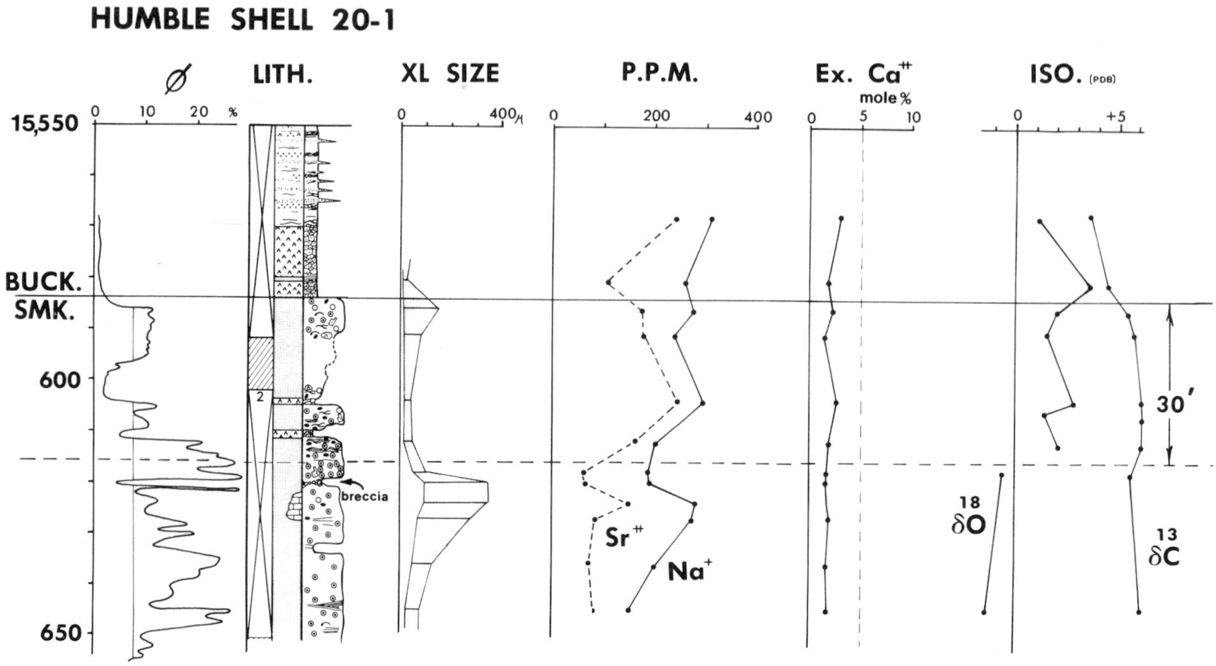

Figure 4. Detailed lithology and geochemistry of the Buckner, uppermost Smackover, and upper Smackover of the Humble No. 1 Shell 20-1 well. Porosity plotted from core analysis with shaded area oil productive. Note anhydrite layers above the discontinuity and the breccia below.

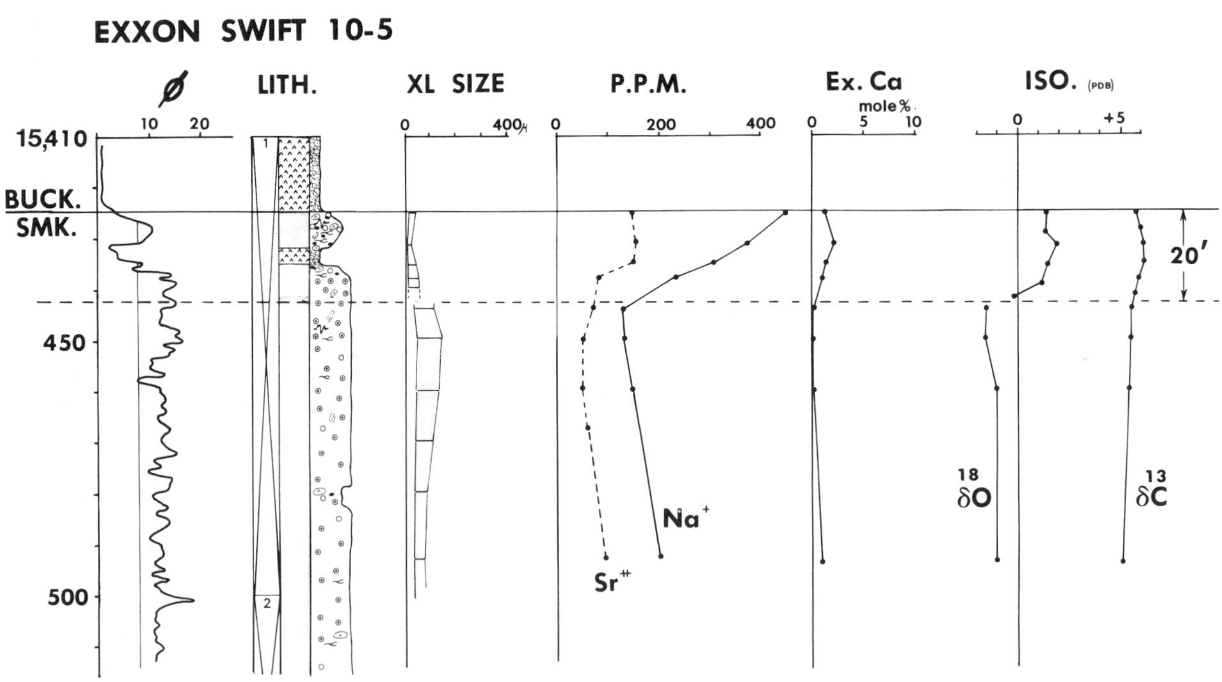

Figure 5. Detailed lithology and geochemistry of upper and uppermost Smackover, Exxon No. 1 Swift 10-5 well. There is no physical evidence of subaerial exposure at the chemical discontinuity.

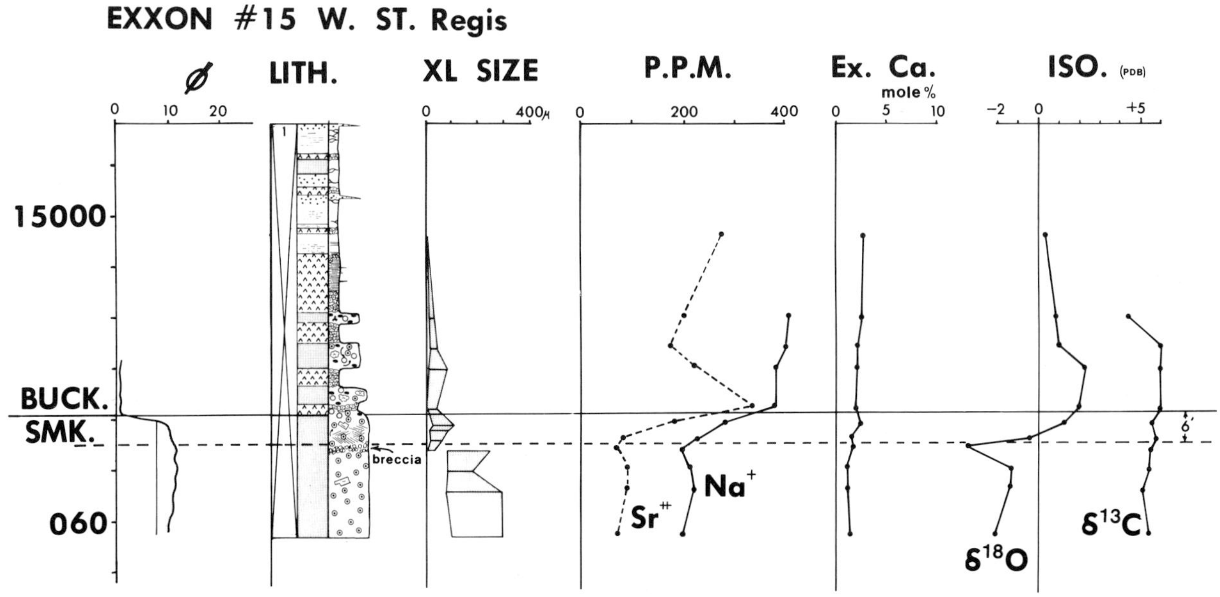

Figure 6. Detailed lithology and geochemistry of the Exxon No. 15 W. St. Regis well. Note breccia at the chemical discontinuity and how it always occurs within porous section.

clearly marks the top of a zone of extensive meteoric diagenesis which is shown by the pervasive moldic fabrics, breccias, microspar and other diagnostic features.

DISCUSSION

The combination of stratigraphic and geochemical evidence indicates that the dolomites of the Buckner and uppermost Smackover are of a supratidal origin. The stratigraphic evidence is the shallowing-upward sequence of fine grained, nodular anhydritic carbonates which cap the grainstone complexes (Figs. 4, 5, and 6); occurrence of abundant displacive, felted crystal, nodular mosaic anhydrite layers characteristic of arid climates; and restricted fossil assemblages characteristic of supratidal-intertidal conditions. The $\delta^{18}O$ values of +0.3 to +3.4‰ (PDB) are similar to Holocene sabkha dolomites described by Kinsman and Patterson (1973) and McKenzie (1981) (Fig. 3). Other dolomites of evaporative origin with heavy $\delta^{18}O$ signatures are described from Baffin Bay, Texas (Behrens and Land, 1972) and the "Solar Lake", Gulf of Elat (Aharon and others, 1977). Ancient sabkha and sabkha-related dolomites also have similar $\delta^{18}O$ values (Rudolph, 1978; Clark, 1980; M'Rabet, 1981). Therefore, the combination of stratigraphic and geochemical evidence suggests arid, intertidal-supratidal environments of deposition of the Buckner and uppermost Smackover dolomites, with preservation of their original $\delta^{18}O$ values. The uppermost Smackover is most likely a lateral equivalent of the massive Buckner anhydrite. Although Na^{+} and Sr^{++} trace-element compositions are lower than in Holocene sabkha dolomites, it is possible that these elements are semi-mobile under deep burial conditions, and/or that their concentrations are controlled by crystal growth kinetics.

The Smackover has long been regarded as a perfect opportunity to employ the reflux model. It can be eliminated as a possible explanation because gradients would be expected at the top of the Smackover and not an abrupt chemical change within a porous and permeable interval.

If dolomitization of the upper Smackover was an early, near surface phenomenon, and reflux does not apply, the upper Smackover was dolomitized prior to uppermost Smackover and Buckner deposition. Assuming, like the Buckner and uppermost Smackover dolomites, that the original $\delta^{18}O$ composition of the upper Smackover dolomites is preserved, a mixed marine-meteoric water mechanism best explains the origin of upper Smackover dolomites. Dolomitization must have occurred during subaerial exposure, the surface of which is now represented by the chemical discontinuity. Recent and Pleistocene dolomites that are most likely of a mixed-water origin (Land and Epstein, 1970; Land, 1973; Magaritz and others, 1980) have $\delta^{18}O$ values similar to those of the upper Smackover (Figs. 3). Ancient dolomites which are suspected to be of a mixed-water origin (Badiozamani, 1973; Land and others, 1975) also have comparable values (Fig. 3). Trace-element compositions of Sr^{++} and Na^{+} are lower (than Buckner values) and conform to the mixed water interpretation.

Another possible explanation for the origin of the upper Smackover dolomite is late burial dolomitization by upward-migrating Louann brines. The upper Smackover must resist early dolomitization either by mixed waters or by refluxing uppermost Smackover and Buckner brines. It is possible that subaerial exposure produced sufficient meteoric diagenesis to stabilize the metastable mineral suite of the grainstones to low magnesian calcite prior to uppermost Smackover deposition. Dolomitization favors aragonite and magnesian calcite over low magnesian calcite (Kinsman, 1971; Patterson, 1972; Gaines, 1980; Sibley, 1980). The low magnesian calcite oomoldic grainstones may then resist early dolomitization and, upon sufficient burial, react with warm Louann brines to form dolomite. In this case, thermal fractionation could account for the lighter (than Buckner) $\delta^{18}O$ values. Although this scenario cannot be disproved, it is difficult to explain the relative scarcity of dolomite in south Arkansas and Mississippi if Louann brines are responsible.

Other similar deep burial theories should result in chemical gradients - not sharp discontinuities. Therefore, early mixed-water dolomitization of the upper Smackover followed by deposition and sabkha dolomitization of uppermost Smackover and Buckner sediments is the preferred explanation for the origin of these rocks.

The lower Smackover and uppermost Smackover-Buckner have both lithologic character and crystallographic character in common. Similarities in lithology are the displacive nodular anhydrite and fine dolomite crystal size which preserves the pellet fabric. Like Buckner dolomites, the lower Smackover dolomites are poorly crystalline as indicated by diffuse XRD peak shapes. Trace-element and oxygen isotopic values intermediate between upper Smackover and Buckner dolomite values may indicate dolomitization by evaporative brines with lower salinities than Buckner brines. Petrographic and crystallographic similarity of lower Smackover to Buckner dolomites, collapse breccias described by Sigsby (1976), and dolomite geochemistry suggest that the lower Smackover sediments were deposited in arid, high intertidal-low supratidal environments with nearly penecontemporaneous dolomitization analogous to unit 6 of McKenzie and others (1980, Fig. 3).

CONCLUSIONS

Stratigraphic and stable isotopic data strongly suggest that dolomites of the Buckner Formation and uppermost Smackover are supratidal in origin and that the uppermost Smackover sediments (up to 30 feet in thickness in some areas) are stratigraphically equivalent to the lower part of the Buckner anhydrite. The presence of a sharp discontinuity in the chemistry and crystal size of dolomites at the base of the uppermost Smackover (Fig. 4, 5, and 6) within a porous and permeable section indicates that the upper Smackover was dolomitized prior to uppermost Smackover deposition. This timing of dolomitization eliminates the reflux of Buckner brines as an explanation for the upper Smackover dolomites. In fact, the upper Smackover in this case cannot serve as a test for reflux. The lower trace-element content and lighter $\delta^{18}O$ (compared to the Buckner), extensive meteoric diagenesis, and exposure breccias at the discontinuity suggest dolomitization of the upper Smackover by a mixed marine-meteoric water system. Petrographic similarity of the lower Smackover dolomites to those of the Buckner; association with abundant, displacive, nodular anhydrite; and trace element and oxygen isotopic values intermediate between upper Smackover and Buckner dolomite values indicate dolomitization by evaporative brines with a lower salinity than Buckner brines. Therefore, the dolomites probably formed in arid, high intertidal-low supratidal environments penecontemporaneous with deposition.

ACKNOWLEDGMENT

This study was conducted under the supervision of Clyde H. Moore, Applied Carbonate Research Program, Louisiana State University, Baton Rouge, Louisiana. Assistance was also received from Eytan Sass, Hebrew University, Jerusalem, Israel; Yehezkeel Druckman, Geological Survey of Israel; J. A. D. Dickson, University of Cambridge, England; Gill Harwood, Louisiana State University - Baton Rouge and Michael Stamatedes, Amoco, New Orleans, Louisiana.

REFERENCES

Aharon, P., Y. Kolodny, and F. Sass, 1977, Recent hot brine dolomitization in the "Solar Lake", Gulf of Elat. Isotopic, chemical and mineralogical study: Jour. Geology, V. 85, p. 27-48.

Badiozamani, K., 1973, The Dorag dolomitization model-application to the Middle Ordovician of Wisconsin: Jour. Sed. Pet., v. 43, p. 965-984.

Behrens, E. W., and L. S. Land, 1972, Subtidal Holocene dolomite, Baffin Bay, Texas: Jour. Sed. Pet., v. 42, p. 155-161.

Carpenter, A. B, 1978, Origin and chemical evolution of brines in sedimentary basins: Oklahoma Geological Circular, v. 79, p. 60-77.

Clark, D. N., 1980, The diagenesis of Zechstein carbonate sediments: Shell Research BV, publ. 562, 27 pages.

Gaines, A. M., 1980, Dolomitization kinetics: recent experimental studies, *in* Zenger, D. H., Dunham, J. B., and Ethington, R. L., (eds.), Concepts and models of dolomitization. Soc. Econ. Paleontologists and Mineralogists, Spec. Pub. 28, p. 81-86.

Katz, A., 1971, Zoned dolomite crystals: Jour. of Geology, v. 79, p. 38-51.

Kinsman, D. J. J., 1971, Discussion of: Subsurface brines and the formation of Mississippi Valley-type ore deposits: Trans. Inst. Min. Metall., v. 80, p. B61B63.

__________ and R. J. Patterson, 1973, Dolomitization process in sabkha environment (abs.): Amer. Assoc. Pet. Geol. Bull., v. 57, p. 788-789.

Land, L. S., 1973, Holocene meteroic dolomitization of Pleistocene limestone, North Jamaica: Sedimentology, v. 20, p. 411-424.

__________, and S. Epstein, 1970, Late Pleistocene diagenesis and dolomitization, North Jamaica: Sedimentology, v. 14, p. 187-200.

__________, M. R. I. Salem, and D. W. Morrow, 1975, Paleohydrology of ancient dolomites-geochemical evidence: Amer. Assoc. Pet. Geol. Bull., v. 59, p. 1602-1625.

Magaritz, M., L. Goldenberg, U. Kafr, and A. Arad, 1980, Dolomite formation in the seawater-freshwater interface: Nature, v. 287, p. 622-624.

McKenzie, J. A., K. J. Hsu and J. F. Schneider, 1980, Movement of subsurface waters under the sabkha, Abu Dhabi, U.A.E., and its relation to evaporative dolomite genesis, *in* Zenger, D. H., Dunham, J. B., and Ethington, R. L., (eds.), Concepts and Models of Dolomitization: Soc. Econ. Paleontologists and Mineralogists, Spec. Publ. 28, p. 11-30.

__________, 1981, Holocene dolomitization of calcium carbonate sediments from the coastal sabkhas of Abu Dhabi, U.A.E.: A stable isotope study: Jour. of Geology, v. 89, p. 185-198.

M'Rabet, A., 1981, Differentiation of environments of dolomite formation, Lower Cretaceous of Central Tunisia: Sedimentology, v. 28, p. 331-352.

Ottman, R. D., P. L. Keys, and M. A. Ziegler, 1973, Jay Field - a Jurassic stratigraphic trap, Gulf Coast Assoc. Geol. Soc. Trans.,

v. 23, p. 146-157.
Patterson, R. J., 1972, Hydrology and carbonate diagenesis of a coastal sabkha in the Persian Gulf Ph.D. Diss., Princeton Univ., Princeton, New Jersey, 498 p.
Rudolph, K., 1978, Diagenesis of backreef carbonates: an example from the Capitan Complex: unpublished Masters thesis, Univ. of Texas, Austin, 159 p.
Sibley, D. F., 1980, Climatic control of dolomitization, Seroe Domi Formation (Pliocene) Bonaire, Netherlands Antilles, *in* Zenger, D. H., Dunham, J. B., and Ethington, R. L., (eds.), Concepts and models of dolomitization: Soc. Econ. Paleontologists and Mineralogists Spec. Publ. 28, p. 247-258.
Sigsby, R. J., 1976, Paleoenvironmental analysis of the Big Escambia Creek-Jay- Blackjack Creek Field areas, Gulf Coast Assoc. Geol. Soc. Trans., v. 26, p. 258-278.

DEPOSITION, POROSITY EVOLUTION, AND MODES OF SULFATE EMPLACEMENT IN THE UPPER SMACKOVER, EDGEWOOD FIELDS, VAN ZANDT COUNTY, TEXAS

DAVID C. WICKLAND

Chevron U.S.A., Inc.
P.O. Box 8200
Concord, California 94520

ALONZO D. JACKA

Department of Geosciences
Texas Tech University
P.O. Box 4109
Lubbock, Texas 79409

ABSTRACT

Cores and cuttings from the upper Smackover Formation, Edgewood fields, northern Van Zandt County, East Texas, were analyzed to determine depositional facies, diagenesis, and porosity evolution. Edgewood fields lie on an anticlinal structure formed by a salt pillow in the East Texas Basin. Structural growth began during deposition of the Smackover Formation and continued until the end of deposition of the Cotton Valley Formation.

Major lithofacies in the Smackover Formation consist of oograinstone, oopisograpestone grainstone, oobiopelwackestone-packstone, anhydrite, quartzose sandstone, and sandy carbonate, with grainstone and packstone being predominant.

Secondary oomolds and intercrystalline pores constitute the major types of preserved porosity. Although large volumes of primary intergranular porosity initially existed in the lime sands, nearly all of this porosity has been occluded first by submarine cements and later by dolomite, anhydrite, and calcite cements.

Evidence indicates that most of the anhydrite in the upper Smackover Formation is of a diagenetic origin and nonevaporitic. Some thin anhydrite layers are present in the formation and may represent sabkha deposits, but they lack certain characteristics that are typical of modern sabkhas, such as pseudomorphs of gypsum crystals, enteroliths, mud crack casts, flat pebble conglomerates, stromatolites, and fenestral cavities. No example of subaqueous evaporites was noted.

Diagenetic anhydrite occurs in the following modes and morphologies: (1) sparry or felted-lath anhydrite, which was precipitated as a cement in primary and secondary voids; (2) abundant anhydrite, which was precipitated as cement subsequently replaced outwardly, invading the surrounding matrix or grains to form nodules, porphyroblasts, or intensively replaced layers; (3) felted-lath anhydrite which occurs as scattered replacement nodules, or which coalesce to form dense mosaics or massively replaced layers; and (4) replacive sparry anhydrite which was emplaced as scattered porphyroblasts and which locally coalesced to form massively replaced intervals.

Most of the anhydrite was emplaced in grainstones and packstones which had been deposited in relatively high-energy subtidal environments. Dolostones that have been replaced intensively by anhydrite contain abundant inclusions of corroded dolomite crystals and relict fabrics of the carbonate. Initially, much anhydrite nucleated in organic-rich grains, such as ooids, shells, or peloids. In some intervals, thalassinoid burrows have been replaced selectively by anhydrite.

In most carbonate intervals, dolomitization preceded anhydritization. In many intervals, emplacement of anhydrite was followed by influx of a lower salinity fluid that dissolved the anhydrite, precipitated a calcite cement, replaced the anhydrite with calcite, and/or brought about dedolomitization.

INTRODUCTION

Exploration of the Jurassic Smackover Formation was initiated in southern Arkansas during the mid-1920's and major oil fields were found in upper Smackover oolitic limestones on numerous large anticlinal structures. In the 1960's, Smackover exploration shifted to East Texas where attention was focused on finding anticlinal structures in the Smackover at greater depths (10,000 to 13,000 feet). This study examines an anticlinal structure in the East Texas Basin (Fig. 1), where twelve gas and gas-condensate wells, now abandoned, formed the Edgewood and Northeast Edgewood fields (Figs. 2 and 3). The Edgewood anticlinal structure is formed by a salt pillow, which is a feature common throughout the Gulf Coast Basin. The hydrocarbons apparently were structurally trapped; however, it has been demonstrated that diagenesis controlled the type and distribution of porosity within the upper Smackover and was probably as important as structure in localizing the accumulation of hydrocarbons present in the Edgewood fields.

Objectives

1. To determine the depositional environments, facies relationships, and geologic history of the Jurassic upper Smackover within the Edgewood fields.
2. To interpret the diagenesis and examine the porosity distribution that results from deposition and diagenesis.
3. To show how diagenesis has affected hydrocarbon production.

Location

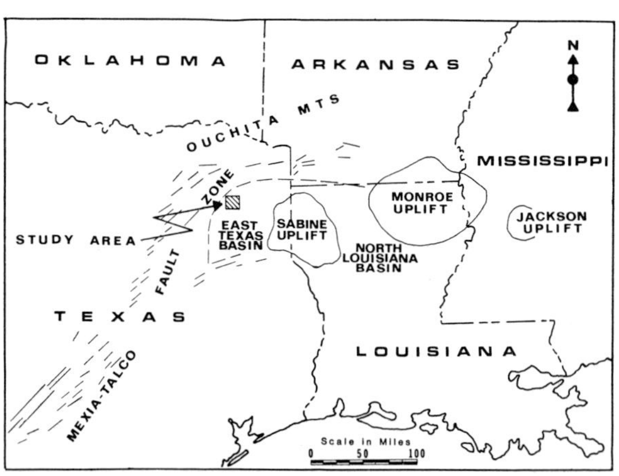

Figure 1. Geologic setting of the northern Gulf Coast Province, showing major structural features and study area (adapted from Imlay, 1940).

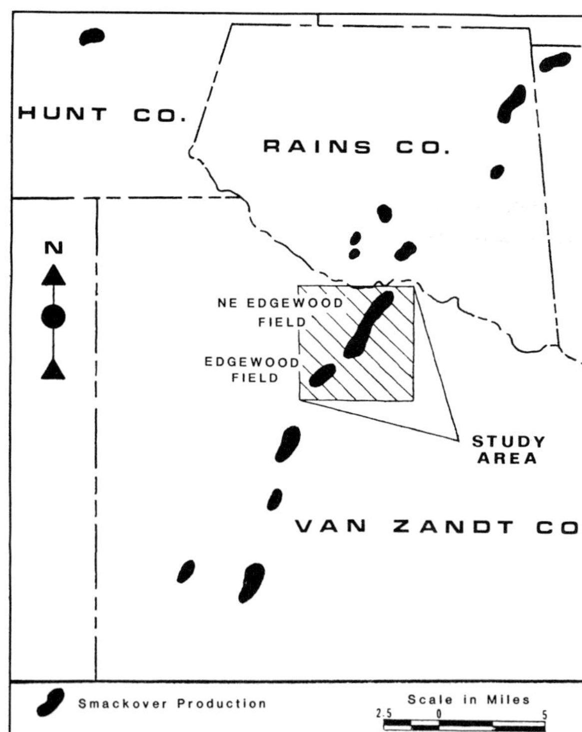

Figure 2. Area of study in Van Zandt County, Texas.

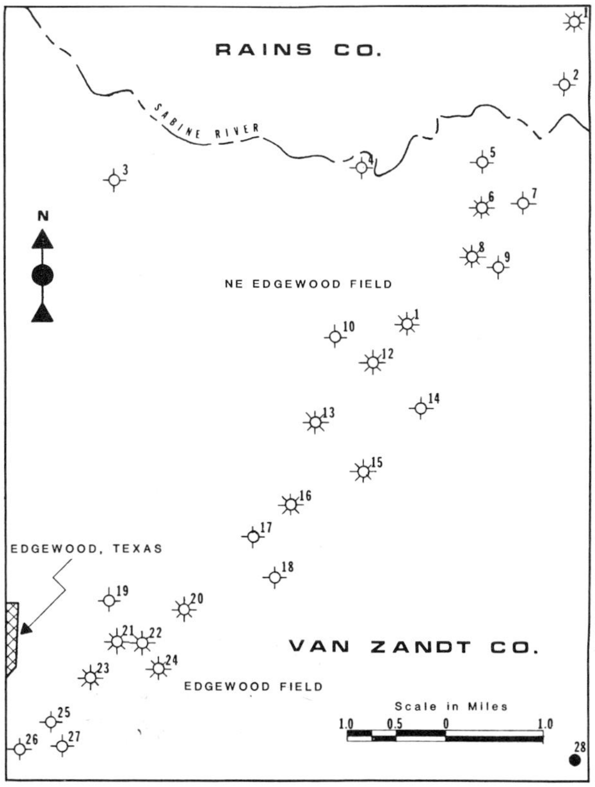

Figure 3. Detail of study and showing well locations and production in the Edgewood Fields area.

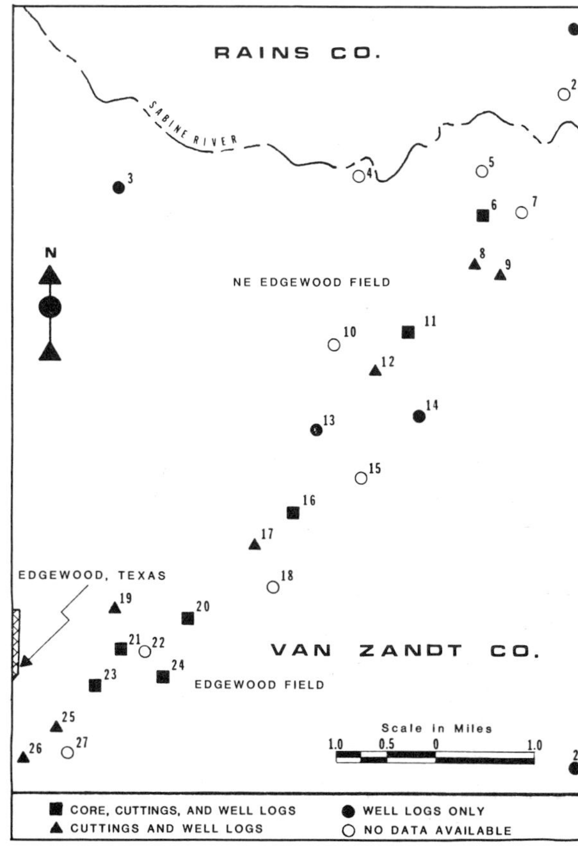

Figure 4. Subsurface data utilized within the study area.

Van Zandt County, in eastern Texas (Figs. 1, 2, and 3) and encompasses 40 square miles that include the Edgewood and Northeast Edgewood fields. Data availability within the Edgewood fields is the major reason that this area was chosen for study. To date, nearly all published diagenetic studies of the upper Smackover are of fields in southern Arkansas, northern Louisiana, and southern Alabama. Thus, a different geographic area and geologic province are presented in this study.

Methods of Study

The structural and stratigraphic framework of the study area is based on the study of 15 electric logs. Approximately 500 feet of core from seven wells within the Edgewood fields were described for gross lithologies, facies changes, and sedimentary structures. Well cuttings and core chips from seven additional wells were utilized where conventional cores were not available. Approximately 400 petrographic thin sections were made from the cores, core chips, and well cuttings, and each was stained according to the method of Lindholm and Finkelman (1972), with a combination Alizarin Red S-Potassium ferricyanide solution to facilitate calcite-dolomite and ferroan-nonferroan carbonate distinction. Thin sections were studied to determine lithofacies, diagenesis, and porosity relationships. Figure 4 and Table 1 show well locations and summarize the data.

GEOLOGIC SETTING

The central Gulf Coast region is a site of Triassic-Jurassic rifting associated with the opening of the Gulf of Mexico (Pilger,

Table 1. Formation Tops and Intervals Described

MAP NUMBER	COMPANY AND LEASE	SUBSEA FORMATION TOPS (FEET) L.C.V.L.	BUCKNER	SMACKOVER	INTERVALS DESCRIBED (FEET) CUTTINGS	CORE
1.	ARK LA, No. 1 Brimer	-11,588	-12,178	-12,828	---	---
2.	HAMON, No. 1 Greene	-11,685	-12,315	TD	IN	BUCKNER
3.	FARMLAND, No. 1 Wingo	-11,023	-11,538	-12,248	---	---
4.	HUMBLE, No. 1 Hazel	-11,395	---	TD	IN	BUCKNER
5.	PRESIDIO, No. 1 Cheatham	NO	DATA	AVAILABLE		
6.	PAN AM, No. 1 Vines	-11,298	-11,858	-12,343	-12,178--12,378	-12,389-12,451
7.	MCBEE, No. 1 Presley	-11,407	-12,000	TD	IN	BUCKNER
8.	PAN AM, No. 1 Lay	-11,232	-11,772	-12,162	-12,062--12,472	---
9.	UNION, No. 1 Albright	-11,400	-11,850	-12,660	-12,660--13,400	---
10.	RUDMAN, No. 1 Read	-11,254	-11,900	TD	IN	BUCKNER
11.	PAN AM, No. 1-C Parker	-11,188	-11,768	-12,218	-12,098--13,108	-12,168-12,229
						-12,255-12,277
						-12,303-12,315
						-12,700-12,712
12.	PAN AM, No. 1-C Parker	-11,140	-11,730	-12,170	-12,224--12,467	---
13.	PAN AM, No. 1-D Parker	-11,145	-11,730	-12,230	---	---
14.	CARAWAY, No. 2 Parker	-11,305	-11,890	-12,510	---	---
15.	CARAWAY, No. 1 Parker	NO	DATA	AVAILABLE		
16.	PAN AM, No. 1 Forbes	-11,186	-11,776	-12,376	-12,265--12,365	-12,365-12,472
17.	PAN AM, No. 1 Knowles	-11,309	-11,884	-12,464	-12,234--12,444	---
18.	ARK LA, No. 1 Holland	NO	DATA	AVAILABLE		
19.	PAN AM, No. 1 Brown	-11,107	11,702	-12,512	-12,422--13,262	---
20.	AMOCO, No. 3 Nichols	-11,147	-11,732	-12,447	-12,367--12,957	-12,857-13,893
21.	PAN AM, No. 1 Nichols	-11,120	-11,868	-12,410	---	-12,356-12,468
22.	ARK LA, No. 1 Estes	NO	DATA	AVAILABLE		
23.	PAN AM, No. 1-A Parker	-11,124	-11,869	-12,369	-12,278--13,238	-12,378-12,460
24.	PAN AM, No. 2 Nichols	-11,200	-11,780	-12,430	-12,329--12,549	-12,474-12,509
25.	PAN AM, No. 1 Simmons	-11,155	-11,730	-12,455	-12,325--12,425	---
					-12,525--12,535	---
26.	PAN AM, No. 1 McDonald	-11,178	-11,808	-12,488	-12,408--12,738	---
27.	SAMEDAN, No. 1 Hamilton	NO	DATA	AVAILABLE		
28.	CONOCO, No. 1 Shields	NO	DATA	AVAILABLE		

1978). The northern edge of this rifting is marked by extensive block faulting. The Mexia-Talco System is one group of these block faults in eastern Texas. Just south of the Mexia-Talco System was a region of active subsidence -- the East Texas Salt Basin (Fig. 1). The East Texas Salt Basin is bounded on the east by the Sabine Uplift, on the north and west by the Ouachita Mountains, and connects to the south into the Gulf Coast Salt Basin (Fig. 1).

Within the East Texas Salt Basin, the Smackover Formation was deposited across a gently inclined ramp in a dominantly regressing sea. Upper Smackover grainstones generally grade basinward into sediments rich in lime mud; this is best explained by a gradual increase in water depth and the consequent lowering of energy in a basinward direction. Present-day regional dip on top of the Smackover is approximately two degrees. However, this regional dip exceeds that which characterized Smackover deposition due to greater post-Jurassic subsidence basinward (Becher & Moore, 1976). In addition, in the Edgewood fields, salt tectonism locally modified the regional dips.

REGIONAL STRATIGRAPHY

The Jurassic System forms a thick wedge of strata in the Gulf Coast basin that generally dip and thicken to the south. Within the United States, the Late Jurassic sediments are known only from the subsurface. However, the Zuloaga Formation, which is stratigraphically equivalent and lithologically very similar to the Smackover Formation, is exposed in the mountains of northeastern Mexico (Oivanki, 1974).

The Late Jurassic rocks of eastern Texas (Fig. 5) include, from oldest to youngest, the Werner Formation, the Louann Salt, the Norphlet Formation, the Smackover Formation, the Buckner Formation, and the Cotton Valley Group, which consists of the Bossier and Schuler formations. A pre-Oxfordian age is assigned to the Werner-Louann sequence. The Norphlet and Smackover formations were deposited during Oxfordian time. The overlying Buckner and Bossier formations are considered to be Kimmeridgian in age and an age of Portlandian is assigned to the Schuler Formation (Dickinson, 1968). Only the Norphlet, Smackover, and Buckner formations are described in this paper.

The Norphlet Formation conformably underlies the Smackover and unconformably overlies the Louann Salt. It consists of approximately 50 feet of red and gray shale and red-to-gray conglomeratic sandstone (Dickinson, 1968). Only two wells, the Union No. 1 Albright and Pan Am No. B-1 Brown, penetrate the Norphlet Formation within the Edgewwod Field area.

The term Smackover Formation is used here for the lithologic unit originally defined by Weeks (1938) as the Smackover Limestone and it consists of three members. The

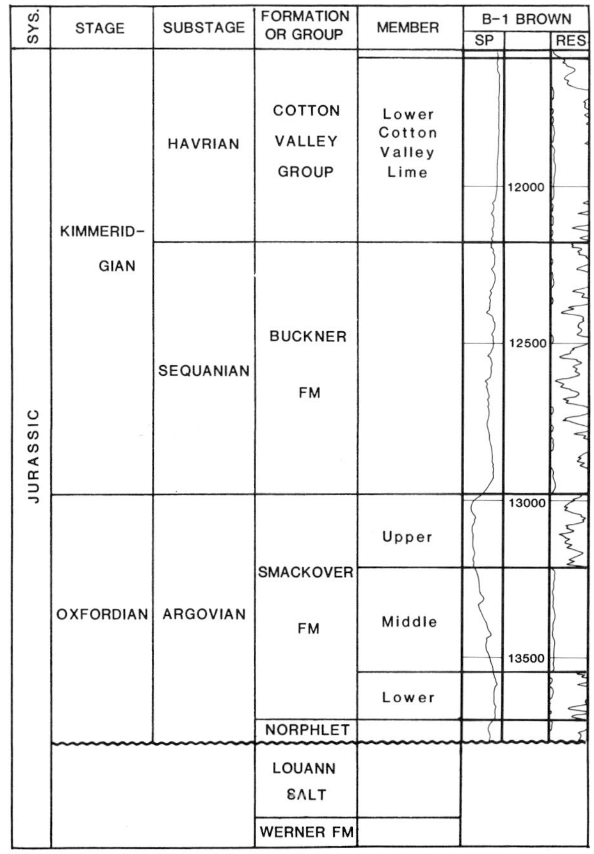

Figure 5. Stratigraphy for eastern Texas and log response from Pan Am B-1 Brown (after Dickinson, 1968).

lower member contains dark-brown and gray-colored, dense, silty-to-clayey limestone and calcareous siltstone which often is laminated and contains peloids. This member ranges from 200 to 300 feet in thickness, and it is penetrated by only four wells within the study area. The middle member consists of dense, brown-to-gray colored limestone which usually contains dolomite, peloids and oolites, and replacement anhydrite (Fig. 10a). Dickinson (1968) states that the middle member forms two-thirds to three-quarters of the total 900- to 1000-foot thick Smackover. The upper member, approximately 200 feet thick, consists primarily of dolomitized ooid grainstones containing pisoliths intraclasts, bioclasts, grapestones, algal grains, and peloids (Fig. 10b). The primary focus of this study is on this member, the Reynolds Oolite

(Weeks, 1938), from which the hydrocarbons are produced.

The Buckner Formation (Figs. 10c and 10d) conformably overlies the Smackover and is considered to be a partial updip equivalent of the Smackover (Dickinson, 1968). In the study area, the Buckner is composed of anhydrite, dolomite, red and tan sandstones, and red siltstones. The sandstones may contain anhydrite nodules and occasionally oolites.

LOCAL STRUCTURE

Gentle folding in the Late Jurassic beds of eastern Texas is caused by structures present in the Louann Salt. Hughes (1968) described such structures as anticlines, and named them low-relief salt pillows. Salt pillows are located near the updip limit of the Louann Salt and are arranged in linear ridge fashion with their long axes oriented approximately parallel to regional strike.

Within the study area, the Edgewood Fields are located on a large, elongate salt anticline (Fig. 6). The top of the Smackover Formation occurs from 13,500 feet on the flanks of the structure to 12,138 feet on the crest of the structure. There are approximately 400 feet of structural closure on the top of the Smackover at Northeast Edgewood Field and approximately 200 feet of closure at Edgewood Fields. A structure map of the top of the Lower Cotton Valley Massive Lime shows progressively less anticlinal structure through geologic time (Fig. 7). An isopachous map of the Buckner Formation and Lower Cotton Valley Massive Lime to the top of the Smackover Formation shows that the sediments thin over structural highs, and thicken in structural lows (Figs. 8 and 9). Therefore, structural growth began during deposition of the Smackover Formation and continued until the end of Cotton Valley deposition.

LITHOFACIES AND ENVIRONMENTS OF DEPOSITION

The upper Smackover is part of a shallow-marine, dominantly-regressive, upward-shoaling carbonate deposit. The upper Smackover possesses many depositional features that are similar to those of the Late Jurassic Arab Formation lime sands of Saudi Arabia (Wilson, 1975), to modern carbonates of the Bahaman Platform, and to carbonates of the Trucial Coast in the Persian Gulf. Regionally, the upper Smackover at Edgewood fields lies within a grainstone facies belt (Bishop, 1968). Lithofacies root names employed in the present study are based on Dunham's terminology (1962), and are modified by Folk's terminology (1965). Primary depositional facies are described and presented in order of decreasing energy; diagenetic modifications are treated in a subsequent chapter. Intensive early dolomitization of the upper Smackover at Edgewood obscured, and in many instances obliterated original rock fabrics.

Oograinstone Facies

Ooids comprise 95 percent of the allochems in the oograinstone facies, accompanied by minor amounts of bioclasts, peloids, and quartz grains (Figs. 12a, 12b, 13a, and 13b). The ooids, usually well sorted, range in diameter from 0.2 mm to 0.5 mm, and probably had a structure very similar to modern ooids. They nucleated around quartz grains, bioclasts, and peloids. The oograinstone facies is widespread at the Edgewood fields and comprises 60 to 90 feet of the 150 to 200 feet total thickness of the upper Smackover rocks. In addition, the oograinstone facies represents the highest-energy subtidal environment at Edgewood.

Oopisograpestone Grainstone Facies

This facies is present in only a few wells at Edgewood. Ooids form approximately 50 percent, pisoliths approximately 25 percent, and grapestones approximately 25 percent of the allochems (Fig. 12c). Minor amounts of bioclasts, peloids, and algal-coated grains also occur. The grapestone grains are composed of aggregated ooids and peloids. The composite grains range from one

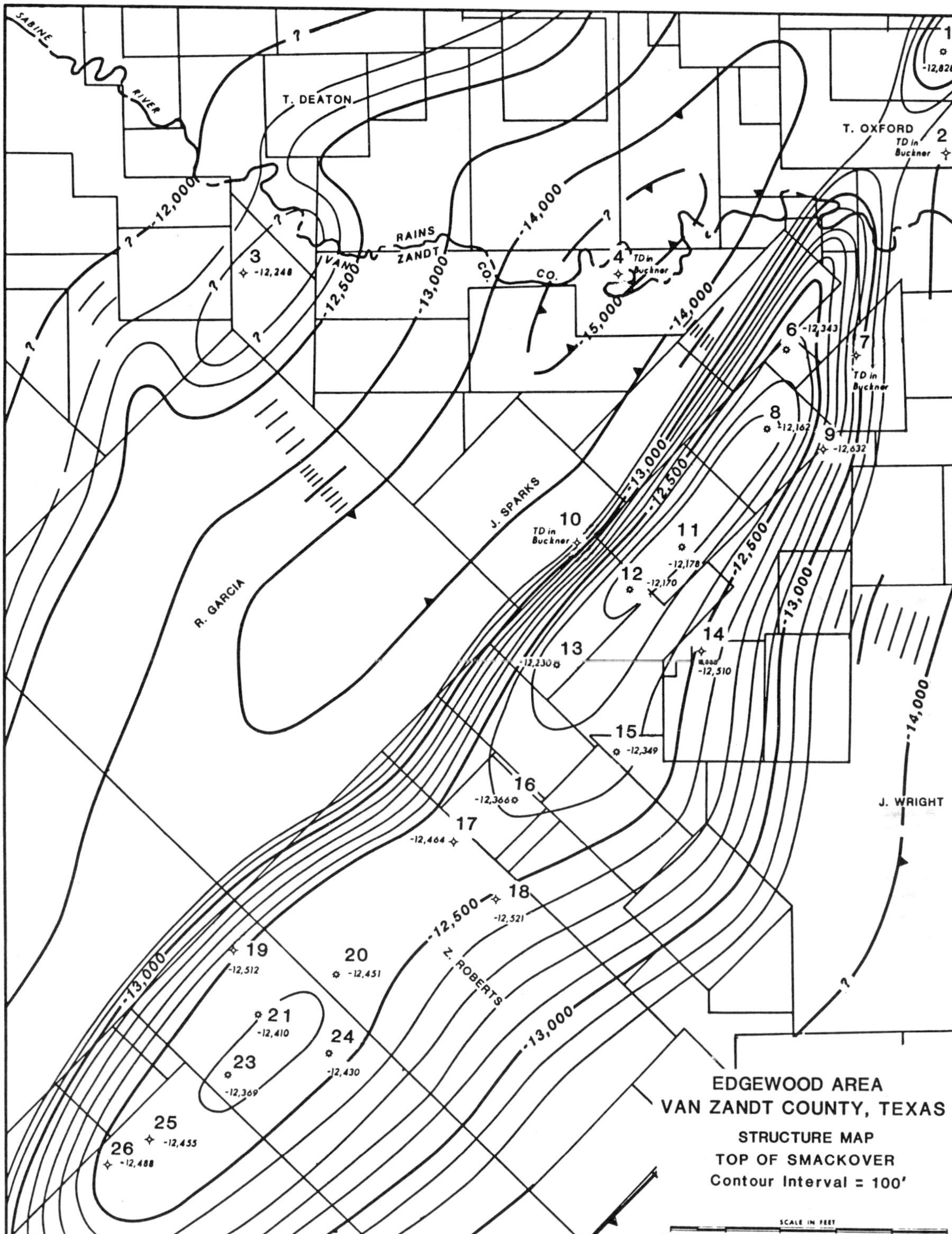

Figure 6. Structure map, datum: top of Smackover, contour interval = 100 feet; Edgewood area, Van Zandt County, Texas.

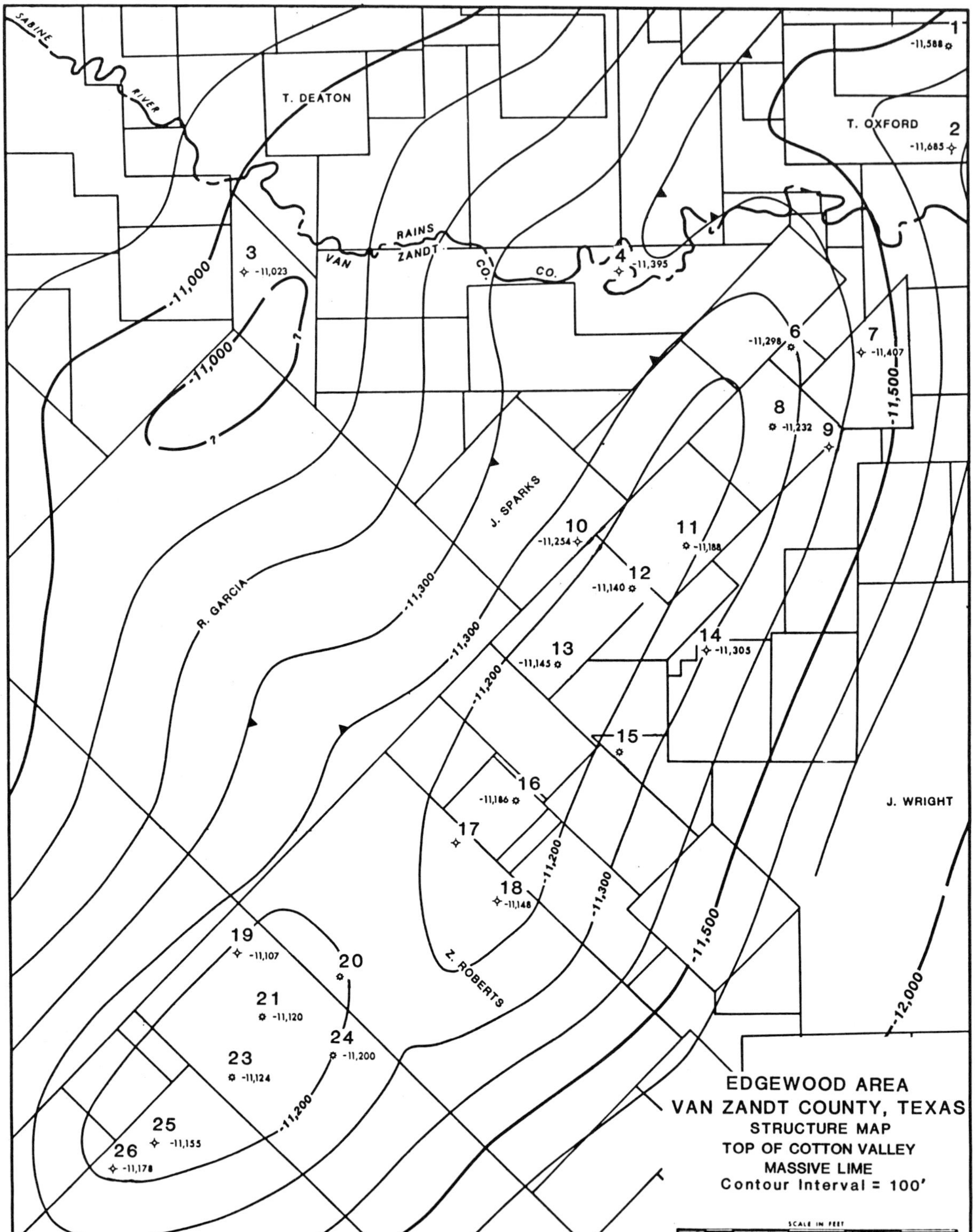

Figure 7. Structure map, datum: top of Cotton Valley Massive Limestone, contour interval = 100 feet; Edgewood area, Van Zandt County, Texas.

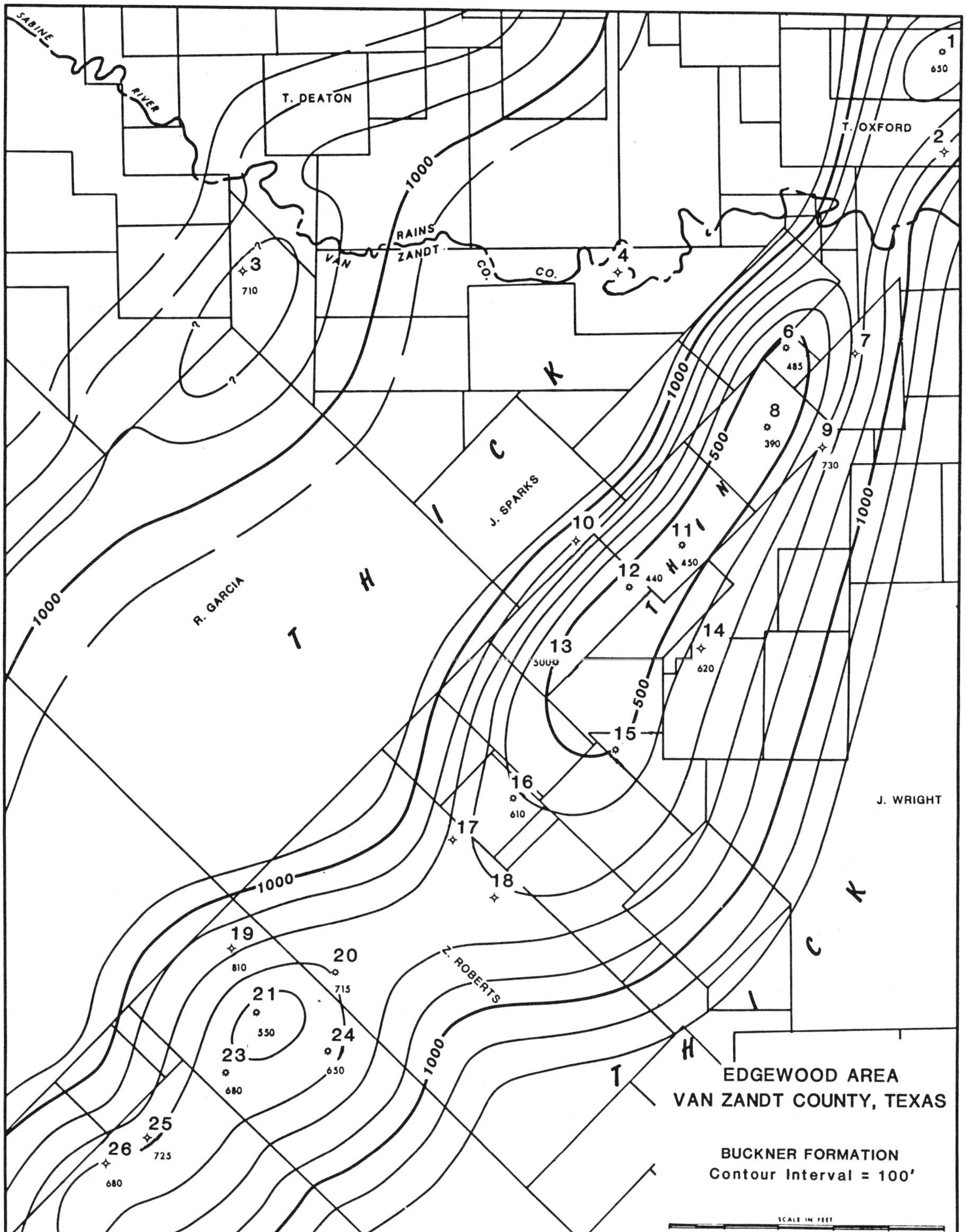

Figure 8. Isochore map, Buckner Formation, contour interval = 100 feet; Edgewood area, Van Zandt County, Texas.

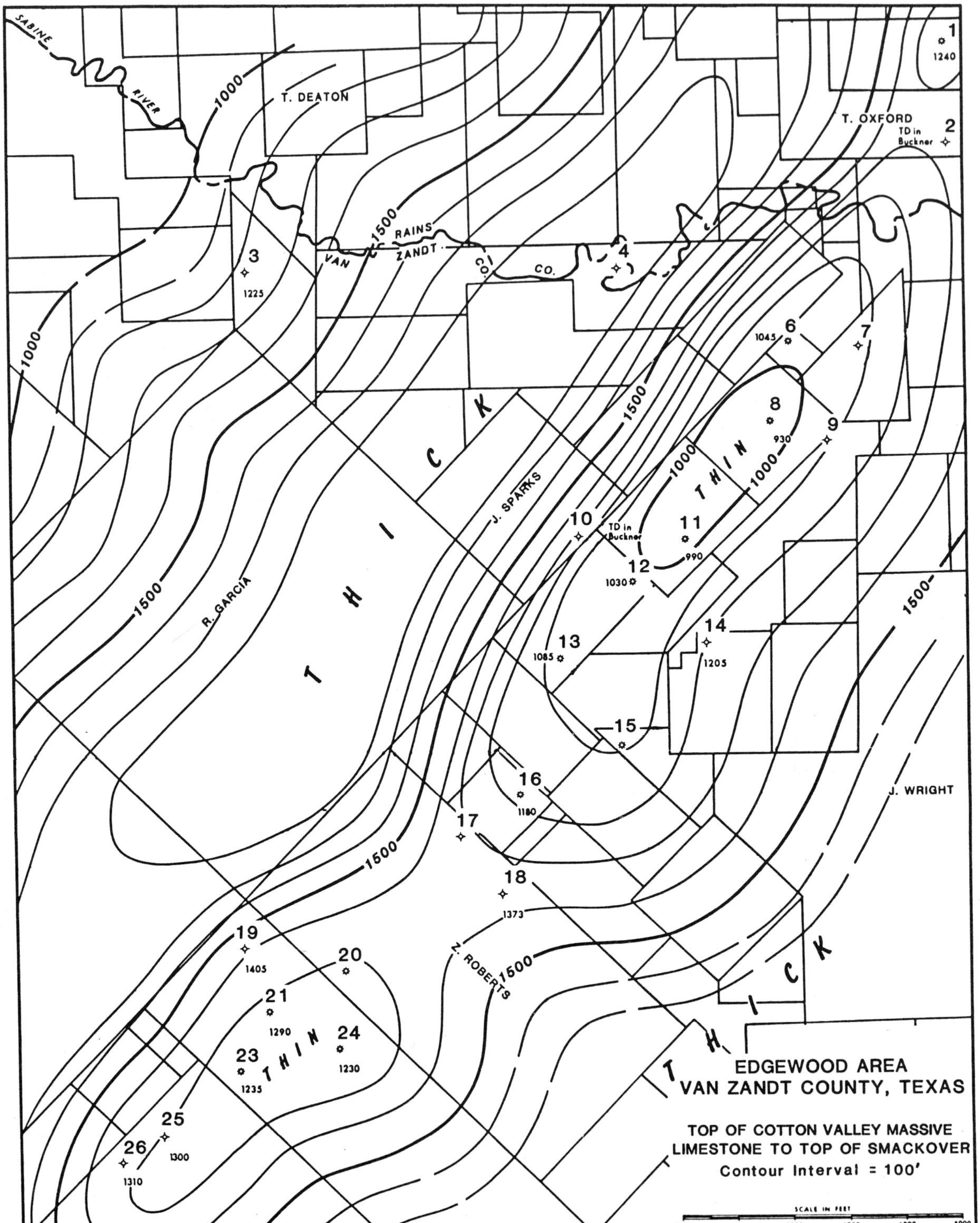

Figure 9. Isochore map, top of Smackover, contour interval = 100 feet; Edgewood area, Van Zandt County, Texas.

to four mm in diameter and are usually spheroidal in shape; the pisoliths are approximately two to three mm in diameter and are spherical. This facies is concentrated into layers 0.5 to 2.0 feet thick, and is evidence of minor bottom agitation because such grains can only be cemented together where there is an intermittent lack of grain movement (Illing, 1954; Purdy, 1963). The abundance of ooids and the absence of lime mud suggest a certain amount of current induced agitation, but during periods of decreased grain movement the grapestones formed. Therefore, this facies represents a slightly lower-energy subtidal environment than the oograinstone facies.

Oobiopelgrainstone Facies

In this facies, ooids constitute 40 to 50 percent, bioclasts 30 to 40 percent and peloids approximately 10 percent of the allochems (Figs. 12d and 13c). Minor amounts of grapestones and algal-coated grains are present. Peloids are spherical to ellipsoidal in shape and range from 0.2 to 0.5 mm in diameter. The peloids result predominantly from micritization of ooids, and only a small proportion represents fecal matter. The most common bioclasts are fragments of pelecypod and brachiopod shells and occasionally gastropod shells. It is possible that the ooids were transported into this facies belt during storms. If so, this facies was deposited in a slightly lower-energy, subtidal environment than the oograinstone facies, with intermittent periods of decreased bottom agitation.

Oobiopelwackestsone-Packstone and Biopelwackestone-Packstone Facies

A predominance of carbonate mud is the primary characteristic of this facies, and peloids, bioclasts, and ooids are the dominant allochems (Fig. 13d and 14d). In addition, silt-sized quartz grains are common and algal-coated grains occur in some intervals. The ooids which occur were transported from neighboring higher-energy areas into this environment during storms. The abundance of carbonate mud suggests that this facies was deposited in a relatively low-energy subtidal environment and was affected only by occasional storms or unusually high tides.

Quartzose Sandstone Facies

This sandstone facies is common at the top of the Smackover Formation and in the lower part of the Buckner Formation. It is composed of well-sorted, subrounded, quartz grains which range in size from 0.1 to 0.3 mm (Fig. 14b). These grains occur in a lime-mud matrix which was replaced by finely crystalline dolomite. Replacement anhydrite nodules also are common, and some ooids occur. This facies probably was deposited on a tidal flat or in shallow water as a result of increased clastic supply or lowered sea level.

Anhydrite Facies

Even though the anhydrite facies is diagenetic in origin, it is described briefly in this chapter to complete the facies discussion.

The anhydrite facies consists of nodular mosaics of felted-lath anhydrite, and fine dolostones containing quartz grains and occasionally ooids (Fig. 14c). The anhydrite occurs at the top of the Smackover and in the lower Buckner in units less than three feet thick. These thin anhydrite units may represent sabkha deposits, but evidence for this interpretation is not conclusive. This type of anhydrite should not be confused with the thicker, subtidal anhydrite units described in the following section.

FACIES RELATIONSHIPS AND DEPOSITIONAL MODEL

It was not possible to delineate the lateral facies relationships because data were not available that would permit a comparison of equivalent intervals in the Smackover Formation throughout a relatively large area. All facies changes observed are vertical, based on core and cuttings evaluations. In some areas, lateral facies relationships are

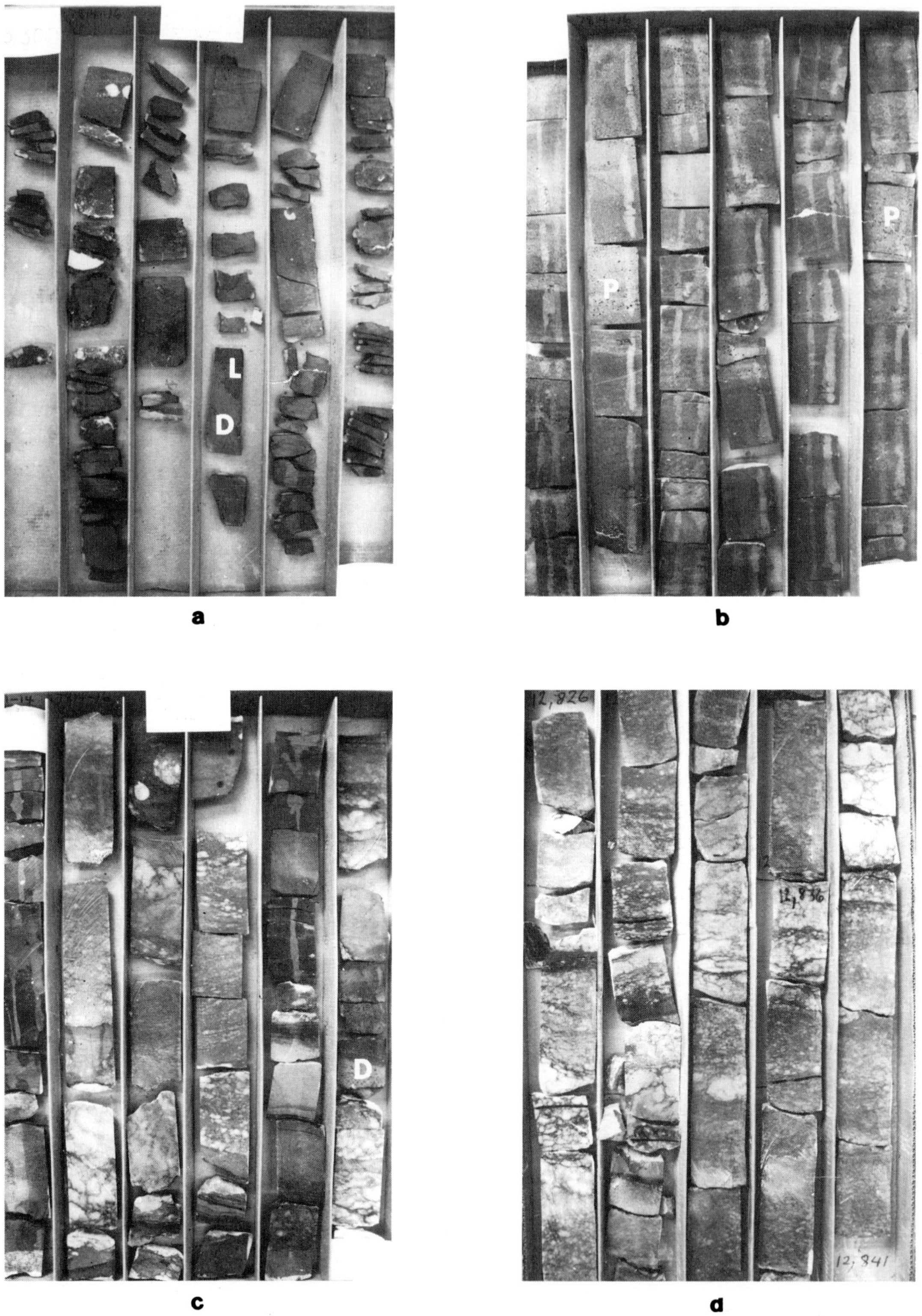

Figure 10

inferred, based on vertical relationships.

The upper Smackover at the Edgewood fields is dominantly a shallow-water marine, subtidal deposit, containing thin intertidal and supratidal intervals at the top and is capped by the thick supratidal-clastic Buckner Formation. The subtidal facies presented previously in this section represent a broad spectrum of wave- and current-energy levels. In order of increasing energy they are: the oobiopel- and biopelwackestone-packstone facies, oobiopelgrainstone facies, oopisograpestone grainstone facies, and oograinstone facies.

Within the upper Smackover at Edgewood, the vertical distribution of facies indicates a general trend of increasing energy upward. During Smackover time, the depositional strike was approximately parallel to the present-day Gulf Coast shoreline (Bishop, 1968). General facies belts probably developed parallel to depositional strike in a northeast-southwest alignment. From northwest to southeast, high- to low-energy environments should have existed due to the change from shallow to deep water. Grainstones were deposited toward the northwest while wackestones were deposited toward the southeast. Early in upper Smackover deposition, dominantly wackestones were deposited across the study area in water depths below the normal wave base. During most of the upper Smackover deposition at Edgewood, grainstones were deposited under moderate- to relatively high-energy levels. Minor sea-level fluctuations and sediment progradations caused some interbedding of wackestones and packstones with grainstones. It is possible that local topographic highs caused by salt tectonism affected the local facies distribution. Near the end of Smackover time, quartz sandstones, wackestones, and thin anhydrite intervals were interbedded with the grainstones. Smackover sedimentation closed with the progradation of the Buckner anhydrite-clastic sequence across the area.

DIAGENESIS

The upper Smackover at Edgewood was composed primarily of oograinstones containing various amounts of peloids, bioclasts, pisoliths, and grapestones. Probably most grains consisted of metastable polymorphs of aragonite and magnesium calcite (Stehli and Hower, 1961). Initial depositional porosities ranged from 26 to 37 percent for a well-sorted oograinstone (Graton and Fraser, 1935), and up to 70 percent for a lime mud-rich sediment (Bathurst, 1971). Diagenesis of the upper Smackover obliterated some of the original fabrics, changed the mineralogic composition of the sediments, and was largely responsible for the occurrence and present distribution of porosity.

Lateral facies relationships in the Smackover Formation are inferred from the observed vertical relationships. The term proximal will be used to indicate that the average shoreline was relatively close to the Edgewood area, and the term distal will refer

Figure 10. Core-slab photographs illustrating facies and porosity relationships in the Smackover of Edgewood Fields.

a. Top of middle member of Smackover Formation, showing subtidal wackestones and packstones. A contact between limestone (L) and dolostone (D) is shown. Isolated nodules of replacement anhydrite (A) are present. Amoco No. 3 Nichols Well, 13,300-13,232 feet.

b. Upper Smackover dolostone, showing alternating porous (P) and nonporous zones. The carbonates predominantly consist of oobiograinstones which were deposited in relatively high-energy subtidal environments. Pan Am No. 1 Forbes Well, 12,823-12,844 feet.

c. Upper Smackover to Buckner transition zone containing alternating porous dolostone (D), nodular anhydrite and dense nodular mosaics of anhydrite. Pan Am. No. 1 Forbes Well, 12,760 -12,778 feet.

d. Lower Buckner Formation, consisting of silty dolostone and mosaics and dense mosaics of anhydrite. Pan Am No. 1 Nichols Well, 12,826-12,841 feet.

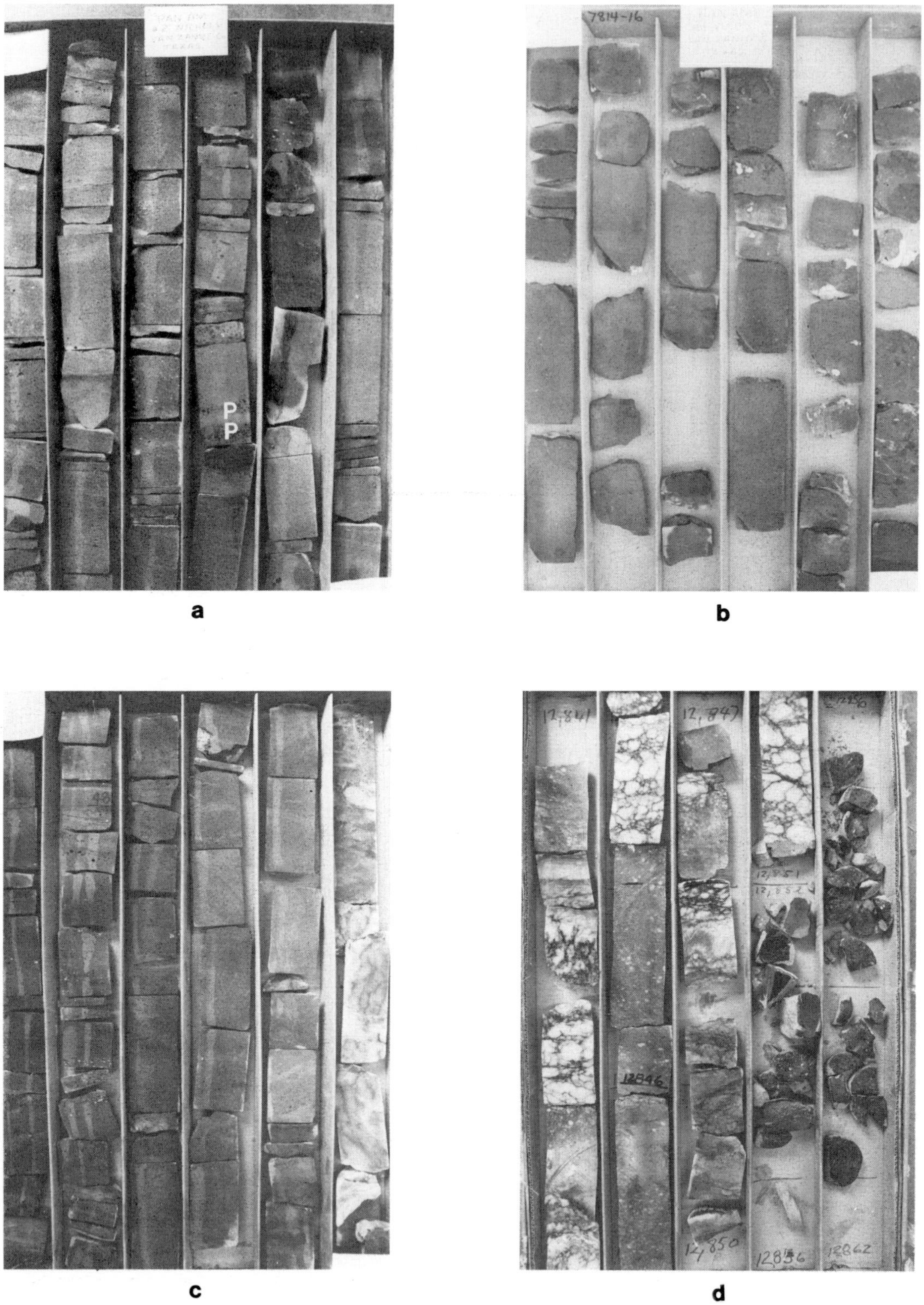

Figure 11

to times when the average shoreline was farther northwest of the study area.

Early Diagenesis in the Marine Environment

Early micritization is common in upper Smackover sediments. Frequently, centripetal micritization only affected the outer surfaces of grains (Figs. 12a and 12b). If micritization went to completion, whole allochems were converted to structureless grains and peloids (Fig. 13d). Bathurst (1966) explained how blue-green algae bore into grains and the bores become filled with micritic carbonate cement. Jacka (1978) suggested that peloids, by virture of their micritization, are highly resistant to later diagenetic alteration.

A few examples of early submarine cementation in the upper Smackover at Edgewood were noted (Figs. 13b and 13c). In Figure 13c, polygonal structure is preserved where cement films thicken toward pore throats. In several samples, marine-cement rims were neomorphically dolomitized (Fig. 13b).

Sabkha Diagenesis

The discovery in the mid 1960's that anhydrite occurs within marginal sediments of the Persian Gulf and the synthesis of this discovery into the sabkha model of evaporite accumulation and diagensis (Shearman, 1963, 1966; Kinsman, 1966, 1969; Butler, 1969) were important developments in understanding carbonate diagenesis. Shearman (1963, 1966) described the formation of sulfate deposition along the Trucial Coast of the Persian Gulf. Here, he found an abundance of nodular anhydrite, gypsum, and dolomite in the sabkha environment. Apparently, some units at the top of the Smackover and in the lower Buckner may represent sabkha deposits. They are usually less than three feet thick and are characterized by nodular to dense nodular mosaics of fine lath-crystal anhydrite, finely crystalline dolostones, variable amounts of quartz grains, and occasional quartz sandstones (Fig. 14b). Anhydrite nodules range in size from 0.1 to 3.5 inches and dolomite crystals range from 1 to 30 microns. Evidence indicates that the dense nodular anhydrite mosaics (chicken-wire anhydrite) were produced predominantly by replacement of dolostone and sandstone or sandy dolostone (Fig. 16b), rather than by the displacive growth of anhydrite. Corroded dolomite rhombs commonly occur in the anhydrite nodules (Figs. 15c, 15d, 16c, and 17a), and corroded quartz grains occur occasionally.

It cannot be stated with certainty that these upper Smackover intervals were deposited in sabkha environments. Butler (1969) suggested that dolomite replaced aragonite when gypsum precipitation raised the Mg/Ca ratios to more than a value of 35, from normal marine values of approximately three, and also showed that anhydrite replaced gypsum landward of the zone of gypsum precipitation. Within the upper Smackover in the Edgewood area, no gypsum or gypsum morphologies were preserved, and it cannot be proven that gypsum was present. In addition, the Smackover Formation lacks characteristics that are com-

Figure 11. Core-slab photographs illustrating lithofacies and porosity in the upper Smackover of Edgewood Fields.

a. Upper Smackover core showing thin pisolith (P) and grapestone intervals and typical upper Smackover dolostone. Pan Am No. 2 Nichols Well, 12,895-12,919 feet.

b. Isolated nodules of anhydrite in the upper Smackover appear to have nucleated in organic-rich burrows and subsequently to have replaced outward, invading matrix of surrounding subtidally deposited carbonate. Pan Am No. 1 Vines Well, 12,841-12,873 feet.

c. Massive replacement of subtidal intervals by sparry and lath-crystal anhydrite in core box on right. Anhydritized dolostones consist of oolitic grainstones and packstones. Pan Am No. 1 Forbes Well, 12,844-12,866 feet.

d. Nodular and dense nodular mosaics of lath-crystal anhydrite replacing upper Smackover dolostone intervals. Pan Am No. 1 Nichols Well, 12,841-12,862 feet.

mon to modern sabkhas, such as pseudomorphs of gypsum crystals, enteroliths, mud-crack casts, flat-pebble conglomerates, stromatolites, and fenestral cavities. However, it appears that these intervals were deposited in proximal environments because of the relatively high percentage of clastic quartz grains and the relatively small size of dolomite crystals.

Dolomitization and Dolomite Cementation

Three basic characteristics were used in describing Smackover dolomite. First, dolomitization may produce neomorphic or paramorphic fabrics (Jacka, 1975); second, neomorphic dolomitization produces rhombic crystal fabrics; and third, paramorphic dolomitization preserves the original calcite and magnesium-calcite crystal fabrics (Jacka, 1975). Jacka (1975) postulated that neomorphic dolomite forms predominantly as a replacement of aragonite. Second, neomorphic fabrics of dolostones may be classified using terminology developed by Friedman (1965), who proposed the terms idiotopic, hypidiotopic, and xenotopic to describe euhedral, subhedral, and anhedral crystal fabrics, respectively. Neomorphic replacement of original lime mud initially produces idiotopic fabrics of loosely packed euhedral crystals with high intercrystalline porosities. Continued flux of Mg-rich brines forms hypidiotopic fabrics of subhedral crystals, which were produced by epitaxial cementation on earlier formed rhombs. Third, dolomite crystal sizes were observed throughout the Smackover section. Smaller dolomite crystals are formed from supersaturated fluids (Berner, 1971) and may be indicative of proximal environments which are closer to the origin of dolomitizing fluids than are the distal environments. Large dolomite crystals form from fluids with lower concentrations of magnesium (Berner, 1971) which may occur far from the origin of dolomitizing fluids as a result of magnesium depletion.

Most Smackover dolomite is neomorphic, indicating that a large amount of Smackover carbonate was deposited as aragonite and re-

Key to Photomicrograph Captions
Figures 12–19

All photomicrographs were taken with a reticle which imprints a scale across the center of the field. The equivalency in mm is given for each smallest division of the scale, d, which ranges from 0.010 mm to 0.110 mm. Those micrographs taken in plane transmitted light are indicated by U, while those taken under crossed polarizers are indicated by X.

Figure 12. Photomicrographs illustrating facies and diagenesis of the upper Smackover.

a. Oograinstone facies. Ooids have been dolomitized neomorphically to a hypidiotopic fabric. Some partial oomolds have been occluded by calcite (light gray) while others remain open (white). Micrite envelopes were dolomitized more finely than ooid cores. Pan Am No. 1 Forbes Well, 12,795 feet, d=0.110 mm, U.

b. Oograinstone - micrite envelopes were dolomitized more finely than ooid cores. Some partial oomolds remain open (white). Pan Am No. 1 Nichols Well, 12,900 feet, d=0.050 mm, U.

c. Oopisograpestone grainstone facies. Dolomite was emplaced as cement in intergranular voids and as partial replacement of grains which now consist of calcite (dark gray). Doubly-terminated quartz porphyroblasts (white) were emplaced within pisoliths. Pan Am No. 2 Nichols Well,12,908 feet, d=0.110 mm, U.

d. Oobiopelgrapestone grainstone. Dolomite occurs as cement in intergranular voids and as partial replacement of grains which now consist of calcite. Pan Am No. 2 Well, 12,923.2 feet, d=0.110 mm, U.

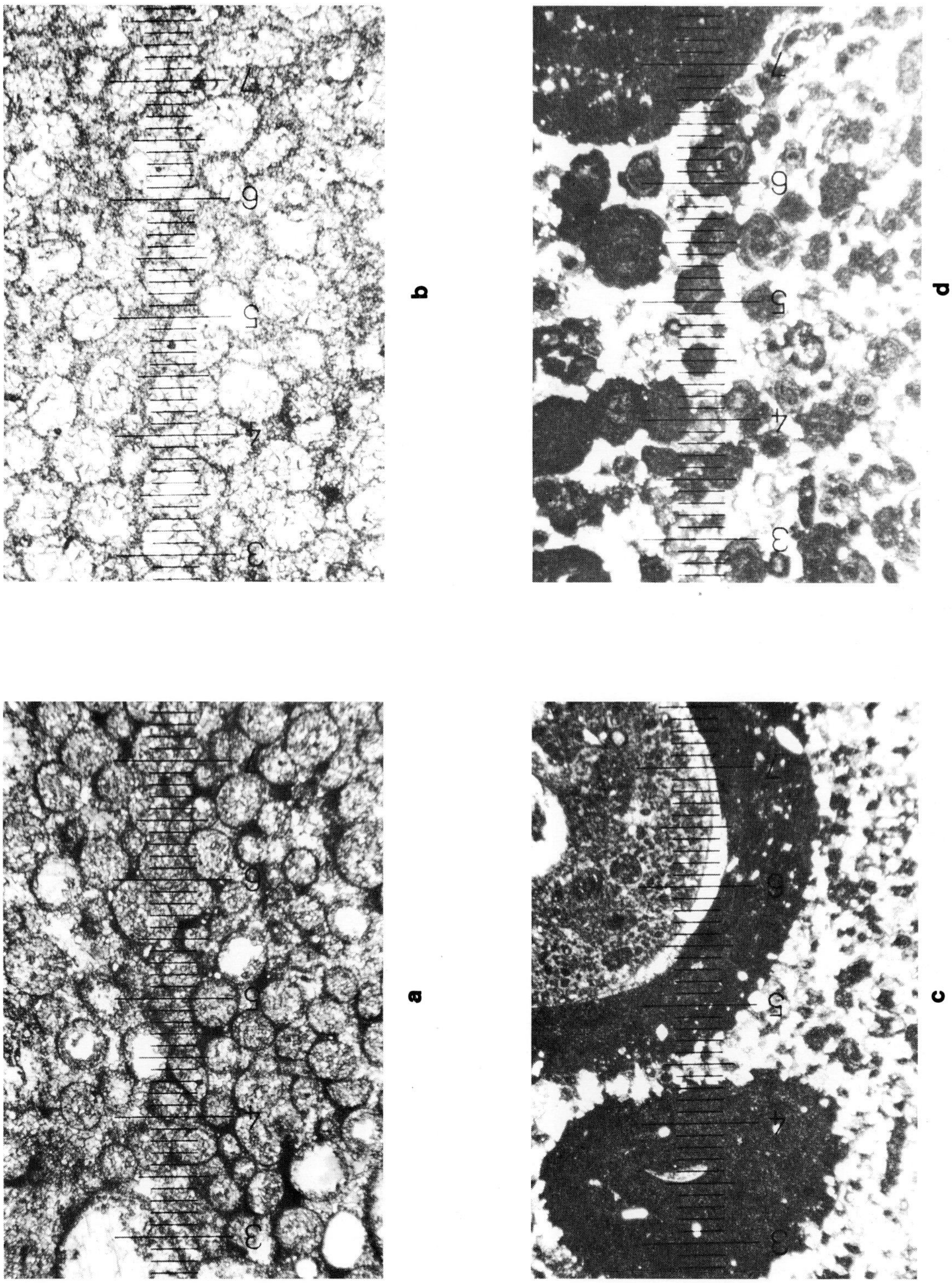

Figure 12

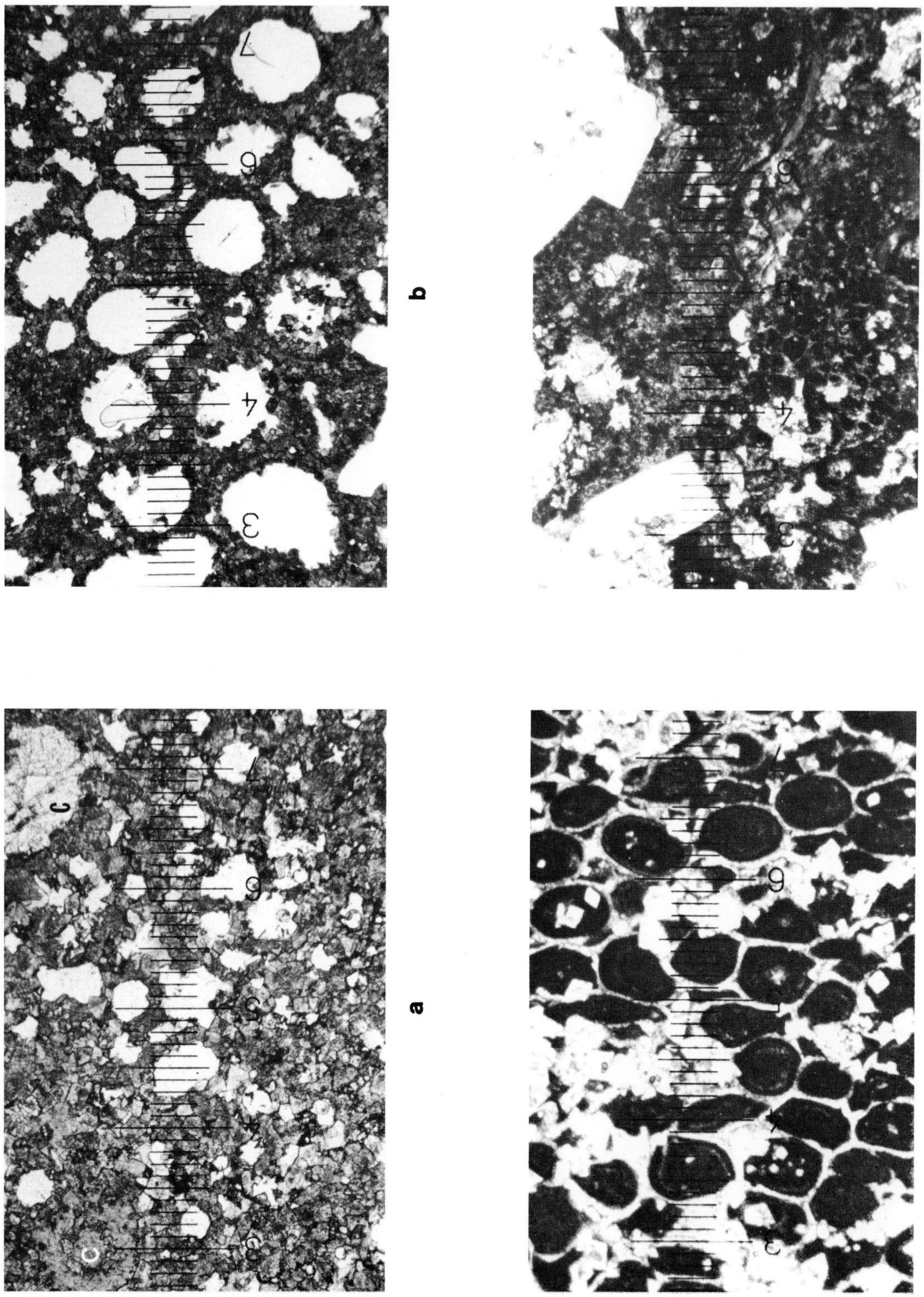

Figure 13

placed by dolomite before the aragonite stabilized to calcite in a meteoric environment.

Dolomitization of subtidal sediments produced idiotopic to hypidiotopic dolostones, which are composed of rhombic crystals ranging in size from 50 to 200 microns and which make up 75 to 85 percent of the upper Smackover at Edgewood. Dolomitization frequently has partially to completely destroyed the original rock fabric (Fig. 13A).

Much dolomite from the Smackover Formation was precipitated as cement, and is characterized by clear rhombs (Figs. 12d, 13c, and 15a). Epitaxial dolomite rims are common in intercrystalline pores.

Many micrite envelopes were neomorphically dolomitized with fine, (one to five microns) hypidiotopic dolomite, while the allochems were dolomitized with coarse, 40 to 50 microns idiotropic, crystals (Figs. 12a and 12b).

Neopmorphic dolomite also occurs in oobiopel- and oopisograpestone grainstones which now represent dolomitic limestones and limy dolostones. In these cases, dolomite initially was precipitated as cement and then began to replace aragonitic allochems, the remainder of which later stabilized to calcite (Figs. 12c and 12d). Scattered dolomite rhombs, which range in size from 100 to 200 microns, have replaced the original submarine cement and ooids.

Paramorphic dolomite is rare in the upper Smackover and the predominant example consists of monocrystalline echinoderm components.

Evidence for flux of second generation, Mg-rich, potentially-dolomitizing fluid was observed in the Upper Smackover in the form of minor amounts of late dolomite cement. In some instances, clear rims of cement occur on corroded dolomite crystals (Fig. 14d), and, in some intervals, oomolds contain dolomite cement crystals (Figs. 15d). These crystals have undulose extinction, and only partially fill the voids.

Sulfate Emplacement

Within the upper Smackover, anhydrite comprises approximately five percent of the total rock, but in some zones the content of anhydrite exceeds 95 percent. Anhydrite has been emplaced in all facies, depositional environments, and rock types. Anhydrite occurs in the following modes and morphologies within the upper Smackover: (1). felted-lath anhydrite, which occurs as scattered, replacive nodules and dense nodular mosaics or massively replaced layers, which formed by the coalescence of nodules (Figs. 10a, 10c, 10d, 11b, 11c, 1d, 14c, 15c, 16a, 16b, and 16c). (2). Sparry or felted-lath anhydrite which was precipitated as cement in secondary molds, intercrystalline voids, and fractures (Figs. 17d and 18a). (3). Anhydrite which initially was precipitated as cement, subsequently replaced outwardly, and invaded surrounding matrix (Figs. 16a). (4). Replacive sparry anhydrite which was emplaced as scattered porphyroblasts with rectangular stairstep outlines (Figs. 13d, 17b, and 17c) and

Figure 13. Photomicrographs illustrating facies and diagenesis of upper Smackover.

a. Neomorphic dolomitization has obliterated oograinstone fabric almost completely. Some partial oomolds and intergranular and intercrystalline voids within dolomitized ooids remain open (white), and some have been occluded by calcite cement (C). Pan Am No. 1 Forbes Well, 12,766.5 feet, d=0.050 mm, U.

b. Much oomooldic porosity has been preserved in oograinstone. Rims around oomolds may represent dolomitized marine cement. Pan Am No. 1 Forbes Well, 12,825.8 feet, d=0.110mm, U.

c. Submarine cement fills intergranular voids of an oobiopelgrainstone. Neomorphic dolomite partially replaced some grains and cement. Cement rims thicken toward pore centers to form polygonal patterns. Pan Am No. 2 Nichols Well, 12,910 feet, d=0.050 mm, U.

d. Oobiopelpackstone facies was micritized, dolomitized partially and replaced partially by prophyroblasts of sparry anhydrite (white). Pan Am No. 1A Parker Well, 12,904 feet, d=0.110 mm, U.

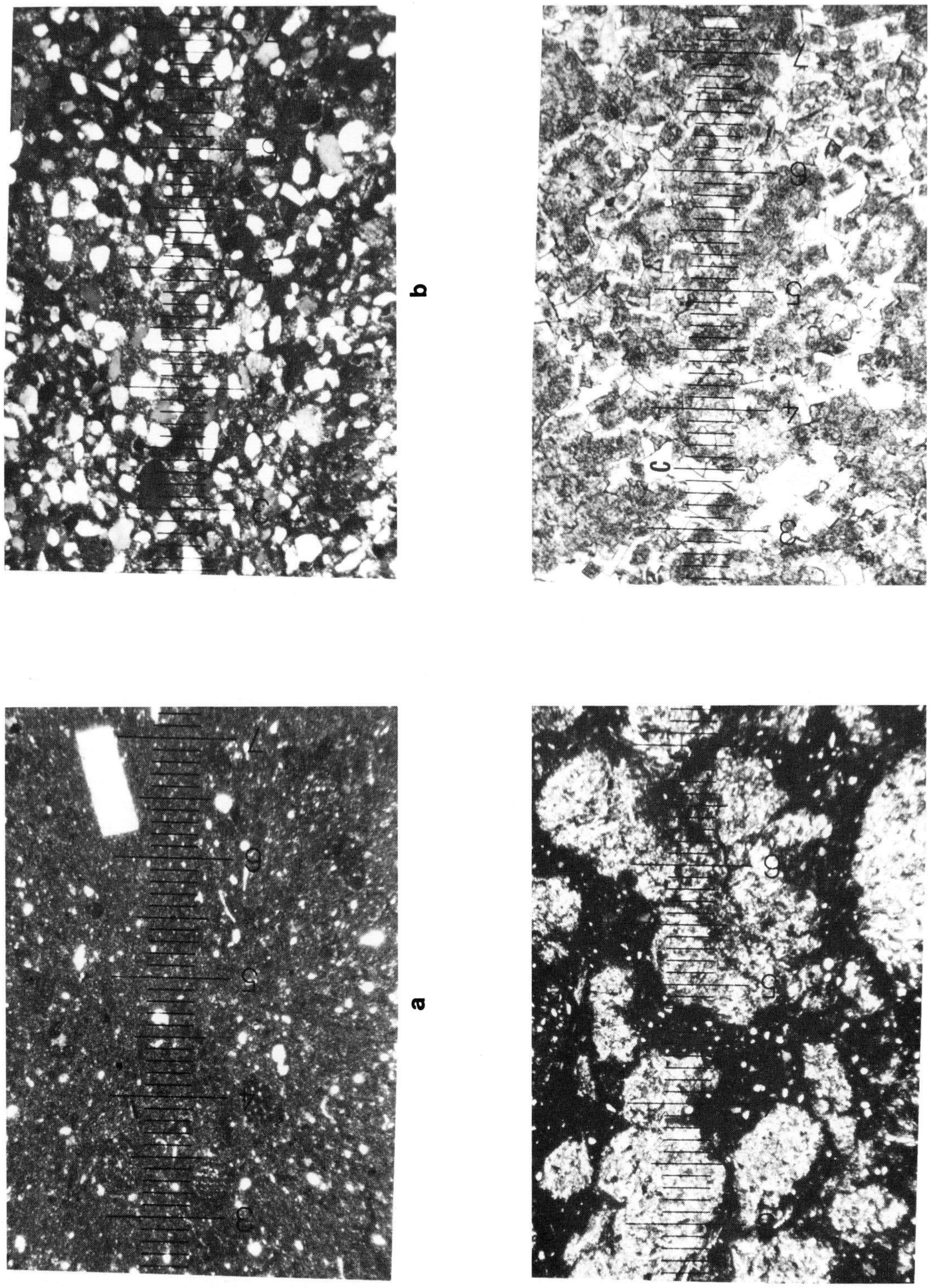

Figure 14

rarely coalesced to form massively replaced layers. Sparry anhydrite also occurs as replacement of ooids and calcite shells (Figs. 18b and 18c). In many dolomitic limestones and limy dolostones, sparry anhydrite tends to replace calcite selectively and may engulf dolomite (Figs. 17b, 17c, 18b, and 18c).

Most Smackover anhydrite was emplaced in subtidal deposits after deposition of sediment and subsequent dolomitization. In most instances, nodules of felted-lath anhydrite were observed to replace dolostones (Figs. 10a, 10c, 10d, 11b, 11c, 11d, 14c, 15c, 16a, 16b, and 16c). In dolomitic limestones, rectangular porphyroblasts one to three mm long selectively replaced calcitic matrix and allochems first and dolomite second (Figs. 17b and 17c). In some instances, lath-crystal anhydrite has nucleated in and selectively replaced burrows (Fig. 11b)

Celestite ($SrSO_4$) and barite ($BaSO_4$) are present in small amounts in the upper Smackover. They are always associated with anhydrite and occur in the same modes and morphologies as anhydrite. These minerals are distinguished by their relatively low birefringence and morphologies which are similar to anhydrite (Figs. 18a). Celestite and barite can be discriminated only by x-ray or SEM-EDAX analyses.

Early Calcitization

Within the upper Smackover at Edgewood, only isolated cases were found where the rocks displayed early stabilization in a meteoric environment. In these cases, predominantly aragonitic oopelpisograpestone grainstones were converted to calcitic grainstones. Furthermore, these calcitized grainstones were much less susceptible to dolomitization than aragonitic materials, and only scattered replacement dolomite rhombs are present (Figs. 13c), which probably were emplaced prior to calcitization. In some cases, dissolution of sulfates in other intervals may have occurred during these times of fresh-water influx.

Late Calcitization and Dedolomitization

A second generation of calcite emplacement caused partial dedolomitization of some dolostones. Dedolomitization occurs when dolomite loses Mg^{2+} ions and converts to calcite, but retains its rhombic fabric. These dedolomites usually retained their intercrystalline porosity, and some were redolomitized by a second dolomitizing fluid (Fig. 15b). In addition, a second generation of calcite replaced the sparry or fel-

Figure 14. Photomicrographs illustrating facies and diagenesis in the upper Smackover.

a. Biopelpackstone (limestone) contains scattered quartz grains (light) and a replacement lath of anhydrite (upper right). This facies is common at the base of the upper Smackover. Amoco No. 3 Nichols Well, 13,300 feet, d=0.110 mm, U.

b. Sandy carbonate interval. Sand grains were deposited within a lime-mud matrix which subsequently became dolomitized. Pan Am No. 1 Forbes Well, 12,778 feet, d=0.050 mm, U.

c. Possible sabkha deposit in which nodules of anhydrite were emplaced within a finely crystalline dolostone containing scattered quartz grains. Pan Am No. 1 Forbes Well, 12,763 feet, d=0.110 mm, X.

d. Oopelgrapestone wackstone. Both matrix and grains were dolomitized. Anhydrite porphyroblasts were emplaced as replacement of grains and matrix and then dissolved. Clear dolomite cement rims then were precipitated on corroded rhombs which appear cloudy because of organic inclusions. Many voids then became filled by calcite cement (c-light gray). Amoco No. 3 Nichols Well, 13,313.2 feet, d=0.050 mm, U.

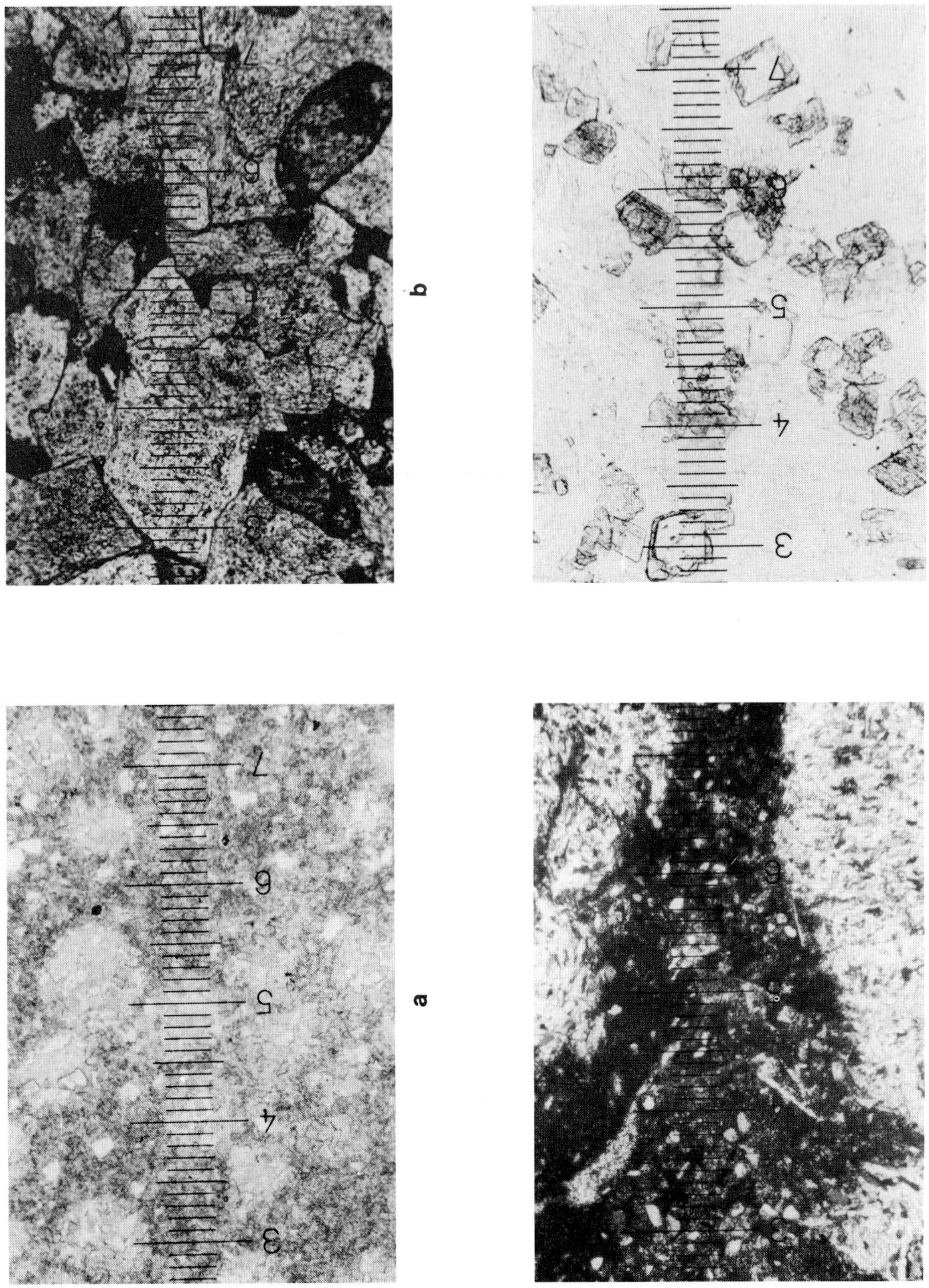

Figure 15

ted-lath anhydrite which was emplaced earlier in the dolostones. Corroded inclusions of anhydrite in the calcite are evidence that calcite replaced the anhydrite (Figs. 18D). Late calcite also occurs as cement in anhydrite molds (Figs. 14d, 15a, 19a, and 19b).

Silica

Silica replacement is uncommon in the upper Smackover at Edgewood. Occasionally, doubly terminated quartz crystals replaced calcitic allochems (Figs. 12c and 19c). The quartz probably was emplaced by a low-salinity fluid during meteoric diagenesis.

Fluorite

Small amounts of fluorite, characterized by isotropic extinction and high relief, are present in many intervals of the Smackover (Figs. 19d). Fluorite generally occurs as a replacement of anhydrite or as a precipitate within anhydrite molds.

POROSITY EVOLUTION

An important aspect of Smackover diagenesis is the formation, occlusion, and preservation of porosity. Primary intergranular porosity is depositionally controlled, whereas secondary and tertiary porosity are developed and occluded by diagenesis. The amount of preserved primary, secondary, or tertiary porosity usually determines the potential quantity of hydrocarbons that will be found in a reservoir rock. The types and relationships of preserved porosity determine the permeability of the rocks, and therefore the accessibility of the hydrocarbons. Percentages of porosity in the Smackover rocks were visually estimated from thin sections, however, it was difficult to visually evaluate the permeability within the Upper Smackover.

Primary Porosity

Primary intergranular porosity in the upper Smackover developed in moderate- to high-energy subtidal environments. Predominantly oolitic grainstones which contain various amounts of pisoliths, peloids, bioclasts and grapestones were deposited with intergranular porosities ranging from 26 to 37 percent (Graton & Fraser, 1935). The scarcity of bioclasts is a major reason why negligible amounts of primary porosity are found within intrabiotic cavities.

Soon after the lime sands of the Smackover were deposited, some primary porosity was occluded by submarine cementation (Figs. 13b and 13c). Nearly all primary porosity in Smackover grainstones has been occluded by

Figure 15. Photomicrographs depicting diagenesis and sulfate emplacement in the upper Smackover.

a. Oopackstone. Dolomitization was accompanied by, or followed by, partial to complete dissolution of many ooids. Coarse dolomite cement then was precipitated within oomolds. Anhydrite then was emplaced as cement and replacement, and centers of many dolomite crystals were replaced selectively. Dissolution of anhydrite created hollow dolomite "pill box" structures (15D). Calcite cement then was precipitated and some dedolomitization took place. Pan Am No. 1 Coleman Parker Well, 12,573 feet, d=0.050 mm, U.

b. Relatively coarse dolomite crystals have redolomitized a more finely crystalline dedolostone. Dead oil is present in intercrystalline pores of dedolostone. Pan Am No. 2 Nichols Well, 12,897.8 feet, d=0.010 mm, U.

c. Nodules of felted-lath anhydrite have replaced a finely crystalline silty dolostone. Many corroded dolomite cyrstals occur within the nodules. Pan Am No. 1 Forbes Well, 12,763 feet, d=0.110 mm, X.

d. Closer view of anhydrite nodule (15c). Corroded remnants of many dolomite crystals can be seen. Anhydrite has replaced cores of many dolomite crystals selectively to form "pill box" structures. Pan Am No. 1 Forbes Well, 12,763 feet, d=0.010 mm, U.

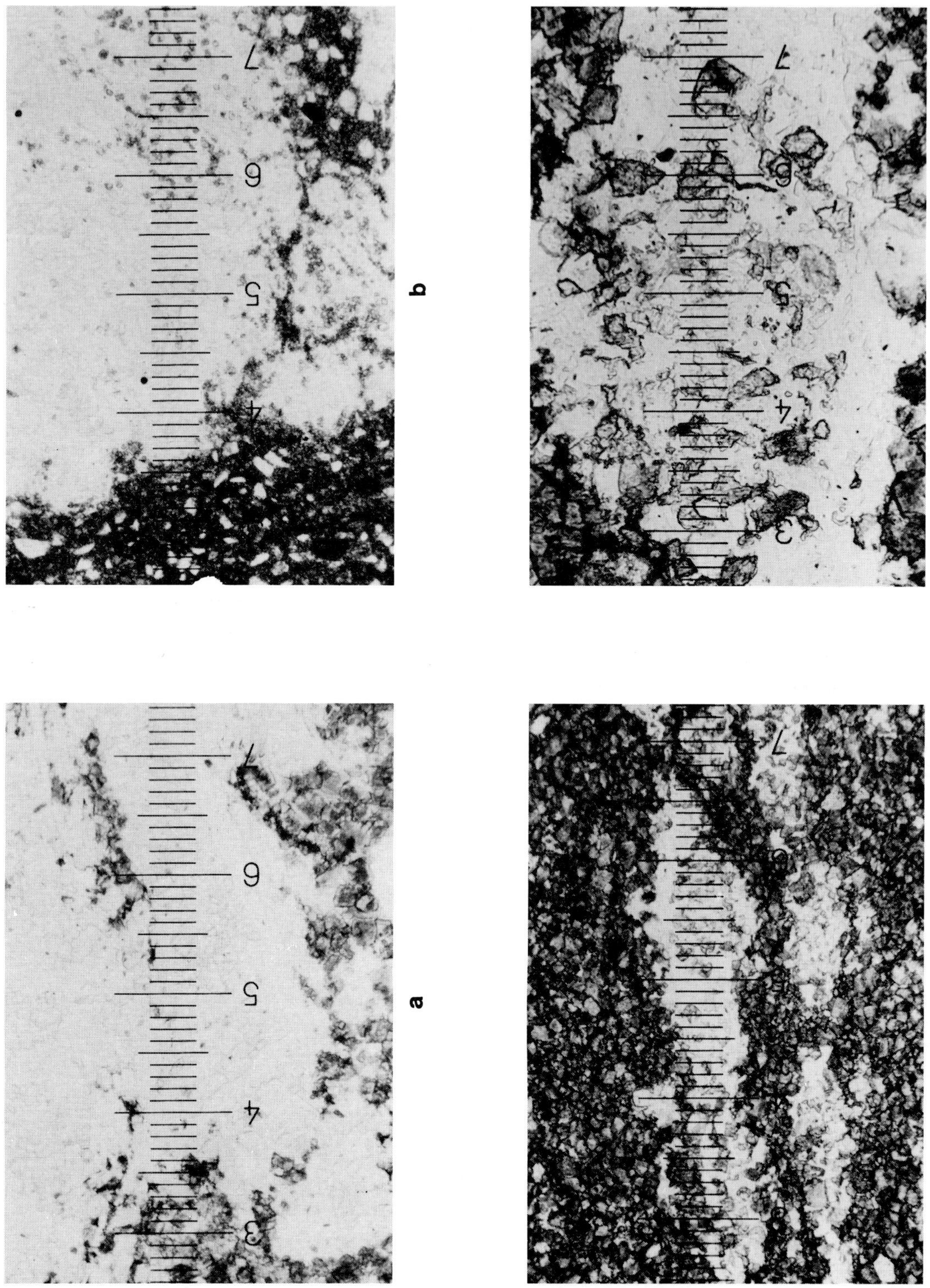

Figure 16

dolomite and calcite cements (Figs. 12a, 12b, 12c, and 12d). In some instances, anhydrite cement occurs in primary voids.

Secondary Porosity

The predominant porosity preserved within the upper Smackover at Edgewood is secondary and consists of intercrystalline voids and oomolds. Dolomitization of aragonitic materials produces a fabric of rhombic crystals which initially do not fit well together; thus pore space remains between the crystals. Most intercrystalline porosity has been preserved within dolomitized ooids and original lime-mud matrix of some wackestones and packstones (Fig. 13a). The intercrystalline porosity in many fine- to medium-crystalline dolostones has been occuluded by anhydrite.

In terms of volume, most of the preserved porosity in the upper Smackover is oomoldic (Figs. 12a, 13a, and 13b). The formation of large amounts of oomoldic porosity is an enigma. Originally, the ooids were composed of relatively unstable aragonite. It is possible that while the outer portions of ooids were being dolomitized, the cores dissolved, creating partial molds. However, evidence indicates that, following dolomitization of outer portions of ooids, ooid cores or the entire ooids were replaced by anhydrite in many cases. Then, dissolution of anhydrite resulted in the formation of partial or complete oomolds. Three observations support this theory. First, examples were observed where anhydrite partially to completely replaced ooids and other allochems following dolomitization (Figs. 18b and 18c). Second, dolomite "pill boxes," are common around the cores of many oomolds, (Fig. 15a) formed after dolomitiztion, when anhydrite replaced the organic-rich dolomite crystal centers (Figs. 15d and 17a). In the Smackover Formation, anhydrite emplaced in dolomite centers was replaced subsequently by calcite or intracrystalline molds were filled by calcite cement (Figs. 19a and 19b). Third, in many instances, rims around an oomold interior are corroded (Figs. 12a, 13a, and 13b). This corrosion suggests that when anhydrite replaced the ooids it may have partially replaced the interiors of the dolomite rims and later was dissolved.

In some cases, oomolds are occluded by lath and sparry anhydrite (Figs. 17d and 18a), calcite (Figs. 12a and 13a), and occasionally by dolomite. Usually sparry calcite occludes the oomoldic porosity. However, there is evidence that much of the calcite occurs as a replacement of anhydrite (Fig. 18d), which is indicated by the corroded inclusions of anhydrite within calcite crystals.

Tertiary Porosity

Tertiary porosity in the Smackover was created through dissolution of anhydrite porphyroblasts and anhydrite which had been emplaced as cement and replacement (Fig. 14d). Most of these anhydrite molds were occluded by dolomite and calcite cements (Figs. 14d).

Dissolution of anhydrite, which had re-

Figure 16. Photomicrographs illustrating modes of sulfate emplacement in the upper Smackover.

a. Oopisopelgrainstone. Felted-lath anhydrite was precipitated as cement in primary voids, and then it replaced surrounding dolomitized grains to form nodules. This core interval can be seen in Fig. 11B. Pan Am No. 1 Forbes Well, 12,862 feet, d=0.050 mm, U.

b. Sandy pelwackestone. Neomorphic dolomitization was followed by emplacement or replacement nodules of felted-lath anhydrite. Both dolomite and sand grains have been replaced by anhydrite. Pan Am No. 1 Forbes Well, 12,763 feet, d=0.050 mm, U.

c. Oopelgrapestone wackestone. After dolomitization, felted-lath anhydrite was emplaced as replacement nodules. Corroded dolomite cyrstals occur within the anhydrite. Pan Am No. 1 Nichols Well, 12,798 feet, d=0.050 mm, U.

d. Closer view of 16c showing corroded dolomite crystals in anhydrite nodule. Pan Am No. 1 Nichols Well, 12,798 feet, d=0.010 mm, U.

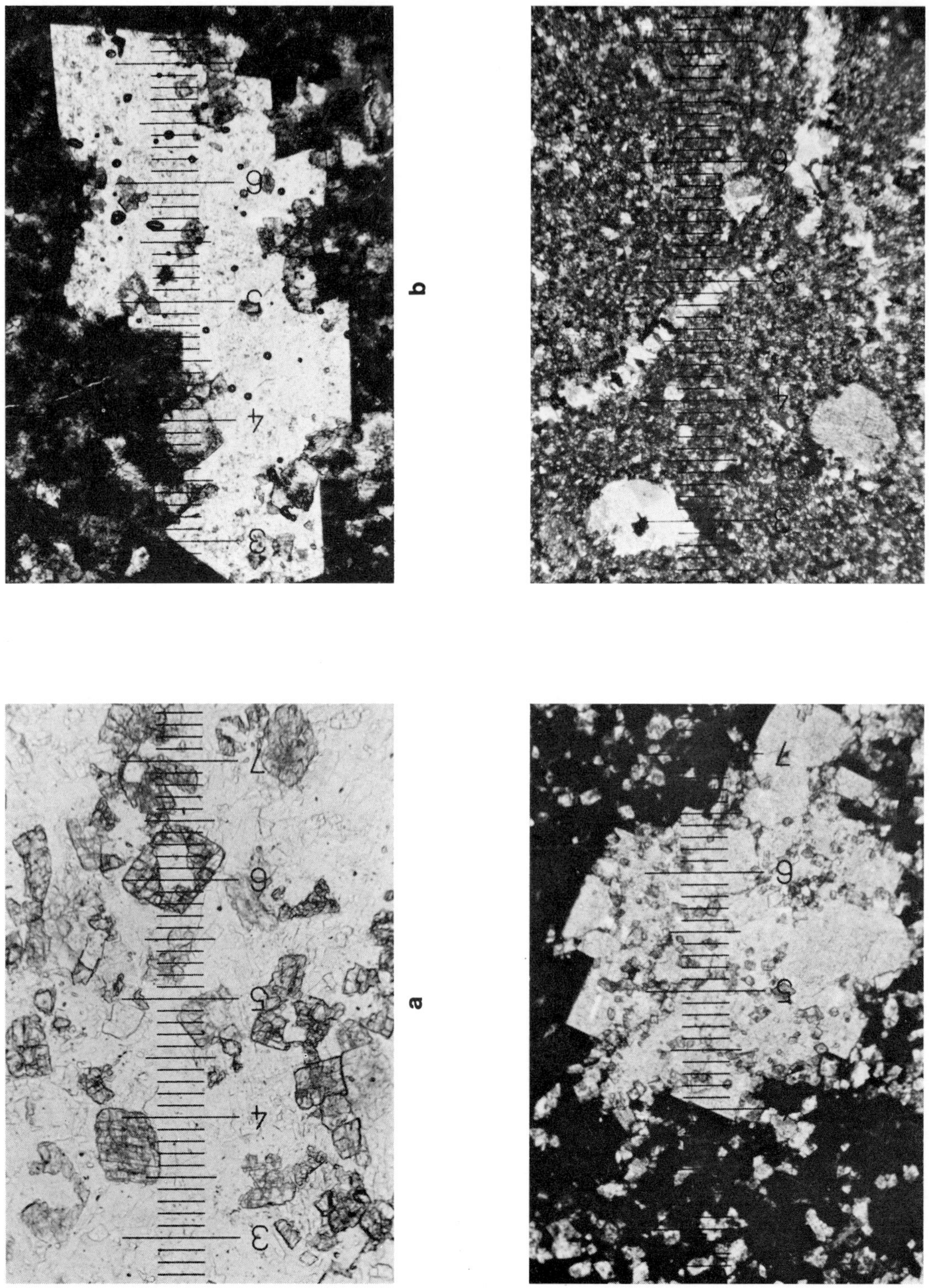

Figure 17

placed organic-rich cores of dolomite crystals, produced "pill box" structures or intracrystalline pores (Figs. 15a, 15d, and 17a) which represent another type of tertiary porosity. Although much intracrystalline porosity once existed in some intervals, nearly all of it has been occluded by calcite cement (Figs. 19a and 19b). The occurrence of calcite cement within intracrystalline voids of dolomite crystals has traditionally been interpreted as an example of dedolomitization. However, it can be seen under crossed polarizers that the calcite cement in the Smackover samples we studied is not syntaxial with the dolomite rims (Fig. 19b) as it would be in the case of dedolomitization.

Porosity Trends

The original primary porosity of the upper Smackover grainstones was controlled by the energy levels within individual environments of deposition. Present porosity values are either 15 to 20 percent or less than three percent and fall short of the original porosity estimates, and preserved primary porosity is almost negligible.

Nearly all of the preserved porosity of Edgewood consists of secondary intercrystalline and oomoldic voids in dolograinstones. The distribution of porous and non-porous intervals seems to be independent of the subtidal facies subdivisions. A single lithofacies frequently exhibits alternating porous and non-porous intervals (Figs. 10, 11, and 14). The highest porosity zones are where oomoldic pores in the dolostones contain minor amounts of cement or lack cement. In many instances, it appears that oomolds are not well connected, and such intervals may have relatively low permeabilities. Zones which contain moderately high intercrystalline porosity probably have the highest permeabilities. High porosity zones are common on top of the Edgewood structure, but they are not limited to the structural highs. Zones of high secondary porosity also are present in dry wells on the flanks of the structure.

PARAGENESIS

Smackover sediments originally were deposited as aragonite and magnesium calcite in predominantly shallow-water marine, subtidal environments which played a dominant role in determining the relative proportions of carbonate mud to allochems, and types of allochems deposited. In high-energy environments, grainstones were deposited, and in low-energy environments wackestones were deposited. Allochems, particularly those deposited in relatively low-energy zones, were centripetally to completely micritized. Submarine cements precipitated on the surface of some of the allochems, so that some or all of the primary porosity was occluded. Evidence is not conclusive as to whether or not some thin zones at the top of the Smackover represent anhydrites and dolostones deposited in a sabkha environment.

Figure 17. Photomicrographs depicting modes of sulfate emplacement within the upper Smackover.
a. Another example of corroded dolomite remnants occurring within a replacement nodule of felted-lath anhydrite. Note that cores of several dolomite crystals have been replaced selectively. Pan Am No. 1 Nichols Well, 12,799 feet, d=0.010 mm, U.
b. A porphyroblast of sparry anhydrite has replaced this biopelgrapestone packstone. The carbonate is a dolomitic limestone. Calcite is more susceptible to replacement by anhydrite than dolomite, and inclusions within anhydrite consist of dolomite. No. 1 J. H. Parker Well, 12,904 feet, d=0.050 feet, U.
c. Rectanguloid prophyroblast of sparry anhydrite has replaced a dolomitic limestone. Again, calcite was more susceptible to replacement by anhydrite and most dolomite cyrstals have been engulfed. Pan Am No. 1 Nichols Well, 12,886 feet, d=0.050 feet mm, U.
d. Sparry and lath crystal anhydrite cement fill oomolds and fracture in a dolostone. Pan Am No. 1 Forbes Well, 12,862 feet, d=0.110 mm, X.

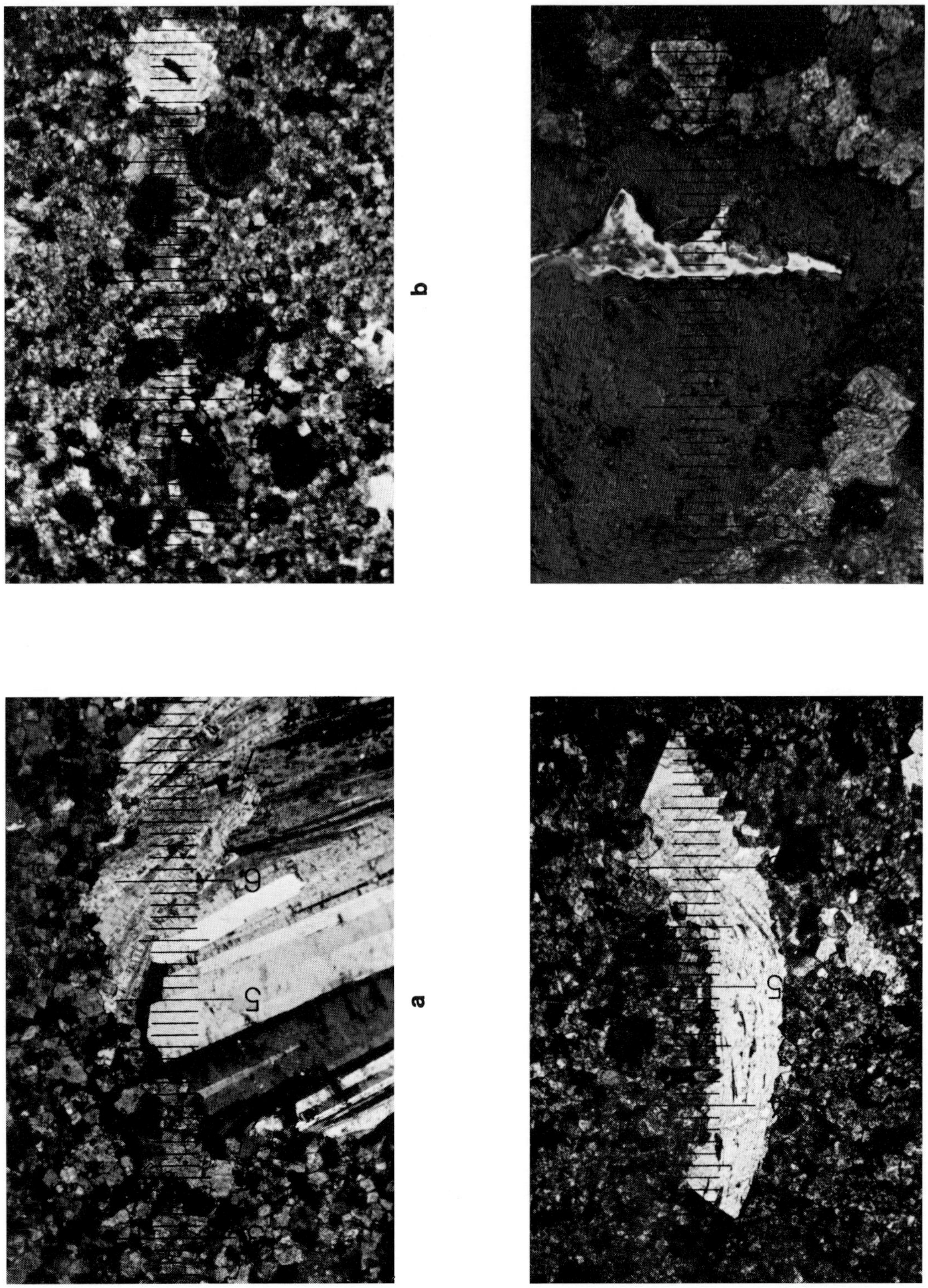

Figure 18

The following evidence indicates that the upper Smackover was affected by a regular sequence of diagenetic events subsequent to burial below the depositional interface: (1) dolomitization (Mg-rich fluid), (2) anhydritization (hypersaline brine), and (3) calcitization (low-salinity fluid).

Subtidal sediments were dolomitized neomorphically, initially producing idiotopic fabrics. Continued exposure to dolomitizing fluids caused epitaxial dolomite cementation and changed idiotopic fabrics into hypidiotopic fabrics, which, in turn, were converted to xenotopic fabrics. Possibly, in extreme distal sediments, magnesium depletion lowered the Mg/Ca ratios substantially, resulting in calcitization of some intervals. However, it is more likely that these sediments were stabilized in fresh-water diagenetic environments.

Following dolomitization, Smackover sediments were subjected to an influx of hypersaline fluids which emplaced anhydrite and small amounts of barite and celestite. In proximal areas, felted-lath anhydrite cemented and massively replaced dolostones. In distal areas, sparry and lath anhydrite were emplaced less extensively as cement, replacement of allochems, and replacement of dolostone. Farther seaward, rectangular replacement porphyroblasts of anhydrite were emplaced in dolomitic limestones. Where calcite and dolomite occur together, anhydrite has selectively replaced the calcite before the dolomite.

Next, the Smackover rocks were exposed to a fluid which was less concentrated than the previous hypersaline fluid. This less-saline fluid caused dissolution of anhydrite, dedolomitization, calcite replacement of anhydrite, calcite cementation, and minor emplacement of silica and fluorite. It is possible that the formation of many oomolds occurred at this time. Subaerial exposure and an increase in precipitation may explain the formation of this relatively less-saline fluid.

The sequence of dolomitization, anhydritization, and calcitization represents one cycle of diagenesis. It appears that much of the Smackover was subjected to more than one cycle of diagenesis. Second-generation dolomite precipitated in oomolds and tertiary anhydrite molds and partially replaced some dedolostones. The presence of anhydrite cement in first generation anhydrite molds represents an example of second-generation anhydrite. No more than five stages of diagenesis could be recognized with confidence in any Smackover interval within the study area.

PETROLEUM GEOLOGY

Four factors are required for the existence of hydrocaron accumulations: a reservoir rock, a trapping mechanism, a source rock, and the proper geologic timing for oil migration. At Edgewood, the reservoir rocks are upper Smackover dolograinstones, which contain 2 to 25 percent oomoldic, intercrystalline, and intergranular porosity. It is

Figure 18. Photomicrographs illustrating anhydrite diagenesis in the upper Smackover.

a. Lath-anhydrite and celestite cement fill an oomold in a dolostone. Pan Am No. 1 Forbes Well, 12,861 feet, d=0.050 mm, X.

b. Sparry anhydrite (white) has replaced a calcitic ooid selectively over the dolostone matrix on right side of micrograph. Pan Am No. 1 Nichols Well, 12,886.5 feet, d=0.050 mm, U.

c. A fibrous calcitic pelecypod shell in a dolostone has been replaced selectively by sparry anhydrite. Some fibers remain as inclusions. Pan Am No. 1 Nichols Well, 12,886.5 feet, d=0.050 mm, U.

d. Sparry anhydrite, which was emplaced as cement in an oomold, subsequently replaced outward into a dolostone. Anhydrite later was replaced by calcite (dark gray) and anhydrite inclusions (light) are present in the calcite. Pan Am No. 1 Forbes Well, 12,783 feet, d=0.110 mm, X.

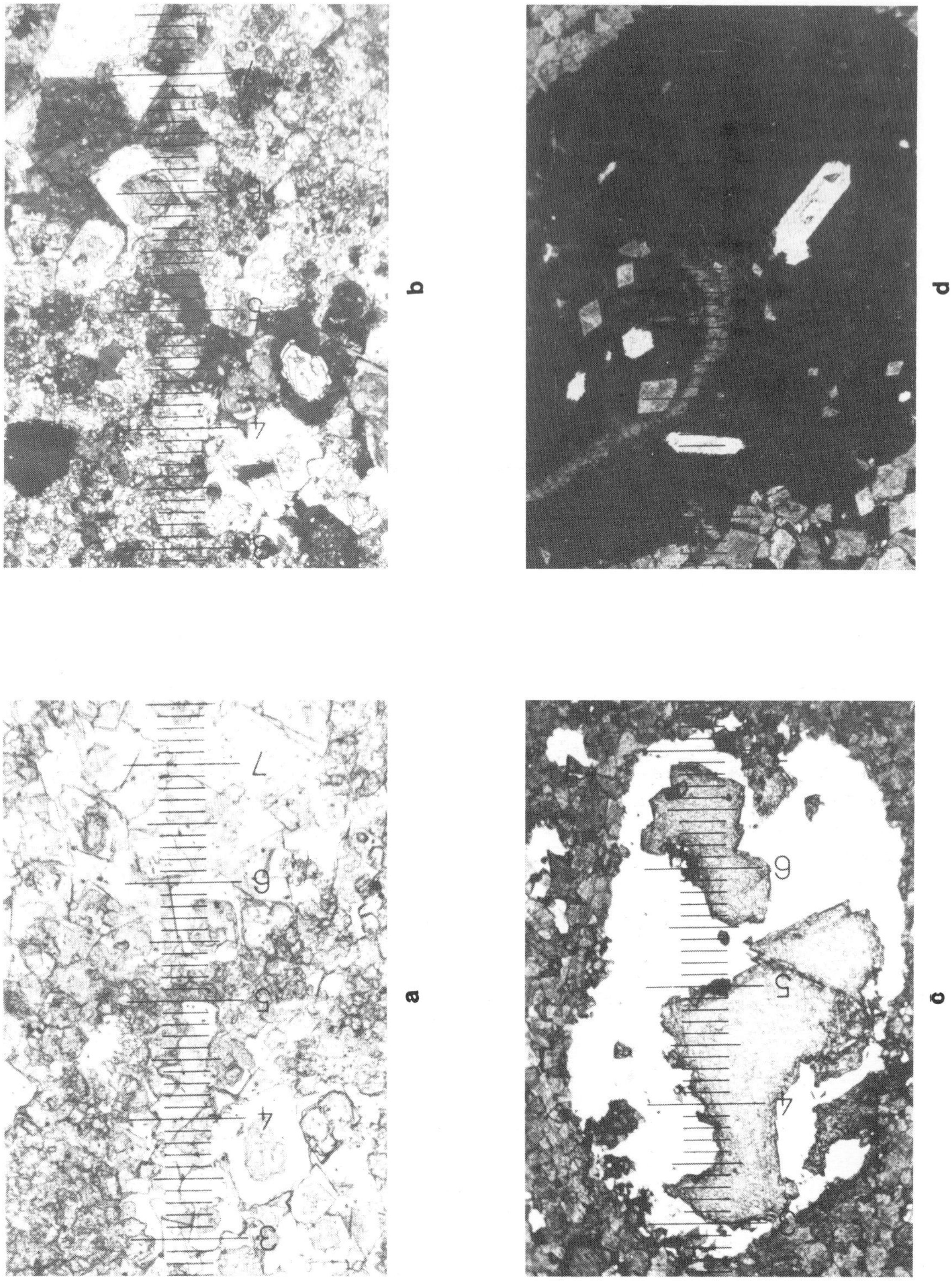

Figure 19

common to see dead oil coating oomolds and intercrystalline pores. However, the hydrocarbons produced at Edgewood were gas and gas condensate. Here, the gas-water contact is at a depth of approximately 12,470 feet subsea (Gratton, 1965). Table 2 summarizes production data for the Edgewood fields.

Edgewood is an example of a classical anticlinal trap (Figs. 6) in which the Buckner Anhydrite provides the seal over the upper Smackover reservoir. The Bossier Shale and the lower lime mudstone member of the Smackover are commonly thought to be the source of Smackover hydrocarbons (Moore and Druckman, 1981).

The fourth requirement, proper timing of hydrocarbon migration, is often the most difficult to evaluate because geologists cannot directly observe hydrocarbon maturation and migration. At Edgewood, it is apparent that the structural closure was present, that the Buckner capped the upper Smackover, and that oomoldic and intercyrstalline pores were open before gas was emplaced in the upper Smackover reservoir. However, the presence of dead oil, which coats some voids, suggests that oil migration occurred prior to the emplacement of gas and the existence of the trap, but after there was effective porosity in the reservoir. Thus, diagenesis was relatively complete at the time of oil and gas migration. In some intervals, dead oil overlies cements in open pores and partially open pores, and in completely cemented voids dead oil generally is absent. Indications are that nearly all cementation and dissolution occurred prior to hydrocarbon emplacement. A few examples were observed where dead oil coated pores that were completely occluded by calcite cement. Two explanations are offered for this. First, it is possible that the emplacement of calcite occurred after moderate burial before, during, and after oil migration. Second, it is possible that the last generation of calcite was emplaced completely prior to hydrocarbon emplacement, but this calcite was not connected tightly to some of the pore walls. During oil migration, oil was forced into some spaces between the pore walls and the calcite cement.

SUMMARY AND CONCLUSIONS

1. The Edgewood fields are located on a large, elongate anticline, probably caused by salt tectonism, in the East Texas Basin.

2. At Edgewood, the Smackover Formation conformably overlies the Norphlet Formation and conformably underlies the Buckner Formation. The Smackover consists of three members; the upper member was the focus of this study. The lower and middle members consist of dense brown to gray, silty limestone which may contain dolomite, ooids, pellets and replacement anhydrite nodules.

3. The upper Smackover at Edgewood is predominantly an upward shoaling, shallow-marine deposit. At the top, thin intertidal and supratidal zones may be inferred.

Figure 19. Photomicrographs illustrating diagenesis in the upper Smackover.

a. Anhydrite selectively replaced cores of many dolomite crystals as shown in Figs. 15D and 17A. Anhydrite later was dissolved and dolomite "pill box" structures were filled by calcite cement (medium gray). Such occurrences of calcite within dolomite cyrstals commonly have been misinterpreted as dedolomitization. Pan Am No. 1 Coleman Parker Well, 12,573 feet, d=0.110 mm, U.

b. Same field of view as 19A, but under crossed polarizers. It can be seen that the calcite cement within the dolomite "pill boxes" is not syntaxial with the dolomite rims as it should be in dedolomite. Pan Am No. 1 Coleman Parker Well, 12,573 feet, d=0.110 mm, X.

c. Fluorite was emplaced in an anhydrite mold either as cement or as replacement of anhydrite. It may have been emplaced during fresh water diagenesis. Pan Am No. 1 Coleman Parker Well, 12,610 feet, d=0.050 mm, U.

d. Quartz porphyroblast, emplaced during fresh-water diagenesis was emplaced in a grapestone grain. Pan Am No. 2 Nichols Well, 12,897 feet, d=).050 mm, X.

4. The predominant facies present in the Smackover is the high-energy oograinstone facies, which contain varying amounts of bioclasts, pisoliths, peloids, and grapestone. Interbedded with thick oograinstone intervals are zones of lower-energy wackestone and packstone. Interbedding of these intervals probably reflects fluctuations of sea level.

5. Certain intervals at the top of the Smackover appear to be similar to modern sabkha deposits, but they lack many of the characteristic features.

Table 2. Production Data

MAP NUMBER	COMPANY & LEASE	SMACKOVER TOP (FEET SUBSEA)	PRODUCTION INTERVALS (FEET SUBSEA)	INITIAL FLOWING POT. (MMCFGPD/BCFMMCFG)
1	ARK LA, No. 1 Brimer	-12,828	------	------
2	HAMON, No. 1 Greene	------	------	------
3	FARMLAND, No. 1 Wingo	-12,248	NO PRODUCTION	
4	HUMBLE, No. 1 Hazel	------	------	------
5	PRESIDIO, No. 1 Cheatham	------	------	------
6	PAN AM, No. 1 Vines	-12,343	-12,345 - - 12,375 -12,395 - - 12,415	11.0/-
7	MCBEE, No. 1 Presley	------	------	------
8	PAN AM, No. 1 Lay	-12,162	-12,167 - - 12,257 -12,357 - - 12,406	6.2/200
9	UNION, No. 1 Albright	-12,660	NO PRODUCTION	
10	RUOMAN, No. 1 Read	------	------	------
11	PAN AM, No. 1 C. Parker	-12,218	------	------
12	PAN AM, No. 1 C. Parker	-12,170	-12,136 - - 12,166 -12,294 - - 12,314	47.0/181
13	PAN AM, No. 1-D. Parker	-12,230	-12,196 - - 12,220	9.2/171
14	CARAWAY, No. 2 Parker	-12,510	NO PRODUCTION	
15	CARAWAY, No. 1 Parker	------	------	------
16	PAN AM, No. 1 Forbes	-12,376	-12,389 - - 12, 423	10.0/-
17	PAN AM, No. 1 Knowles	-12,464	NO PRODUCTION	
18	ARK LA, No. 1 Holland	------	NO PRODUCTION	
19	PAN AM, No. 1 Brown	-12,512	NO PRODUCTION	
20	AMOCO, No. 1 Nichols	-12,447	-12,415 - - 12,425 -12,429 - - 12,443 -12,454 - - 12,512	6.0/
21	PAN AM, No. 1 Nichols	-12,410	-12,340 - - 12,352 -12,412 - - 12,436 -12,460 - - 12,468	5.1/204
22	ARK LA, No. 1 Estes	------	------	------
23	PAN AM, No. 1-A. Parker	-12,369	------	------
24	PAN AM, No. 2 Nichols	-12,430	-12,425 - - 12,445 -12,456 - - 12,459 -12,464 - - 12,479	2.8/-
25	PAN AM, No. 1 Simmons	-12,455	NO PRODUCTION	
26	PAN AM, No. 1 McDonald	-12,488	NO PRODUCTION	
27	SAMEDAN, No. 1 Hamilton	------	NO PRODUCTION	
28	CONOCO, No. 1 Shields	------	------	------

6. In many intervals, early dolomitization of aragonitic sediments obscured original fabrics by the formation of hypidiotopic fabrics.

7. Anhydrite was emplaced in the following modes and morphologies: (a) scattered to dense mosaics of felted-lath anhydrite nodules replaced dolostones or coalesced to form massively replaced layers; (b) rectangular replacement porphyroblasts replaced dolomitic limestones and occasionally coalesced to massively replace entire units, (c) felted-lath and sparry anhydrite were precipitated in primary voids, oomolds, intercrystalline pores, and fractures; and (d) much anhydrite which was precipitated as cement subsequently replaced outward, invading the surrounding matrix and grains.

8. Large quantities of calcite were emplaced as cement, replacement of anhydrite, and dedolomite. During these periods of calcite emplacement, quartz and fluorite probably were also emplaced, and anhydrite was dissolved.

9. Diagenetic sequences in the Smackover Formation can be explained by the following sequence of fluid invasion: (a) Mg-rich fluid causes dolomitization, (b) hypersaline brine emplaces anhydrite and other sulfates, and (c) low-salinity fluid results in sulfate dissolution, calcitization, silica and fluorite emplacement. It appears that some of the upper Smackover was subjected to one full cycle of diagenesis and the first two stages of a second cycle.

10. Most of the preserved porosity at Edgewood is oomoldic and intercrystalline in hypidiotopic dolostones. The formation of large amounts of partial and complete oomolds possibly occurred during dolomitization, but evidence suggests that many were caused by anhydrite replacement and subsequent dissolution. Calcite occludes the most porosity, although occlusion by anhydrite is common in some intervals.

11. The Edgewood fields are examples of a classical anticlinal trap. The Buckner Formation forms a non-porous seal over the Upper Smackover reservoir rocks. Gas and gas condensate were produced from secondary oomolds and intercrystalline voids in dolostones.

ACKNOWLEDGMENTS

Appreciation is extended to Amoco Oil Company for supplying the cores necessary for this study, to Chevron U.S.A., Inc. for providing financial assistance, and to Superior Oil Company for their contribution of well logs, base maps, and drafting. This paper is adapted from an M.S. thesis completed at Texas Tech University, August 1982.

REFERENCES

Bathhurst, R.G.C., 1966, Boring algae, micrite envelopes and lithification of molluscan biosparites: Geol. Journal, v. 5, p. 15-32.

____________, 1971, Carbonate sediments and their diagenesis: Developments in sedimentology, v. 12, Elsevier, Amsterdam, 620 p.

Becher, J.W., and C.H. Moore, 1976, The Walker Creek field: a Smackover diagenetic trap: Trans. Gulf Coast Assoc. Geol. Socs., v. 26, p. 34-56.

Berner, R.A., 1971, Principles of chemical sedimentology: New York, McGraw-Hill, 240 p.

Bishop, W.E., 1968, Petrology of Upper Smackover limestone in North Haynesville Field, Claiborne Parish, Louisiana: Am. Assoc. Petrol. Geol. Bull., v. 52, p. 92-128.

Butler, G.P., 1969, Modern evaporite deposition and geochemistry of coexisting brines, the sabkha, Trucial Coast, Arabian Gulf: Jour. Sed. Petrology, v. 39, p. 70-89.

Curtis, R., G. Evans, D.J.J. Kinsman, and D.J. Shearman, 1963, Association of dolomite and anhydrite in the recent sediments of the Persian Gulf: Nature, v. 197, p. 679-680.

Dickinson, K.A., 1968, Upper Jurassic stratigraphy of some adjacent parts of Texas, Louisiana, and Arkansas: Geological Sur. Prof. Paper 594E, 25 p.

Dunham, R.J., 1962, Classification of carbonate rocks according to depositional texture, *in* W. E. Ham (ed.), Classification of Carbonate Rocks: Am. Assoc. Petrol. Geol., Tulsa, p. 108-121.

Folk, R.L., 1965, Some aspects of recrystallization in ancient limestone: *in:* Dolomitization and limestone diagenesis: a Symposium, Soc. Econ. Paleontologists and Mineralogists, Spec. Publ., no. 13, p. 14-48.

Friedman, G.M., 1965, Terminology of cyrstallized textures and fabrics in sedimentary rocks: Jour. Sed. Petrology, v. 35, p. 643-655.

Graton, L.A., and H.J. Fraser, 1935, Systematic packing of spheres with particular relation to porosity and permeability: Jour. Geology, v. 34, p. 785-909.

Gratton, P.J.F., 1965, Several ground rules can aid search for Smackover: Oil and Gas Jour., v. p. 258-261.

Hughes, D.J., 1968, Salt tectonics as related to several Smackover fields along the northeast rim of the Gulf of Mexico Basin: Trans. of Gulf Coast Assoc. of Geol. Soc., v. 10, p. 320-330.

Illing, L.V., 1954, Bahama calcareous sands: Am. Assoc. Petrol. Geol. Bull., v. 38, p. 1-95.

__________, A.J. Wells, and J.C.M. Taylor, 1965, Penecontemporary dolomite in the Persian Gulf, *in* Dolomitization and limestone diagenesis: Soc. Econ. Paleontologists and Mineralogists, Spec. Publ., no. 13, p. 89-111.

Imlay, R.W., 1940, Lower Cretaeous and Jurassic formations of southern Arkansas and their oil and gas possibilities: Arkansas Geol. Survey Inf. circ. 12, 64 p.

Jacka, A.D., 1975, Observations on paramorphic and neomorphic dolostones and associated porosity relationships (abs.): Am. Assoc. Petrol. Geol. Annual Convention Book of Abstracts, p. 38-39.

__________, 1977, deposition and diagenesis of the Ft. Terret Formation (Edwards Group), in the vicinity of Junction, Texas: Bur. Econ. Geol., Univ. Texas at Austin, Symposium on Cretaceous carbonates, Report of Invest. No. 89, p. 182-200.

Kinsman, D.J.J., 1966, Gypsum and anhydrite of Recent age, Trucial Coast, Persian Gulf, *in* Second Symposium on Salt, Northern Ohio Geol. Soc., Cleveland, Ohio, p. 302-326.

__________, 1969, Models of formation, sedimentary associations, and diagnostic features of shallow-water and supratidal evaporites: Am. Assoc. Petrol. Geol. Bull., v. 53, p. 830-840.

Lindholm, R.C., and R.B. Finkelman, 1972, Calcite staining: semi-quantitative determination of ferrous iron: Jour. Sed. Petrology, v. 42, p. 239-242.

Moore, C.H., and Y. Druckman, 1981, Burial diagenesis and porosity evolution, Upper Jurassic Smackover, Arkansas and Louisiana: Am. Assoc. Petrol. Geol. Bull., v. 65, p. 597-628.

Oivanki, S.M., 1974, Paleodepositional environments in the Upper Jurassic Zuloaga Formation (Smackover), Northeastern Mexico: Trans. of Gulf Coast Assoc. of Geol. Soc., v. 24, p. 258-278.

Pilger, R.H., 1978, A closed Gulf of Mexico, pre-Atlantic Ocean plate reconstruction and early rift history of the Gulf and North Atlantic: Trans. of Gulf Coast Assoc. Geol. Soc., v. 28, p. 385-393.

Purdy, E.G., 1963, Recent calcium carbonate facies of the Great Bahama Bank. 2. Sedimentary Facies: Jour. Geol., v. 71, p. 472-497.

Shearman, D.J., 1963, Recent anhydrite, gypsum, dolomite, and halite from the coastal states of the Arabian shore of the Persian Gulf: Geol. Soc. London Proc., No. 607, p. 63-65.

__________, 1966, Origin of marine evaporites by diagenesis: Trans. Inst. Mining Met., Section B, v. 75, p. 208-215.

Stehli, F.G., and J. Hower, 1961, Mineralogy and early diagenesis carbonate sediments; Jour. Sed. Petrology, v. 31, p. 358-371.

Weeks, W.B., 1938, South Arkansas stratigraphy with emphasis on the older coastal plain beds: Am. Assoc. Petrol. Geol. Bull., v.22, p. 953-983.

Wilson, J.L., 1975, Carbonate facies in geologic history: New York, Springer-Verlag, 471 p.